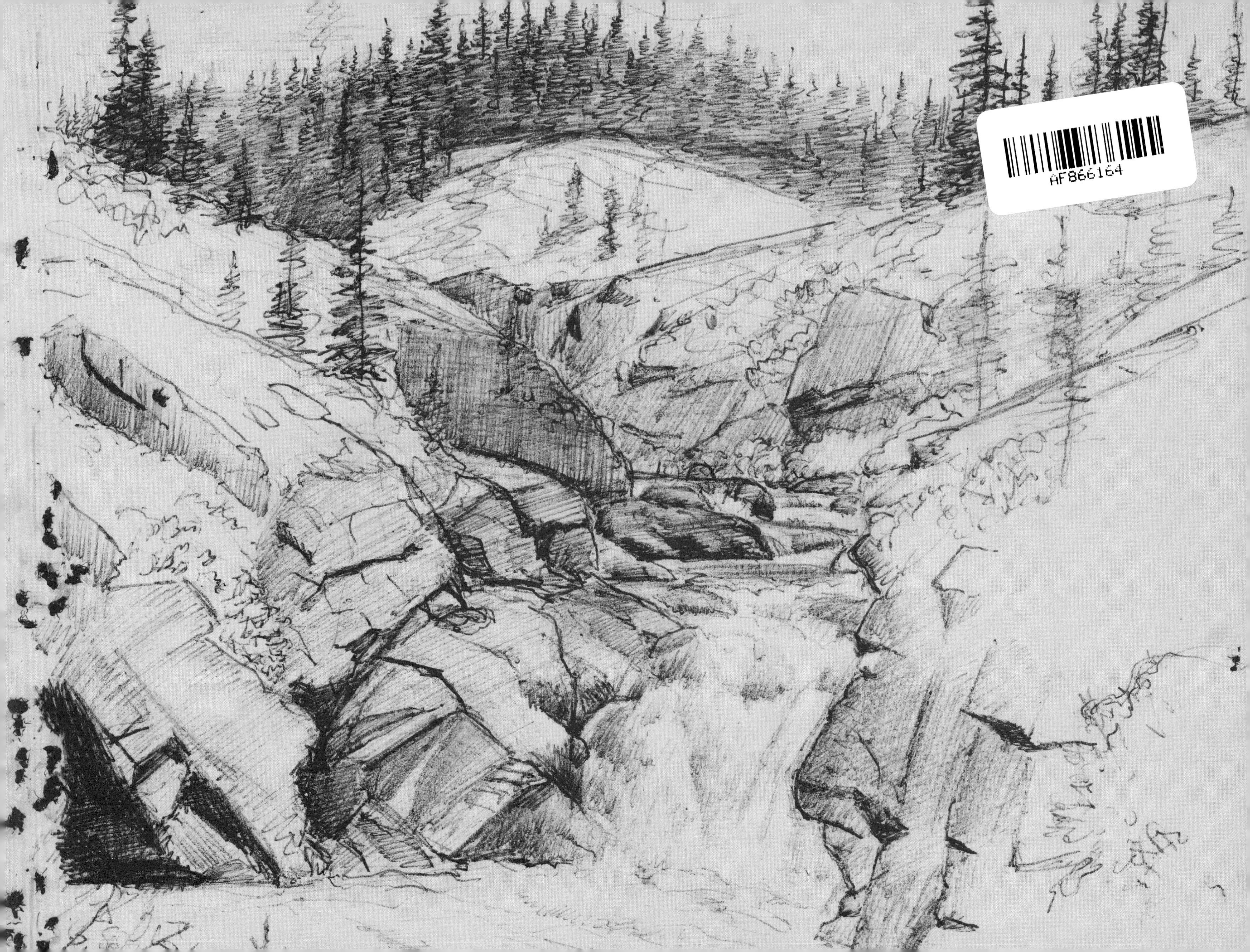

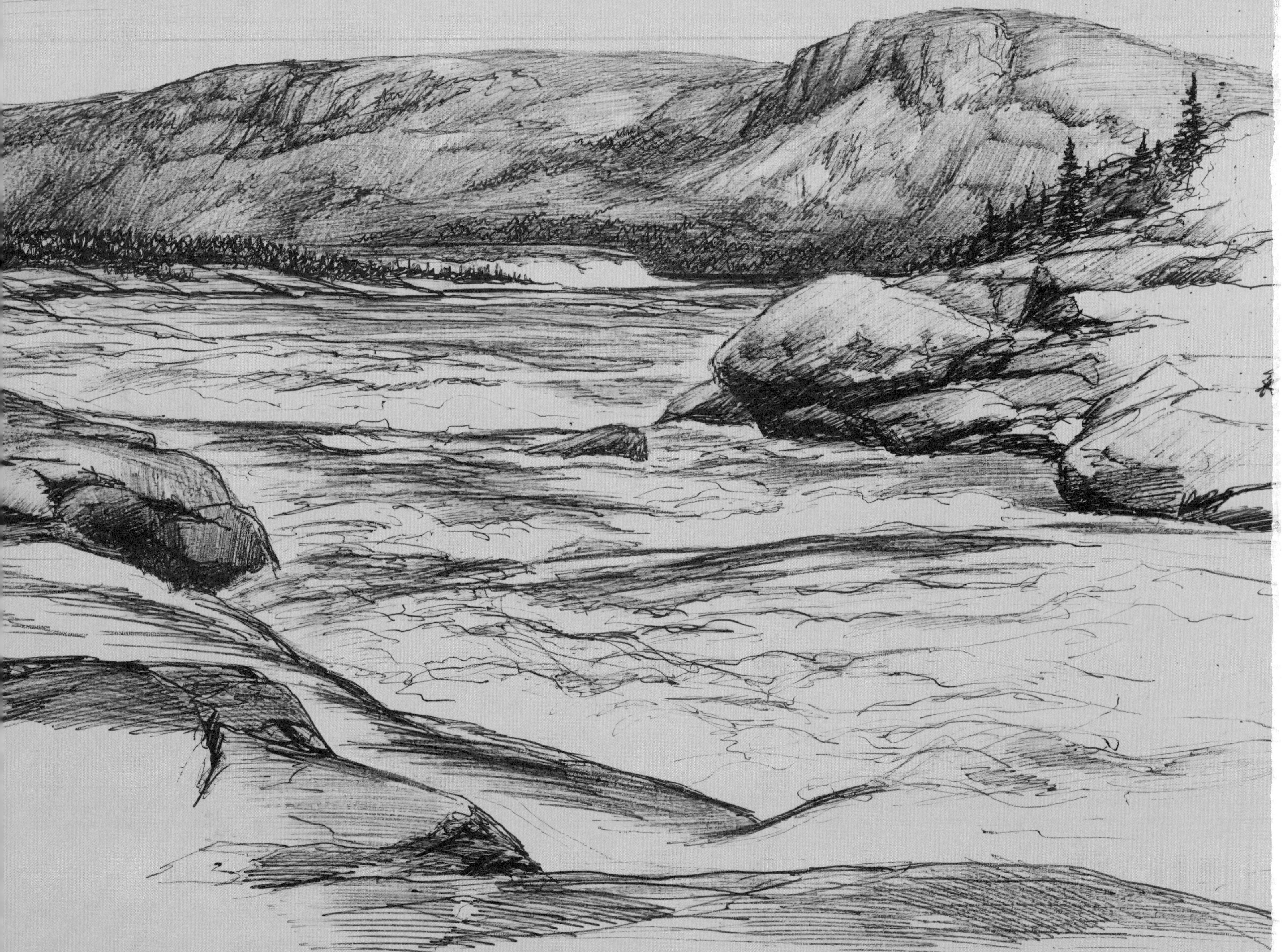

BOREAL WILDERLAND

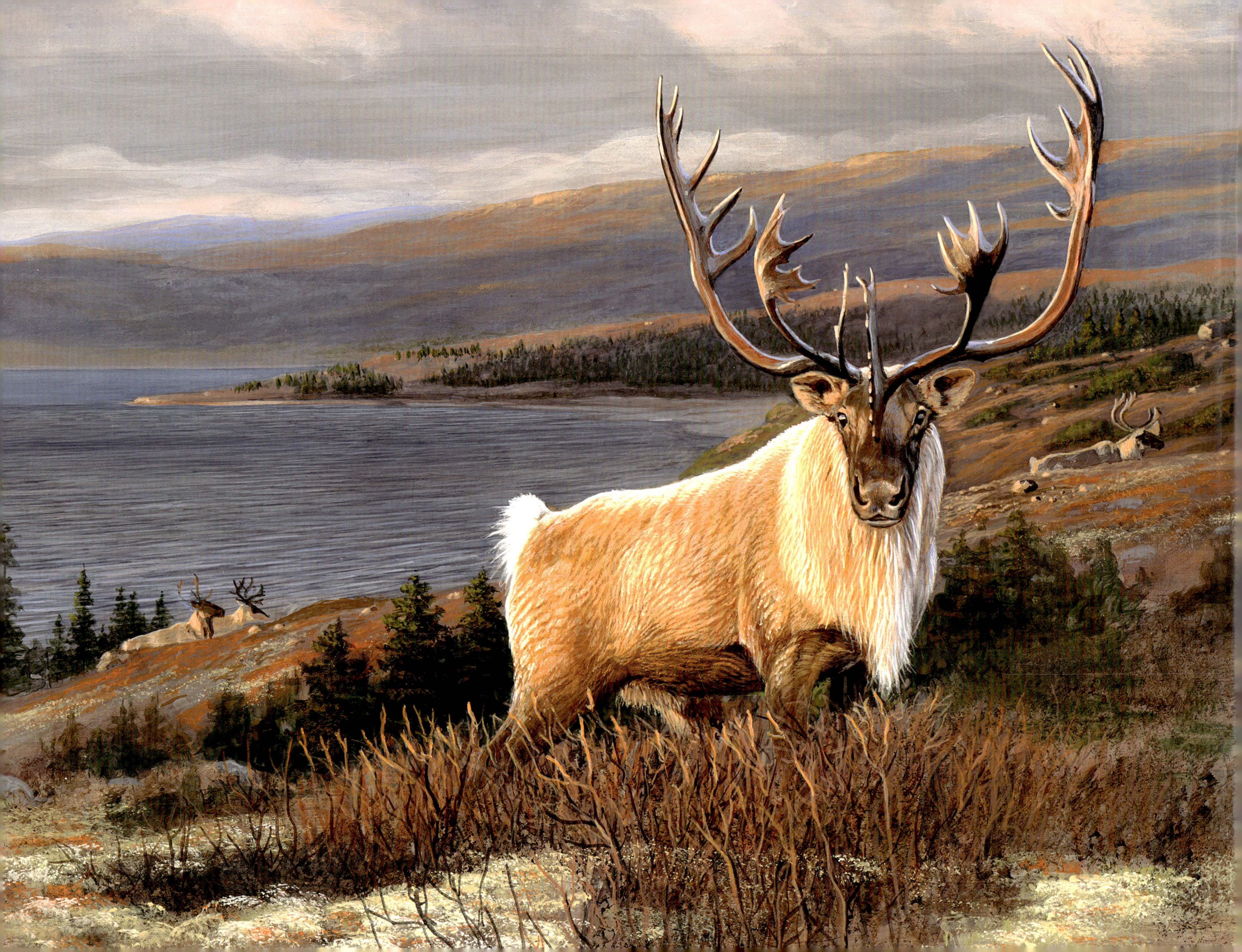

ART, SCIENCE, AND ADVENTURE IN EARTH'S GREATEST WILDERNESS

BOREAL WILDERLAND

35 Artists · 6 Natural History Experts · 20 Expeditions · 1 Vast Ecosystem

ROB MULLEN · FOREWORD BY ROBERT BATEMAN

CONTRIBUTIONS BY DR. WILLIAM FITZHUGH & DR. STEPHEN LORING

SMITHSONIAN ARCTIC STUDIES CENTER

PAGE TWO

FRONTISPIECE: George River caribou, Kamestastin Lake, Labrador. Small acrylic study for a larger painting. *Coming Ashore*, acrylic, Rob Mullen

From the expeditions of the
Wilderness River Expedition Art Fellowship (WREAF)
Director ROB MULLEN

A program of the **Center for Circumpolar Studies**
Director DR. STEVEN YOUNG

and the **Smithsonian Arctic Studies Center**
Director and archaeologist DR. WILLIAM FITZHUGH
Archaeologist/ethnologist DR. STEPHEN LORING

at the **National Museum of Natural History**

WREAF Wilderness River Expedition Art Fellowship

Arctic Studies Center

The Center for Circumpolar Studies, Inc.
www.circumpolarstudies.org + info@circumpolarstudies.org

Is it because of the barrenness of the land that the skies in the
Great North are always such a grand spectacle?
No lazy summer clouds here, with trailing white puffballs taking
shapes of bunnies and cows . . .
No! Here the sky is always at war, always a battlefield for colossal
forces fighting for dominion over the forest, tundra and cold lakes.

JEAN-LOUIS COURTEAU

WREAF 2006 George River Expedition

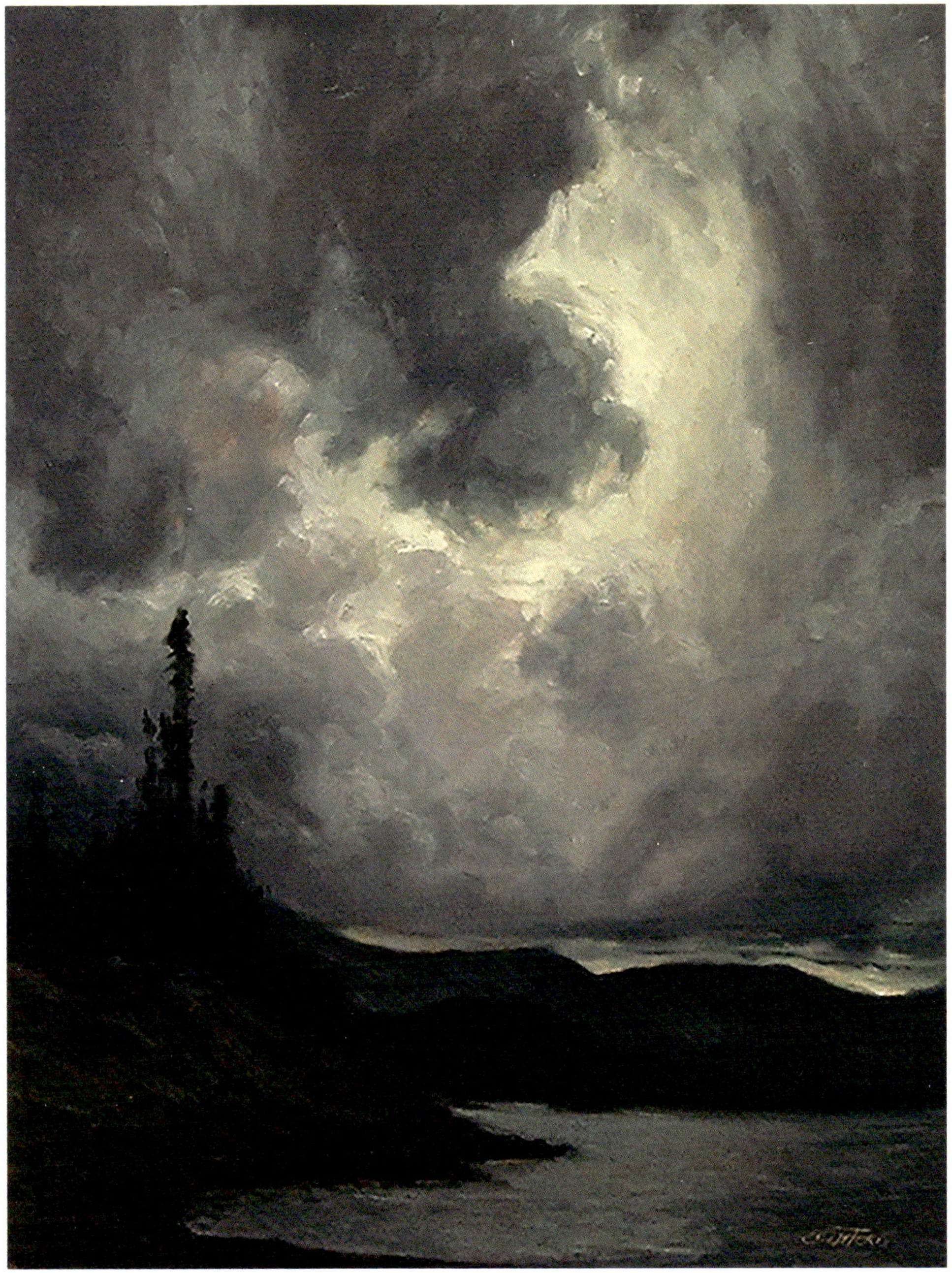

Boreal Watch, oil, Jean-Louis Courteau

Dedication: To my wife Bonnie Rowell

There are people, individually and collectively, without whom this book would not exist. They may have made one critical contribution, or a dozen. There is no easy metric to rank such generosity; "I couldn't have done it without you" need be said only once to earn a full measure of gratitude (see Acknowledgments and others in the text).

Yet maybe because humans respond to patterns, repetition stands out, and, by sheer volume, no one has done more, more often, or in more ways than Bonnie. Large and small, across the wide gamut of the components of this long-suffering project, Bonnie has come through with often brilliant solutions to problems, seen issues I had overlooked, and moreover, anticipated others that I never saw coming. Yet, as many times as she has saved the day, her most inexplicable and treasured trait is her constancy.

From the time we met, she has supported this inexorable background mania of mine. Why, I don't know. But she has. She has had my back no matter what. Despite the stress of my being gone for weeks at a time, a few times alone, she never suggested backing off, but came up with better communications and acted as the homebase coordinator. After setbacks and even when the whole project seemed to have crashed around us (twice), she never suggested giving up, but helped figure out how to move forward. I started this boreal forest project several years before I met Bonnie, and I was determined to see it through. In hindsight, I don't know how I could have persevered without such a steadfast and loving partner.

Besides, she's a brave photographer and a good bow paddler.

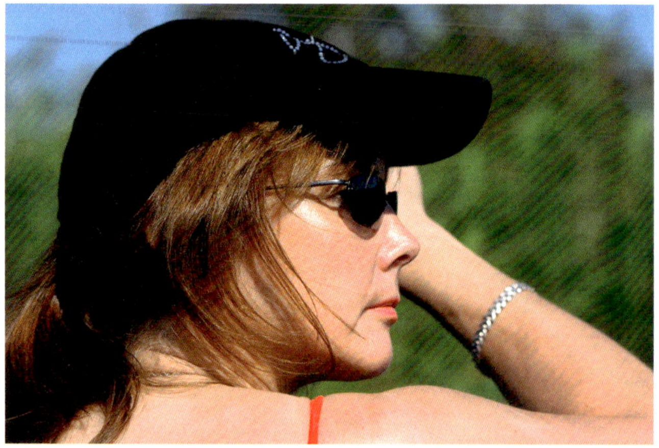

TOP: Bonnie Rowell paddling in Lac Saint-François.

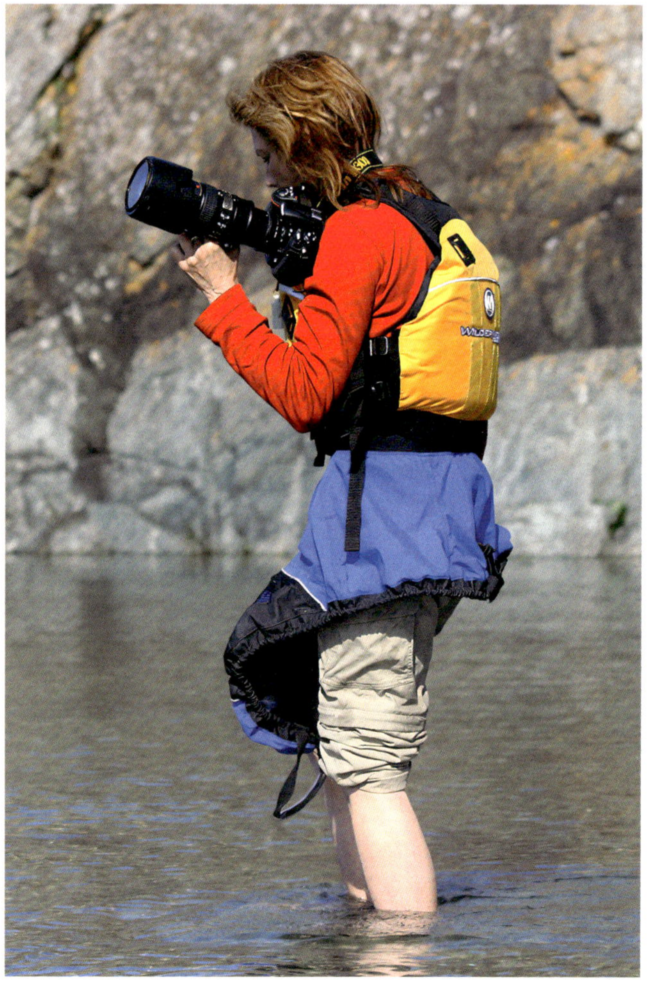

NEAR RIGHT: Bonnie getting into position for a better shot. Slate Islands, Lake Superior.

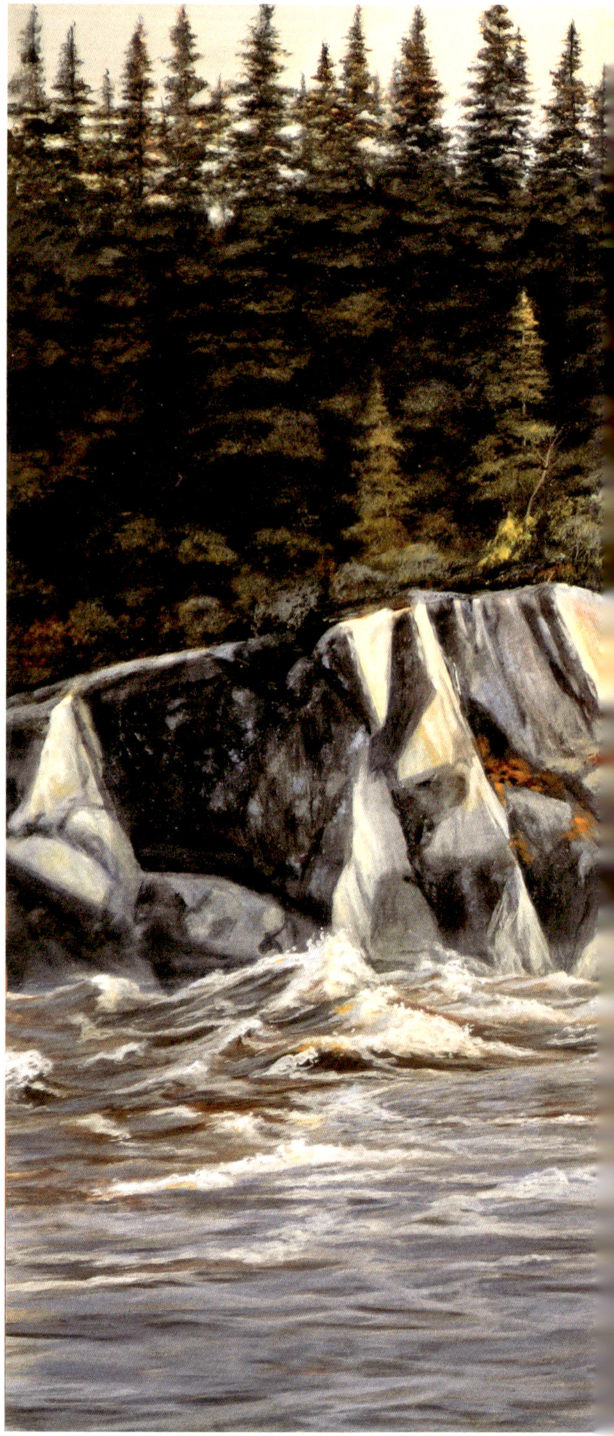

FAR RIGHT: Wolves on the Harricana. *Homeward Bound*, oil, Rob Mullen

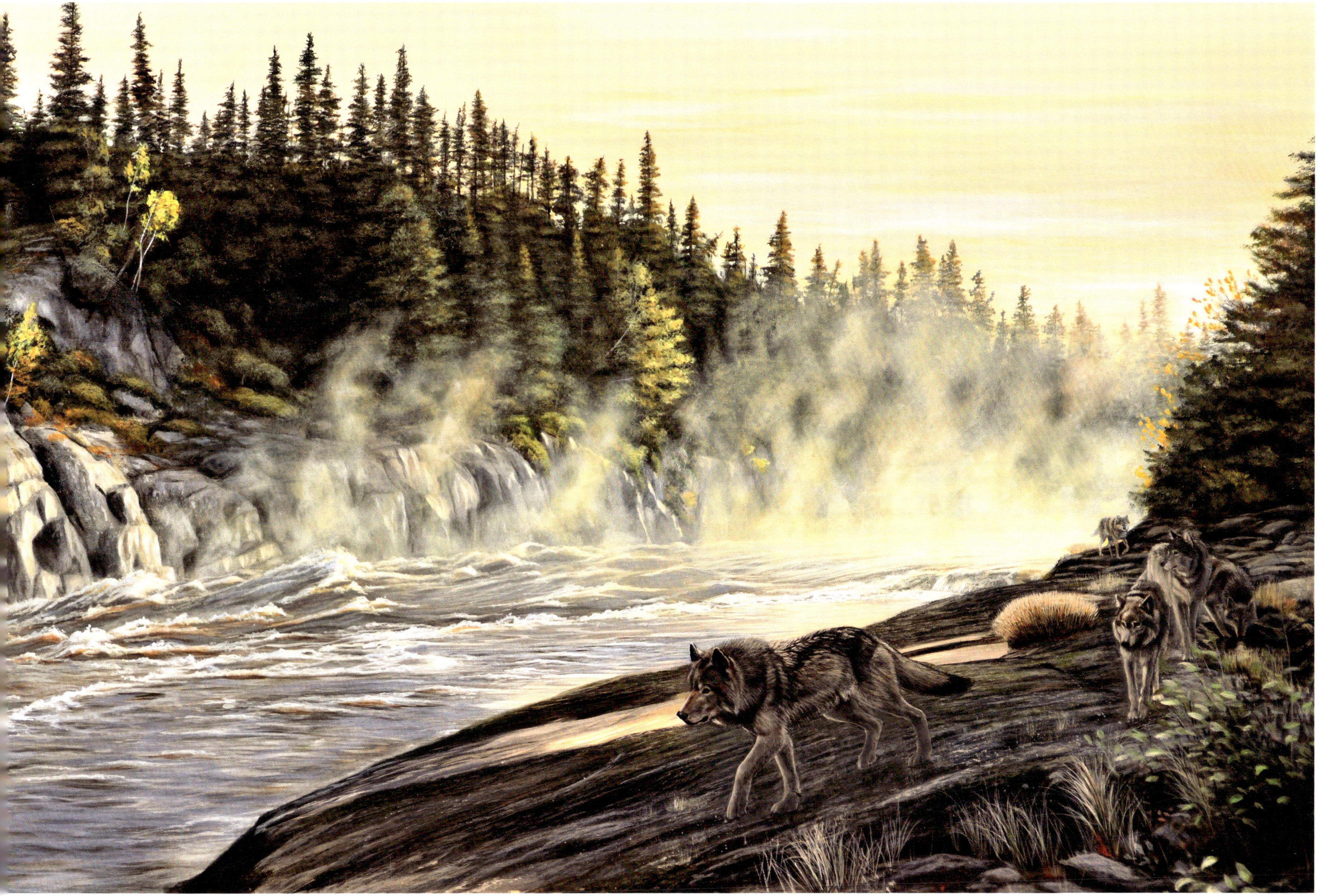

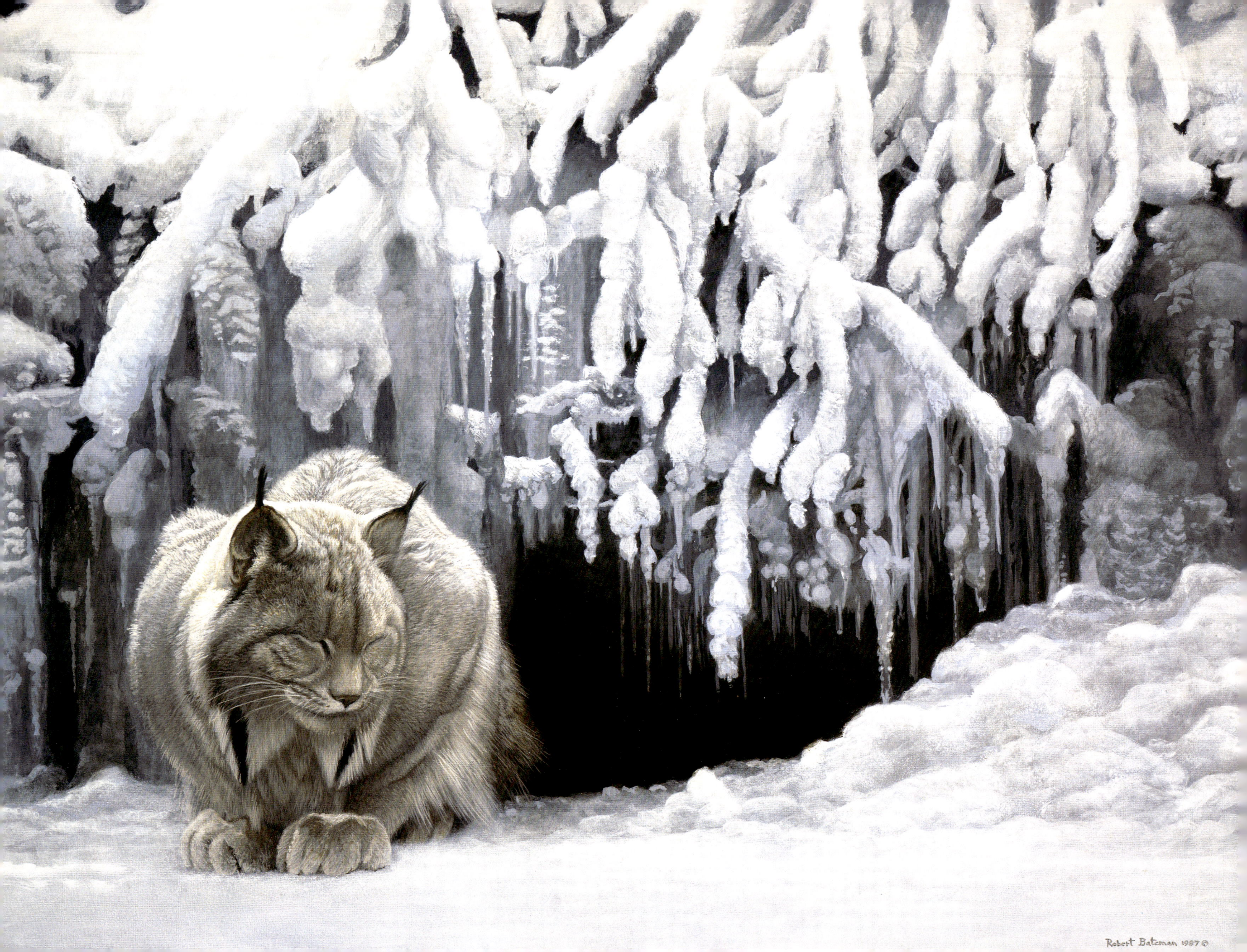
Robert Bateman 1987 ©

Contents

Foreword by Robert Bateman *10*

A Word About the Word "Wilderness" *12*

Prologue: Into the Great Canadian Northwoods *17*

PART ONE **TREES AND TUNDRA** *23*

1 The Boreal Forest and "Wilderland" *27*

2 Architects of the Boreal Forest: Winter, Water, and Fire *41*

3 A Tour of the Boreal Forest *61*

PART TWO **LIONS AND TIGERS AND BEARS (AND MORE!)** *97*

4 Non-Human Mammals *101*

5 Birds *147*

6 Lower Vertebrates: Fish, Reptiles, and Amphibians *215*

PART THREE **HUMANS IN WILDERLAND** *235*

7 Cultures of the Boreal Forest *239*

8 Ecosystem Services: The Other Ecological Value *259*

9 Conservation Threats and Action *275*

A Last Word About the "Wild" in Wilderness *281*

Acknowledgments *284*

Notes *286*

Index *288*

FACING: *Dozing Lynx*, acrylic, Robert Bateman

Foreword

SINCE MY earliest years, I have been enchanted by the boreal forest. These forests are made up of perpetually green coniferous trees that provide cover and protection for many species of the animal kingdom from invertebrates to large mammals such as moose. Also, like other forests, they absorb carbon dioxide and act as a carbon sink, which helps set off our excess industrial carbon production.

Growing up in Southern Ontario, I thought of the great Canadian North as an exciting and somewhat mysterious place. When I was 17 years old, a piece of good luck landed me a job at the Algonquin Wildlife Research Station in Algonquin Park, Ontario. A chap I knew in Boy Scouts was forced by his father to take a job there. He wanted to have a friend join him as he wouldn't know anyone and he asked me; he hated it, but I thought I'd gone to heaven. The researchers were a combination I had never encountered before: outdoorsmen but intellectuals. They wore blue jeans, knew how to "dubbin" a boot, but discussed philosophers such as Immanuel Kant in their conversations.

I might have gone into scientific research, but I have little ability in math, so I majored in geography. This led to several summer jobs in the wilderness besides those in Algonquin Park. I spent a summer in Newfoundland with the Geological Survey; another in Churchill, Manitoba, as a research assistant to a friend doing his master's thesis; and another at the edge of Ungava (far northern Quebec), which, although being north of the boreal forest, provided me with a contrast in ecosystems. In 1954, I travelled to Europe and went as far north as Lapland where the boreal forest is called "taiga."

The artist side of me reveled in the wilderness subject matter of my heroes: Tom Thomson and the Group of Seven. My earliest serious artworks were oil paints on board such as they used.

A large solo exhibition of my paintings, *Portraits of Nature,* was held at the Smithsonian National Museum of Natural History in 1987. It was the first time that a fine arts exhibition was mounted in the natural history museum. As I understand, this exhibit was one of the inspirations that, in 2004, led artist Rob Mullen and Smithsonian Arctic Studies Center anthropologists Dr. William Fitzhugh and Dr. Stephen Loring to start a project on the boreal forest that led to this book.

In 2006, I went on an expedition that Rob arranged to the George River in Northern Quebec along with Dr. Loring. The George River is a key research area for Dr. Loring and home to the George River caribou herd that, even in 2006, was starting what has in recent years become a drastic decline in numbers. Rob has not only knowledge of this forest, its inhabitants, and its history, but also an appreciation of its ecological importance. He advocates for wise management, holding that wildlife and ecosystems have intrinsic, existential value far beyond any economic returns from extractive industries or energy production.

The expedition artists, scientists, naturalists, and photographers whose work is included in this book portray a stunning part of our planet Earth. This is the main reason I paint: I want to engage others with the natural world. This book, *Boreal Wilderland*, continues that goal.

ROBERT BATEMAN
Salt Spring Island, British Columbia

Author's note

The *Portraits of Nature* exhibition at the National Museum of Natural History was indeed an inspiration behind the collaboration that Dr. Fitzhugh, Dr. Loring, and I began in July of 2004.

Robert Bateman's work and environmental advocacy have been and are inspirations to me and most of the other artists that have been on WREAF expeditions. He is one of the giants on whose shoulders we are perched as we strive to highlight how essential healthy ecosystems and wildlife are—not just for wildlife artists, outdoors enthusiasts, and animal lovers, but to the very survival of our civilization, if not our species.

ROB MULLEN
West Bolton, Vermont

A Word About the Word "Wilderness"

THIS BOOK is about a region typically called a "wilderness." It is a more evocative word than "woods," "city," or "vacant lot," conjuring a variety of charged images ranging from a frightening wasteland to an idyllic utopia, so an understanding of what wilderness means in this book will be useful.

Probably the most common definition of wilderness is the "natural world" absent signs of human disruption. But that presumes our presence can lessen the world's naturalness, implying that we are separate from nature. That is biological nonsense. However, such a twofold worldview is possibly "natural" for any sentient species. Beavers may see the woods, beyond a short distance from their pond, as some nebulous, non-verbal notion of "wilderness." Moreover, while beavers' activities (like ours) can have profound effects on ecosystems, unlike beavers, our effects can be, and often are, destructive. Therefore, areas largely unaltered by humans may be useful comparisons for examining our effects on the world around us.

But how absent need we be? The San Gorgonio Wilderness in Southern California (where I used to live) is about 70 miles from the seven-thousand-acre Laguna Coast Wilderness Park. Both are near LA, and the latter accepts Mastercard. Vermont's Long Trail, known as the "footpath in the wilderness," is crossed by roads and much of it has cell reception (I hiked it end to end in 2020, and it is a short walk through the woods from our house). To purists, these are not real wilderness. However, in Labrador, what such purists might breathlessly revere as wilderness traditional Innu hunters call *nutshimits* ("the country" in Innu-aimun), which they and their ancestors have known intimately as home for millennia.

It's complicated, but we need a working definition. Given that settled life, agriculture, and, more recently, the industrial revolution probably refined the concept of wilderness out of the age-old distinction between the area around the home territory and unfamiliar land far away, it stands to reason that signs of their presence could be benchmarks. Therefore, for the purposes of this book, bearing in mind that we are all part of nature, "wilderness" will be the general absence of agriculture and industry: no big towns, clear-cuts, mines, roads, or dams (not human ones anyway). Canoes are allowed. Whether one thinks of it as a wasteland or utopia is a personal call. However, I will attest that it can flip from enchanting to terrifying in a psyche-shredding instant.

Online appendix

A lot of material needed to be left out of this book so that it would be lighter and take up less space in your bookshelf than a canoe. Scan the QR code to access that material, such as a bonus chapter (Being There), abridged contributions to the book, expedition trip reports, annotated maps, satellite images, videos, essays, new paintings, and more. The landing page has a table of contents, and throughout this book there are notes when there is additional information in the online appendix at robmullen.com/boreal-wilderland.

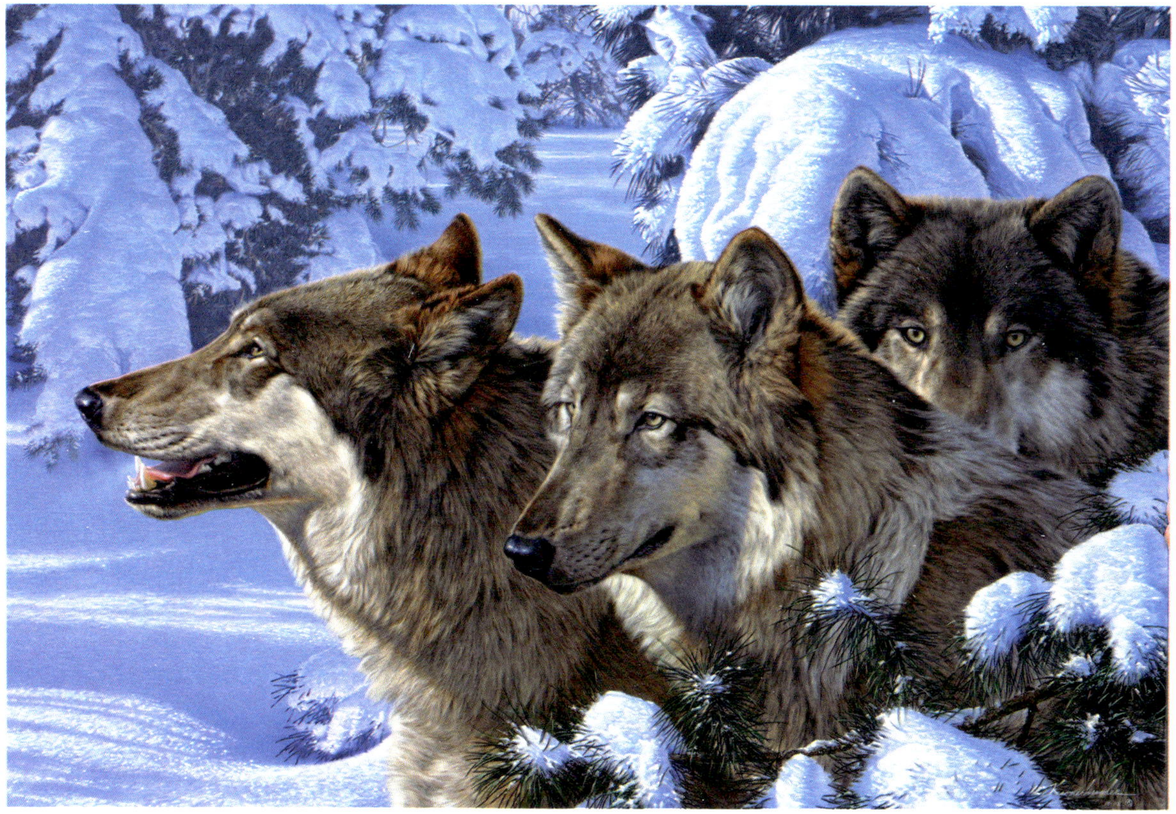

Gray wolves. *Winter's Warmth*, acrylic, Lee Kromschroeder

OVERLEAF: Woodland Caribou Provincial Park, 2011. *Red Canoes*, oil, Dawn Banning

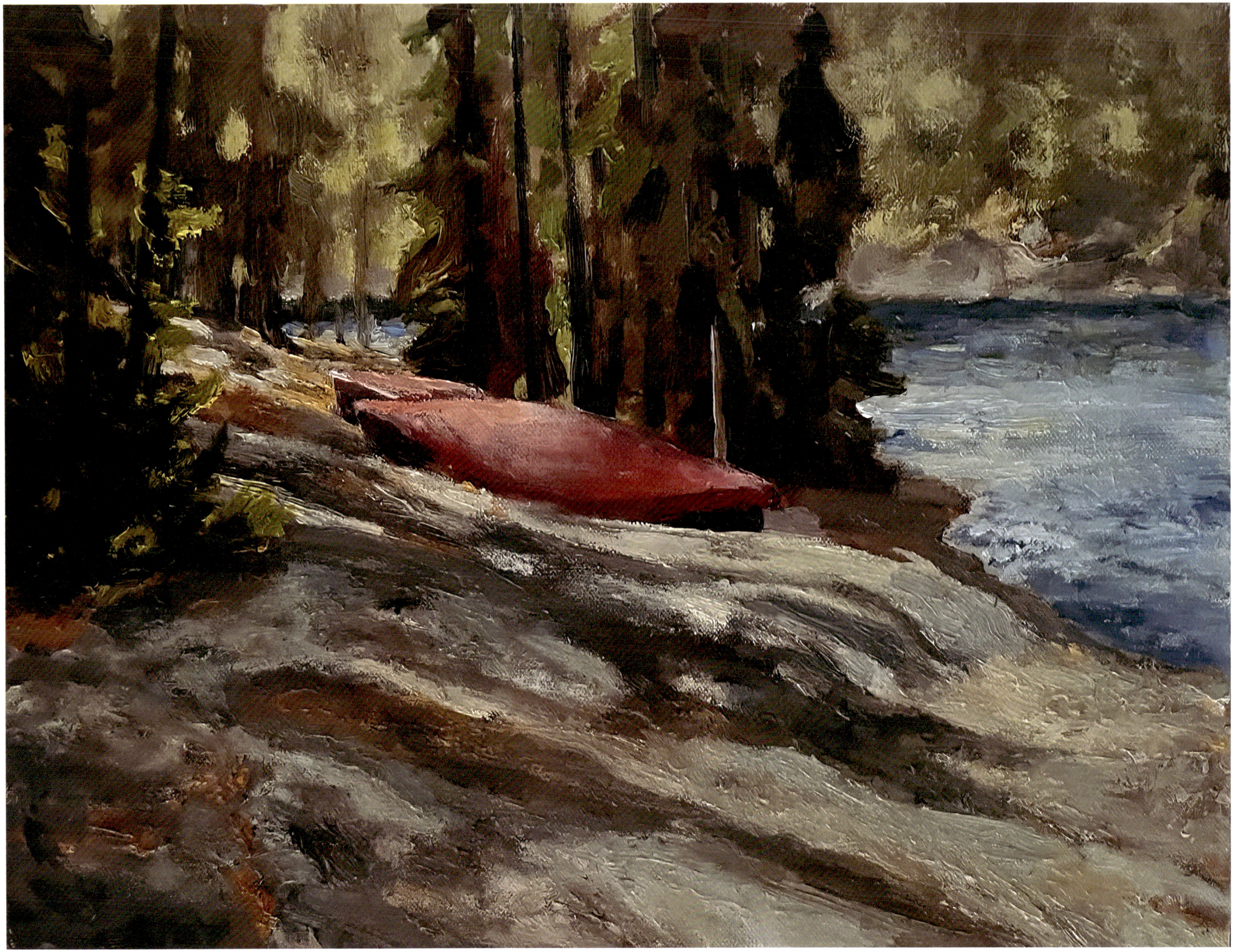

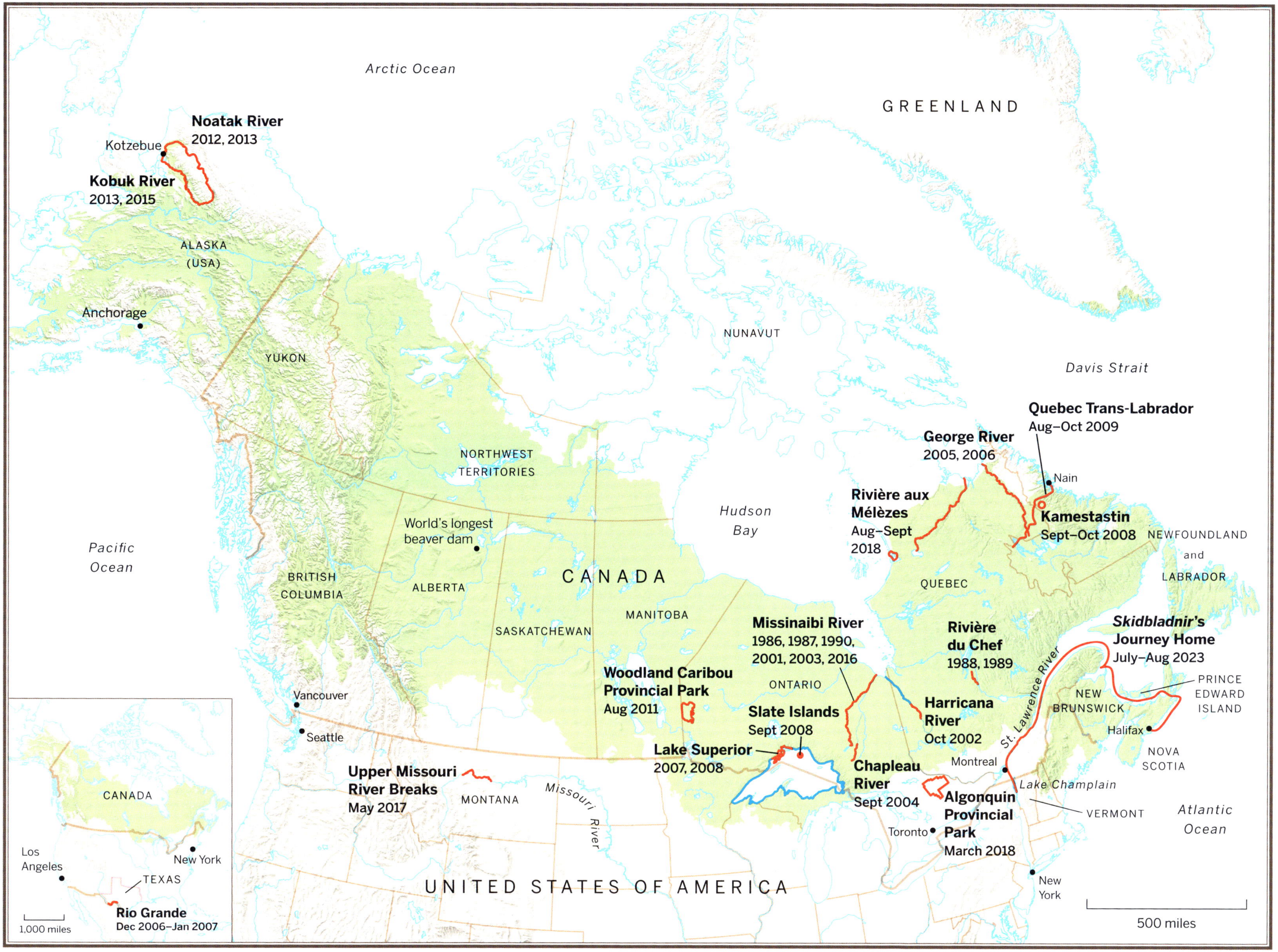

Arctic Ocean
GREENLAND
Noatak River
2012, 2013
Kotzebue
Kobuk River
2013, 2015
ALASKA
(USA)
Anchorage
YUKON
NUNAVUT
Davis Strait
Quebec Trans-Labrador
Aug–Oct 2009
George River
2005, 2006
Nain
NORTHWEST
TERRITORIES
Rivière aux
Mélèzes
Aug–Sept
2018
Kamestastin
Sept–Oct 2008
Hudson
Bay
World's longest
beaver dam
NEWFOUNDLAND
and
LABRADOR
Pacific
Ocean
BRITISH
COLUMBIA
ALBERTA
CANADA
QUEBEC
MANITOBA
SASKATCHEWAN
Missinaibi River
1986, 1987, 1990,
2001, 2003, 2016
Rivière
du Chef
1988, 1989
Skidbladnir's
Journey Home
July–Aug 2023
Woodland Caribou
Provincial Park
Aug 2011
ONTARIO
PRINCE
EDWARD
ISLAND
St. Lawrence River
NEW
BRUNSWICK
Vancouver
Harricana
River
Oct 2002
Slate Islands
Sept 2008
Halifax
Seattle
Lake Superior
2007, 2008
NOVA
SCOTIA
Montreal
Upper Missouri
River Breaks
May 2017
Chapleau
River
Sept 2004
Lake Champlain
MONTANA
Missouri River
Algonquin
Provincial
Park
March 2018
VERMONT
Atlantic
Ocean
CANADA
Toronto
Los
Angeles
New York
TEXAS
New
York
UNITED STATES OF AMERICA
Rio Grande
Dec 2006–Jan 2007
1,000 miles
500 miles

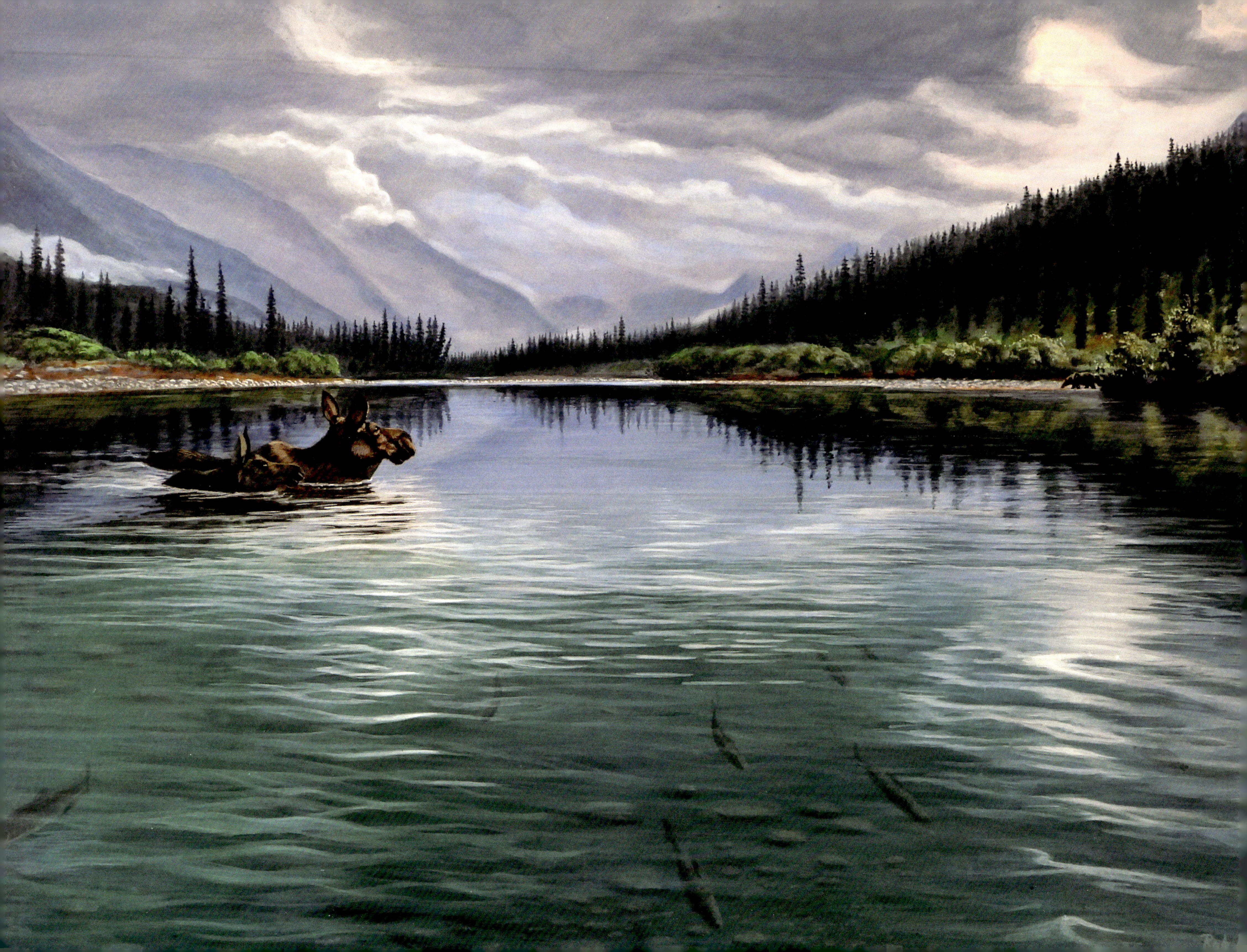

FACING: Reed River to the Kobuk, Alaska. *Crossing Paths*, acrylic, Rob Mullen

PROLOGUE

Into the Great Canadian Northwoods

IN JUNE 1978, I arrived in Manhattan with $250 and a biology degree from the University of Vermont, hoping to become the Audubon of bugs. That didn't work out. I slid into advertising art, married, became a father—and, as the famed "over the hill" age of 30 loomed, felt the pull of my adventurous youth. My brother Frank, then a 28-year-old Coast Guard officer, was in the same boat, so to stave off becoming "old, settled toads," we planned a wilderness canoe trip. Duty called Frank elsewhere, but Dad joined me.

A close encounter of the big, hairy kind

June 1986: Upper Missinaibi River, Northern Ontario

With no suitable campsite for miles, Dad and I wedged our tent among trees at the edge of a cut bank, four feet above the river. Long after dinner, Dad snored while I lay out of the doorway, meditatively tending a twig fire and marveling at the still faint midnight glow of the sunset.

Ah, the serenity of a wilderness evening . . . shattered with an explosion in the water below. Five feet away, a huge black mass loomed, dimly lit by the small flames. To my growing alarm, another shape rose beside it, water streaming from wide—aha!—velveted antlers. The bull moose's beady eye glinted at me, and I realized the "explosion" had been possible commentary, delivered directly into the river. He moved on, leaving us grudgingly accepted into the neighborhood—though clearly below him in the forest pecking order.

That began a tradition of annual Canadian canoe trips, a tonic for my city life. The next summer, Frank joined me and a colleague for 130 miles on the Missinaibi to Mattice, Ontario—the last road crossing on the river's journey north to James Bay.

June 1988: Rivière du Chef, Northern Quebec

We finally arrived at the river near sunset, and a minute after the poor driver left, I realized my camera was still in his truck. He was a fill-in and didn't know the way and was anxious to get home. Distracted by trying to reassure him in our best

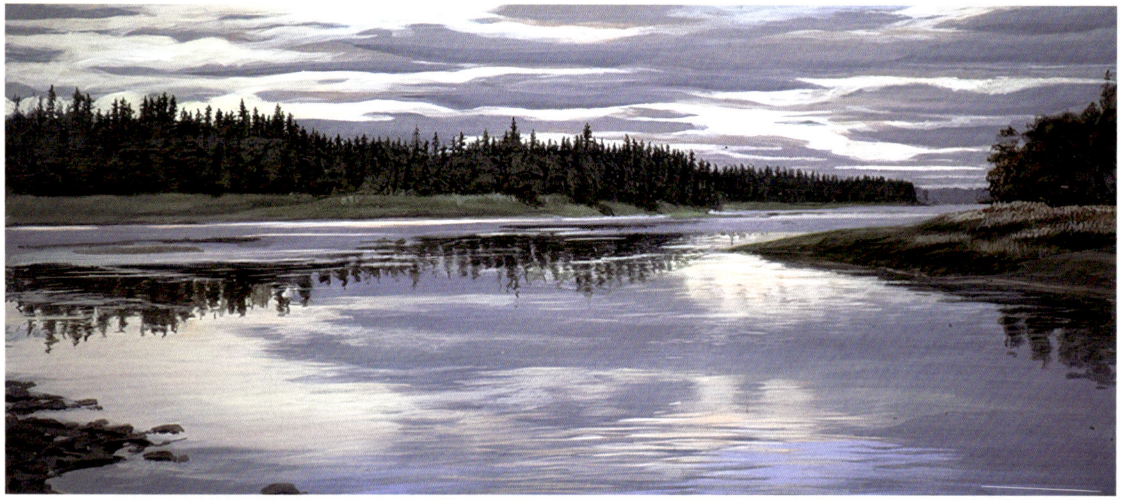

Field study from Portage Island, Missinaibi River, September 11, 2001. *River's End*, acrylic, Rob Mullen

FACING: Portage Island, Missinaibi River, September 11, 2001. *River's End*, acrylic, Rob Mullen

French (not very good), the camera went unnoticed. The drive from Chibougamau had already been a grim ride through miles of abandoned logging roads and shredded forest. Yet behind us was the river and the edge of intact wilderness, the sharp edge of where the maw that feeds civilization was chewing its way north.

July 1989: Rivière du Chef, Northern Quebec

Short on time to research a new stream, Frank and I returned for an encore run of the Chef. This time Joe St. Clair of Chibougamau drove us in, laughing with us all the way as we tried each other's languages. Joe hated the destruction but knew the road—he was helping to build the bridge across the Chef. Indeed, at the river, there was an unfinished bridge, and on the far side of what had been untrameled wilderness, a 250-yard semi-circle of clear-cut. Today, satellite images show log roads crisscrossing the region, with no trees left along most of the river. In fact, for over a hundred miles around, the forest is a tattered patchwork of massive clear-cuts (see "Maps and Satellite Images" in the online appendix).

In 1994, I left advertising, returned to Vermont, and dove into wildlife art. Six years on the show circuit left me feeling more like a teamster than an artist. In 2001, I returned to the Missinaibi for a "source to salt" painting trip to the Arctic Ocean—just to recharge my creative batteries—and found more than I bargained for.

Finding my muse (or "catching hold of a dragon by the tail")

September 11, 2001: Lower Missinaibi River

I awoke to silence. After four days of storm, no rain! And light glowed through the tent—sunrise! I raced to the riverbank and froze. Under the glowing vault of the dawn, three sandhill cranes fed beside my canoe. In that moment, I understood how sun-worshipping was born (*First Light*—the jacket art).

Later that day, after 31 miles of exhilarating paddling (including the aptly named "Deception Rapid"), I reached the end of the Missinaibi at Portage Island, where it joins with the Mattagami to form the Moose River. I cooked dinner, fed peanuts to Canada jays, and painted the last stretch of river (*River's End* field study, above left). I was excited to get home. Not only for the studio paintings I was keen to do, but for the bigger idea that had taken shape as I paddled north. It was 40 easy miles to James Bay. I didn't make it.

What? The boreal forest?!

That "source to salt" journey on the Missinaibi River in August–September 2001 was to be solo only because, with an artist's itinerary, I assumed no one else would want to come. However, at a June art show reception, wildlife painter John Pitcher surprised me by asking to join. On August 24, with a

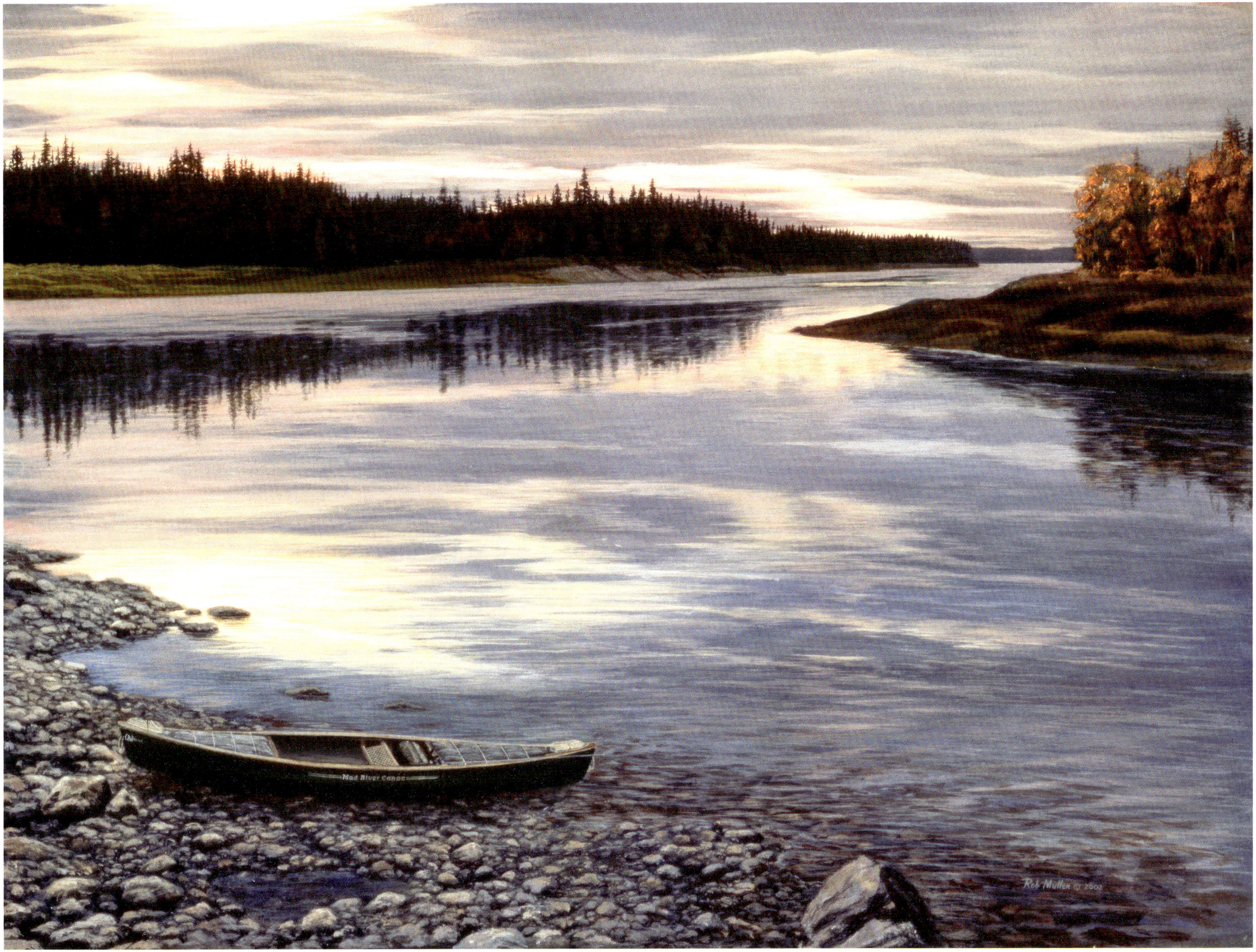
Mad River Canoe

second solo canoe supplied by Mad River Canoe Company—our first and "accidental" sponsor—we headed north from Barclay Bay on Missinaibi Lake. On the river we became fast friends, plotting future trips with other artists.

A wrist injury forced John home after 130 miles, at the last road crossing in Mattice-Val Côté, Ontario. I had never been north of Mattice. Talking about bravely venturing alone into the wilderness over wine and cheese at art shows was one thing; facing it for real with the Missinaibi flowing inexorably into the north past the back door of the Empire Inn (since closed and replaced by a bar and grill) was quite another. After debating my odds over poutine and beer, I set off on September 4 for the 200-mile run to Moosonee, the largest Canadian town unconnected by road to anywhere. Camped that night at Kettle Falls, I was nearly overwhelmed by loneliness but focused on setting a secure camp and starting dinner. While the water heated, the lowering sun created such wonderful light that, hoping for an additional distraction, I started a painting. As the painting took shape, an almost

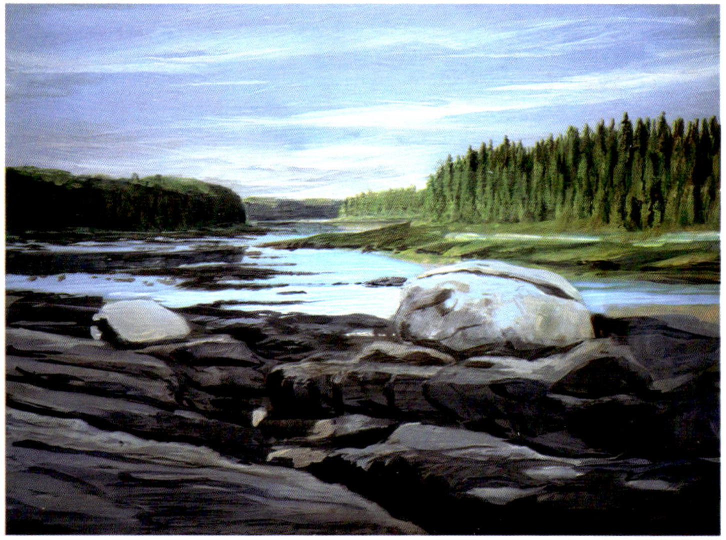

Camped at Kettle Falls portage, Missinaibi River, September 4, 2001. *View North*, acrylic field painting, Rob Mullen

frightening euphoria overtook me. The loneliness vanished and I was wrapped in a profound sense of connection—to people even more than my surroundings. I was never lonely again on the river.

North of Mattice, I was again in intact wilderness. Alone, I was keenly aware of how easily a small mistake could cascade into calamity. Solitude has an odd edge: After a few days, I found myself talking to passing rocks, a few gregarious Canada jays, and two sandpipers (*Solo* and *Back to Business,* facing page), who raised my spirits after a bout of hypothermia. They were all very supportive.

On the other hand, an oblivious moose nearly killed me by accident, and a large bear looked like he might have been considering doing it on purpose.

From the river and forest, however, I sensed only indifference. There was sound, but no noise—no urgency—and nothing I did mattered. It was humbling. Earth's age and scale pressed in, carrying a faint thrum of latent danger. Thinking of all the species that had come and gone, I realized Mother Earth births millions of marvels but is an impassive parent.

That thought circled back to the outsized cutting, mining, and damming I'd seen over the years. Such destruction now seemed rash as well as ugly. I was just one artist, but I knew many more. My earlier talks with John had focused on art. Paddling north with 200 miles of river and nothing but my thoughts for company, I realized we needed to add science and conservation.

On September 12, I reached Moose River Crossing, a small Cree village where the river is crossed by a bridge for the Ontario Northland's Polar Bear Express (the tourist train—no canoeists allowed) and the Little Bear (the train for locals, fishermen, and canoeists that no longer runs; now everyone is allowed on the Polar Bear Express). There, I caught up with three canoeists who had left Mattice a few days ahead of me but had run out of food and were waiting to flag down the

Little Bear in the morning. I had enough to share, so we made a night of it under the northern lights.

The next morning, as I dithered about continuing to Moosonee, an Elder shocked me with the news of 9/11. He told me the village had kept silent as we set camp the night before to spare us needless worry. My brother Frank's first day of duty at the Pentagon had been September 11, and my cousin Peter's office was near the top of the World Trade Center. An hour later, I was on the southbound Little Bear too.

By incredible luck, Frank was spared by the Pentagon's size, and Peter, by turning back home after a canceled meeting. Still, like many, I was too stunned to focus. It took a year and a half to begin the studio paintings I'd envisioned in my isolated ignorance on the river.

However, I soon began calling artist friends about my idea and it caught on. By summer 2004, we had a fourth expedition set for the Chapleau River, continued backing from Mad River Canoe Company, and some of the biggest names in wildlife art involved—including Robert Bateman. Just before leaving for Chapleau, I pitched the Smithsonian Arctic Studies Center's Director Dr. William Fitzhugh on combining art, science, and adventure to spotlight the boreal forest. Bill introduced me to his colleague Dr. Stephen Loring, and they added Indigenous perspectives to the interdisciplinary mix. We've been working together ever since.

A Smithsonian Institution Traveling Exhibition Service show on the boreal forest finally opened in 2023, though it morphed away from our original concept. Bill suggested I put that concept—a natural history of the boreal forest, combining artistic, scientific, and Indigenous perspectives set within a background of wilderness adventure travel—into print. This book is the result.

Forests are living systems: Remove the trees, fungi, or wildlife, and the whole collapses—just as we can't function without our own circulatory or nervous systems. Similarly, this book has three interdependent parts. Part 1 focuses on the trees and the rooted life of the forest, and its seamless blending into the tundra. Part 2 covers wildlife, leaning toward vertebrates. This isn't bias—just a time and space issue. Invertebrates are no less vital or interesting (recall my early dream of becoming the Audubon of bugs). Wild animals keep energy and nutrients cycling through ecosystems, though in a healthy ecosystem rarely is a single species essential. Earth managed without us for billions of years. Our presence is optional. Our understanding of our place is not. Therefore, we are the focus of part 3.

An important yet often underappreciated term that you will see often is "biodiversity." In its simplest essence, biodiversity is the number of species in an ecosystem. Its importance can be thought of in the context of a sports team. Its first role is having enough team members (species) to fill all the positions (or niches in an ecosystem). Its second role is to have even more team members (species) to fill a deep bench for every position (niche). When this level of biodiversity is maintained, ecosystems are at their peak of productivity and stability.

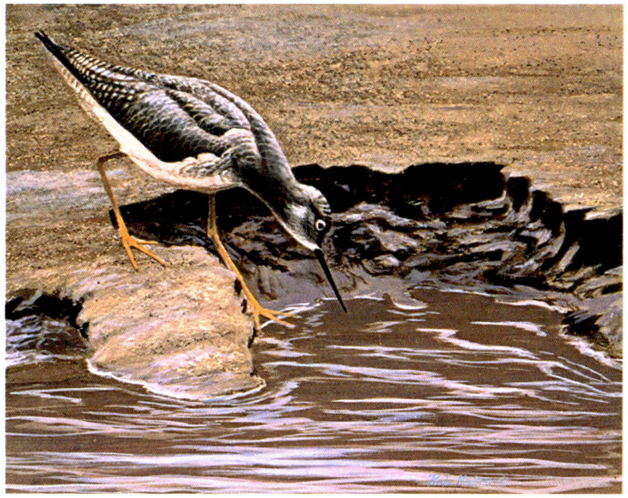

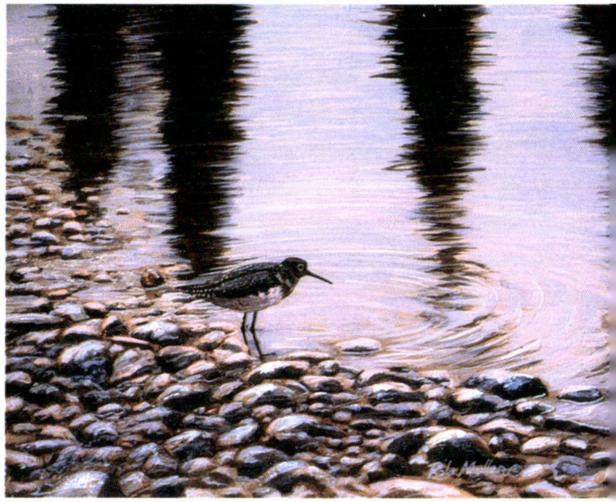

LEFT: Lesser yellowlegs, Missinaibi River, September 2001. *Back to Business*, acrylic field painting reworked in the studio, Rob Mullen

RIGHT: Solitary sandpiper, Missinaibi River, September 8, 2001. *Solo*, acrylic field painting, Rob Mullen

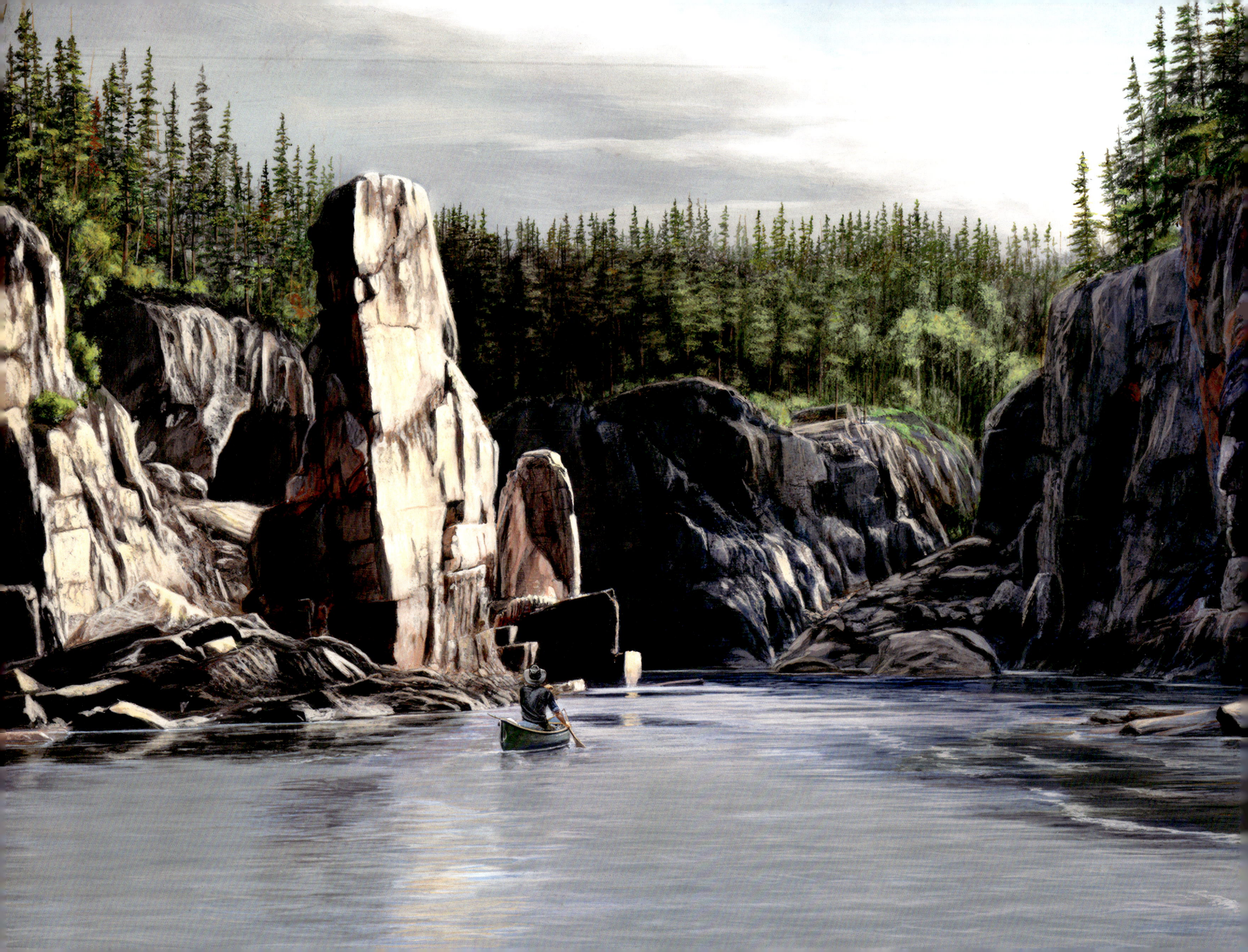

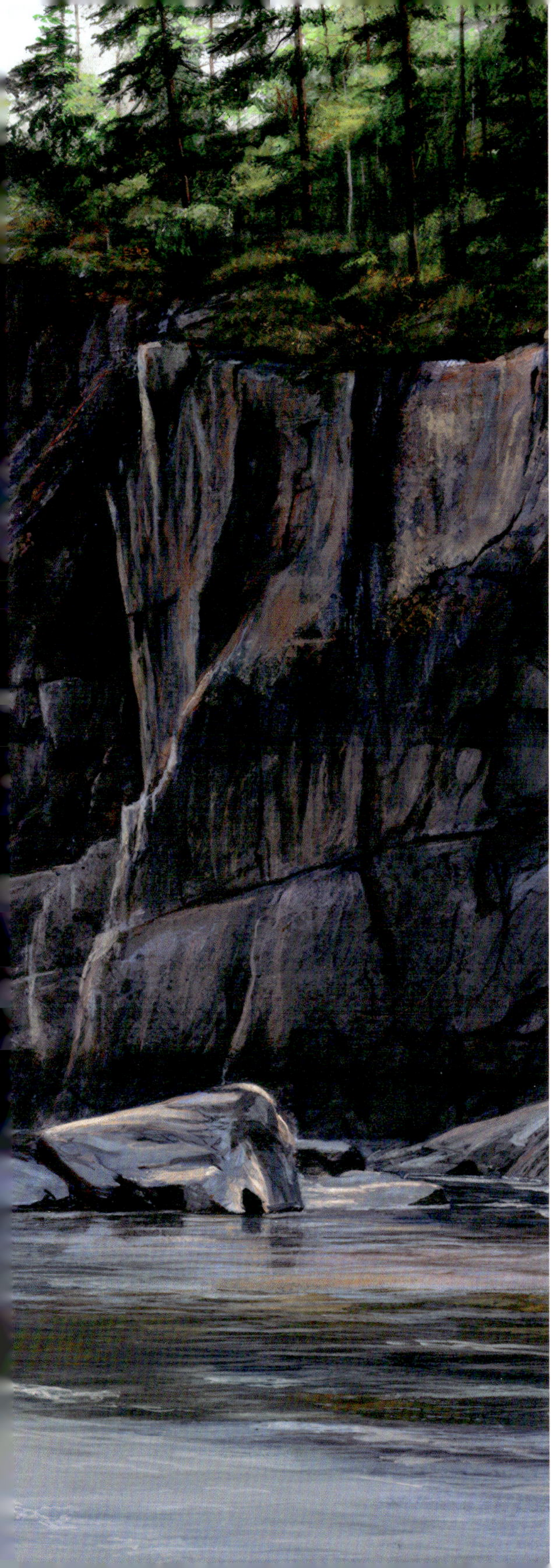

Thunderhouse Gorge, Missinaibi River. A self-portrait taking advantage of my good side. *Conjuring House Rock*, acrylic, Rob Mullen

PART ONE

TREES AND TUNDRA

FACING, CLOCKWISE FROM TOP LEFT:

Mount Igikpak, the tallest peak in the Gates of the Arctic National Park and Preserve, photographed August 7, 2013, 1:42 a.m. local time.

Angiaak Pass bear chase composite, Angiaak Pass, Brooks Range, Alaska. My two young companions chasing and playing with each other in the rain around my camp.

"Entmoot" gathering of waters and trees, Angiaak Pass, Reed River. Moving out of the pass, the trees grew more densely within the shelter of the river valley.

With my trusty .45-70 and mismatched boots. Hendrick Luckey was a German soloist on a fishing trip I met after reaching the Kobuk River, a day after an event that will be recounted later. My mismatched boots were met with polite amusement from the posse of young Iñupiaq boys who toured me around the village of Kobuk several days farther down the river from where I met Hendrick.

The first spruce, August 11, 2013.

August 10, 2013: Alone in Angiaak Pass, Brooks Range, Arctic Alaska

I had just dropped the gear pack as I leapfrogged my kit down Angiaak Pass when I noticed the black bear again—the second time in three days. The canoe pack (tarped in orange while not being carried) held all my boating gear and art supplies. I double-checked that it was secure before hauling the gear pack onward.

Rain returned as I reached a side stream. Dropping the pack and sitting on it, I pulled my green tarp over my trekking poles as a makeshift tent. Listening to the rain, I had drifted off with my Marlin .45-70 lever-action carbine across my lap when a chuffing noise near my right ear jolted me awake. Hoping it was just my hat scuffing the tarp, I turned my head slightly. Not the hat.

I pulled back the tarp's edge and stared into the wet face of the bear. My brain oddly registered the soaked triangles of fur under its eyes before I levered a round into the rifle and sent a warning shot into the hills. The bear hopped back, but not far.

That was when bear number two arrived: a cinnamon phase, bolder, and probably a sibling. After some fruitless attempts to shoo them away, the two decided I was no fun and began wrestling and chasing each other nearby. Eventually, I went to check the canoe pack a quarter mile back.

That got their interest, and they followed. The torn orange tarp flapped in the wind like a distress signal as the reprobates innocently scarfed blueberries nearby. Fortunately, the folding PakCanoe was intact. They had done only minor damage but something important was missing: my right waterproof neoprene Chota boot. Losing it meant glacier-fed misery ahead. I yelled and did a wild and crazy dance. The cinnamon phase gave me a quizzical look and went back to the blueberries. They escorted me back down but respected my personal space. As I set up camp, they continued playing and disappeared at times, one of which I used to scarf down dinner. That night, under my tarp shelter, they disconcertingly bedded down just 20 yards away. I decided to let sleeping bears lie. I made sure the rifle was close at hand but fatalistically fell asleep immediately and slept well.

Waking to heavier rain in the morning, with odd disappointment, I noticed my bears were gone. The rain lightened after breakfast, and figuring that they'd probably been chasing each other all over the tundra with it, I set off to search for the boot. Suddenly, while zigzagging through the sodden hummocks of sedge, I stopped short. After two weeks in Arctic and Arctic-alpine tundra, there it was: not the boot, but a white spruce. Less than two feet tall, it was the first outlier of the boreal forest.

Three days later, after another night with my ursine fellow lonely hearts (yes, they returned), I was deep in it. The boot never turned up.

I believe in having bear spray. In the rush to depart the airfield it was forgotten. For a discussion of bear spray versus firearms, see "Being There" in the online appendix available at robmullen.com/boreal-wilderland.

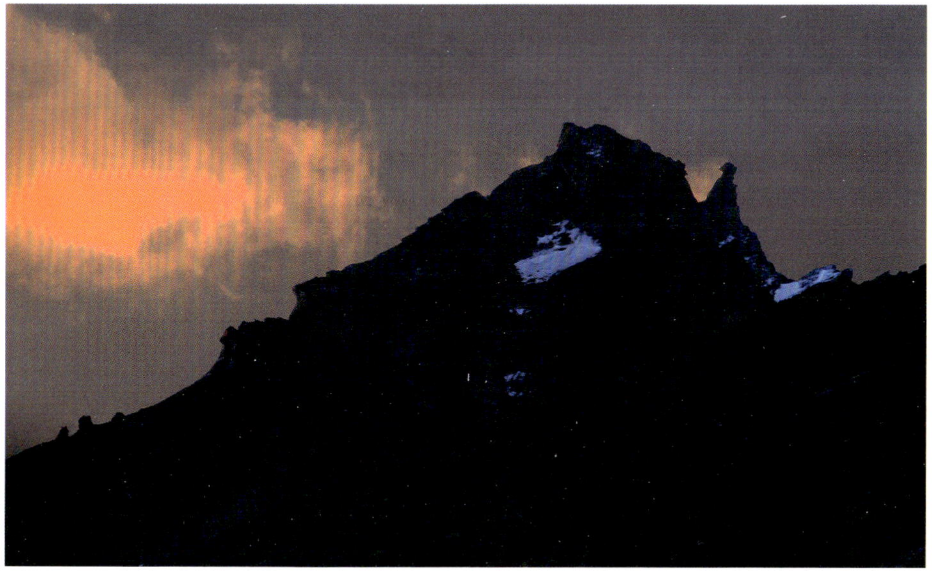
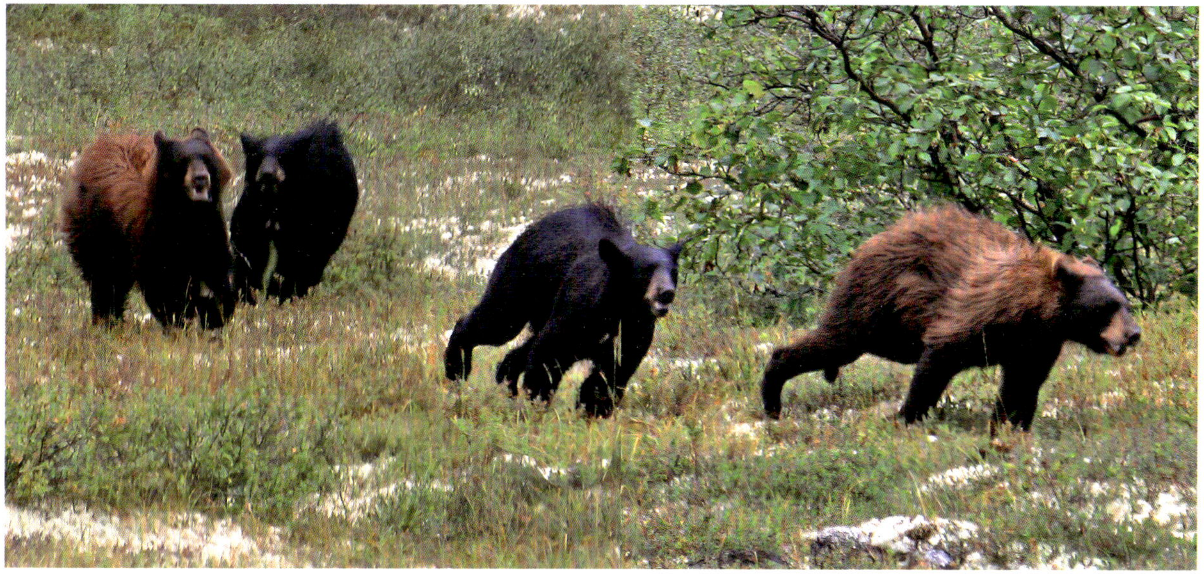
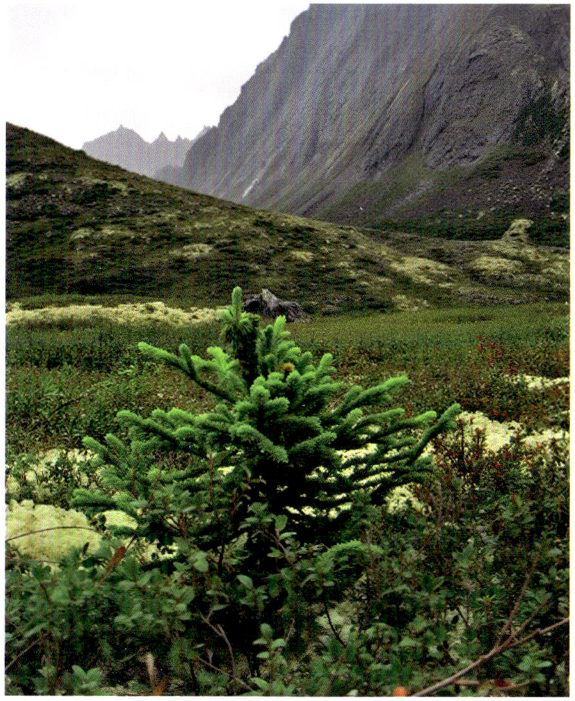
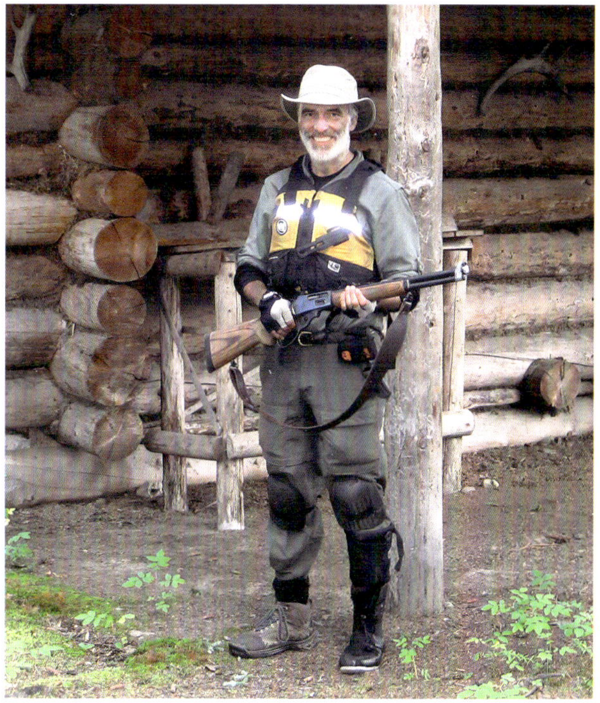
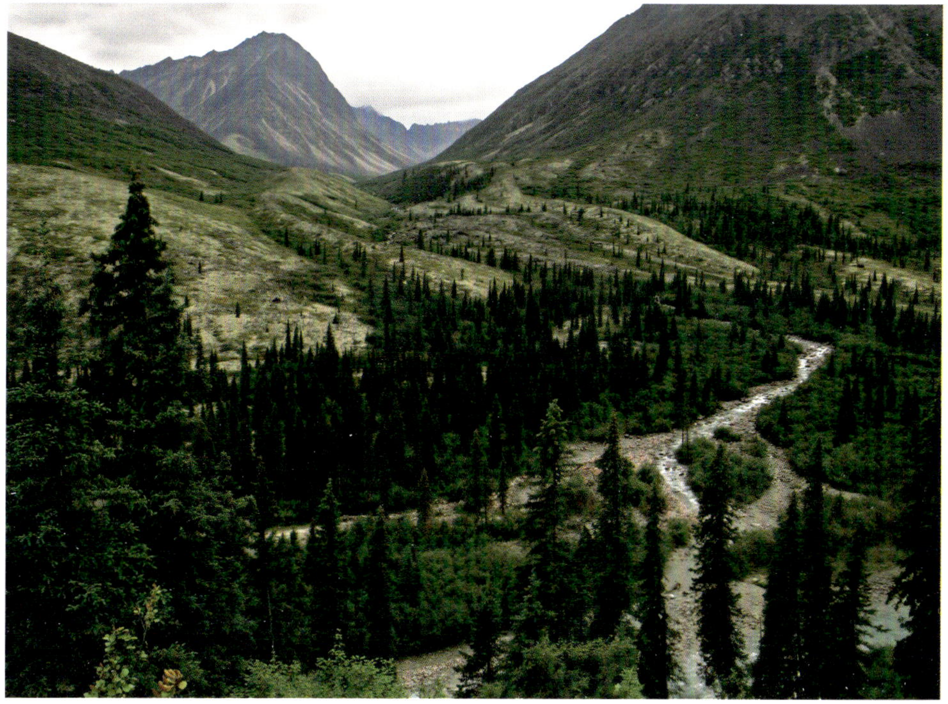

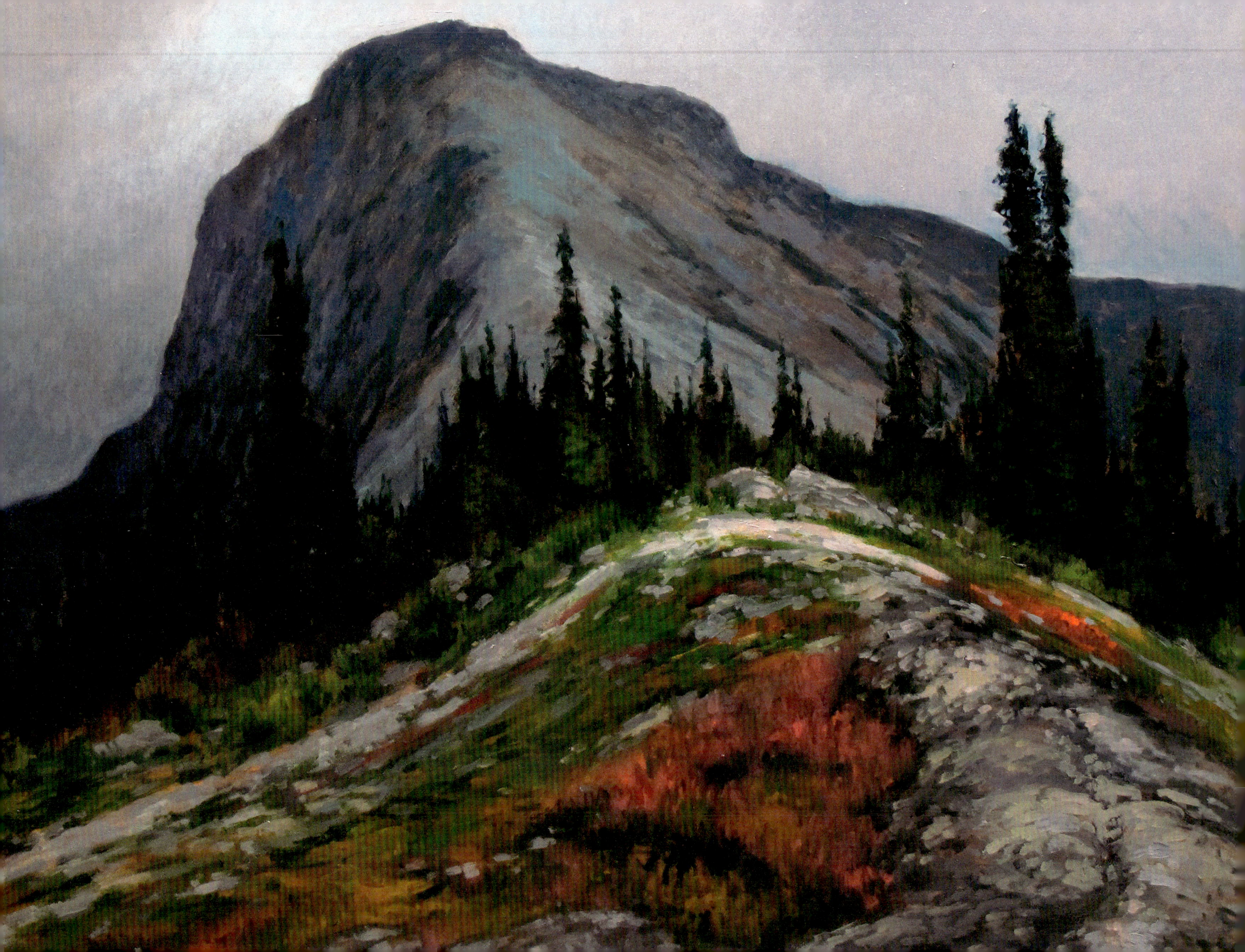

FACING: La Colline Misurtuq, George River, Nunavik. *Eagle's Peak*, oil, Jean-Louis Courteau

1

The Boreal Forest and "Wilderland"

DIGGING INTO conservation issues after my Missinaibi trip, I stumbled onto an embarrassing—and inspiring—truth. Those remote rivers I'd been paddling for years in Ontario, Quebec, and New Brunswick weren't just wild waterways; they were all part of one vast global biome: the boreal forest, Earth's largest wilderness ecosystem.

I was floored to learn that the planet's greatest forest wasn't some far-off tropical jungle, but within a day's drive of New York City, Chicago, Vancouver, and Moscow, and barely an hour from Montreal. Three times the size of the Amazon, the boreal forest wraps the Arctic in horizons of evergreens, bogs, lakes, rivers, and wetlands—from Labrador to Alaska, across Siberia and Scandinavia, even brushing Iceland and Greenland. It holds most of Earth's liquid surface freshwater and plays a critical role in carbon storage, climate, oxygen production, and wildlife survival. Here roam some of the largest free-ranging ungulate herds in the world (caribou), the last wood bison, the endangered Siberian tiger and Amur leopard, and population reservoirs of moose, wolves, grizzlies, and wolverines. Indigenous cultures have thrived here for millennia, keeping ancient knowledge alive. Still largely intact, it is now under accelerating threat from industrial development. Dr. Joshua Reichert of The Pew Charitable Trusts called the boreal forest "the greatest conservation opportunity in the world."

I'd grown up on its fringes, earned a biology degree, paddled and painted in it for nearly 20 years—and had never heard of it.

For decades, the boreal forest was overshadowed. Conservation groups focused their campaigns on small regions with many threatened species, branded with charismatic wildlife like pandas. Termed "biodiversity hotspots," they are sensible in the short term, but the strategy leaves large ecosystems, like the boreal forest, out in the cold. And, as important as biodiversity is (and it is critical as you will see), it is exactly their size that enables intact global biomes like the boreal forest to keep our world habitable. Dr. Loring summed up the typical boreal forest image: "When thought of at all, it is usually as an inhospitable wasteland, aswarm with hordes of blood-sucking insects in summer or hard-frozen under deep winter snows."

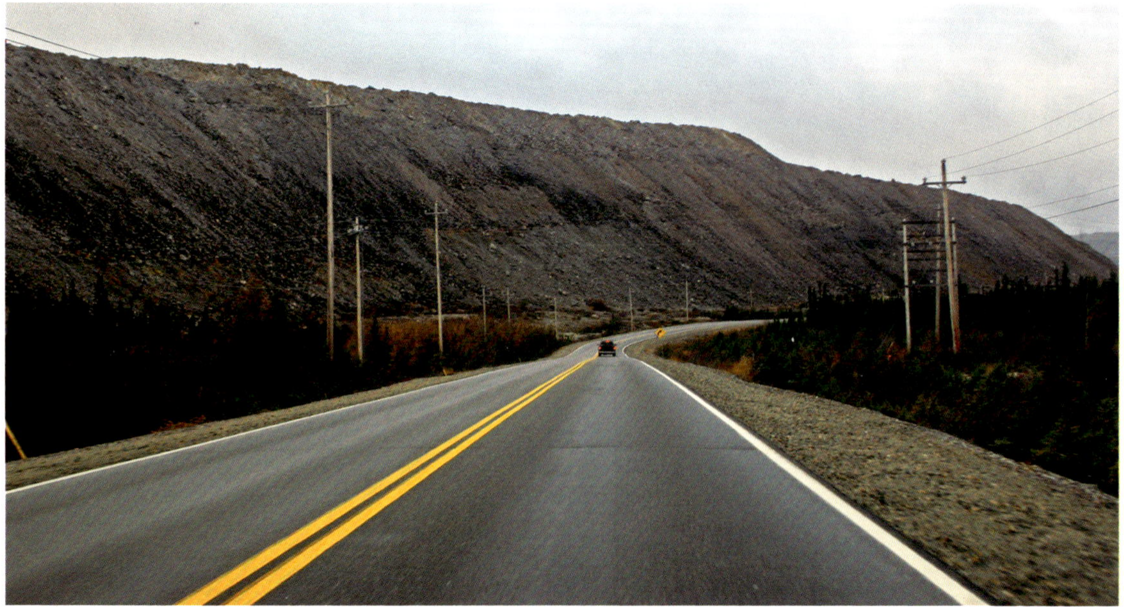

Iron mining slag mountains, Fermont, Quebec.

The insects and snow can be memorable, but this is no wasteland. Life in the boreal forest is not stunningly diverse, but it is staggeringly prolific. Caribou migrations here can surpass wildebeests in eastern and southern Africa. And while their predators may not have the primordial aura of crocodiles, a pack of wolves on the chase under a northern moon loses nothing to crocs lurking in a muddy river. And the forest is a "bird nursery," providing nesting habitat for half of North American bird species, including migratory warblers and 60 percent of its waterfowl, producing billions of new birds each year. It's also the most fungi-diverse biome on Earth, with mushrooms feeding plants and animals and quietly driving the carbon cycle. And then there are the trees.

All healthy ecosystems perform "environmental services": producing food and clean water, decomposing waste, regulating climate, and giving us breathable air. As the largest and most intact biome on land, the boreal forest is one of the planet's main life-support systems—second only to the oceans. It's done this for millennia without our management. And all for free. All we need to do is not foul it up. And there's the rub.

In 2003, Michelle Ann Marvier and Peter Kareiva termed such large, less diverse, but critical ecosystems "biodiversity coldspots."

First things first: The trees

Trees in the boreal forest number in the trillions—an order of magnitude more than the stars in our galaxy. This is fortunate, because we need them for far more than two-by-fours, junk mail, and toilet paper. Beyond producing oxygen, storing carbon, stabilizing soils, feeding wildlife, and shaping weather and climate, trees provide other, less obvious benefits. In *The Treeline*, Ben Rawlence describes how they clean the air, yield medicines, and release therapeutic aerosols. He even suggests we may one day consider lumber one of the least essential uses for them.

Another surprise: Even scientists can't agree on the forest's boundaries. Botanist Dr. James Vogelmann, senior ecologist at the U.S. Geological Survey and a friend since first grade, confirmed there's no single definition. Everyone agrees it's a cold, conifer-dominated northern forest, but soils, wetlands, latitude, altitude, and even politics shift the lines. Given the wide disagreements among reputable maps, my own estimate of the boreal's extent is likely as accurate as any. These maps aren't from an "official" panel of experts, but they're informed by decades of travel, research, and picking experts' brains.

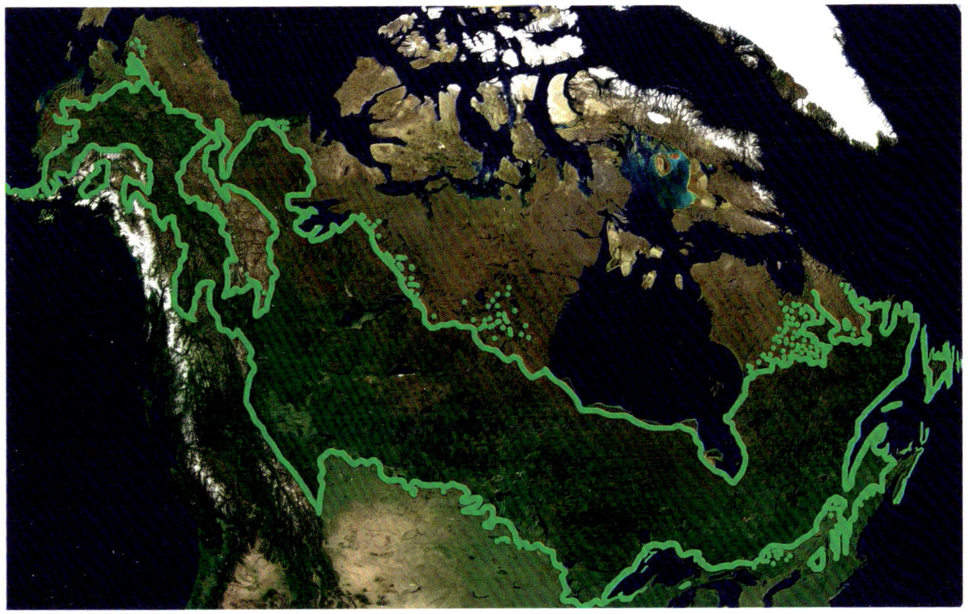

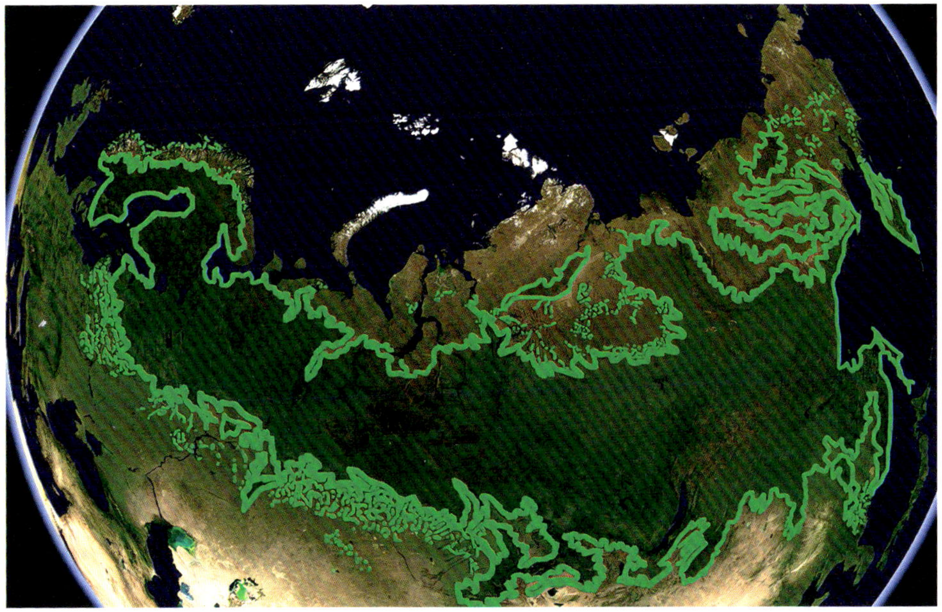

LEFT: North American boreal forest. NASA WorldWind/USGS Landsat

RIGHT: Eurasian boreal forest. NASA WorldWind/USGS Landsat

While warmer than the polar regions, the boreal forest is still one of the coldest places on Earth. Biodiversity generally decreases with temperature, so instead of teeming multitudes of tree species, as in temperate or tropical forests (the Amazon has around 16,000), the boreal forest's trillion-plus trees come from only about 20 species.

Most are conifers: spruces, larches, pines, firs, and cedars. In Eurasia, Norway spruce (*Picea abies*) and Scots pine (*Pinus sylvestris*) dominate the western boreal, giving way to Siberian (*Larix sibirica*) and Dahurian larches (*Larix gmelinii*) farther east into Asia. Across the Bering Strait grows the American larch, or tamarack (*Larix laricina*), but in North America spruces rule. White spruce (*Picea glauca*) dominates in the west, black spruce (*Picea mariana*) in the east, though both range across the continent. Balsam fir (*Abies balsamea*) and jack pine (*Pinus banksiana*) are also major players in the eastern two-thirds, but spruces most define the forest. Note, conifers are often called evergreens, and generally that fits. Larches (tamarack) are the exception: Their needles turn brilliant yellow in fall and drop for the winter (deciduous).

There are hardwoods here too: modern deciduous, broad-leaf trees (flowering plants, aka angiosperms) such as birches, aspens, alders, and willows (alders and willows are often shrub-sized in the harsh conditions of the boreal forest).

The understory

Speaking of shrubs—trees aren't all there is to the boreal forest. Look down. Blueberries, bilberries, cloudberries, crow-berries, huckleberries, bearberries, snowberries, cranberries, salmonberries, serviceberries, Labrador tea (a rhododendron), mountain ash, dwarf willows, dwarf birch, wild rose, fireweed, grasses, water lilies, pickerelweed, pondweed, water shield,

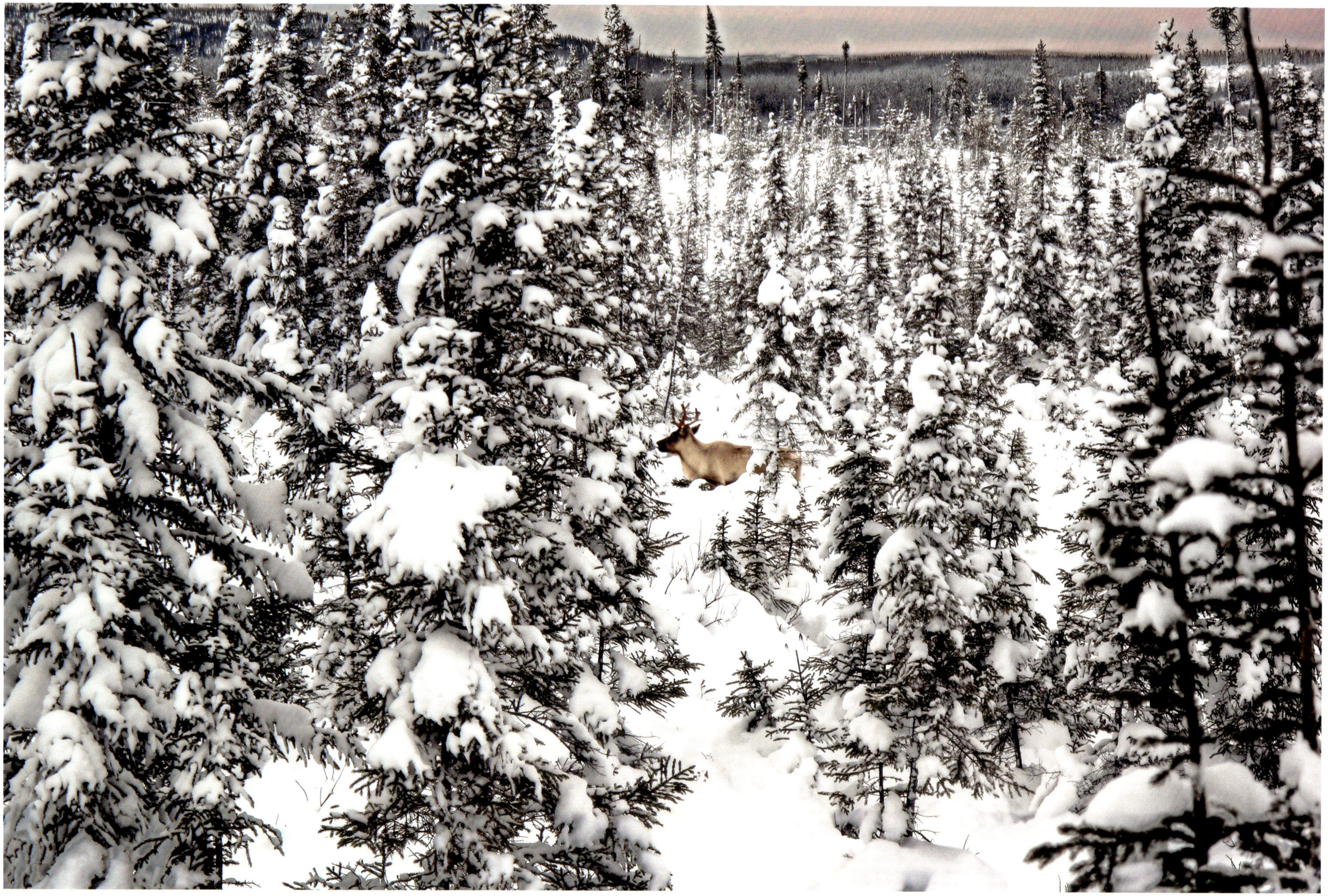

duckweed, cattails, pitcher plants, sundews (plus a profusion of other vascular plants) carpet the forest floor, wetlands, openings, and tundra margins.

Adding a twist to the old saying about not seeing the forest for the trees is the fact that, in the boreal, the greatest biodiversity is on and just below the ground. Much of the forest's carbon storage is there too. And that's just the modern vascular plants, those with internal water and nutrient transport. In addition to them, the following is a brief overview of some of the more primitive vascular plants, legions of non-vascular plants, and whole groups of organisms that aren't plants or animals, all of which add to the understory's role in the forest's functioning, biodiversity, and carbon storage (limiting carbon dioxide in the atmosphere by incorporating it into organic matter—more in chapter 8).

Ferns and horsetails

Ferns are familiar and well represented in the boreal forest and can make it appear like a jungle in places. Reproducing via spores, they are in the same class as horsetails: ancient "living fossils." Common in wet soils, often in dense clusters, horsetails' segmented stems contain silica, giving them a gritty texture. Before I embraced "leave no trace" camping (in my Boy Scout days they still recommended cutting evergreen boughs for tent floors), we used horsetails to scrub burnt bits out of cooking pots. There were always burnt bits.

Clubmosses (Lycopods)

This group includes the ground pines—princess pine and running pine—overharvested in the past for Christmas decorations. They're the most primitive living vascular plants, miniature descendants of the hundred-foot giants that dominated steamy Carboniferous forests 300 million years ago along with some ancestors of the horsetails.

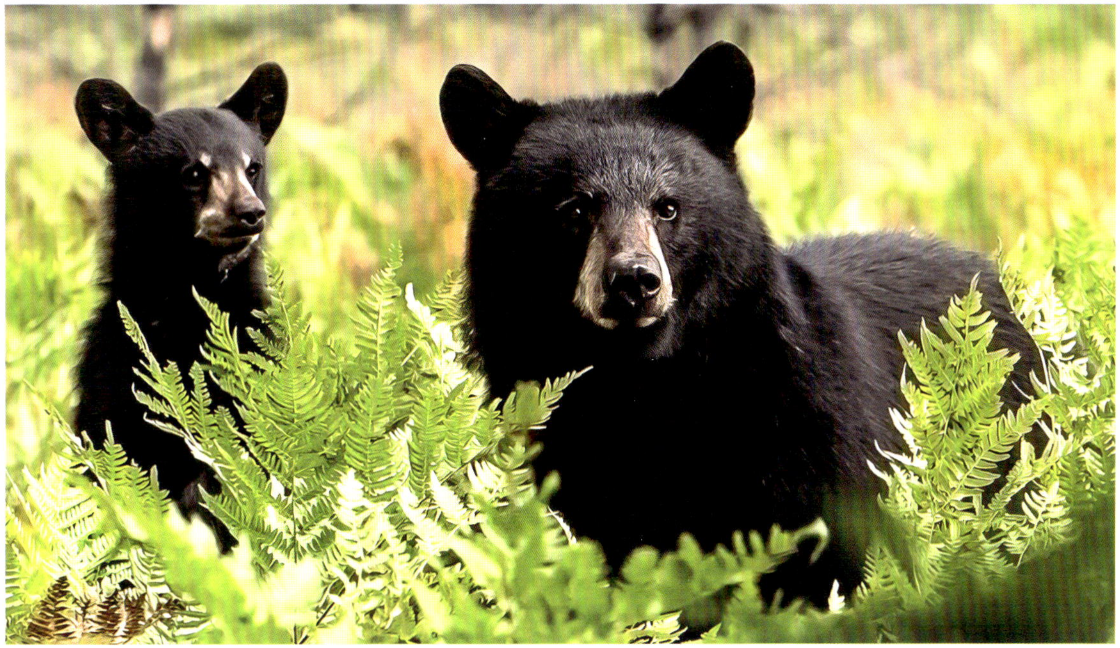

Black bear sow and cub in ferns. Algonquin Provincial Park, Ontario. Steve Bruno

FACING: Lone caribou cow of the Leaf River herd bedded in boreal forest, Nunavik. Susan C. Morse

Bryophytes

Non-vascular plants, mosses, liverworts, and hornworts were among the first plants to colonize land—about 450 million years ago in the Ordovician Period. Today, mosses (especially sphagnum) are the most abundant plants in the boreal forest and are essentially a keystone species. Vast sphagnum beds cover much of the North American boreal floor, absorbing up to 20 times their weight in water, buffering moisture levels, and even overgrowing ponds to create bogs—the main source of peat.

They can also coat tree trunks, though their supposed preference for the north side is weak and, contrary to popular wisdom, an unreliable navigational aid. Lacking true roots, leaves, stems, or vascular tissue, they rely on absorption, diffusion, and osmosis—and partnerships with fungal mycelia—for water and nutrients. They help break down minerals to build soil, paving the way for more advanced vascular plants; provide habitat for small organisms; and, as photosynthetic primary producers, generate oxygen.

Algae

Probably the oldest plants on Earth, possibly dating back 1.6 billion years. While algae are mainly aquatic, some species grow in damp soils, on tree trunks, and on rocks. They remain important as primary producers.

"Blue-green algae" aren't algae at all, but photosynthetic cyanobacteria. About 3.5 billion years old, they were crucial in supplying the oxygen that allowed more complex life, including true algae, to evolve. Today, they still produce food and oxygen, though in warm waters they can bloom excessively and cause trouble.

John Pitcher and Lindsey Foggett examining lichens, George River

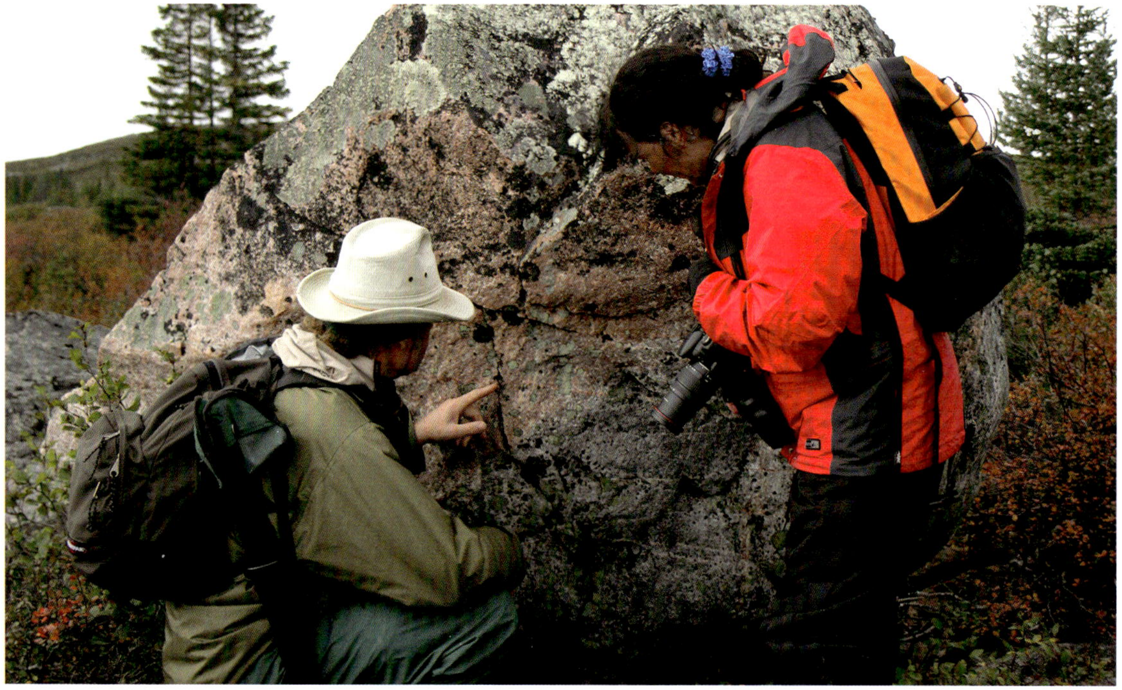

Primary producers are species (most plants and some bacteria) that form the base of food chains, principally by harvesting energy from the sun and carbon from the atmosphere to create sugars (photosynthesis), making carbon and energy available to all other organisms in the ecosystem in the form of the sugar glucose.

Oddballs

Mycoheterotrophic plants: These lack chlorophyll and get food via mycorrhizal fungi from other plants (the meaning of "mycoheterotrophic" in Latin). This can be a permanent or temporary trait. Orchids—nearly all mycoheterotrophic when young—form the largest plant family on Earth. Lacking green pigment, these plants are often mistaken for fungi. A common example is Indian pipe, which I first knew as "Indian pipe mushrooms" as a kid. They're not mushrooms.

Some other botanical outliers are carnivorous, supplementing their diet with insects or other small animals. You will encounter two of them later in chapter 8.

Lichens

Leaving plants behind, we will move next to the lichens. Lichens can be part plant. They are symbiotic partnerships between fungi and either algae or cyanobacteria, and often striking in form, color, and texture.

Because of this partnership, lichens can create food via photosynthesis and feed via decomposition. This dual strategy allows them to thrive where plants fear to tread.

They flourish in tundra and open-canopy forest, where, uniquely among deer, caribou can subsist on some species. This diet lets caribou range far north of other deer—and possibly most wolves.

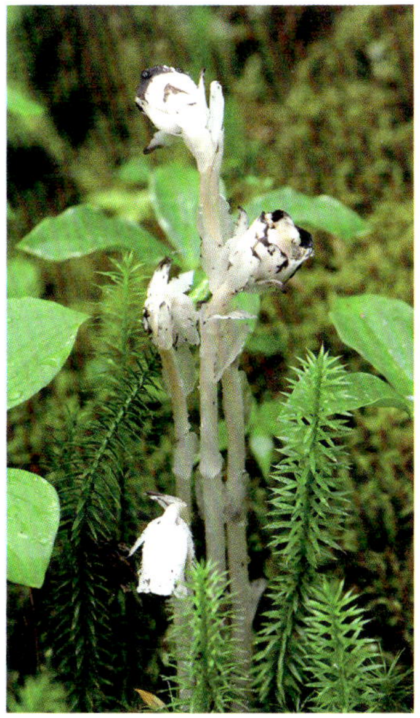
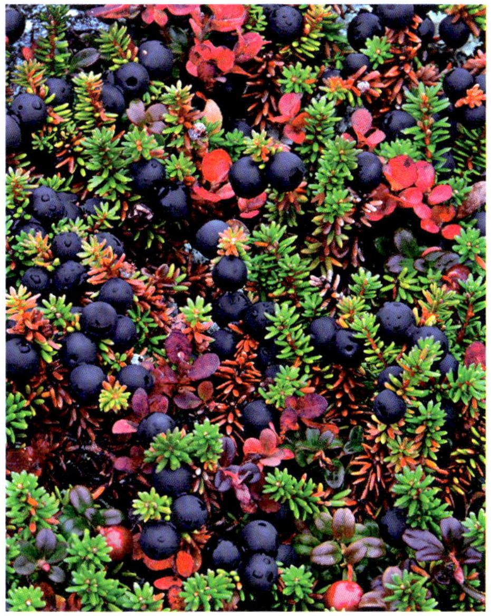
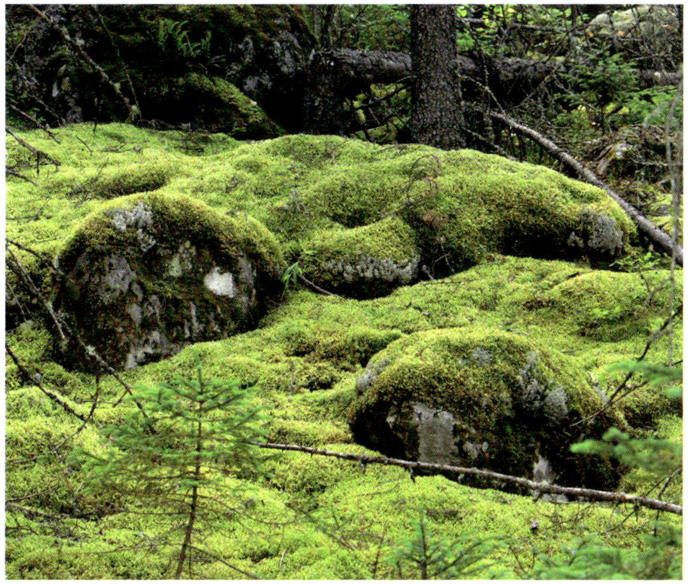
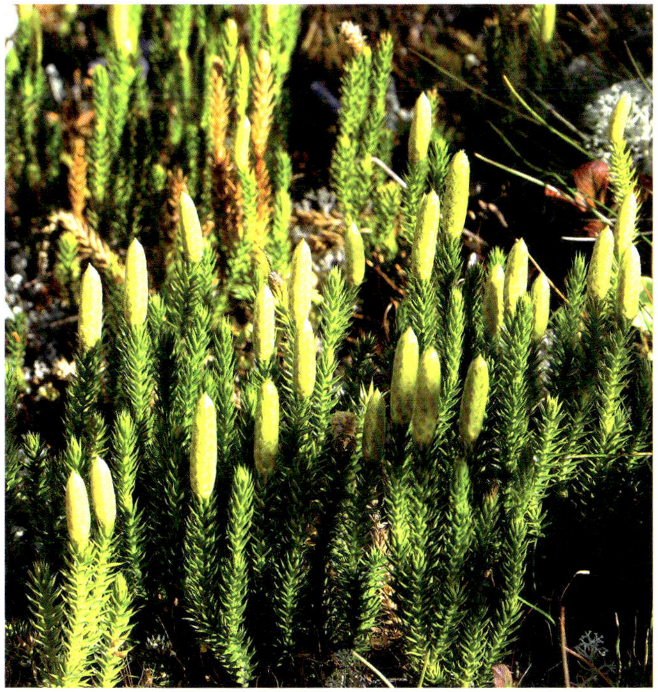
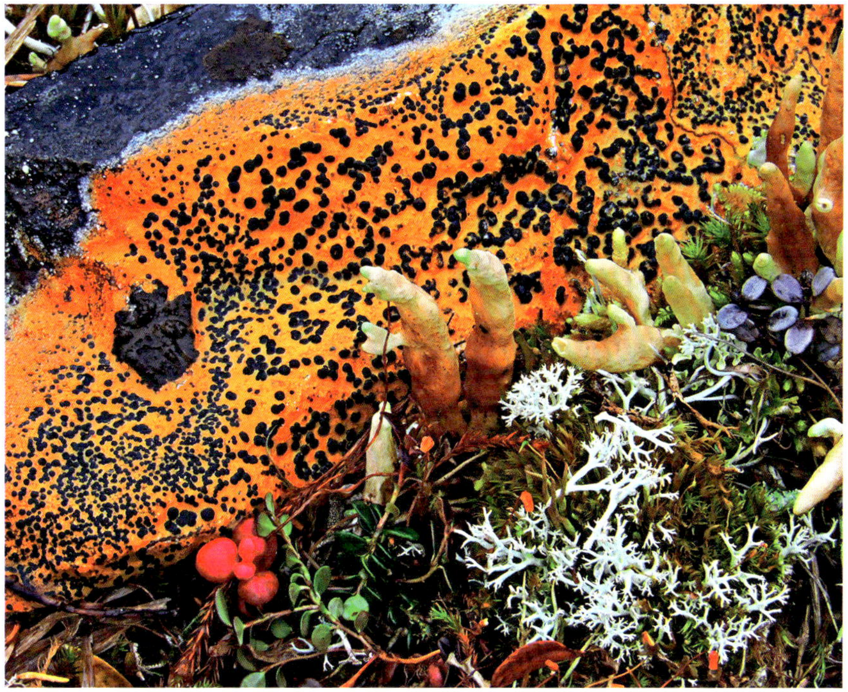

CLOCKWISE FROM TOP LEFT:

Indian pipe, Algonquin Park, Ontario, circa 2004. Bonnie Rowell

Crowberries, bearberries, cranberries, bilberries, ferns, clubmoss, bryophytes, and lichens. Susan C. Morse

Carpet of moss on the forest floor, northwestern Ontario, Smoke on the Water expedition, August 2011. Mosses are the most abundant plants in the boreal forest.

Orange boulder lichen (*Porpidia flavocaerulescens*) and friends, Ungava, Nunavik. Susan C. Morse

Clubmoss on the De Pas River.

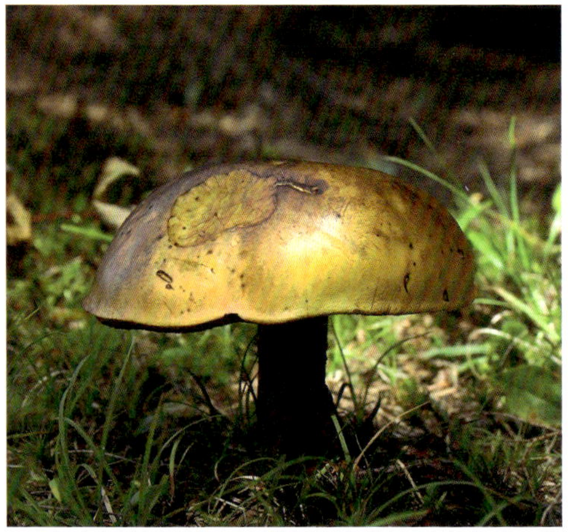
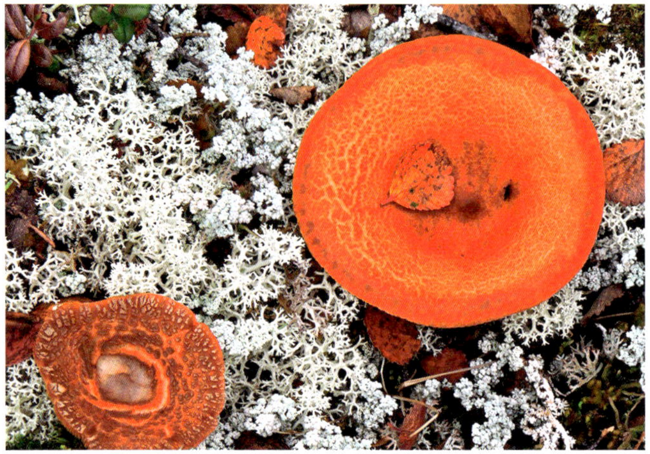
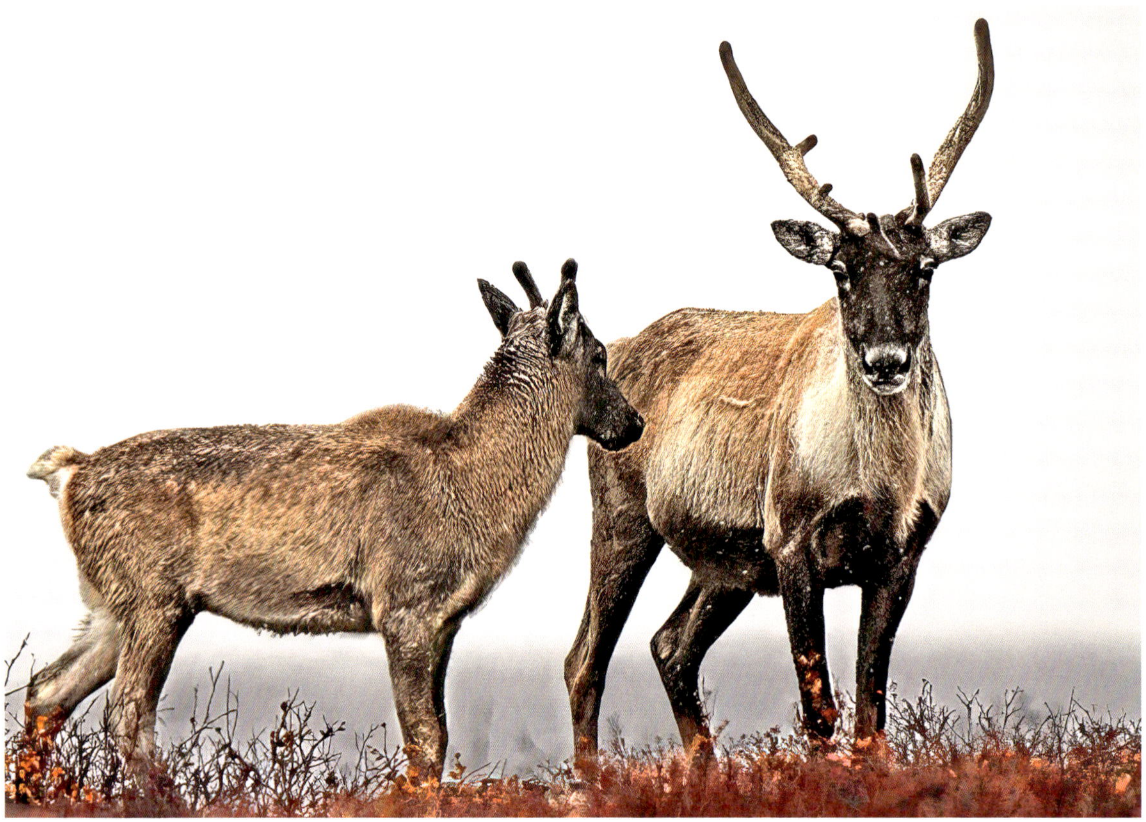
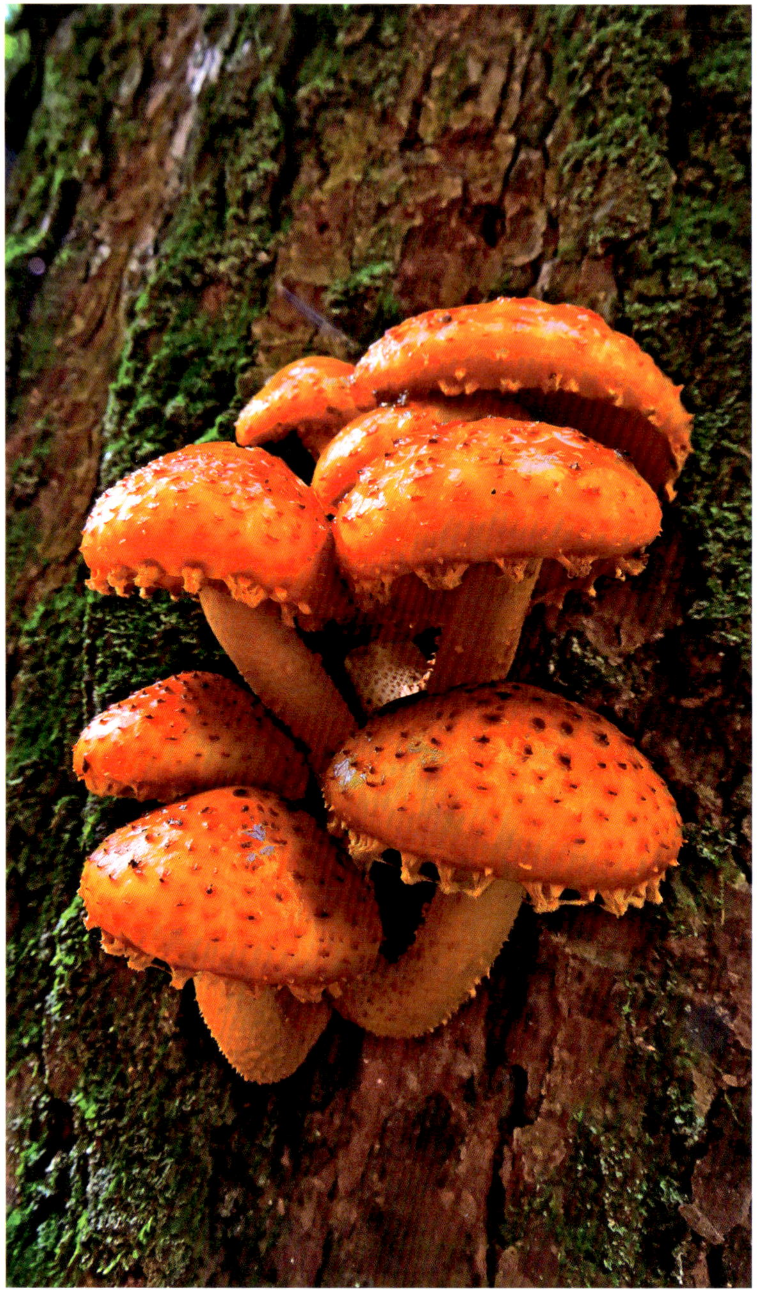

Consumers and decomposers both get carbon, energy, and nutrients from other organisms. The difference: Consumers digest food internally, while decomposers excrete enzymes to digest food externally, then absorb it.

Fungi

Fungi, despite the plant-like look of some mushrooms, are even less plant-like than lichens. They have no photosynthetic ability, feeding entirely on other organisms, living or dead (they are heterotrophic). Nonetheless, despite not being able to move except for their spores and growth, they have thrived for over a billion years in every ecosystem. In the cool, damp boreal forest, they reach their peak diversity and are critical to ecosystem function.

While famous as decomposers, fungi also play another essential role. Many, known as mycorrhizal fungi ("myco" means fungus and "rhiza" means root), spread via fine, densely woven threads called hyphae, which intertwine with plant roots. Formerly considered parasitic, they are now widely recognized as a mutualistic relationship that probably formed as plants colonized land. The fungi break down organic material in the soil, extracting nutrients like phosphorus and nitrogen and providing them to plants. In return, they receive sugars the plants make via photosynthesis.

Most land plants today, including boreal trees, rely on mycorrhizal fungi. The mushrooms we see are just the fruiting bodies (spore dispersers)—most of the organisms' tissues are underground. A handful of forest soil can contain hundreds of miles of fungal hyphae connected to tree roots, trading nutrients and even chemical signals. This "wood-wide web," explored by Suzanne Simard in her revolutionary book *Finding the Mother Tree*, suggests trees may share resources and even "communicate" with each other through these networks—a hot topic in botany.

Once carbon and nutrients travel through the food chain and end up as waste or dead tissue, decomposers (primarily fungi and bacteria) recycle them back into forms plants can use, priming the ecosystem's energy engine. Some fungi are the only organisms that can break down lignin, the tough hydrocarbon polymer that stiffens wood, making them indispensable to forest ecosystems.

Fungi also serve as food—and not just for gourmet humans. Many animals rely on them, including caribou. While in Ungava, naturalist, tracker, forester, and photographer Susan C. Morse witnessed this firsthand:

Early one September morning in Northern Quebec's Ungava Peninsula, several hundred Leaf River herd caribou were migrating by me when a cow and her calf abruptly left the band and entered a copse of black spruce. Their tracks in the new-fallen snow led me to group after group of mushrooms that the animals had consumed. Many decades ago, revered field biologist Olaus Murie made the same observation: "In July and August, when mushrooms become plentiful, these are promptly placed on the caribou menu."

Plant communities

Plants, like all of us, have needs. Where they can grow depends on where those needs are met. The first and foremost factors are temperature and rainfall—climate. On a large scale, this is a matter of latitude: for example, the northern boreal forest at high latitude. On a small scale, it can be a matter of altitude.

The presence—or absence—of other organisms can also determine whether a habitat works for plants. The assemblages of plant species and associated life that coexist under given environmental conditions are called "plant communities." In forests, they're often named for their dominant trees.

CLOCKWISE FROM TOP LEFT:
Fruiting bodies (mushrooms) of ectomycorrhizal fungi. Bonnie Rowell

Mushrooms on the menu, Ungava. Susan C. Morse

Golden scalycap fungi on hemlock, Ungava. Susan C. Morse

Cow and calf after eating mushrooms, Ungava. Susan C. Morse

East end of Kamestastin Lake in north-central Labrador, showing the effect of elevation (and wind). At the top, conditions are Antarctic: bare rock and lichens. Eight hundred feet lower, there are trees. Kamestastin Lake lies at the bottom of a 32-to-38-million-year-old meteorite crater and is deep enough to shelter trees from wind and offer slightly warmer conditions, October 2008

The larch forests of Eurasia and spruce forests of North America are stable, self-sustaining climax communities. But they don't appear out of thin air. Ground scoured and crushed by ice for a hundred thousand years is not the same as soil ready to support trees. Recolonization starts with organisms able to tolerate those bare-bones conditions. As they alter the site, new species can thrive, further changing conditions for yet more species, and so on until the area reaches a climax stage again.

This *succession* process can start from scratch after an ice age; partway along, with soils intact after a fire; or nearly complete, with soils untouched after a windstorm. Disturbance—whether millennia of ice, a day of fire, or a single night's blowdown—sets the cycle in motion.

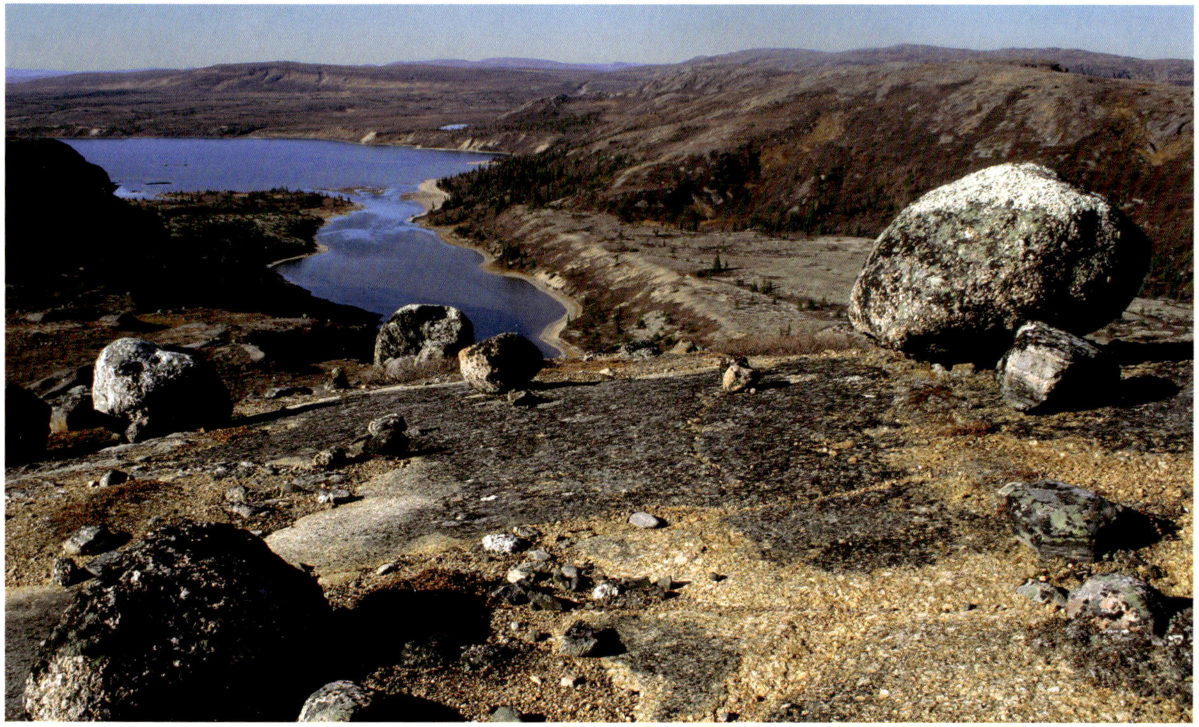

Such disturbances, especially fire on a large scale and storm blowdowns on a smaller one, are the norm in the boreal forest. They create a shifting patchwork of plant communities, each providing different growing conditions and food and shelter opportunities for wildlife. This mosaic adds to the forest's richness, diversity, and long-term stability.

Bogs

Among the boreal forest's most important wetland communities are bogs that create vast carbon stores in the form of partially decomposed peat. Sphagnum moss is the main bog-builder, growing so densely it can choke off groundwater inlets and outlets and reduce atmospheric exchange. Once cut off, the wetland relies solely on precipitation and loses buffering minerals from ground water that keep acidity in check. On bedrock, clay, or permafrost, the area becomes waterlogged, acidic, and low in oxygen—the hallmarks of a bog. Decomposition, ironically quite often of the dying moss on the underside of the growing mat, is slowed and the partially decomposed material accumulates. Dried, it burns readily.

Bogs cover immense areas. The Hudson Bay Lowlands are one of the largest wetland complexes on Earth and mostly bog. In Siberia, the Great Vasyugan Mire, which includes vast expanses of bog, is larger than Switzerland.

Boreal Wilderland

Animals, like plants (over time), go where they will. Many boreal species use the adjoining tundra between the last trees and the Arctic Ocean, while tundra species often use the forest. On a map, it's tempting to draw a hard line—forest here, tundra there—but in the field, that line blurs and disappears.

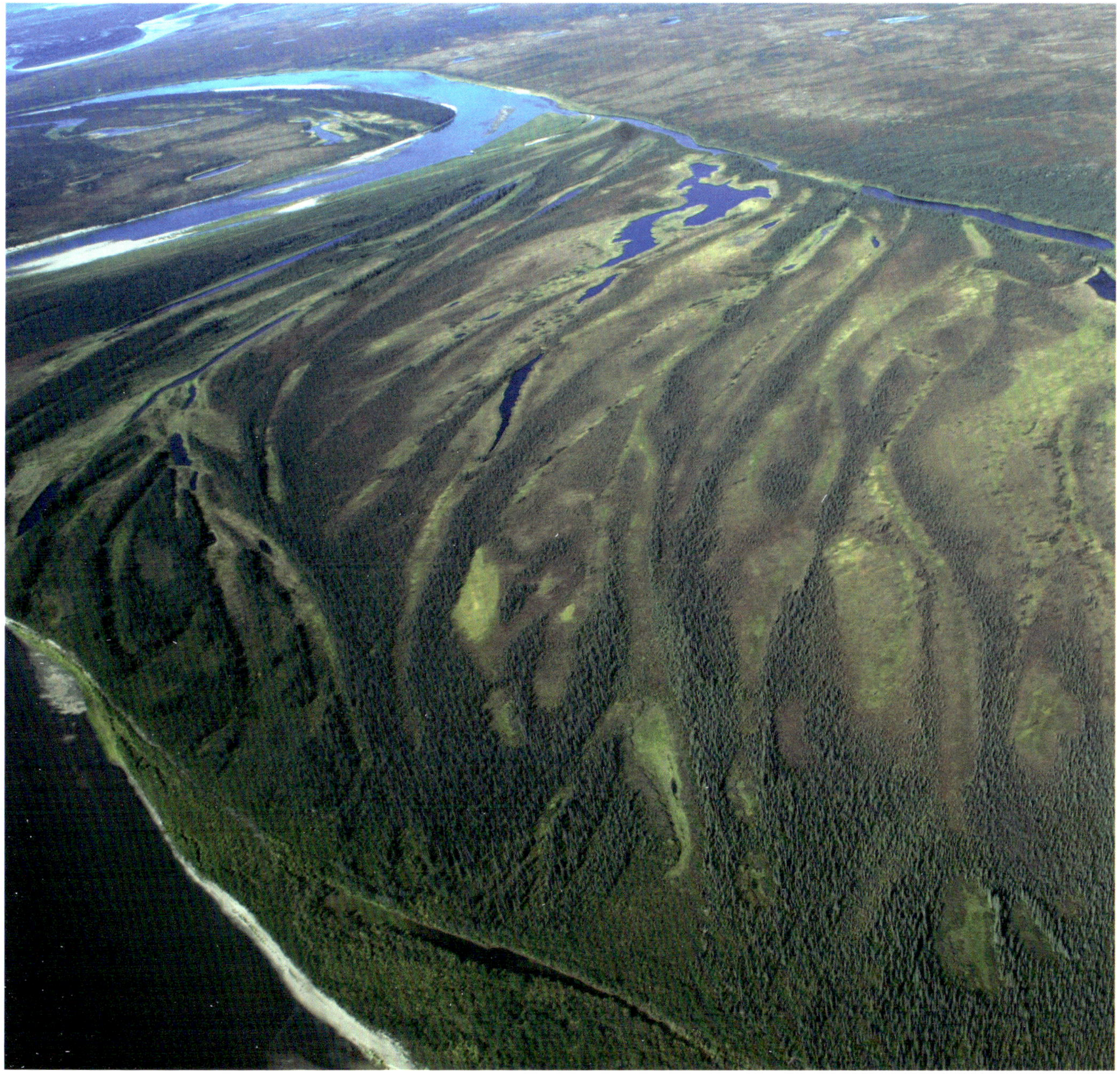

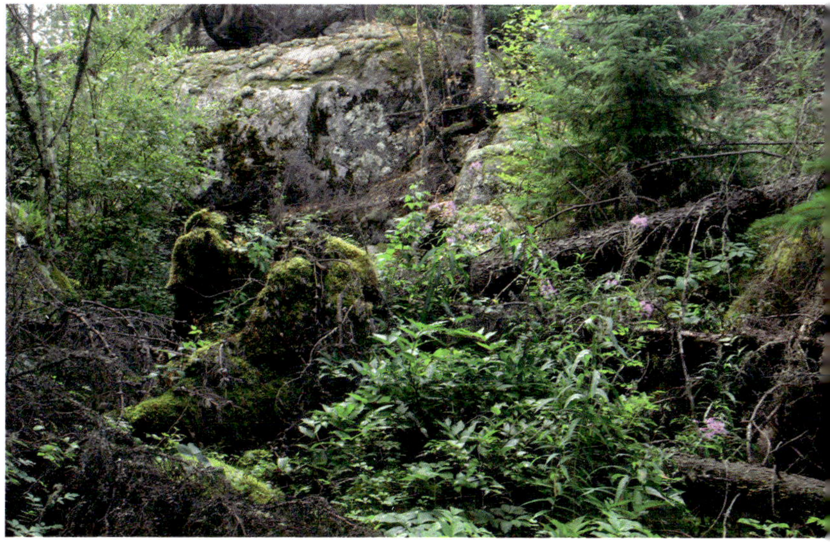

LEFT: String bogs along the Kobuk River, Alaska.

ABOVE: Closed canopy forest blow-down with successional plants such as fireweed recolonizing, Woodland Caribou Provincial Park, Ontario.

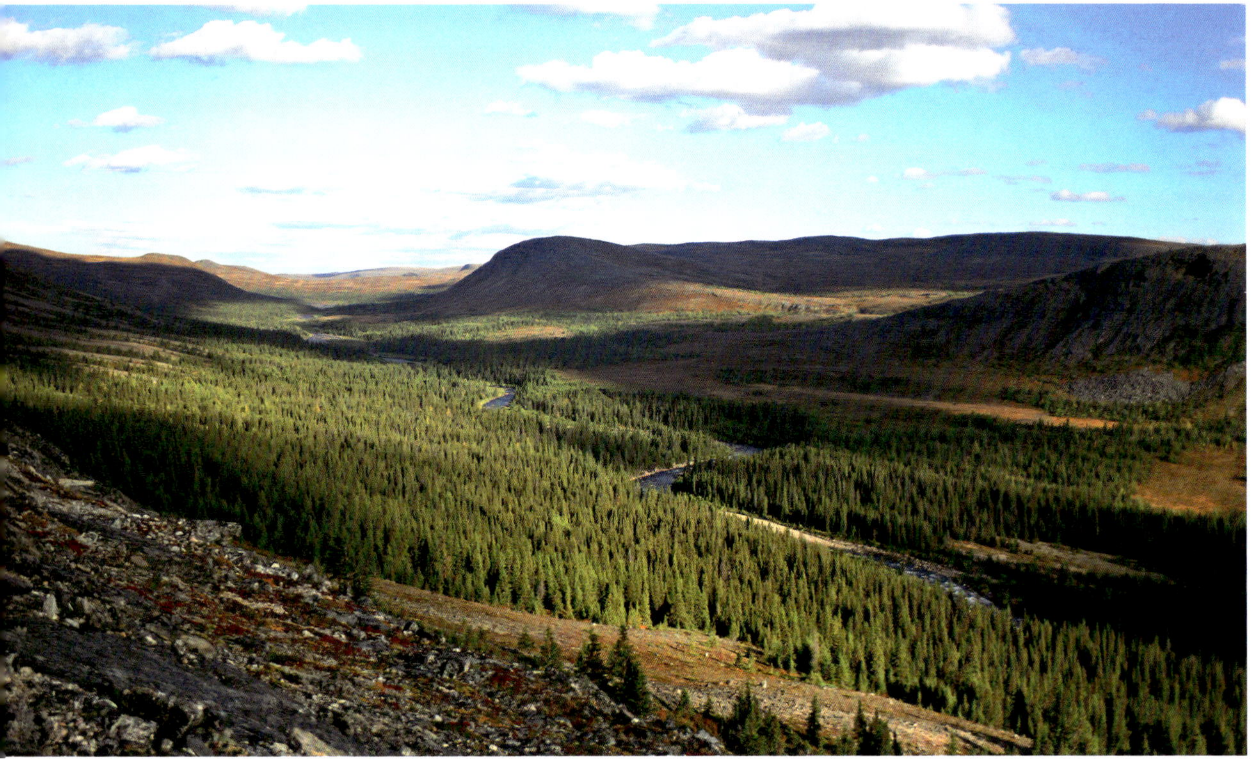

Valley of the Nutillilik River from the Colline Misurtuq, showing the mosaic of forest and tundra, George River expedition, September 2006.

FACING: Northern lights on the George River. *Northern Night*, work in progress (I'm debating whether to keep the owl), oil, Rob Mullen

In the Far North, cold, wind, and dryness challenge tree growth. Young trees take root wherever they can find suitable temperatures, moisture, and shelter. On the Canadian Shield's scarred face, this creates a captivating mosaic of forest fingers reaching into the Barrens along lakes, rivers, and hollows. This interface is both forest and tundra. (Note: The Barrens, also known as the "Barrenlands," or Barren Grounds, with spelling variants, is mostly a Canadian term for the treeless northern tundra.)

On the George River en route to Ungava Bay, we paddled through ribbons of dark spruce flanked by the looming heights of the Barrens. A couple of artist crewmates said what I'd been thinking, "Middle-earth, Tolkienesque": the towering tops standing mute witness to distant ages, the dark forest hiding mysteries, and the vast, lonely silence of the Barrens evoking a world before humanity. This transition zone provides the "edge effect" for ecologists, compositional opportunities for artists, and a more diverse menu for wildlife. In many respects, it is possibly more accurate to consider the two biomes one great boreal "Wilderland" (the name the Hobbits gave to the wilds east of the Misty Mountains, which the elves termed "Rhovanion," in J.R.R. Tolkien's *The Hobbit*).

The boreal's character shifts across its vast range.

July 1989: My brother Frank and I tried to pitch a tent among black spruce no thicker than my wrist and barely two feet apart while a man-eating mass of blackflies dined. That is the boreal forest.

July 1990: Camped at Thunder Falls on the upper Missinaibi, my first wife, father, two friends, and I spent a perfect northern evening under a cathedral of white spruce, Scotch in hand, and no bugs. That is the boreal forest.

September 2009: Up late with logistical worries, I watched the northern lights spill across a pale-green carpet of caribou moss, punctuated by eerie fang-like spires of black spruce. That is also the boreal forest.

Closed canopy, open canopy, taiga, park, bog, tundra—all evoke the wonder I first felt as a child, listening to my father read *The Hobbit* and *The Lord of the Rings* and make it real. Even now, in my studio, I can feel the cool wind, the cold lakes, the looming highlands; I can see the rising moon, the bears, moose, and caribou; I can hear the loons, owls, and the wolves.

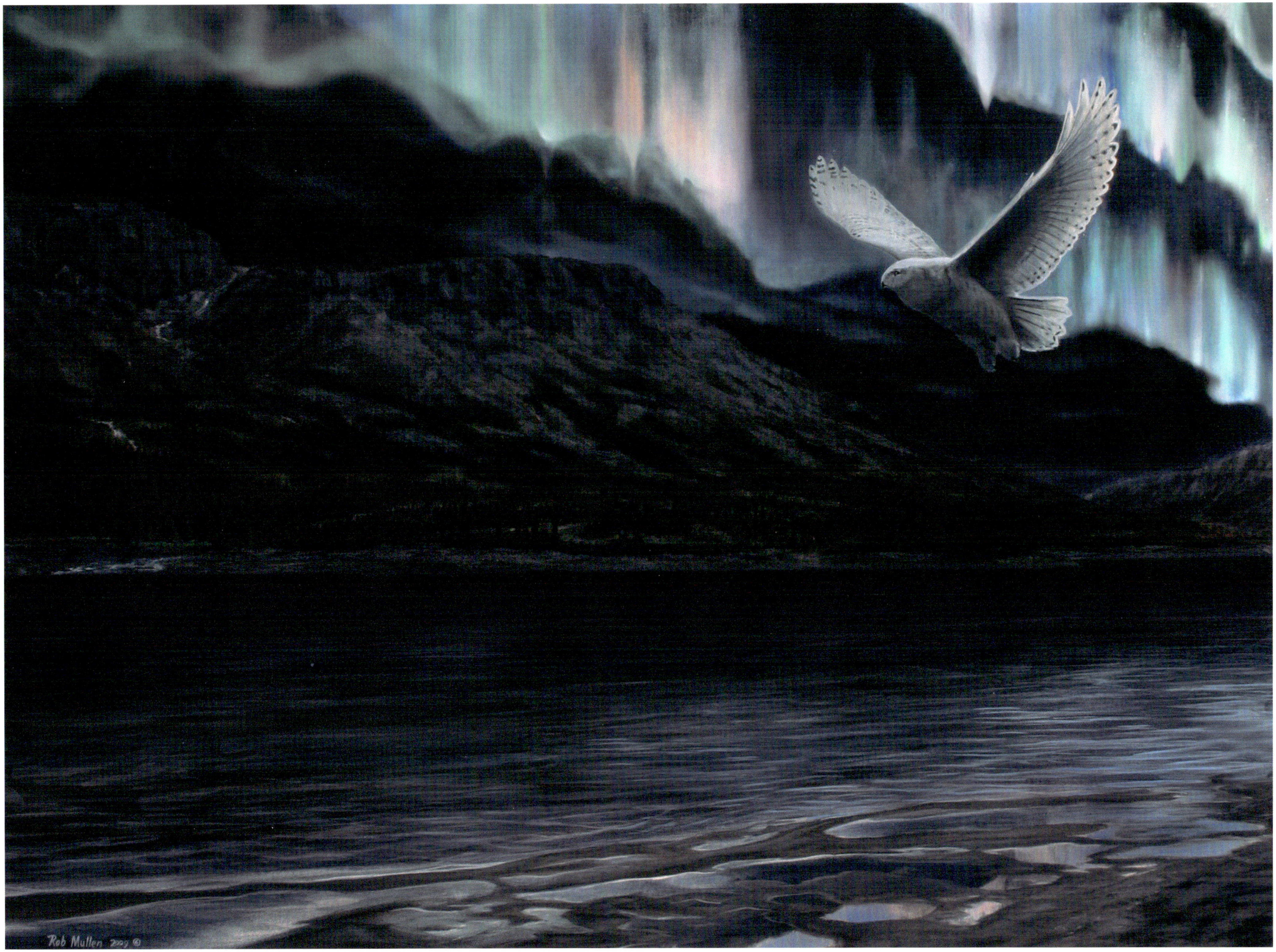
Rob Mullen 2009 ©

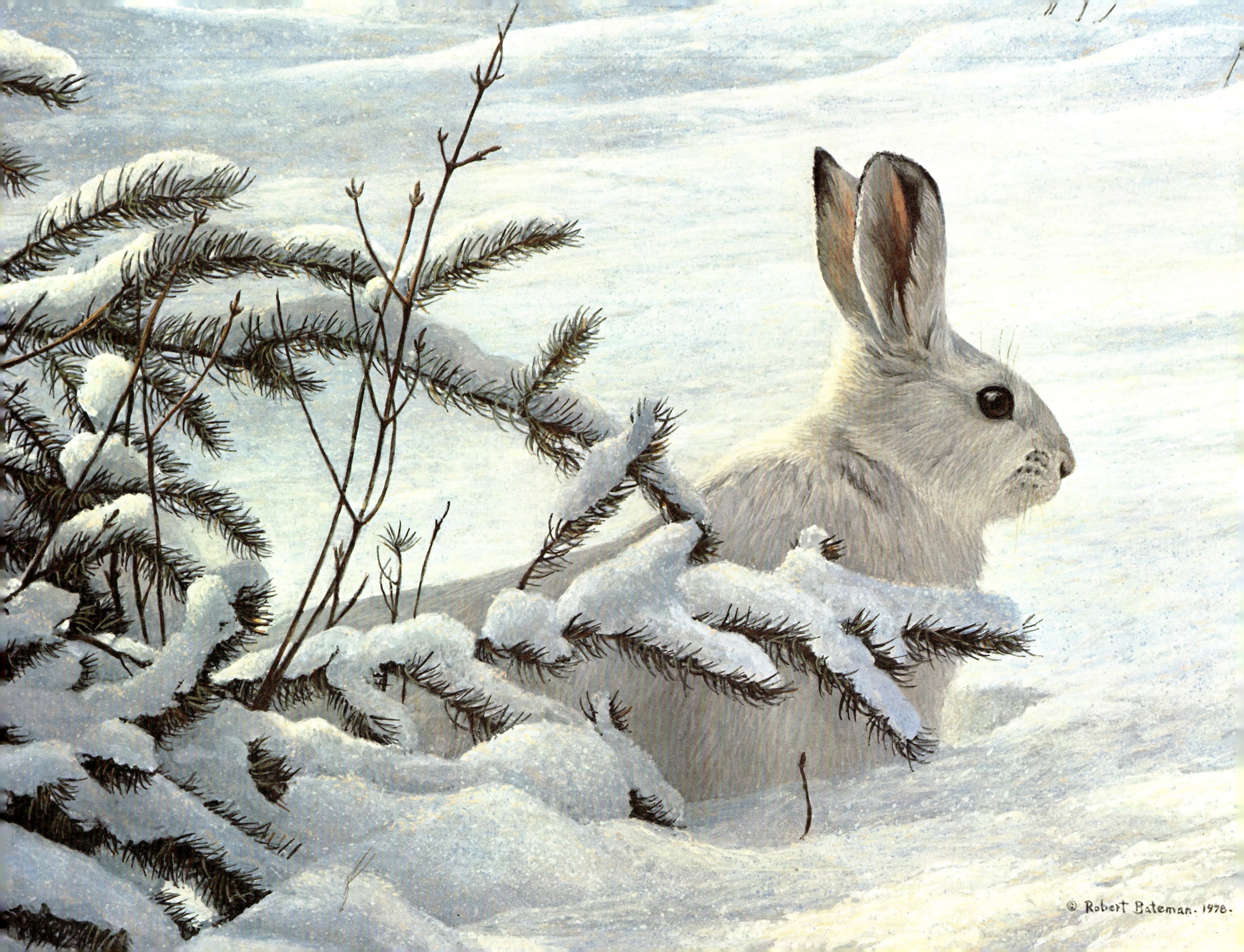
© Robert Bateman. 1978.

FACING: *Winter Snowshoe Hare*, acrylic, Robert Bateman

2

Architects of the Boreal Forest: Winter, Water, and Fire

August 2011: From the small mount we'd climbed, the land stretched away to the east, south, and west. In one sweep of the eye, you could see the work of three forces that shape the boreal forest: winter, water, and fire (see photo on page 43).

Winter's signature was in the dominance of conifers, especially the narrow black spruce. On the sunnier south face of the hill, jack pines spread wider. Water was impossible to miss, everywhere. Fire took a moment to spot: distant patches of lighter green with a softer texture—young jack pines racing to fill the space after a large burn. Fire's legacy was a living quilt of species and ages stitched together.

Winter

February 2008: North shore of Lake Superior: From deep in the five pounds of down of my mummy bag rated –50°F (–45°C), I woke to something cold and wet hitting my lips and chin. I froze, breathlessly listening in the blackness. Eventually, I had to exhale . . . and there it was again.

I cracked the bag just enough to peek out into the –30°F (–35°C) cold of the tent, switched on my headlamp, and exhaled again. A cloud of vapor rose, crystallized in front of my eyes, and fell back as snow. At such moments, the Ice Age doesn't feel so far away.

Technically, it isn't. We're still in the last of Earth's five known great ice ages, in an interglacial phase geologists call the Holocene. The "Ice Age" as it is commonly called was just the latest glacial period, starting about 125,000 years ago and ending roughly 11,000 years ago.

All human civilization has unfolded in this warmer interglacial, so we think of it as "normal." But the ice caps are waiting only for the right combination of orbital cycles, solar output, ocean currents, atmospheric chemistry (now complicated by our interference), and feedback loops in reflectivity and precipitation. Then they'll likely advance again. In the meantime, winter—ice age lite—still dominates the boreal forest and delivers the harshest test any species must pass to use its summer bounty. Winter makes the first cut of "who, how, and when" for all forest life, starting with the most obvious: the trees.

LEFT: Mink hunting a rabbit.

RIGHT: Snow forest, Lake Superior National Marine Conservation Area Winter Walk, February 2008.

FACING: Forest panorama, Woodland Caribou Provincial Park, northwestern Ontario, Smoke on the Water expedition, August 2011.

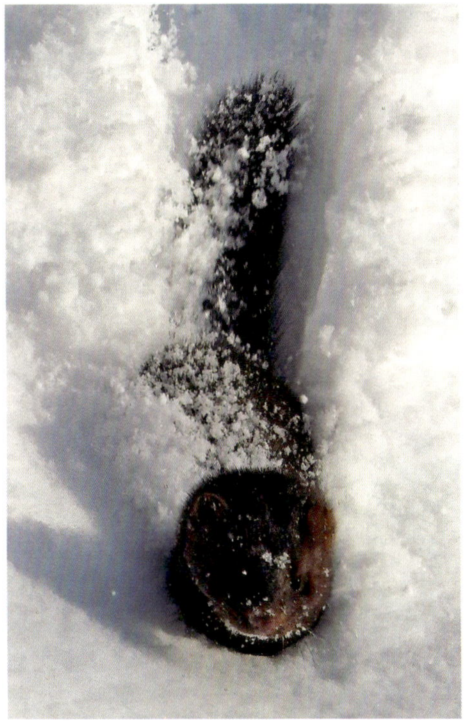

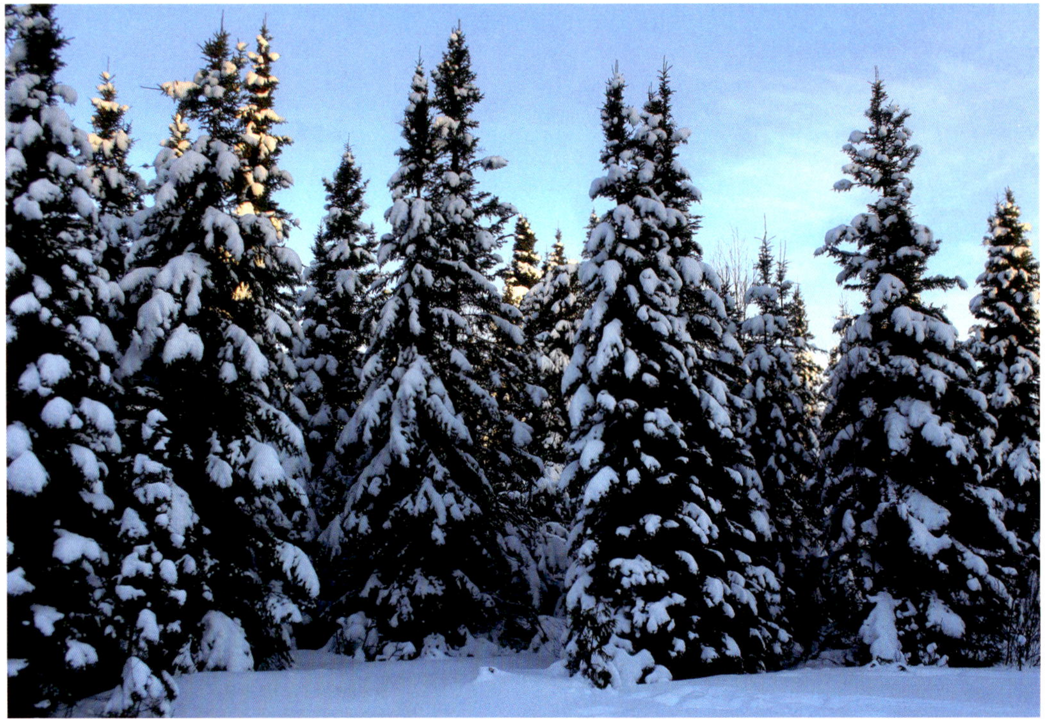

Balsam poplar (*Populus balsamifera*) is a tough broadleaf tree, with a range extending as far into the Arctic as any North American tree. Its range map roughly traces the boreal forest's North American extent. One could imagine it as the forest's signature species.

But winter has handed the title not to a young sprout from the hundred-million-year-old flowering trees, but to the descendants of a three-hundred-million-year-old group: the conifers. Spruces, pines, larches, firs, hemlocks, and cedars are the modern descendants of trees that arose long before the dinosaurs and give the boreal one of its other names: the "needle-leaf forest."

Their most obvious advantage is shape. "Conifer" comes from their reproductive cones, but the trees themselves are often conical—perfect for shedding snow, especially with shorter, more flexible limbs than broadleaf trees. Broadleaves drop their fragile leaves in fall, regrowing them each spring. Conifers keep their tough needles year-round (except larches).

Needles aren't as photosynthetically efficient as broad leaves, but they're tougher: wax-coated to shed snow, ice, and rain, with recessed stomata to reduce moisture loss when winter freezes the water supply. With their snow-shedding shape and armored, year-round leaves, conifers can photosynthesize the moment light and temperature allow, wasting none of the short growing season. And so, they dominate in the "snow forest."

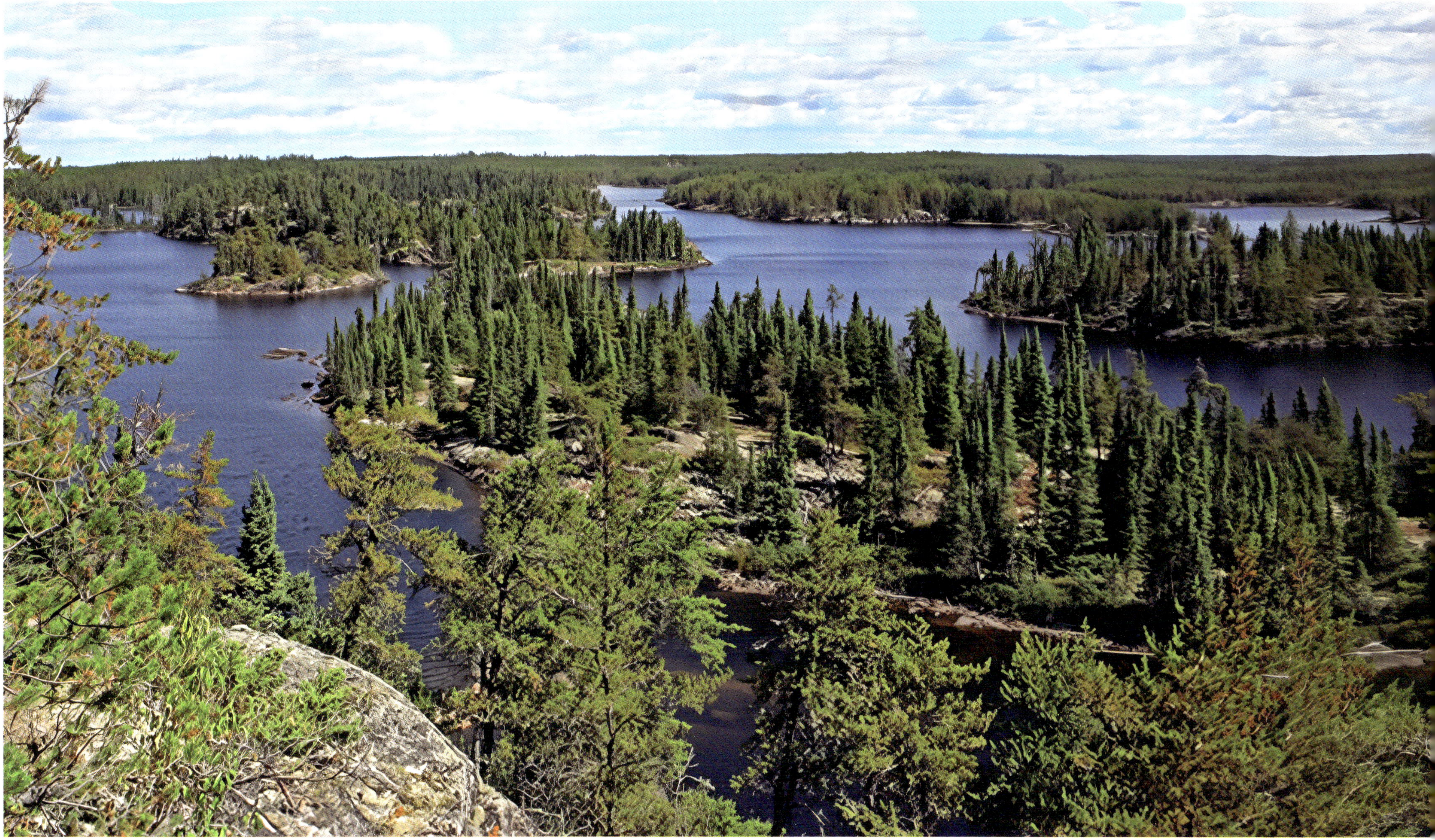

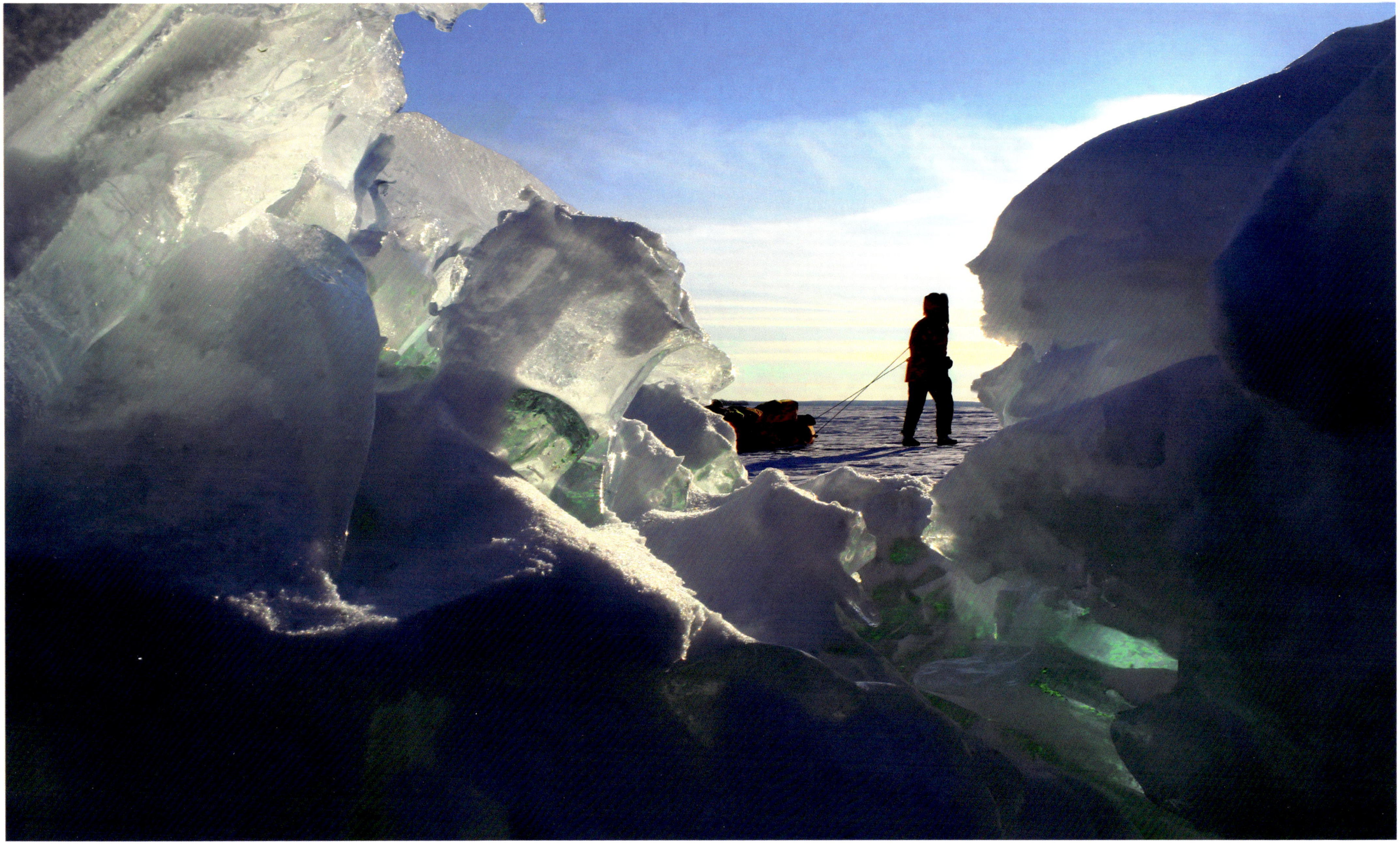

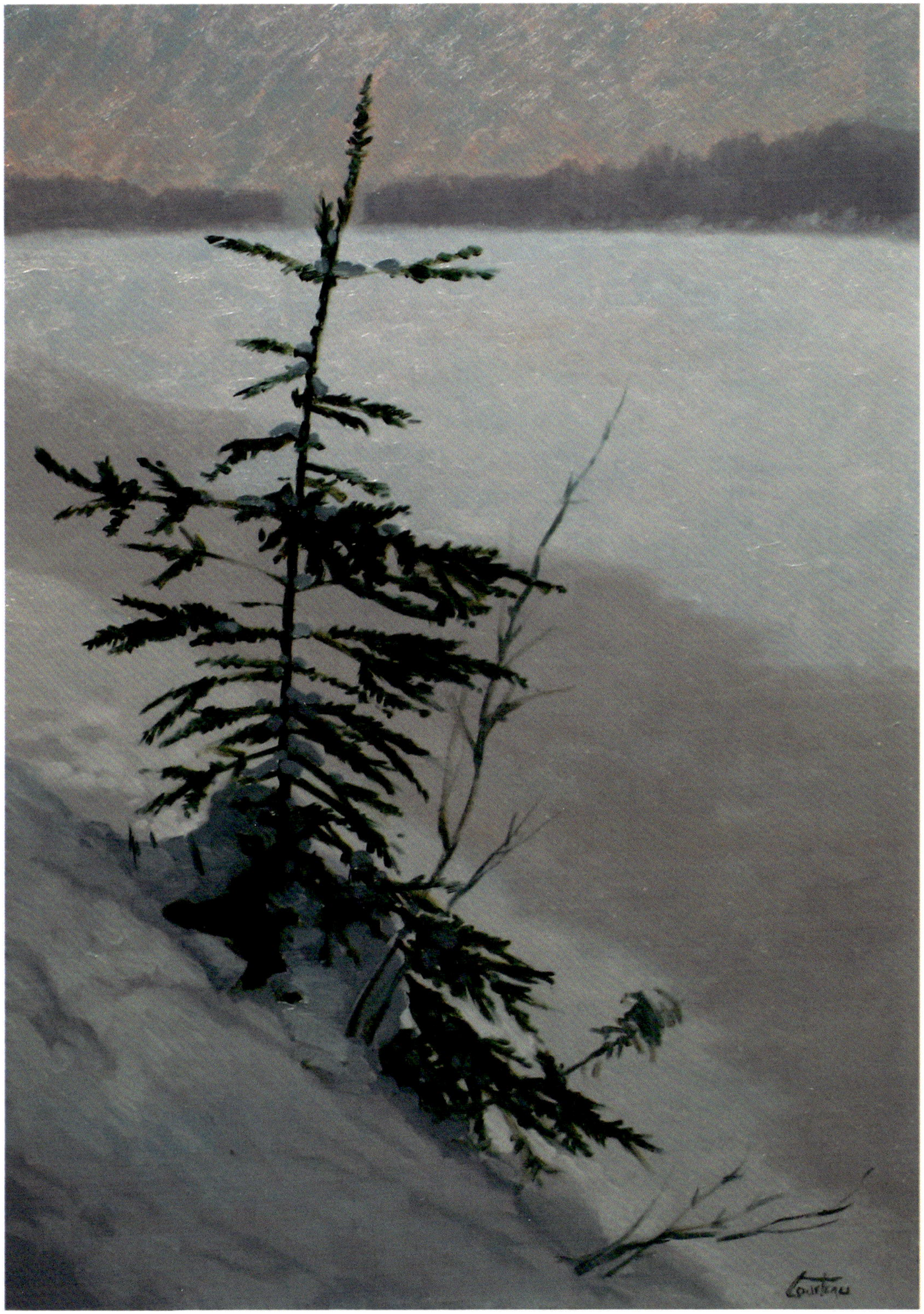

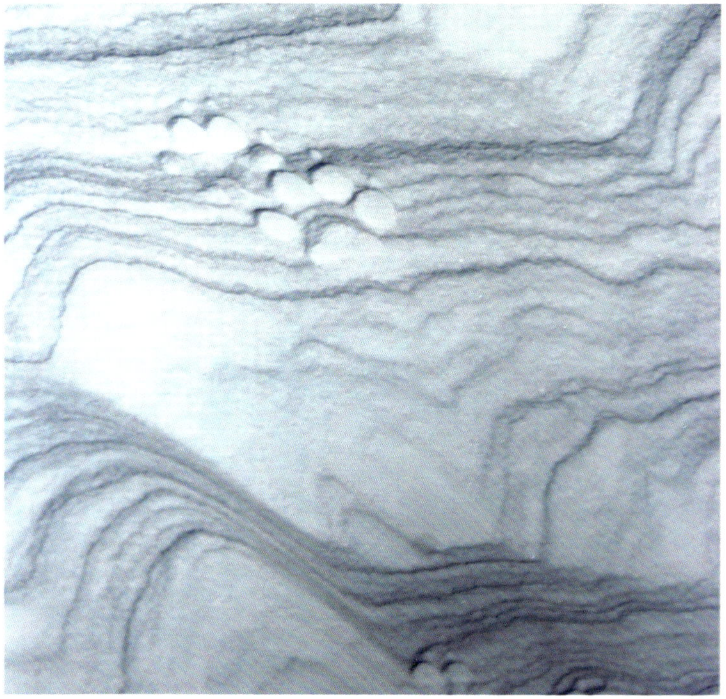

LEFT: *Tenace* (roughly "tenacity"), oil, Jean-Louis Courteau

ABOVE: Elevated wolf tracks in wind-scoured snow, Lake Superior National Marine Conservation Area Winter Walk, February 2008. The compressed tracks resist wind erosion better than the surrounding snow, leaving them raised.

FACING: Toboggan trip leader Jon Farshmin pulling past a pressure ridge, Lake Superior National Marine Conservation Area Winter Walk expedition, February 2008.

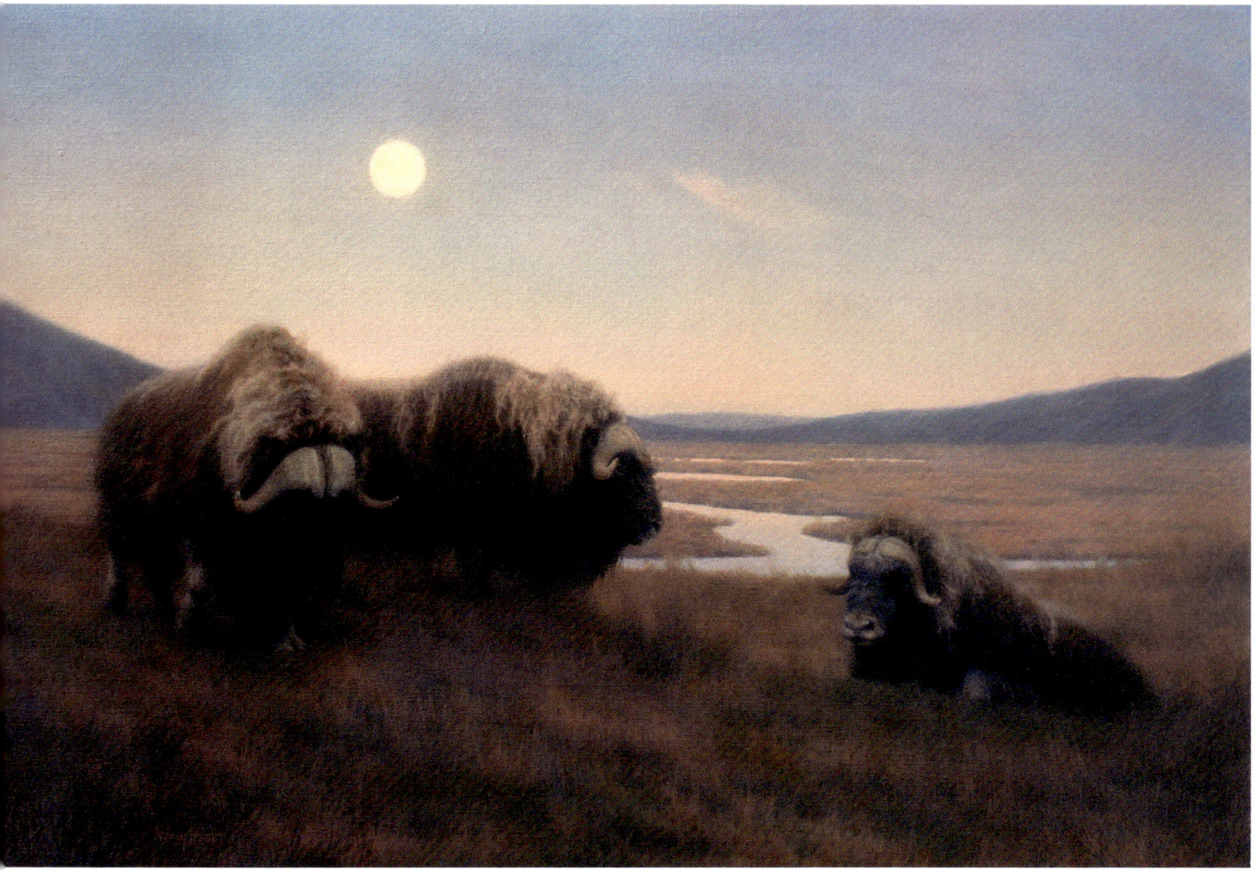

Musk oxen. *Oomingmak on the Rise*, oil, Nancy Howe

Adaptations

Everybody in the boreal forest (plants, bugs, birds, bears, people) must adapt to winter. Animals have more options than plants, but also more complex systems to protect from the cold. Low-lying organisms like lichens, mosses, and other ground huggers can simply go dormant under deep, insulating snow. Animals generally have to work harder, using a mix of morphological (shape/form), physiological (metabolism), and behavioral strategies.

Pelage, plumage, and plumping up

July 1, 1990: Kapuskasing, Ontario: After I called an early end of a third Missinaibi trip due to my concerns of crew fatigue, Dad and I were loaned a truck at a fishing lodge (plus treated to a Canada Day game dinner) to go back and retrieve one of ours from Mattice. Afterward, shuttling our crew out, we were driving back to their camp along a forest road, when I spotted the strangest animal I'd ever seen jogging

HOW BIG?

Size matters. In cold climates, the larger you are, the longer you retain heat. Bigger animals have more warm volume relative to their heat-radiating surface area. This is especially true for warm-blooded animals and underlies Bergmann's rule: Within a species or genus, body mass increases in colder regions.

A close cousin, Allen's rule, deals with appendages. Longer legs and noses only increase their size mostly in one direction and so if they get longer, they don't get the fully multiplied increase in volume as the whole body does when it balloons up. Having a greater amount of radiative surface relative to their slightly increased volume and being farther from the main heat-producing body core, these appendages would lose heat faster than if they stayed short and round. Consequently, in colder climates, appendages tend to be smaller, shorter, and rounder. Bergmann's rule favors bulky bodies; Allen's favors compact extremities—both for heat conservation.

Of course, there are exceptions. Lots of them. Nature is complicated. Humans, being animals, should follow both rules, but our technology (fur and hide clothing, shelters, boats, weapons) and relatively recent arrival in northern latitudes muddy matters.

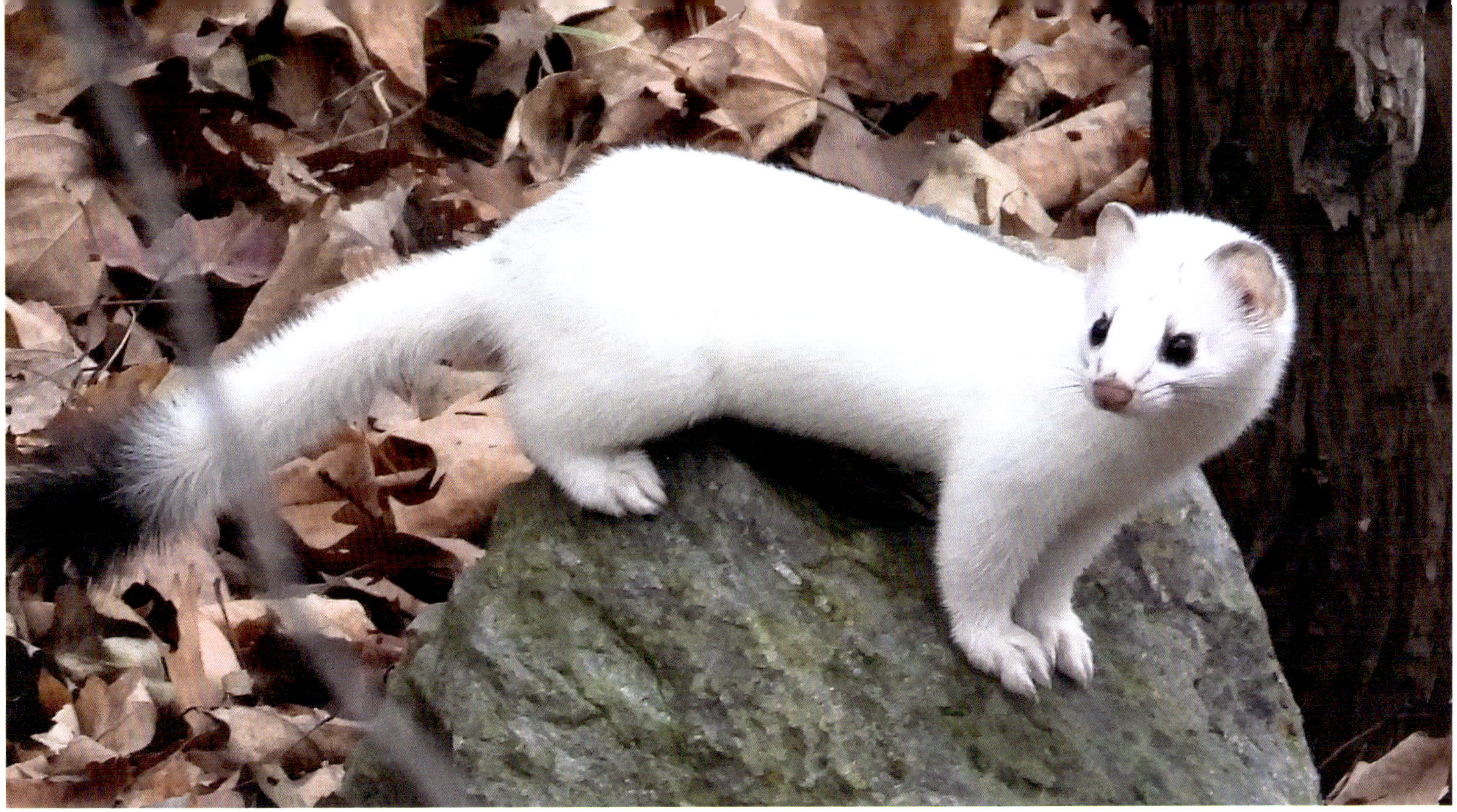

Long-tailed weasel jumping the gun with his molt, West Bolton, Vermont.

between the trees. Sasquatch? No—four legs. Chupacabra? Too far north (and mythical). Then it hit me: lynx—my first. But not the picture-book lynx I'd imagined stalking snowshoe hares through deep powder snow in luxurious winter fur. This one was tall, gangly, long-legged . . . and in its thin summer coat.

For animals that live in the boreal year-round, the first and easiest winter strategy is to put on a heavy coat with a base layer of insulation underneath—plus some extra fat. In mammals and birds, that means growing a water-shedding outer layer over a dense, heat-trapping undercoat (don't forget the fat).

Caribou go further, adding hollow guard hairs for extra insulation and buoyancy—handy for swimming big rivers during migration. Their hides are so effective at shedding water and retaining heat, that they've been prized by northern First Nations for millennia.

Some animals take it up another notch, changing both coat thickness and color. Snowshoe hares, arctic foxes, and four members of the weasel family turn white for camouflage. Among birds, ptarmigan do the same.

Beyond fur, feathers, and fat, shelter is critical. Hares line grass nests; small birds puff up to trap warm air and roost in dense evergreens; grouse burrow into snow at night; moose move to river bottoms; caribou retreat into the forest. Others, like chickadees and Canada jays, hedge their bets by caching food.

If you can't wear a heavy coat or layer on blubber—and you can't remember where you stashed your nuts—two strategies remain.

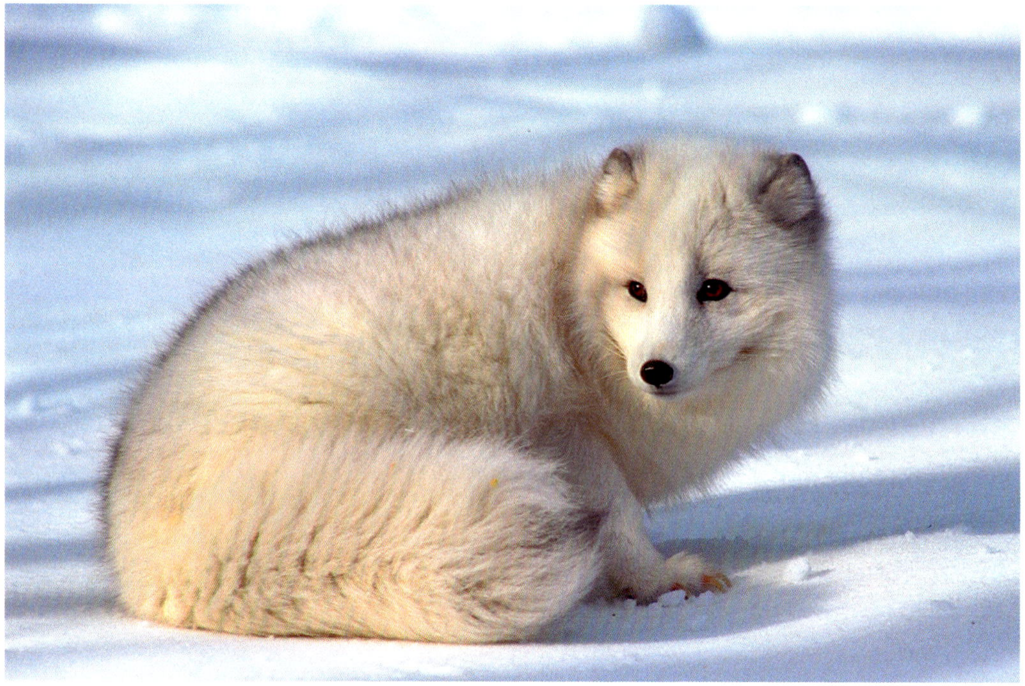

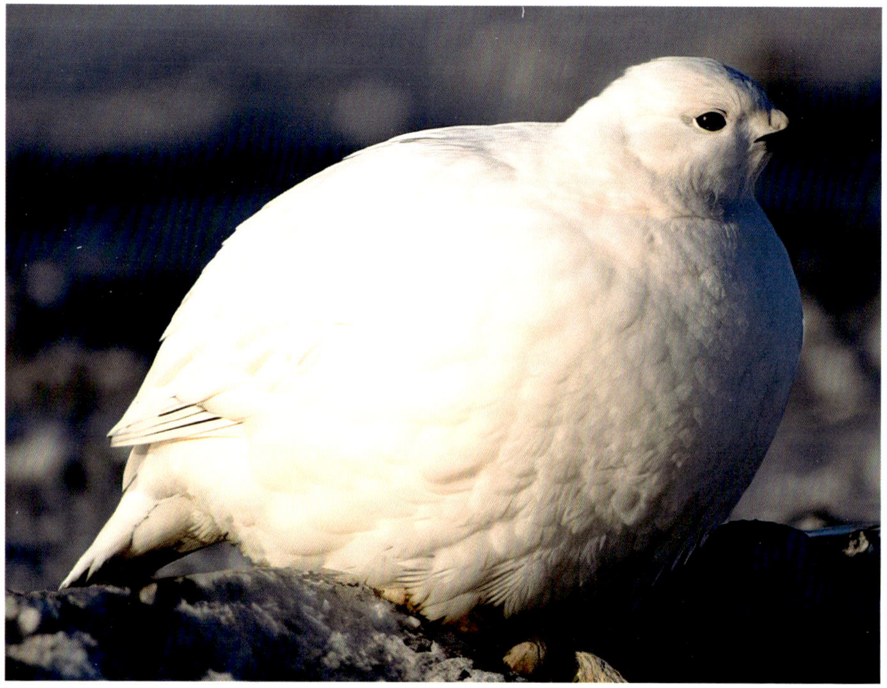

ABOVE LEFT: Arctic fox, Ungava. Susan C. Morse

ABOVE RIGHT: Brrrr... willow ptarmigan puffed up in full winter plumage at −40°F (−40°C), Labrador. Neil Simmons

Sleep it off

From bears to frogs, many species simply wait winter out. "Hibernation" is the common term, though definitions vary. Bears, for example, shut down most functions for months, occasionally stirring, and keep body temperatures relatively high, leading some naturalists to call it "torpor" instead. Arctic ground squirrels, in contrast, drop body temperature dramatically. What is common to them all is a continuum of reduced metabolic states to conserve resources until conditions improve.

Cold-blooded species enter "brumation": hibernation's reptile-and-amphibian equivalent. Some use an antifreeze of sugar-laced body fluids to keep key organs from freezing, aided by insulating leaf litter and snow. Wood frogs, gray tree frogs, spring peepers, and chorus frogs all use this trick.

Flying the coop: Migration

If you don't want to freeze, fast, or get fat all winter, the surest plan is to get out of Dodge. Most boreal birds migrate. Some barely move, staying near their nesting grounds in slightly milder zones. Others head for the tropics. North American boreal forest and Arctic breeding birds that do are often termed "neotropical migrants." It is an important term in conservation due to the dependency such species have on a variety of habitats, often spanning thousands of miles. The term covers any species breeding north of the Tropic of Cancer (roughly the top of Cuba) and wintering south of it. The most famous boreal forest neotropical migrants are the wood warblers, but many vireos, thrushes, shorebirds, waterfowl, and even some raptors all head for southern Mexico, the Caribbean, and Central and South America.

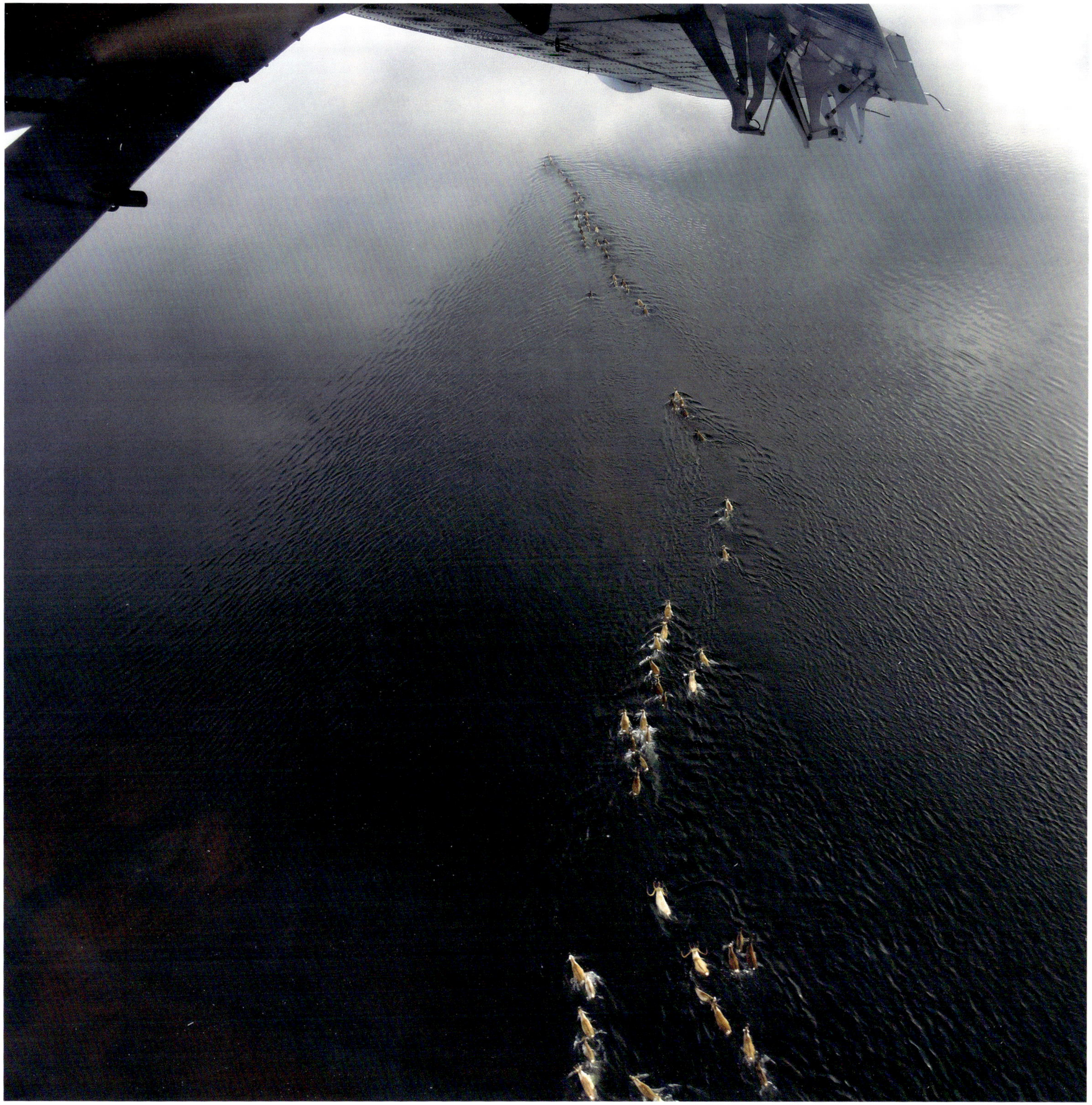

Members of the Leaf River herd swimming across Charlie Lake, Ungava (northern Nunavik). Susan C. Morse

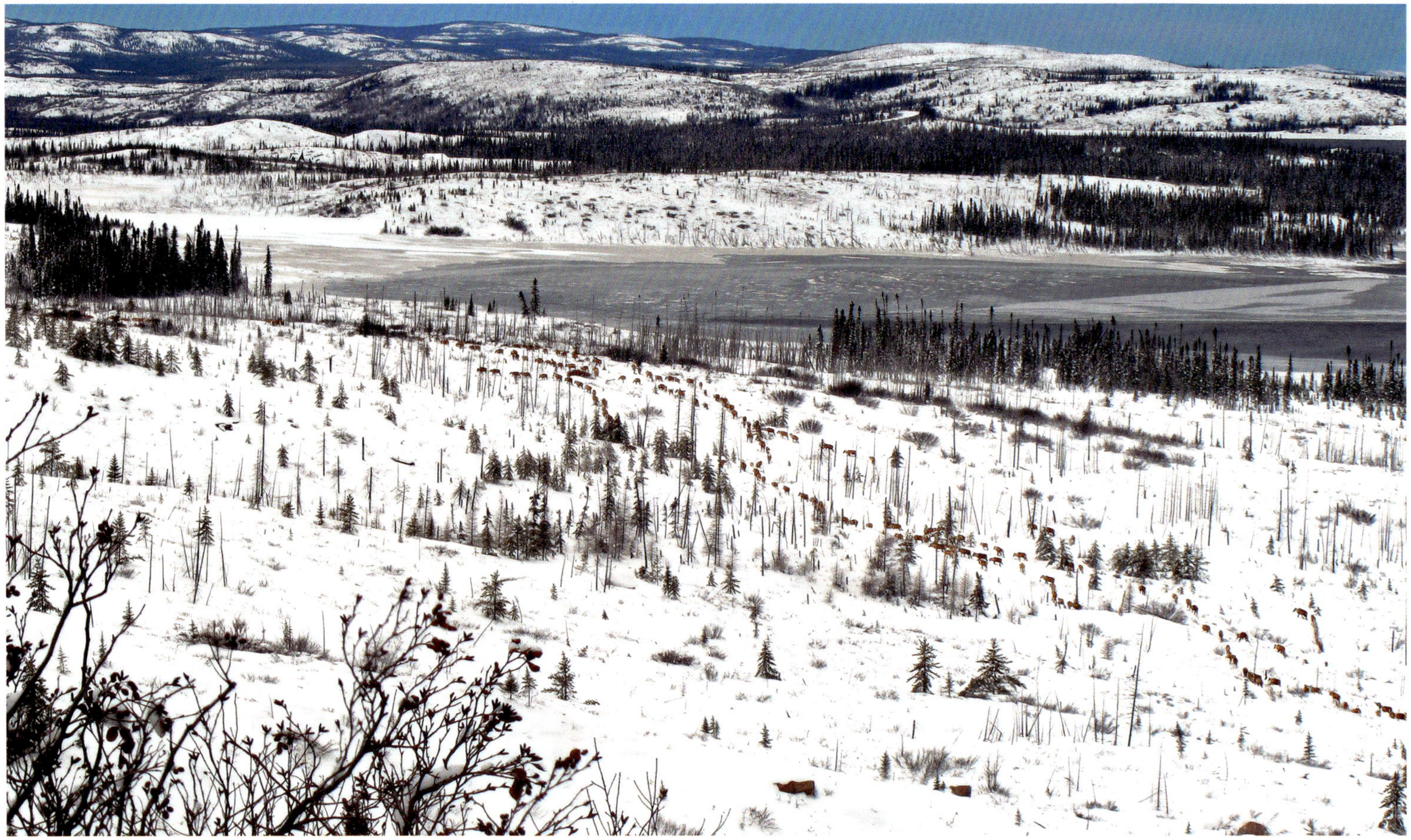

Caribou migrating through the landscape, James Bay. *Evening March*, Susan C. Morse

FACING: *Life's Blood*, oil on canvas, Jean-Louis Courteau

Other birds are opportunists: usually resident, but ready to head south if food runs short. Snowy owls sometimes stage dramatic southward "irruptions" when lemming populations crash or their own spikes.

Food availability drives these choices; breeding success pulls them back north, with long days, abundant habitat, and fewer predators as inducements. Migration offers rewards, but at a high cost in time, energy, and risk, so strategies can vary widely.

The seasonal flow is immense: Billions of birds in the spring arrive for nesting and billions more leave every fall, moving nutrients and resources between ecosystems like the ocean's tides.

Birds aren't alone in migrating, though land animals are generally more limited in range. Some are more like commuters, shifting only a few miles—or just hundreds of feet down in elevation—seeking denser cover, less wind, and more food. And then there are caribou.

Some are resident; making short descents to forested wintering grounds. But others—barren-ground and some woodland populations—undertake the longest and largest land migrations on Earth.

Water

September 26, 2009: Labrador Barrens

Cole Johnson and I dumped our packs in the screaming wind and pushed north toward a hazy blue gash slowly growing across the tundra. It had to be the canyon of the Kogaluk. On the topographic map it was just a dense smudge of contour lines denoting a 1,200-foot sheer canyon wall; in person, even from a few miles away, it felt immense. There had to be water.

Nine days earlier—far to the south—a nameless river we'd been following to the North Atlantic Ocean had vanished into a boulder garden 70 miles from the coast and not bothered to emerge (this was a remote, rarely traveled wilderness area with no river gauges or government monitoring). Now, after dragging our canoe and gear across miles of cold, windswept sedges, bogs, and boulder fields (and sitting out a vicious storm), we were hoping hard.

Cole reached the edge first, throwing his arms up in jubilation as the near-gale tried to tear his anorak off. When I joined him, I felt like Bilbo Baggins looking down into Rivendell: far below, a vibrant river wound between forested banks, around islands, and off toward the sea. We had water, plenty of it. All we needed was a way down.

In the boreal forest, the greater challenge is often too much water. More of this vast region is wetland than any other major ecosystem on Earth. True, you can walk for miles without hip boots in some areas, so a well-drained global northern conifer forest is conceivable—if we wait for a few

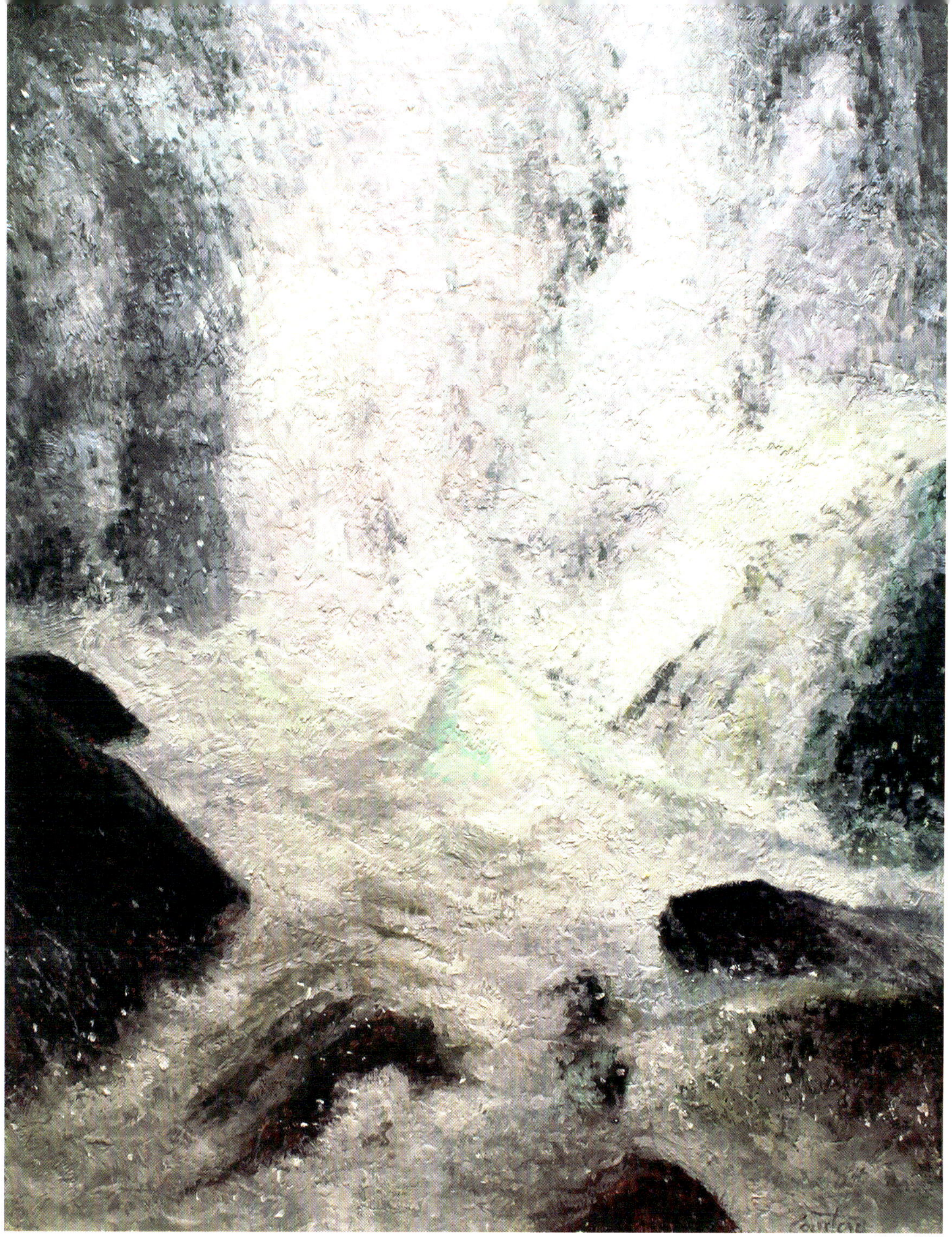

CLOCKWISE FROM TOP LEFT: Canyon of the Kogaluk, Labrador Barrens.

Wetlands east of the George River, Nunavik. Sometimes, when I'm looking at maps or out a bush plane window, the thought of walking this country is dizzying. That Cole Johnson and I once had too little water still amazes me.

Kalon Baughan punching through big waves on the De Pas River, which is rated V (more like IV at our water level—or perhaps boosted for remoteness, as is common). Kalon is a former US National Downriver Canoe Champion. Rod Lawrence and I lined this drop. Quebec Trans-Labrador expedition, September 2009.

Lewis and Clark . . . or Montana artist Kalon Baughan (left) and the author scouting a rapid on the De Pas River, Quebec Trans-Labrador Expedition, 2009. Rod Lawrence

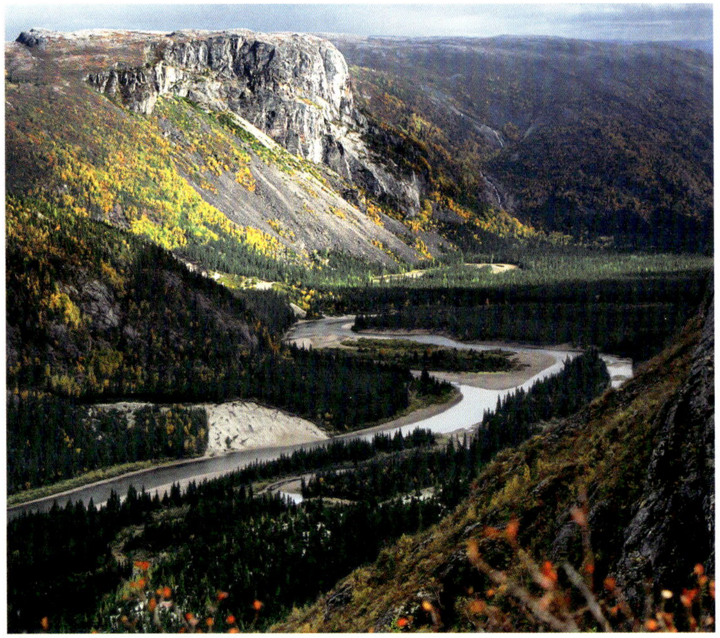

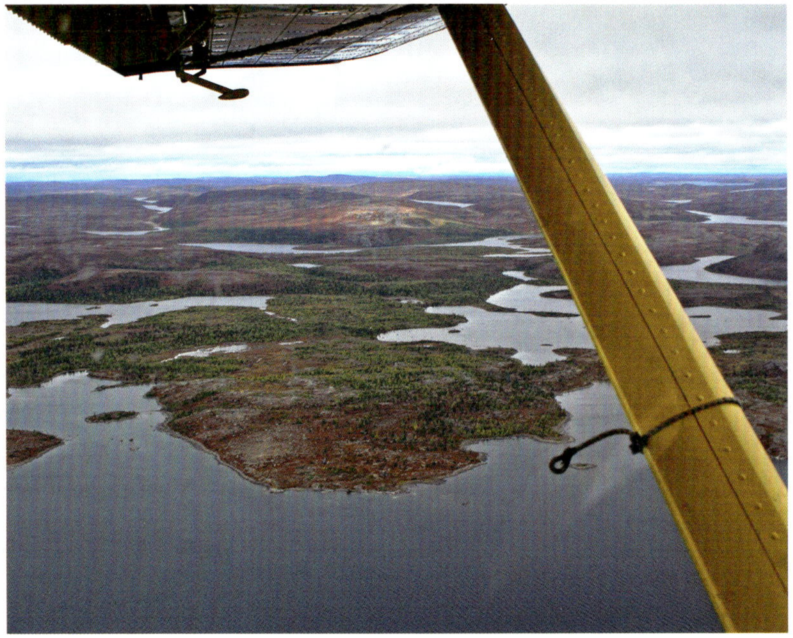

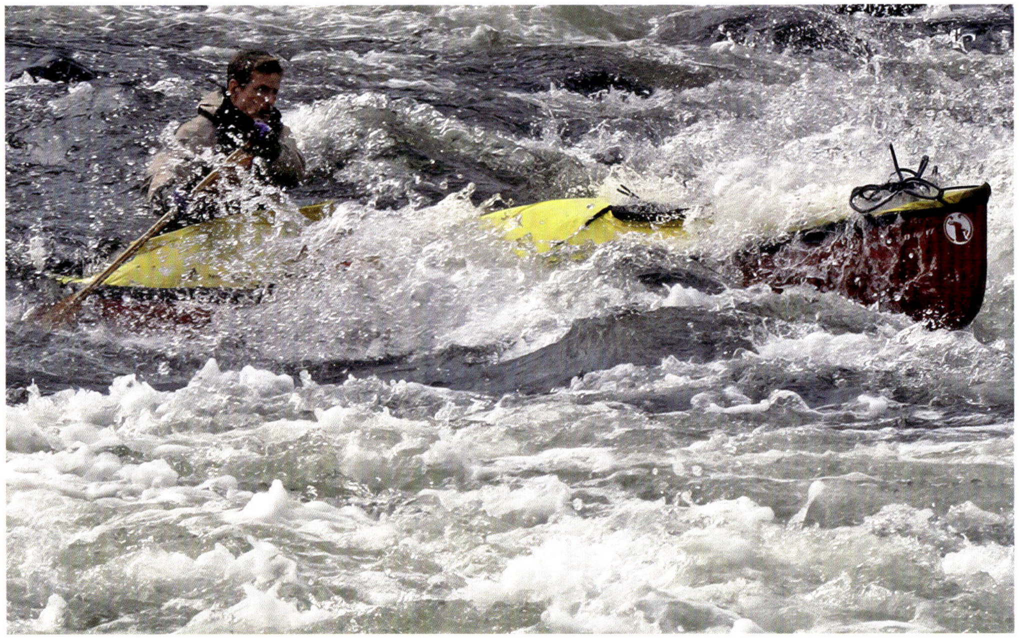

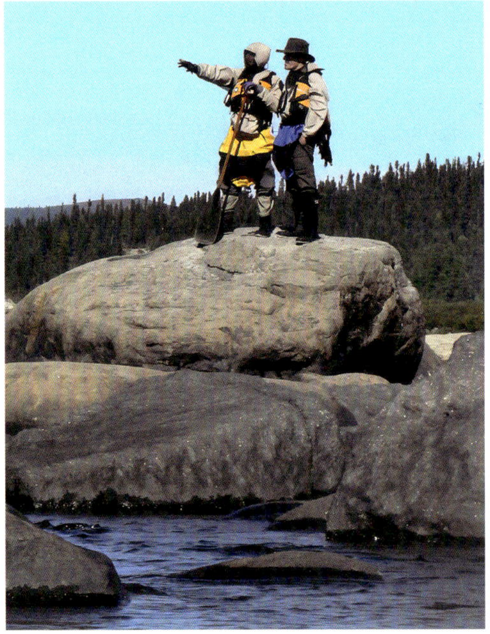

more glaciations and some continental drift. But for now, water is elemental to the forest's character and ecology.

One main reason behind the boreal forest's sogginess is the same force that shaped its winters: the ice sheets. At the height of the last glaciation (25,000 to 18,000 years ago), continental glaciers, two miles thick, covered nearly all of Canada, parts of the northern US, the UK, northern Europe, and much of northern Asia. They didn't blanket every inch of what's now boreal forest, but they scraped and scoured much of it clean.

The glaciers crushed land, rasped mountains, and gouged valleys. As they melted, they dumped untold trillions of tons of gravel, sand, clay, silt, and—every New England farmer's curse—rocks across thousands of miles, along with "erratics": massive boulders carried far from their point of origin, abandoned like the forgotten toys of mountain giants. The glaciers left a big wet mess. And it hasn't dried out much.

One reason for the slow drying is terrain. As their names imply, the North European Plain, West Siberian Plain, and Central Siberian Plateau are all pretty flat. In North America, much of the boreal forest sits atop the Canadian Shield, a vast expanse worn down over billions of years. It too is relatively flat. So, every time the glaciers chew up the ancient land and then retreat, watersheds must sort themselves out all over again. Without the clear guidance of dramatic topography, meltwater spills across the land, wandering through a labyrinth of new wetlands, streams, rivers, and lakes, as it searches out its way to the sea.

Poor soils don't help. Glaciers strip them away. Winter slows the weathering that rebuilds soils, so what little exists saturates quickly. French explorer Jacques Cartier, exploring the north shore of the Gulf of St. Lawrence in 1534, wrote: "I did not see one cart-load of earth and yet I landed in many places." With little to soak it up, water pools and wetlands—especially bogs—are born.

The boreal isn't a rainforest, though on more than 20 expeditions we've never missed rain and/or snow (on one, it stopped raining only to snow). Despite generally modest precipitation, cool temperatures, short summers, and forest cover slow evaporation, so wetlands persist in the glacier-gouged terrain.

All this water shapes the forest's geography, habitats, ecology, and climate. Holding most of the planet's liquid surface freshwater, the boreal forest influences far more than itself, affecting marine fisheries, ocean currents, atmospheric CO_2, and global weather.

Water has also shaped human life in the boreal forest, molding cultures and steering history. In North America, the canoe—a technological masterpiece developed by Indigenous builders and raised to a high art in northeastern North American by Algonquian people—remains the best way to travel long distances through this land (see chapter 7).

There are easy rivers and beautiful lakes in the boreal forest and tundra, and you could spend a lifetime paddling them. But throughout much of the boreal forest, sooner or later you'll face white water. Rapids are classed on a scale of I to VI (easy to insane). If you travel alone or in a small group, know your limits and how to read a river. Lining (floating a canoe down a rapid while controlling it from shore with the bow and stern lines, called "painters"—don't ask *me*) or portaging are usually options, but for safety and enjoyment it's best to be comfortable with Class III. Higher grades on wilderness trips aren't wise—unless you're canoe champion Kalon Baughan (see p. 52). In larger crews with highly experienced companions, even basic canoe skills can take you safely through—and make for the trip of a lifetime.

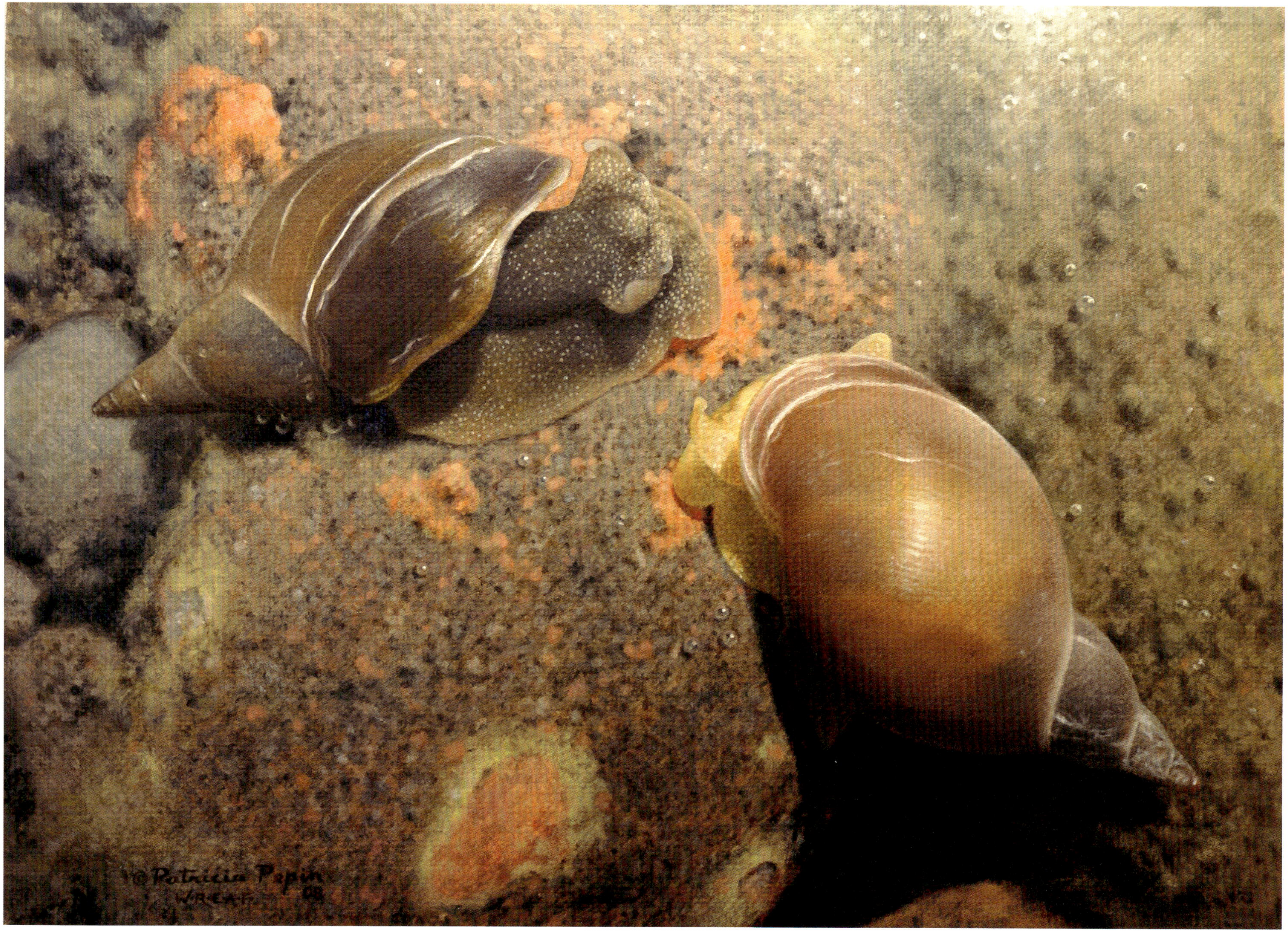

Lake Superior North Shore expedition, July 2007. *Snails*, oil, Patricia Pépin

FACING: Smoke on the Water expedition, August 2011. *Blaze of Dawn*, oil, Dawn Banning

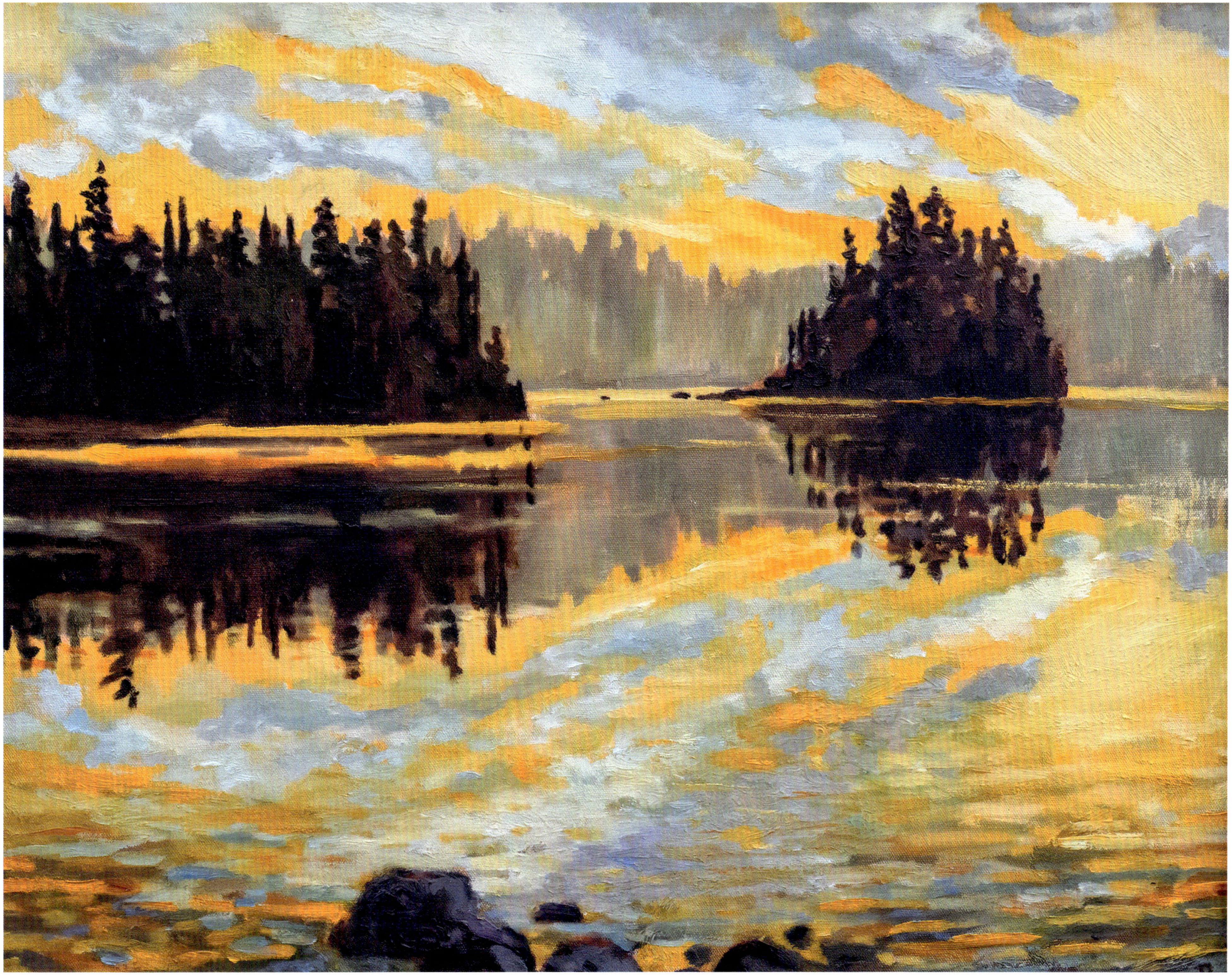

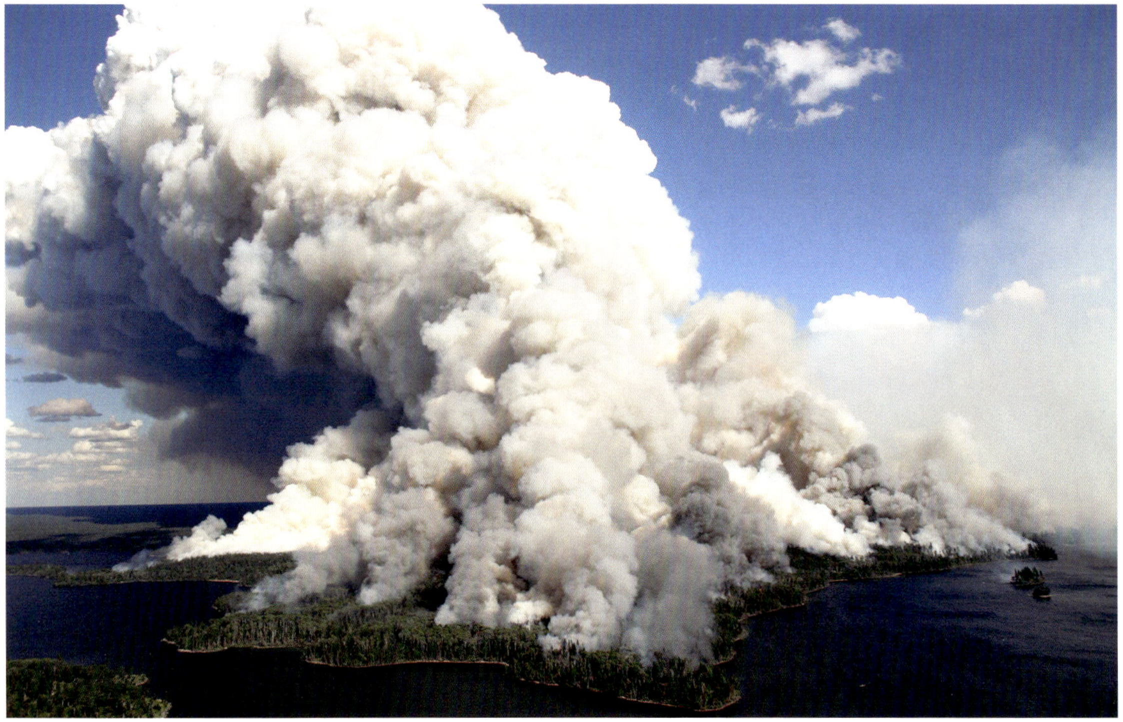

"Red 124" as termed by the Ontario Ministry of Natural Resources (probably shorthand for Red Lake 124) from the other side. Mitch Miller, courtesy of Ontario Ministry of Natural Resources (OMNR)

FACING: Composite of Red 124 and a herd of caribou we'd seen swimming Kamestastin Lake at dawn, Labrador (artistic license). *Belly of the Beast*, oil, Rob Mullen

Fire

August 1, 2011: Woodland Caribou Provincial Park headquarters, Red Lake, Ontario We sat with the park superintendent reviewing our canoe route to the Bloodvein River and on to Lake Winnipeg. Fires were burning in the region, but we were told our route looked safe. One small burn north of our beginning stretch: "Nothing to worry about."

August 2, 2011: West-northwest of Lund Lake After our second long portage, we were greeted by an immense, dirty plume devouring the sky and tinting everything in ominous russet light. Oddly unconcerned, we pressed on and camped on an island.

August 3, 2011: It was still hazy and smelled a bit like smoke in the morning, but it looked better. We were making breakfast when I heard a distant *thwumping*. Channeling Radar O'Reilly, I scanned the horizon. A black dot appeared to the east, growing into a helicopter that roared straight toward us, hovered over the beach, then vanished behind the trees. Moments later, assistant superintendent Claire Quewezence stepped crisply from the brush. Her uniform creases were sharp enough to butter our toast. "Don't bother with breakfast—you need to leave now," she said. A Force 5 (top of the scale) forest fire was only six miles away and if it turned toward us, it could overrun us in minutes. Trucks would meet us at our starting point—evacuation was underway. Later, photographer Mitch Miller shared images from the Ontario Ministry of Natural Resources' fire crews. They were stunning and made me suspect we'd been in greater peril than we realized.

Water and winter are kin: cold, wet, descended from glaciers, and relatively reliably present. Fire is their chaotic opposite. Unpredictable, it's as far from wet and cold as you can get, yet it echoes the Ice Age on a smaller scale, periodically scouring the land and everything that cannot flee.

The boreal forest seems built to burn. The main culprits are its trees.

Conifers, in their trillions, have evolved superb defenses against injury and infestation: thick, resinous sap that seals wounds, poisons pests, wards off fungi—and smells wonderful. For millennia, people have used conifer pitch as medicine, sealant, chewing gum . . . and fire starter.

Resin is highly flammable. Black spruce, dominant in many areas we travel, thrives in poor soils that favor dense, small trees. Their crowns hold green needles, but their trunks are wrapped in dead, pitch-laced branches—perfect kindling. White spruce, tamarack, and most other boreal conifers aren't much different. Outside of bogs, you won't find an easier place to start a fire than the boreal forest.

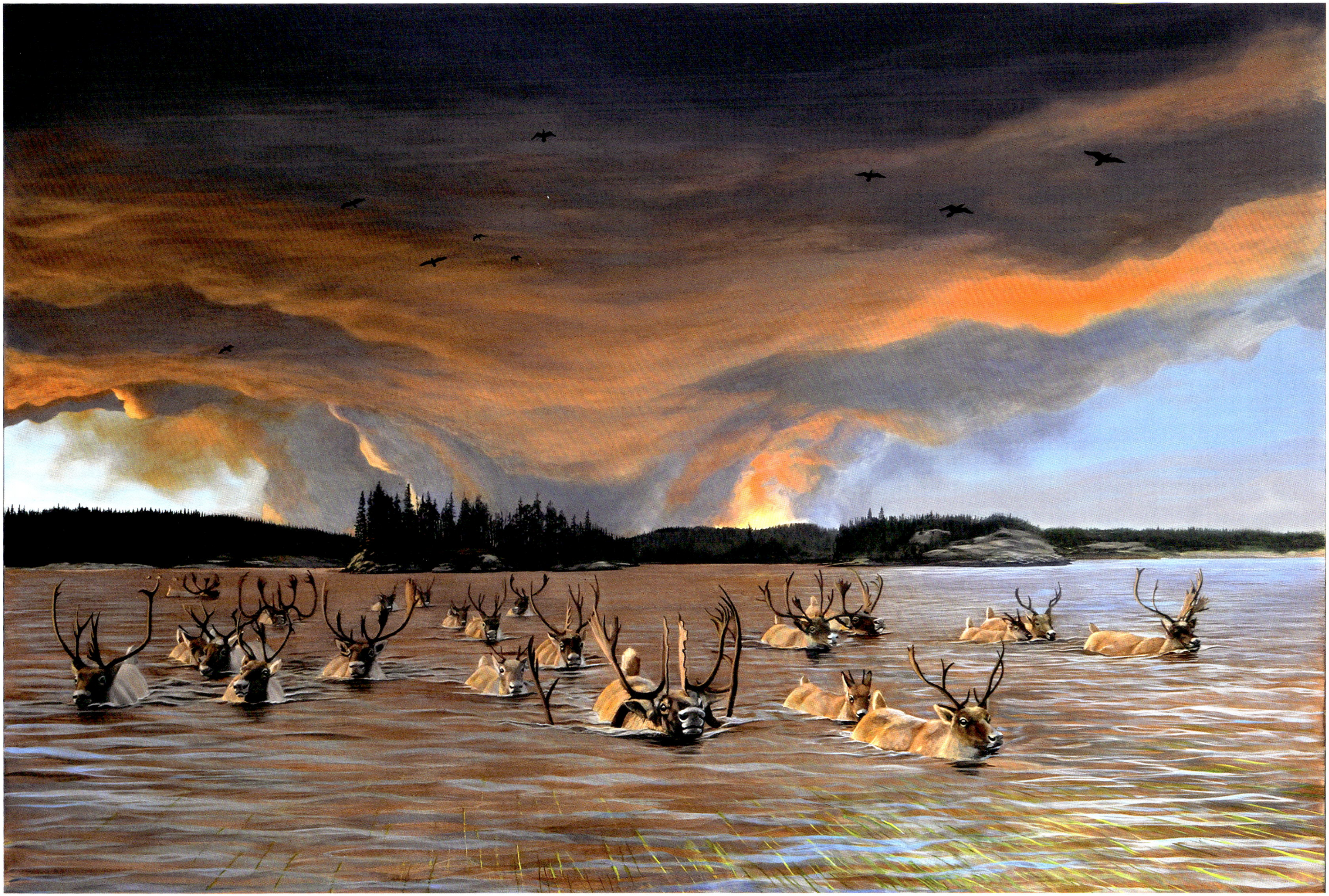

Lady of the Flames, aka Red 124, Woodland Caribou Provincial Park. Mitch Miller, courtesy of OMNR

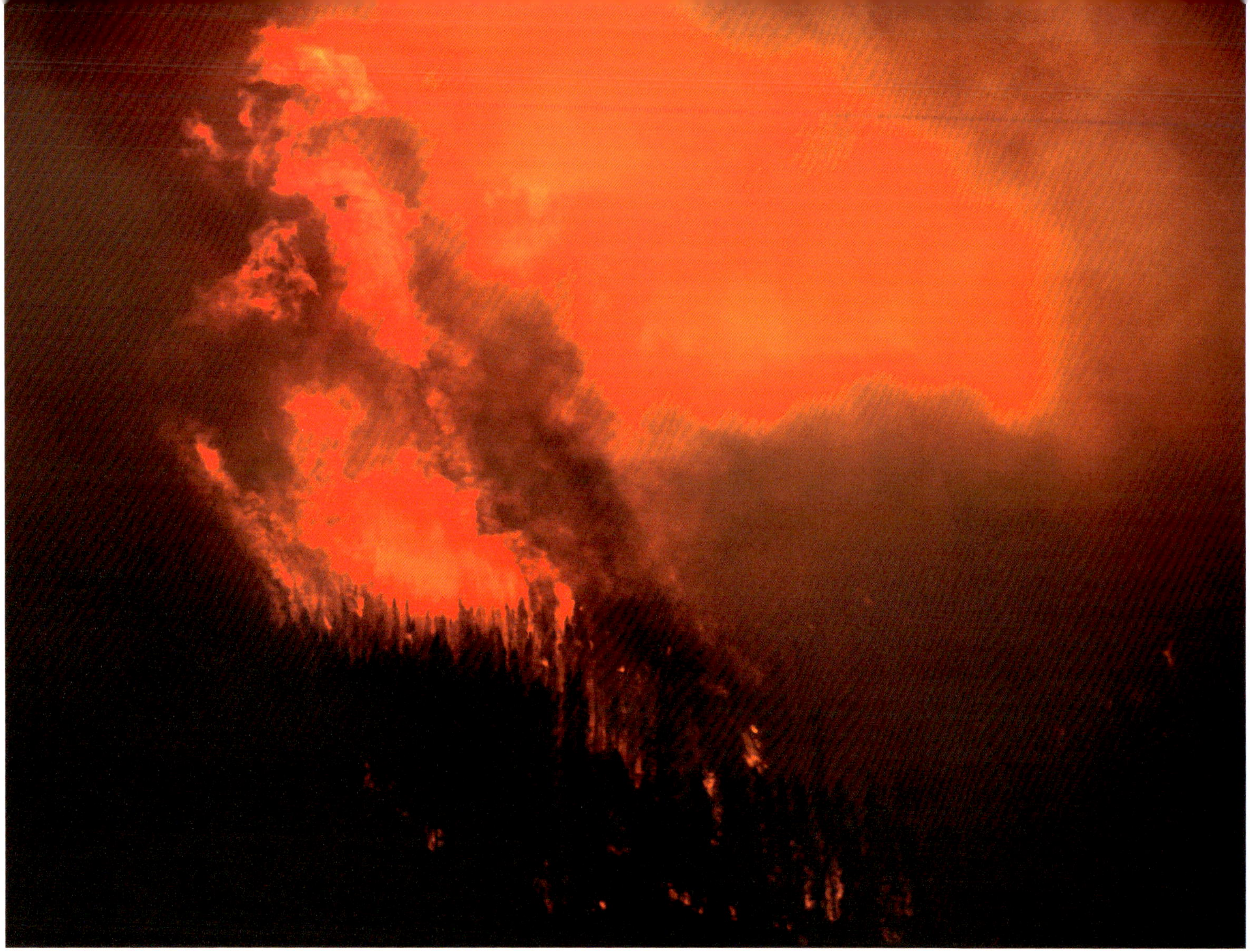

Wildfire is not only a destroyer, but also a creator. Several conifers depend on heat to open their seed cones; many plants and animals flourish in burned-over areas. Burn patterns vary with forest composition, climate, fuel load, weather, and geography. With 30 percent of Canada's boreal forest in wetlands, water is the first line of defense against runaway fires. Therefore, burn areas often create a crazy-quilt pattern in the forest. Moreover, burn severity varies widely, with some areas reduced to bare soil, others scorched only at ground level, and still others untouched. These intertwined variables create a diverse mosaic of ages and species of trees, and habitats for a wide variety of wildlife.

Dr. James Vogelmann, retired senior ecologist, U.S. Geological Survey, says:

Fires have impacted the boreal landscape for thousands of years... Most in northern North America are from lightning, though some were historically set by Native Americans for local management. Although many people view wildland fires negatively, most scientists agree fires are necessary for healthy forests in these latitudes.

Diverse forest types and ages benefit wildlife: moose and caribou browse in open, younger stands; wolverines and lynx prefer dense mature canopies. Severe drought can burn old stands hard, slowing recovery, while younger stands often burn less severely. This patchwork of conditions increases biodiversity.

Decades of fire suppression, especially in western North America, have left many forests dense with young trees, shrubs, and sometimes invasives. When these finally burn, fires can be especially hot and destructive. Yet, if not too severe, wildfire benefits boreal forests by opening cones of jack pine and black spruce, speeding regeneration. Often, the best environmental response to forest fires is to let them burn.

For many years, the pattern of fire behavior across large regions of the boreal ecosystem has been reasonably consistent and predictable. Depending on the forest type and geography, some forests will typically burn every 40 to 60 years, whereas others will burn every 100 to 150 years (or even longer). This has been the case for millennia. Unfortunately, with global climate change and, in many cases, in conjunction with altered forest management practices, not only have the fires been getting bigger and more severe, but the fire return intervals have been skewing toward shorter time periods.

While there are alternative hypotheses to explain the observed forest fire behavior changes, unless proven otherwise, the simplest explanation is likely the most tenable (i.e., Occam's razor). The broad scientific consensus is that fossil fuel burning and secondary effects such as the freeing up and release of methane through permafrost melting are primary culprits. There is also concern within the scientific community that the boreal forest, which has long been considered a carbon sink (where excess global carbon gets stored in the forest ecosystems), is now becoming a carbon source (i.e., it is releasing more carbon than it is taking in). There is concern that the buffering capacity of the boreal forest to take in and store carbon from the atmosphere may be waning. Humans have set into motion a new set of conditions, likely with many unintended consequences, and we can be certain that there will be significant impacts and repercussions in the years to come.

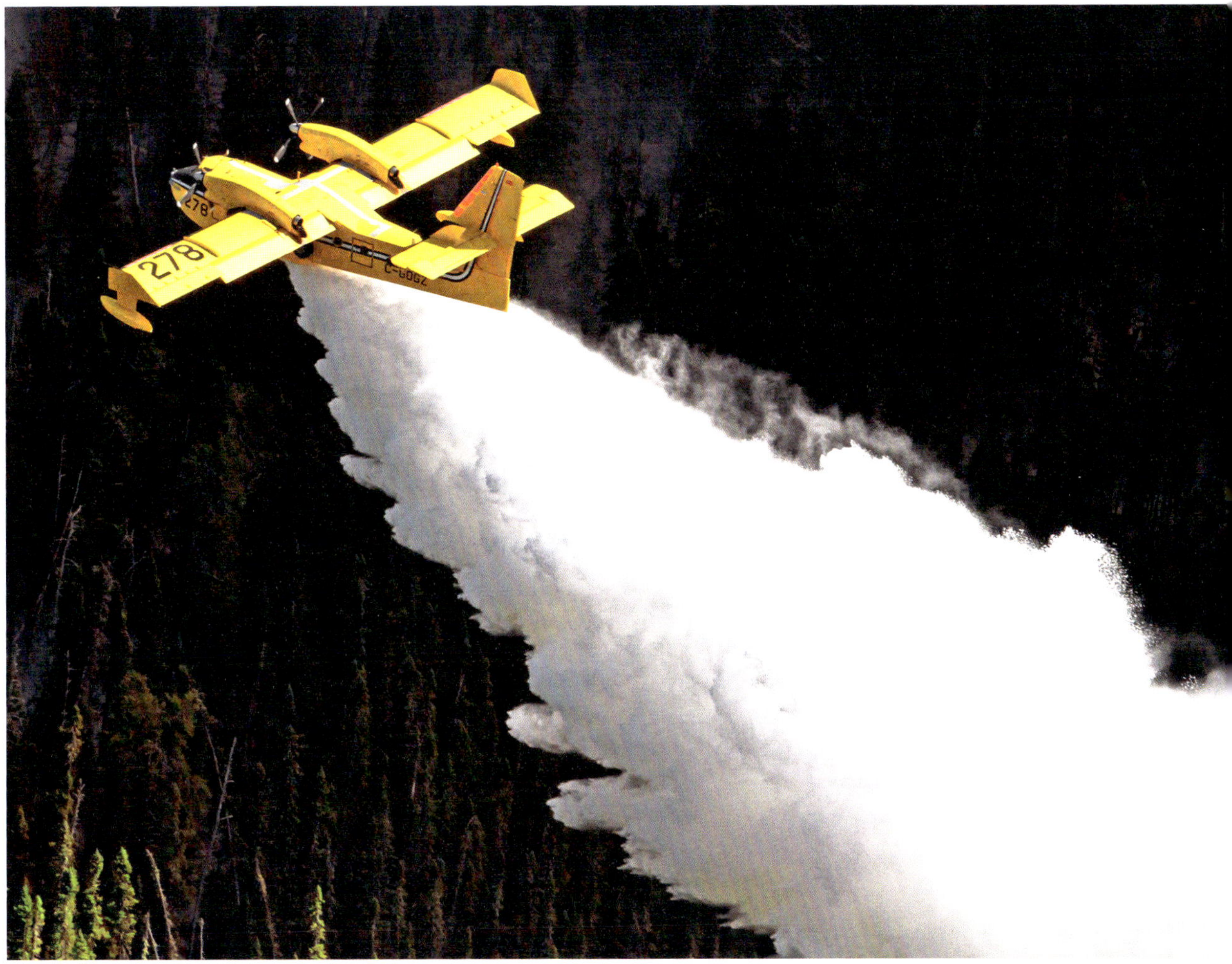

Waterbomber fighting Red 124. Sometimes letting it burn is not an option, especially if infrastructure is threatened. Mitch Miller, courtesy of OMNR

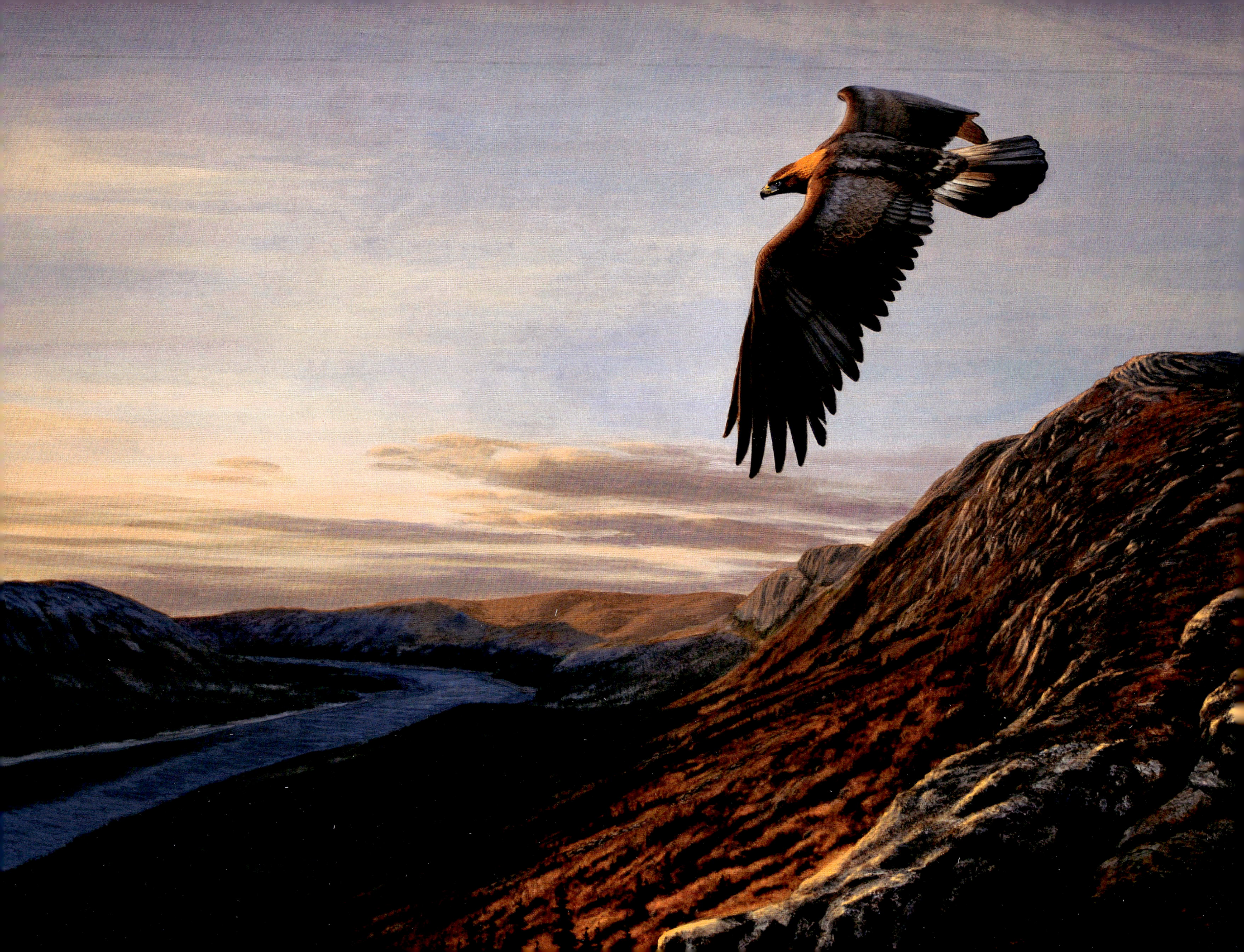

FACING: Golden eagle over La Colline Misurtuq and the George River, Nunavik. *Nunavik Gold*, acrylic, Rob Mullen

3

A Tour of the Boreal Forest

What was before the boreal forest?

By Dr. Steven Young

In relation to the human lifespan, the boreal forest may seem ancient and primeval, but to anyone who looks at it from the perspective of geological time, it has been around for only an instant. When the cooling trend that culminated in repeated ice ages began a few million years ago, temperate forests extended to the northernmost regions of Canada. Our present boreal forest is much younger. If we could have looked at the regions now covered by boreal forest as recently as 15,000 years ago, they wouldn't have been recognizable. In North America, they would have been covered by glacial ice, still a mile or more thick.

The situation in Eurasia was vastly different. While northern Europe and northwestern Asia were covered and alpine glaciers expanded to solidify into an ice sheet in the Alps and the Caucasus, Siberia and Central Asia mostly remained bare, ice-free land. But it was not forested land; it is now understood to have been a vast area of mostly treeless steppe. This stretched from the Atlantic coast of the Iberian Peninsula across central and eastern Europe, virtually all of Siberia, and across the diminished Bering Sea to Alaska and even the Yukon. It also spread northward beyond the shores of the current Arctic Ocean and south to the Gobi Desert and into China, making it the largest biome, or major ecosystem, on Earth.

Scientists have been trying for decades to reconstruct the nature of the "mammoth steppe," as this biome came to be known. There's good information on the large mammals that dominated: woolly mammoths, steppe bison, horses, reindeer, saiga antelope, and, in the Eurasian portion, woolly rhinoceroses. Their numbers indicate that it was a highly productive

> "In Inuktitut, *nuna* means 'the land,' but not in our simple and utilitarian sense. Nuna is more than that, conveying a meaning vastly beyond Nunavik and Nunavut and the Northwest Territories combined. Nuna is the land that includes the caribou, all birds, and all other animals as well as The People—all sustained in intertwined, physically and spiritually, as equal partners in life's journey."
>
> SUSAN C. MORSE

NOTE: The boreal forest is a global biome, yet, as a matter of practicality, every WREAF expedition has been in North America. While the ecological mechanisms and dynamics and many species are certainly global and are acknowledged as much as practicable, our direct experience and imagery are heavily biased to North America.

Immature bald eagle, Missouri River Breaks. Bonnie Rowell

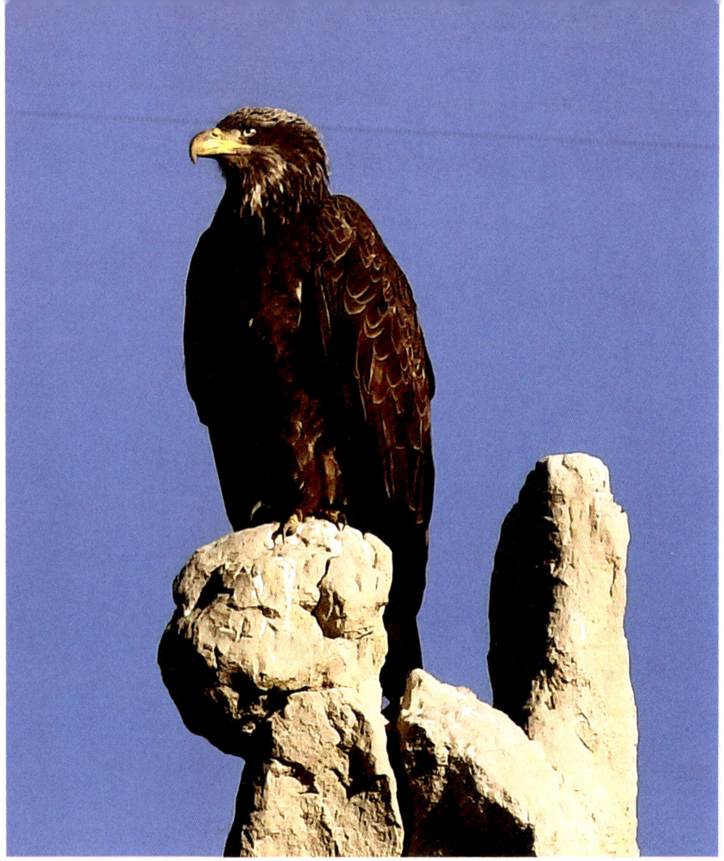

environment, but we are only beginning to uncover details of the nature of the plant cover and the climate. The mammoth steppe appears to have typically had short, relatively warm summers and intensely cold, dry winters.

The mammoth steppe biome came to an abrupt end over most of its range in a remarkably short span of time, geologically speaking. Shortly after 15,000 years ago, as the last ice age was rapidly waning, the vegetation changed radically. The productive steppe was replaced by birch and willow scrub, which was both less nutritious and richer in toxic or indigestible compounds. This "birch rise" seems to have been related to a climate change that may have led to longer but cooler and moister summers. Probably more importantly, increased precipitation, in the form of snowfall, often made grazing impossible because of winter storms. These changes led to population diminishment, range fragmentation, and even the extinction of many of the large mammal species, such as the mammoths. The result of rapid climate change, directly in the case of vegetation and leading to the loss of grazing animals, led to the destruction of the whole mammoth steppe ecosystem. This set the stage for the spread of the boreal forest as we know it.

Where did the trees come from? This is mostly a matter for educated guesswork. The obvious answer is that they lived out the ice ages in areas mostly south of the ice in North America, and less certainly, in pockets of forested environment in Eurasia. These weren't necessarily south of the mammoth steppe, which may well have been treeless because of dry, windy, abrasive conditions rather than extreme cold—or the presence of massive glaciers.

The Eurasian equivalent is much more variable from east to west, unlike the American boreal forest. From easternmost Newfoundland and Labrador to near the Bering Strait in Alaska, the American forest is mostly dominated by two species of spruce. Neither of these species, or any other important tree species of North America, occur in Eurasia. This indicates that the reconstitution of the boreal forest of the two continents took place from different sources. Ice ages have waxed and waned for the past couple of million years, but conditions have apparently never been such as to allow a continuous boreal forest to develop across the Bering Strait region.

Even during the last ice age and the heyday of the mammoth steppe, there must have been enormous differences between its nature in, say, the Iberian and Chukchi peninsulas, almost on the other side of the world. As the mammoth steppe broke up, it would have been colonized by Scotch pine from Europe and possibly Asia; by spruce and birch, perhaps from the Atlantic coastal regions; stone pines from the Alps, Caucasus, Altai, and other mountainous

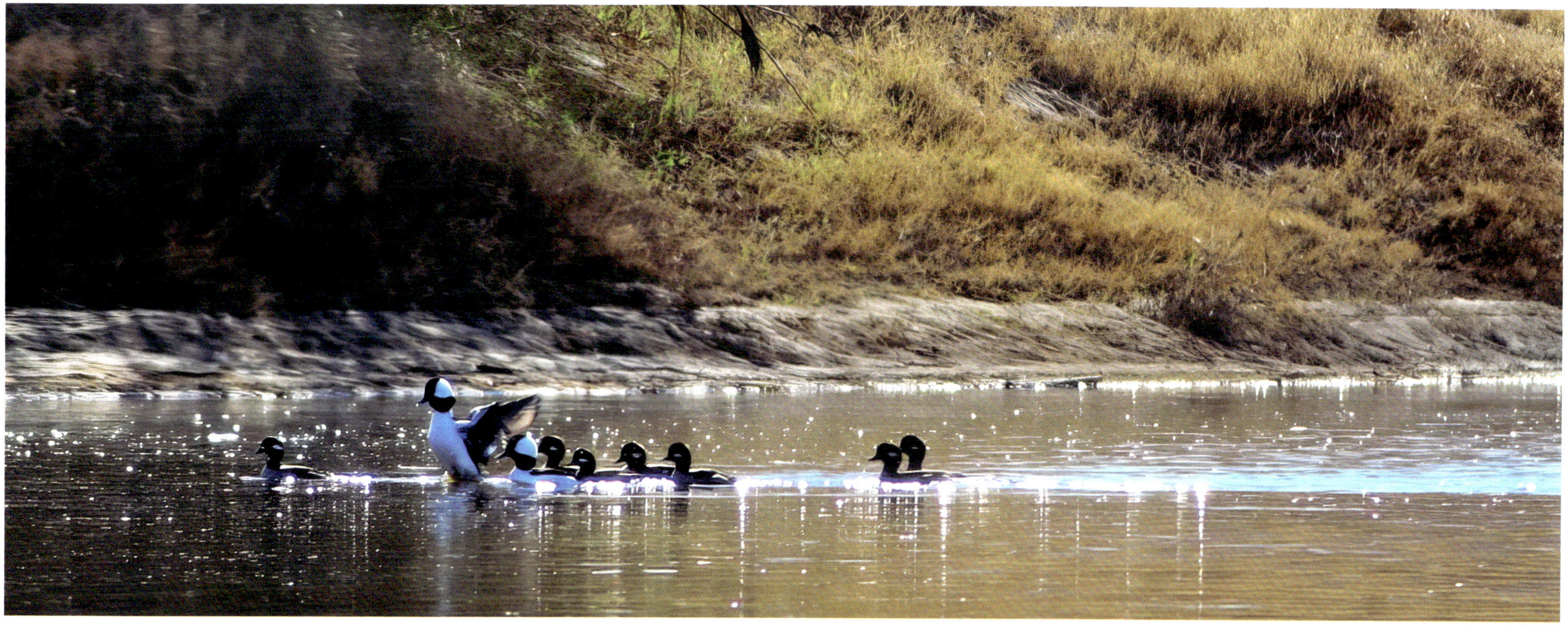

Buffleheads wintering on the Rio Grande, 2006.

regions; and, of course, the larches that now dominate most Asian areas of the boreal forest, or taiga as it's known in Russia.

It's often suggested that the dominance of larch in the Siberian forest is related to its deciduous needles, although the physiology of this is ambiguous. The most extreme winter cold of the Siberian forested region actually occurs in areas in which Scotch pine is often dominant. Larch tends to take over in areas of poor drainage, what would be called muskeg (bog) in Canada.

ON THE TOPIC of regions related to but not boreal forest, WREAF has made two river trips outside of the boreal forest, though both were linked to the North.

The first was the Rio Grande trip in 2006, which was mostly for fun. However, it is an important wintering ground for many boreal-breeding birds, especially ducks. We saw plenty of buffleheads.

The second was the Upper Missouri River Breaks National Monument in Montana in 2017. At first glance, the Missouri doesn't seem tied to the boreal forest, but the Breaks are where the river, blocked from flowing to Hudson Bay by continental ice sheets, spilled over southern tablelands and cut a spectacular new route to the Mississippi. If not for the glaciers, the Missouri would be a major boreal river. (See "A Stolen Boreal Forest River in Bandit Country" in the online appendix under Trip Reports.)

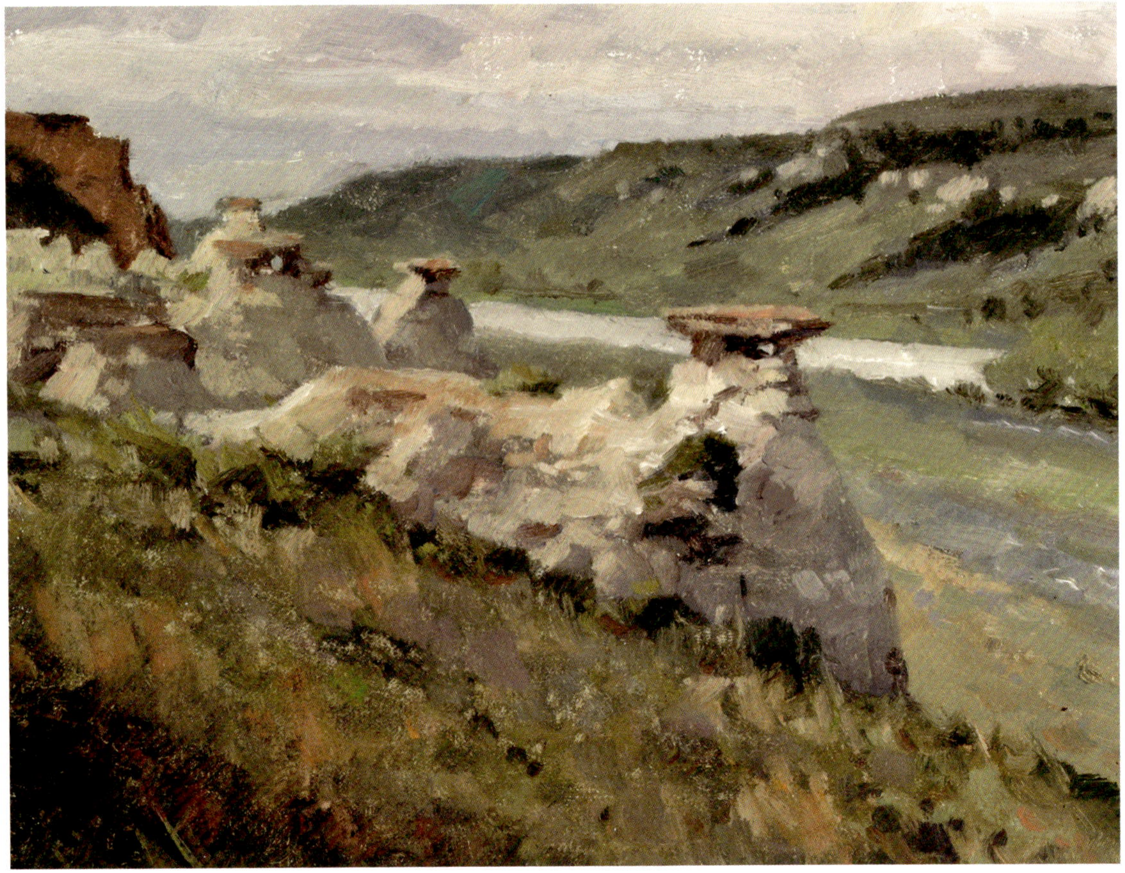

Upper Missouri River Breaks National Monument. *Eagle Rock Along the Missouri*, oil field painting, Ralph Oberg

The far east (of North America): Nunavik, Labrador (Ungava Peninsula), and the Maritimes

Here, Arctic conditions push the tree line far south, creating vast stretches of open-canopy forest carpeted in thick teal-white caribou lichen, snowy in appearance even in summer. Hudson Bay plays a role: Shallow, with restricted exchange with the open ocean, it is low in salinity, and so is ice-covered early, creating a huge reflective surface resulting in a regional coldspot. This may explain why polar bears roam near Nain, Labrador (56°30′N), while in Juneau, Alaska (200 miles farther north) winter temperatures rarely stray much below freezing or stay there long.

Another transition unique to the Far East (and the Mackenzie River delta in the Northwest Territories) is the direct shift from boreal forest to the sea.

We've run six expeditions here: 19 weeks in the field. The first were the 2005 and 2006 George River (Mushuau shipu to the Innu) trips.

Mushuau shipu is the home range of the George River caribou, a keystone species in the ecosystem and a cornerstone of Innu culture for millennia. The river gained fame in 1903 when Leonidas Hubbard died attempting to canoe it to Fort George (Kangiqsualujjuaq) on Ungava Bay. Remarkably, his wife, Mina Benson Hubbard, completed the journey in no-nonsense fashion in 1905, publishing *A Woman's Way Through Unknown Labrador*.

A century later, our first trip to Kangiqsualujjuaq was a reconnaissance to host Robert Bateman the next year. While a practical planning trip, it was unforgettable for the nightly northern lights shows.

In 2006, we returned with art in mind—lingering longer, traveling less. Our outfitter, Norpaq Adventures, flew Robert Bateman to its George River Lodge on Indian House Lake (a 60-mile-long widening of the George River). Smithsonian Arctic archaeologist Stephen Loring joined us as well.

The George River was my first trip into the Barrens: a river mostly forested along its banks, but where, as the land rose, the trees quickly thinned to stunted spruce and tundra bursting with September color and berries. It was my first time seeing a river shift continually between forest and tundra, with the ancient Canadian Shield exposed beneath.

Our next Far East journey was to an Innu hunting site in continuous use for over 7,500 years around Kamestastin Lake ("Mistastin Lake" on most maps), a 38-million-year-old

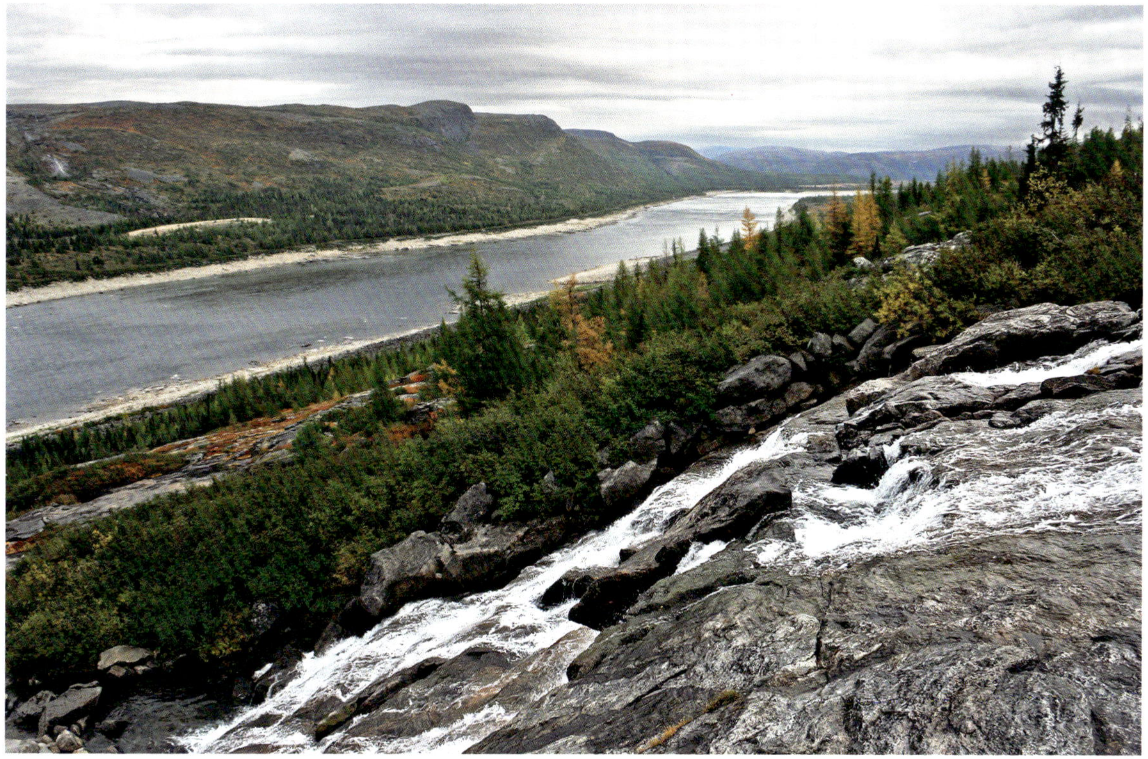

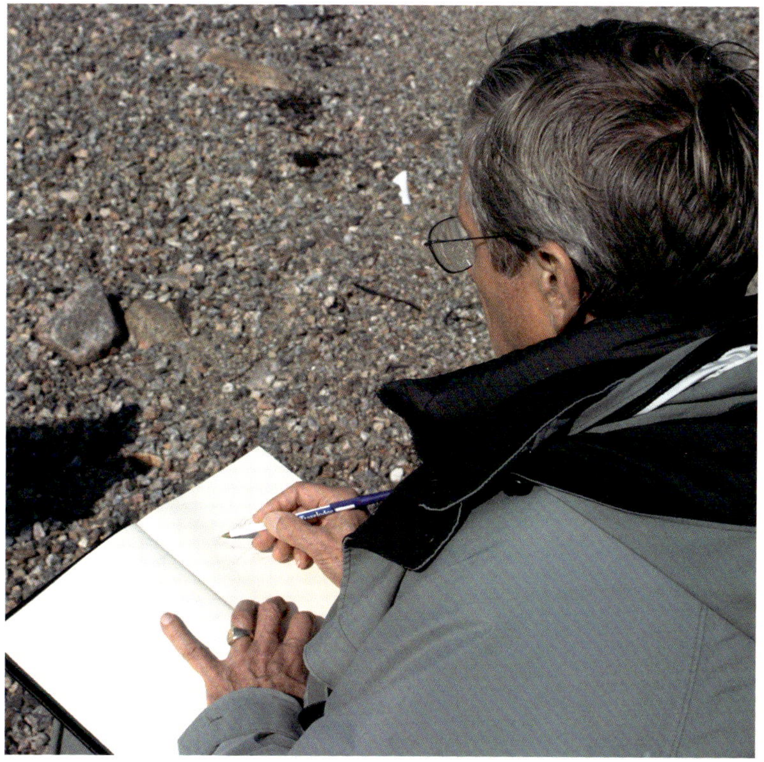

meteorite crater. Sheltered from tundra winds, its southern and eastern shores grow dense spruce, drawing caribou and other wildlife in huge numbers.

In October 2008, invited by the Innu Nation of Labrador and the Tshikapisk Foundation, we headed for Kamestastin (now known as Kameshtashtan and designated as a national historic site). The drive passed through the Charlevoix crater and skirted the Manicouagan crater, the "Eye of Quebec," before arriving at Happy Valley–Goose Bay for our bush flight to Kamestastin.

As we approached the Kamestastin crater (the youngest of the bunch), I pictured a subtropical dawn 38 million years ago, with titanotheres and hyaenodons stirring to greet the new day—moments before a meteorite changed everything. (See "Caribou 'n' Craters" under Papers & Essays in the online appendix.)

Our sense of time is plastic. Fighting a rising wind and frigid building waves across Voisey's Bay in a loaded canoe for over an hour can feel endless; a night storm at sea in a 32-foot ketch can feel longer still. But a quiet hour in camp with a hot toddy (just one; there were polar bears) or dinner with your brother and banged-up old salt of a father in a marina vanishes too soon. Experience and the scale of our lives mold and bind our sense of time. Standing on billion-year-old bedrock, scanning the tundra hills as others have for thousands of years, you can feel those bonds of time start to slip.

ABOVE LEFT: Falls above the George River, September 2006.

ABOVE RIGHT: Robert Bateman working on the George River.

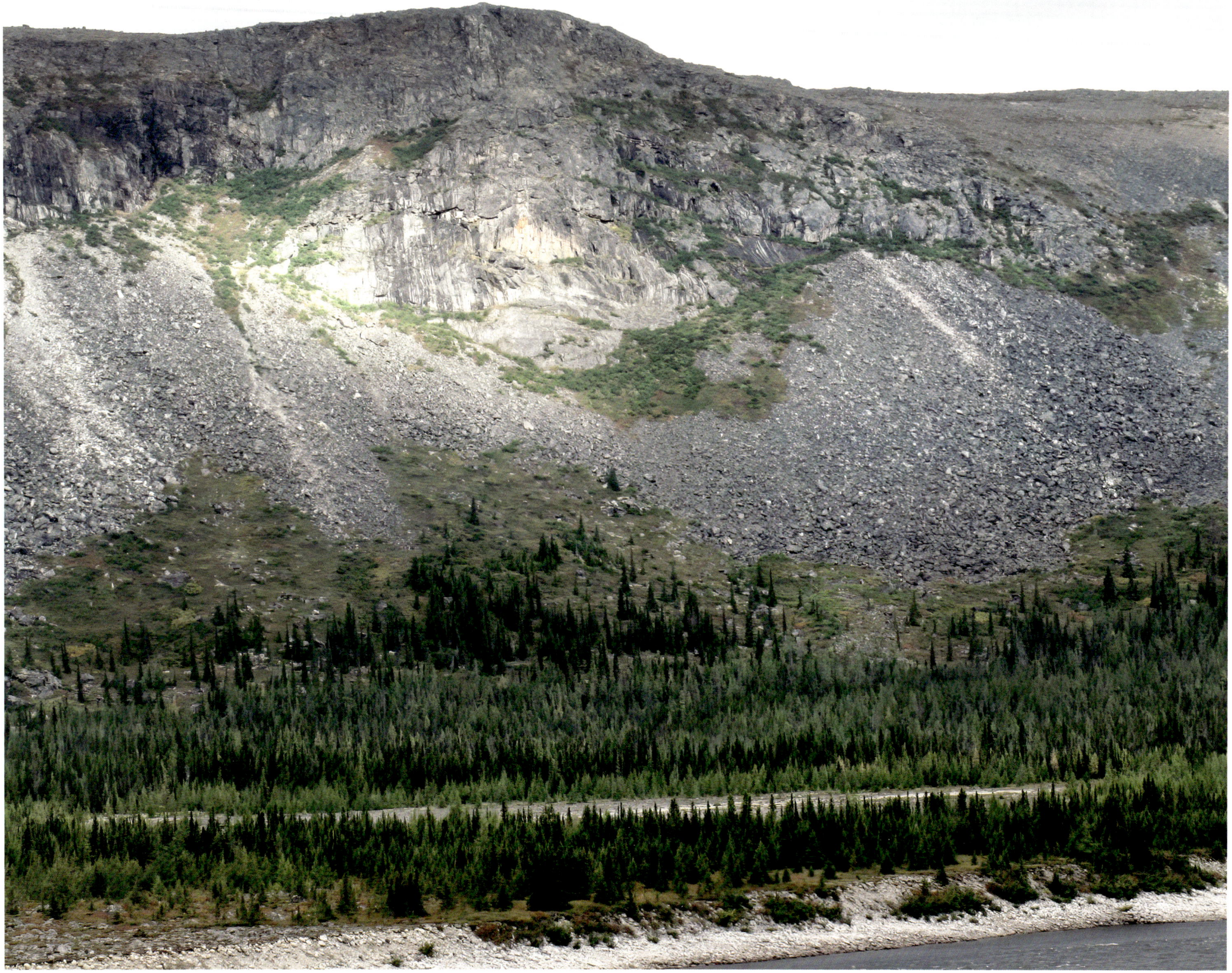

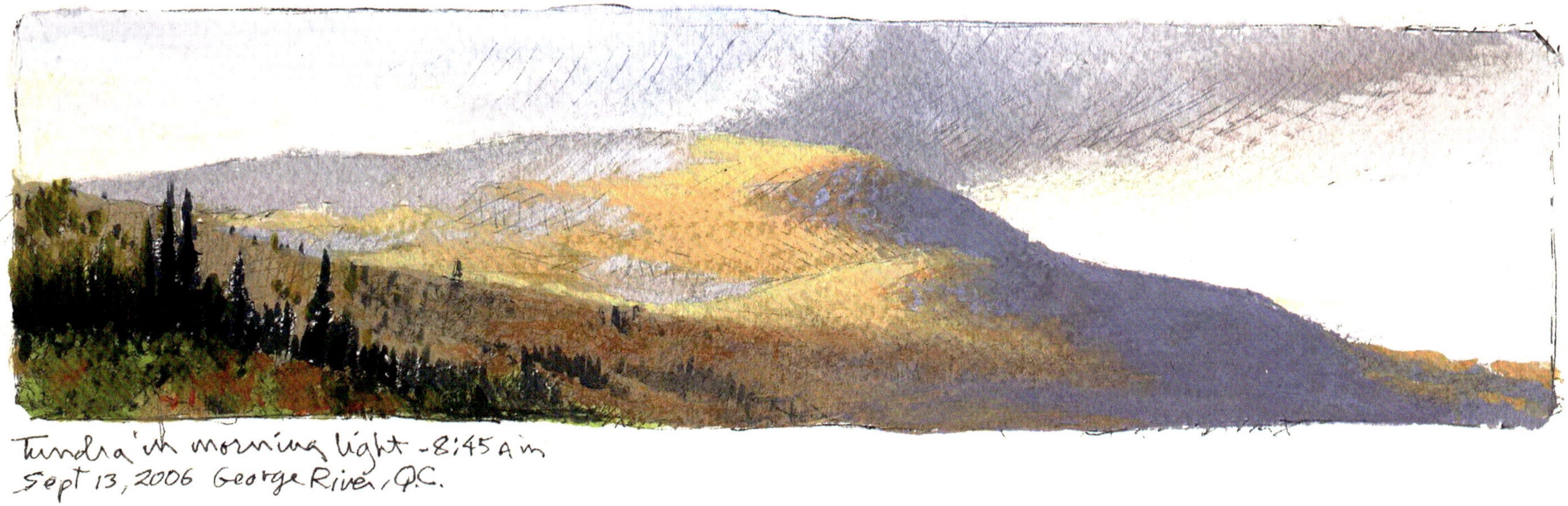

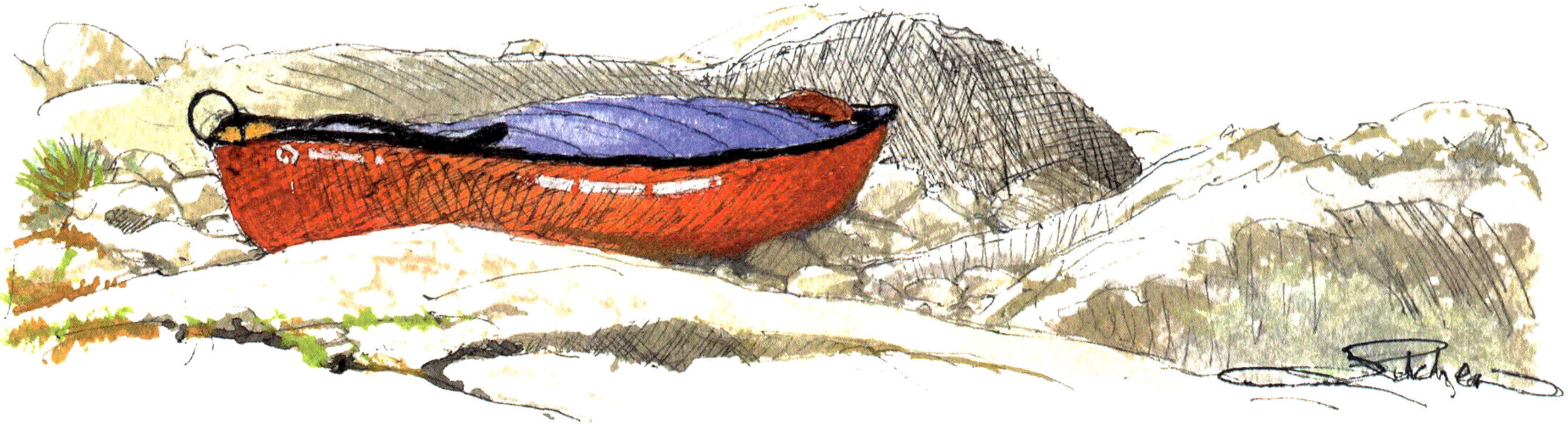

Canoe and tundra in morning light, George River. Watercolor, John C. Pitcher

FACING: Ancient Watcher, George River, Nunavik (if you don't see it, look to the light).

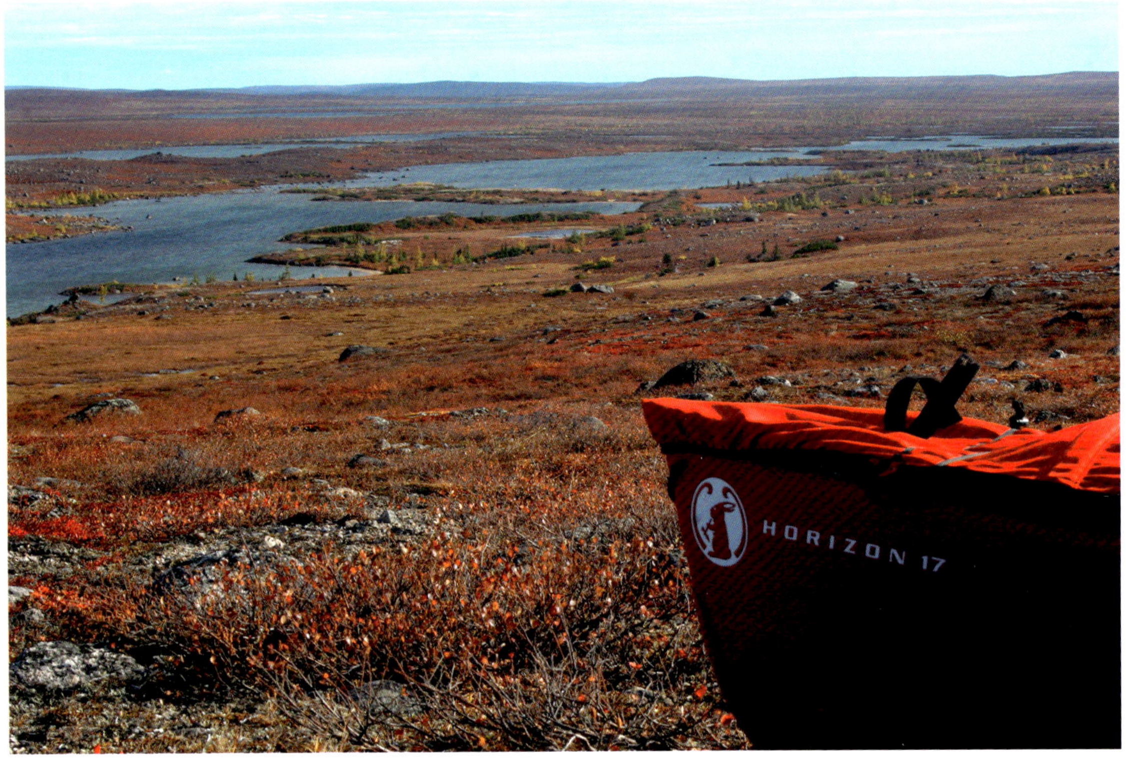

Our trusty canoe overlooking Long Pond from the top of a portage on the way to the Kogaluk, Labrador Barrens.

Following the George River trips and Kamestastin visit, we planned an ambitious Labrador crossing, complete with a team of artists, Innu youth participants, and a rendezvous with Arctic archaeologists to retrace William Brooks Cabot's 1901 route to Nain, the northernmost settlement in Labrador. As the length of that sentence hints, it was complicated—and didn't go as planned.

By departure, we were down to four artists, with Cole Johnson able to come only on the resupply plane. My initial two companions unexpectedly departed on that same plane. That left Cole and me to figure out a way through the rarely traveled country between the George River and Nain on the Labrador Sea.

It became one of my two hardest canoe trips. Some excerpts from "Across Labrador's Spine" in the winter 2010 issue of *Che-Mun: The Journal of Canadian Wilderness Canoeing* will impart the essence.

September 19, 2009: "Fresh snow coated everything with the torn remnants of the overnight storm scudding southeastward. A gyrfalcon swooped in low over our camp, and wolves howled in the distance as we prepared for the portage."

September 21, 2009: "We came to shore for the next portage at what must have been the same place whoever named the region the 'Barrenlands' did."

September 29, 2009: "The water was crystal aqua green, the sea floor clearly visible through 10 to 20 feet. As the east wind strengthened, our progress slowed until the sea urchins on the bottom seemed to be going nearly as fast as we were. The wind continued to build, the waves followed suit and before long, we were in a fight."

The rest: bears, gales, the 1,200-foot canyon of the Kogaluk, repeated white-knuckled descents of the canyon (one with a canoe), huffing-and-puffing ascents to get our kit down, a straight transition from forest to ocean (whales, seals, probably a shark), and the crossing of Voisey's Bay. (Full article can be found in the online appendix.)

This expedition stands out from all the rest for a special reason. Shortly after we arrived in Nain, I was on the satellite phone to Bonnie Rowell when I met a lonely puppy. All Bonnie said, with a laugh in her voice, was "oh no." (There were puppy toys waiting when we got back to Montreal; Bonnie and I are

Maps label it Kogaluk River, but *Kogaluk* means "little river" in Inuktitut. If you say "Kogaluk River," the Inuit will politely smile and maybe chuckle.

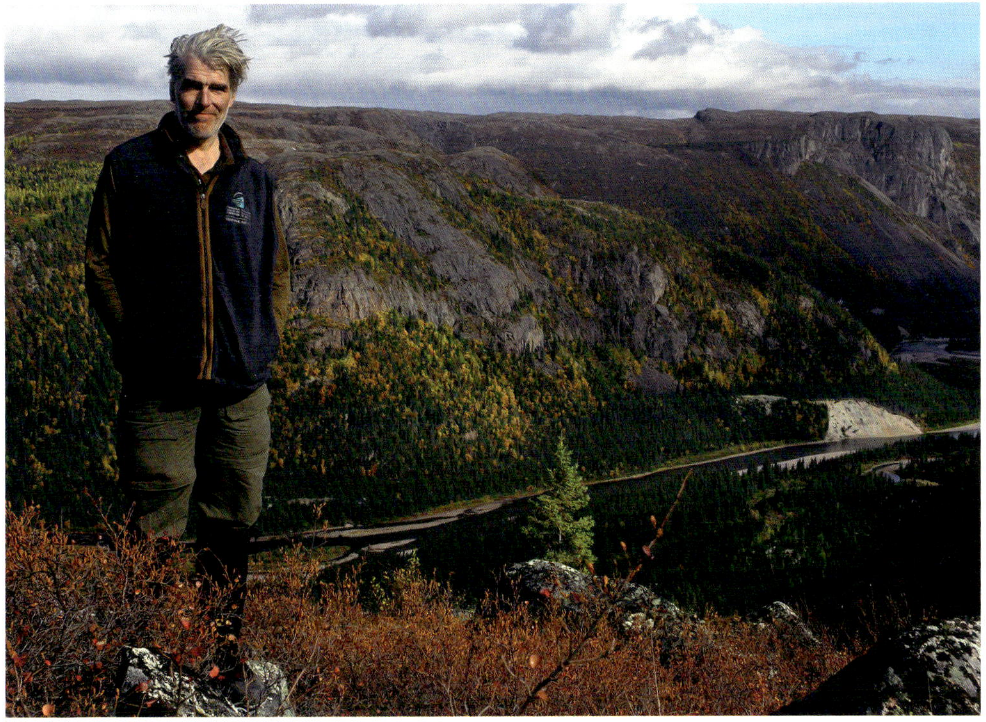

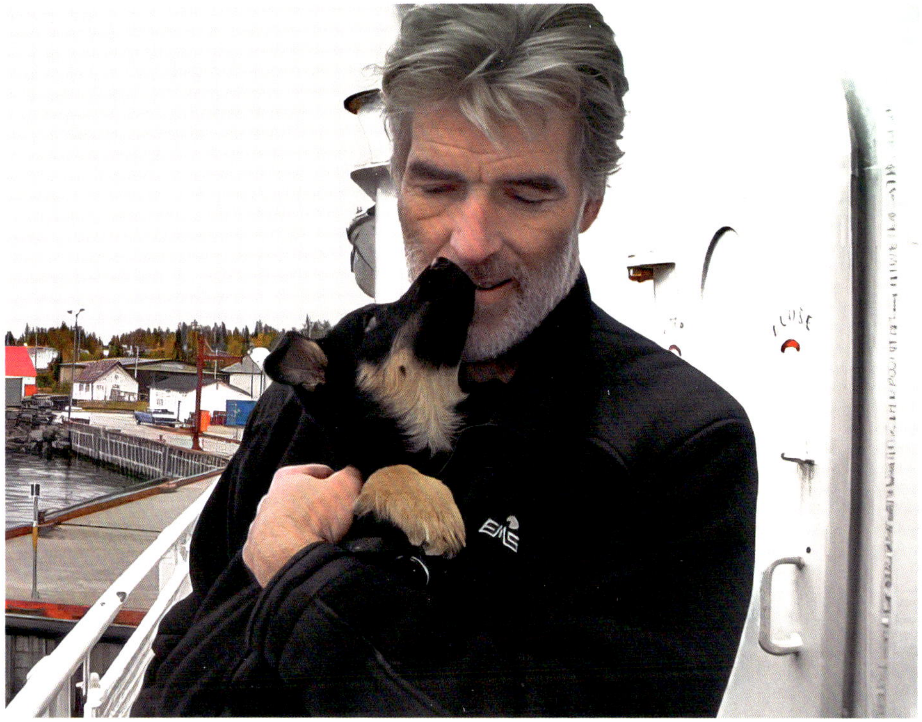

now married.) Henry Webb, the fisherman we stayed with after our hotel money ran out waiting for the weekly coastal freighter, said the little pup was a runt, tossed out to die and would be "raven-picked bones in a snowbank in a month."

As we boarded the freighter, the purser droned that "the dog" would have to be kept in the cages on the stern rail, open to the raging North Atlantic in October. A steward helped me smuggle her to our cabin, and Nainette became the most beloved stowaway aboard the MV *Northern Ranger* for the two-and-a-half-day voyage south to Goose Bay. (I had to put her in the kennels as we came into port in Goose Bay for appearances.)

At the time of writing, Nainette was 16 years old. She was my constant companion in the studio for more than 15 of those years and our perfectly behaved trail hiking dog, never needing a leash unless legally required. She became too old for much more than walking over to the pond or all the stairs up to the studio, but she was next to me for the writing of this entire book. She made it to the end but passed away, peacefully at home in Bonnie's and my arms, on October 17, 2025.

August 1–September 7, 2018: Tursujuq National Park and Trans-Ungava via the Rivière aux Mélèzes

Crew: Steve Gorman (photographer), Stephen Loring (anthropologist/archaeologist), Rob Mullen (artist), Robin Tapley (photographer; Robin did not join the Mélèzes segment). Full trip report with annotated 1:50,000 maps can be found in the online appendix.

ABOVE LEFT: Above the Kogaluk. Cole Johnson

ABOVE RIGHT: With Nainette aboard the MV *Northern Ranger* on the coast of Labrador. Cole Johnson

BELOW: Caribou at Kamestastin Lake. *Outgunned*, oil, Linda Besse

FACING: Caribou cresting a tundra ridge above Kamestastin Lake under the rising moon.

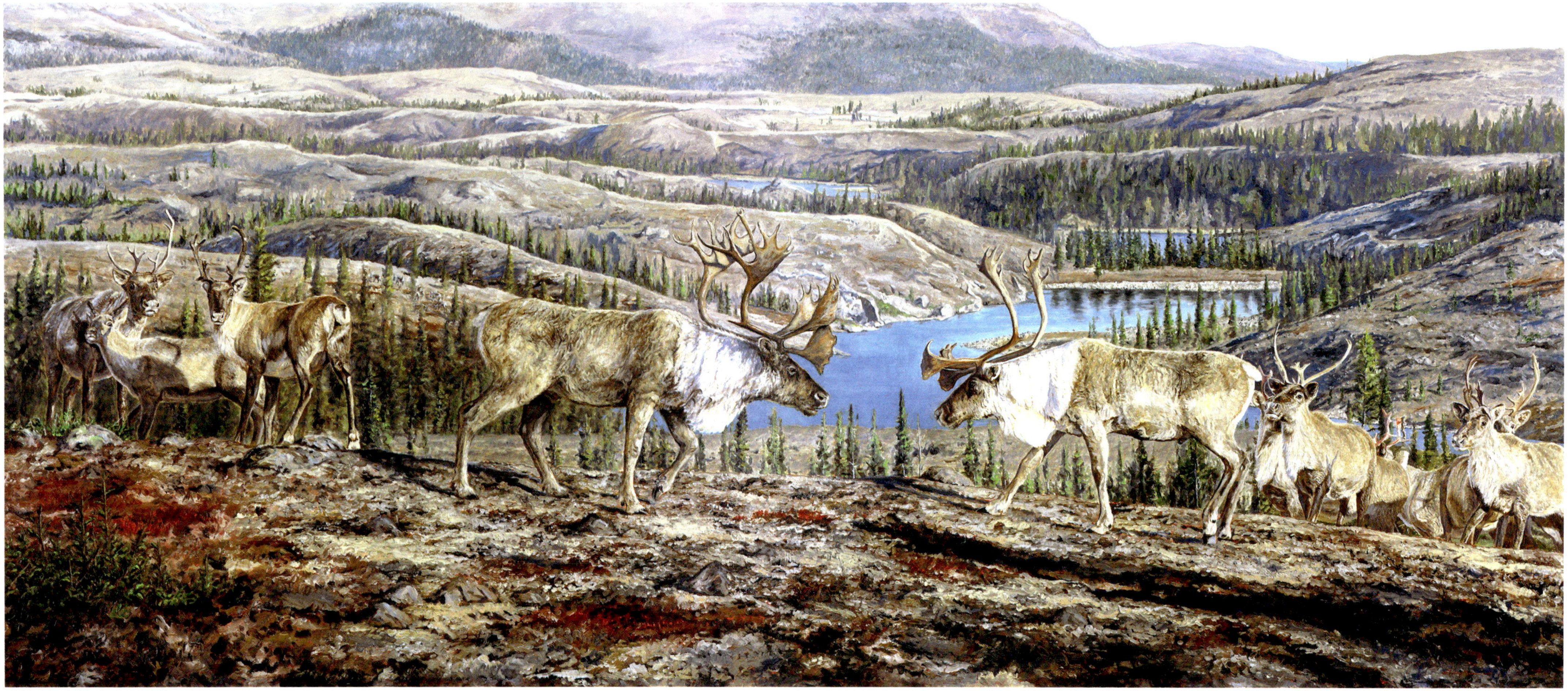

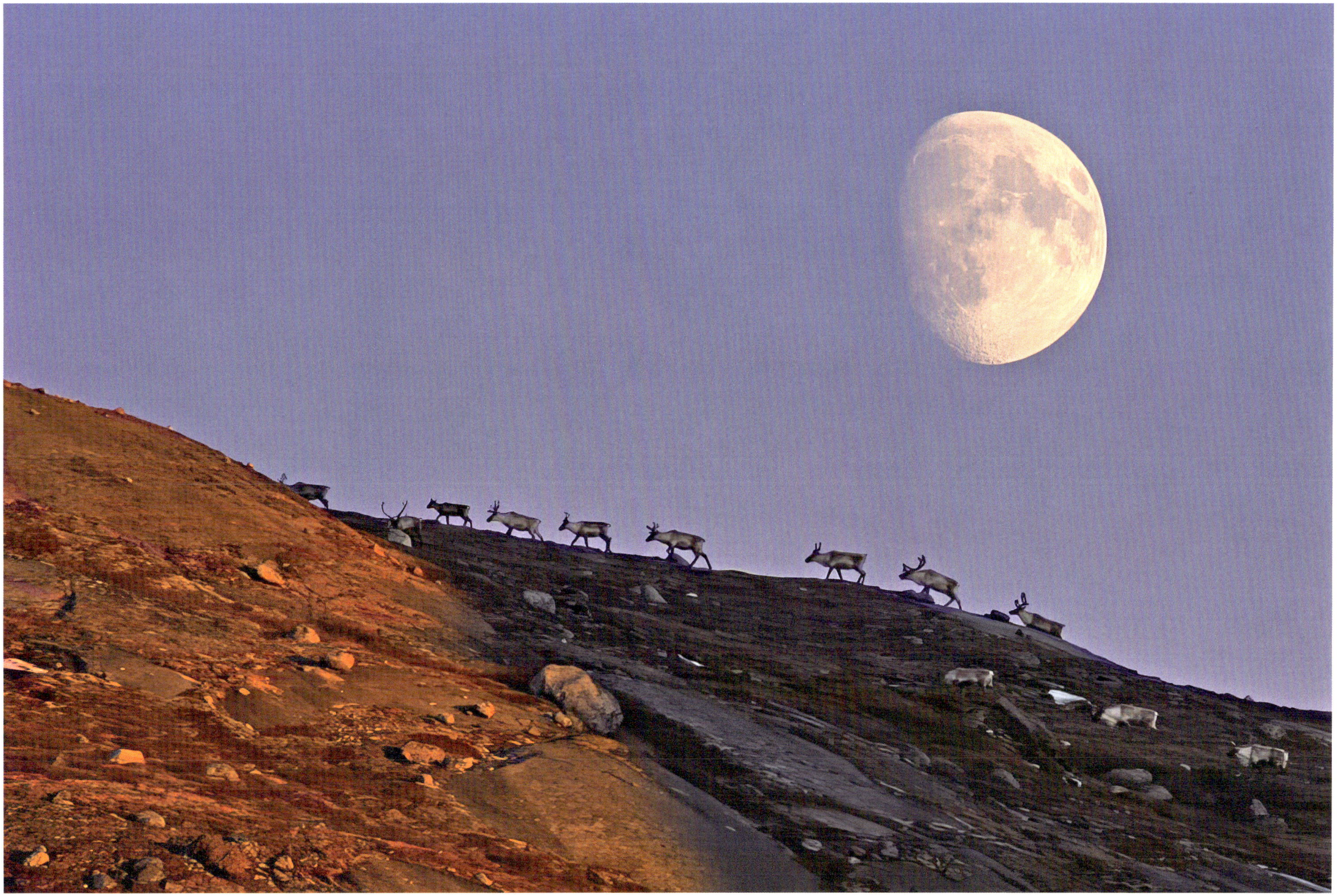

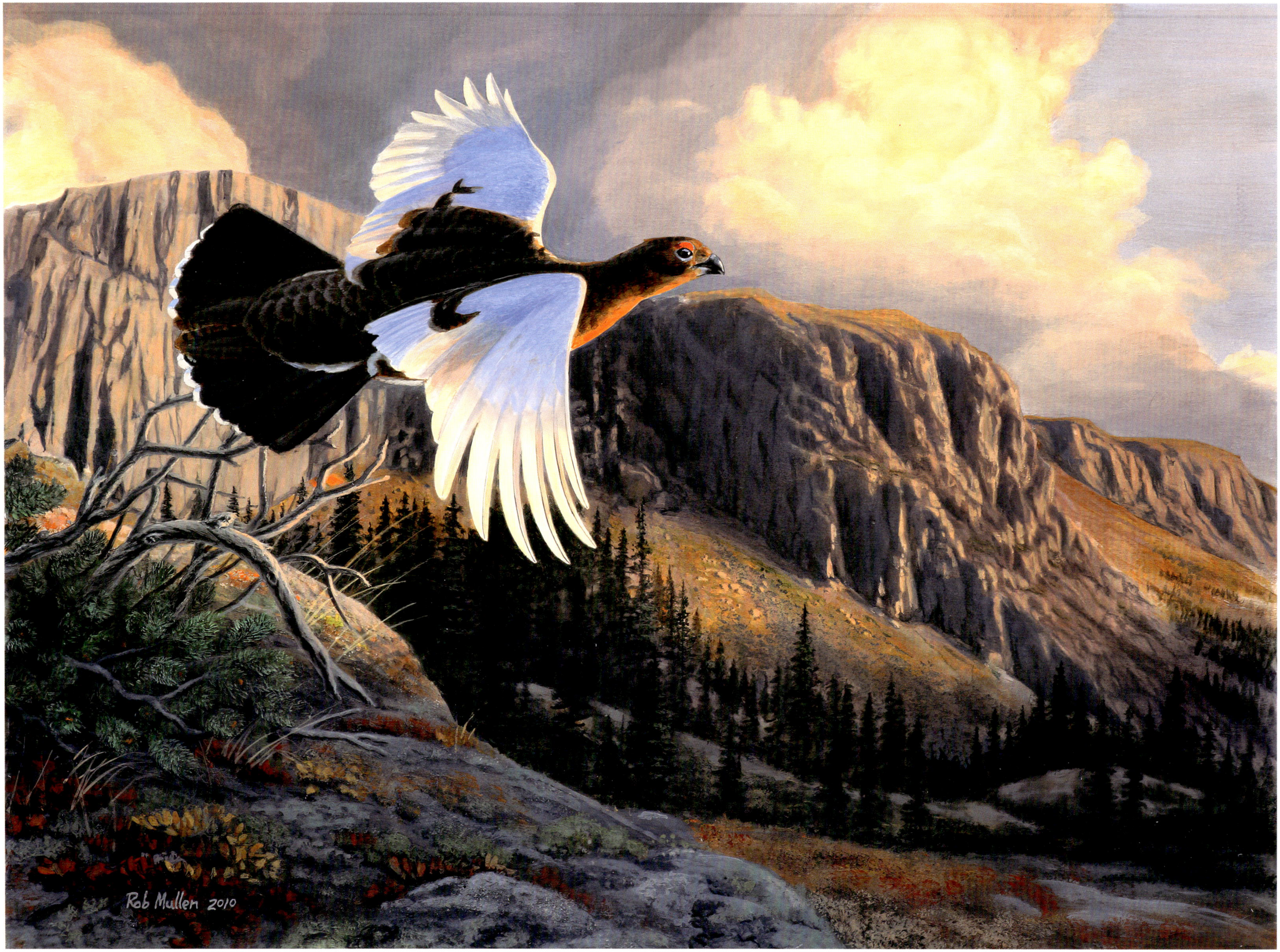
Rob Mullen 2010

The original plan was to traverse the Ungava Peninsula from Hudson Bay to Ungava Bay, roughly following the boreal tree line via a traditional route dating to at least the fur trade, first documented by Albert Peter Low in 1896. Low, a geologist, noted that the Rivière aux Mélèzes was "full of many rapids with boulders dangerous to canoes." He wasn't wrong.

Stephen Loring planned to record Cree and Inuit ethnohistorical sites along the route, which had never been professionally surveyed. But starting in Umiujaq (the headquarters for the new Tursujuq National Park) led to a detour: Bonjour Québec invited us on a nine-day guided tour. To make up for the lost time, they flew us "uphill" to Lac Wiyâshâkimî (the Clearwater Lakes): two overlapping meteorite craters nearly 200 million years apart in age. The "crossing" was shortened, but we swapped 40 miles of hard sweat and toil for catered wilderness glamping. Oh well. The pampered travel finished, we arrived at the north shore of Wiyâshâkimî by floatplane on August 11.

Stephen found ethnohistorical sites almost immediately, often at our camps, since what makes a good camp to canoeists probably hasn't changed much in thousands of years. Some were subtle enough though that, without him, I'd have passed thinking only, "Boreal chickadee?"

There is always a wondrous sense of freedom and anticipation when finally setting off from shore in a fully loaded wilderness canoe. It is at its delicious peak of flavor on that first day when you leave behind the logistics and their clamoring pack of ankle-biting details. However, on Lac Wiyâshâkimî we soon sensed an accompanying presence, an eerie stillness lurking in the surreal landscape: For days, there were no animals beyond the occasional Canada jay, a lonely trio of semipalmated sandpipers, a few loons, and one desperate Canada goose pair raising a late brood.

Tursujuq National Park is an enormous, wild expanse of coastal habitat, forest, and tundra, and can be reached in a few hours from Montreal by air. To learn more, visit nunavikparks.ca/en/parks/tursujuq.

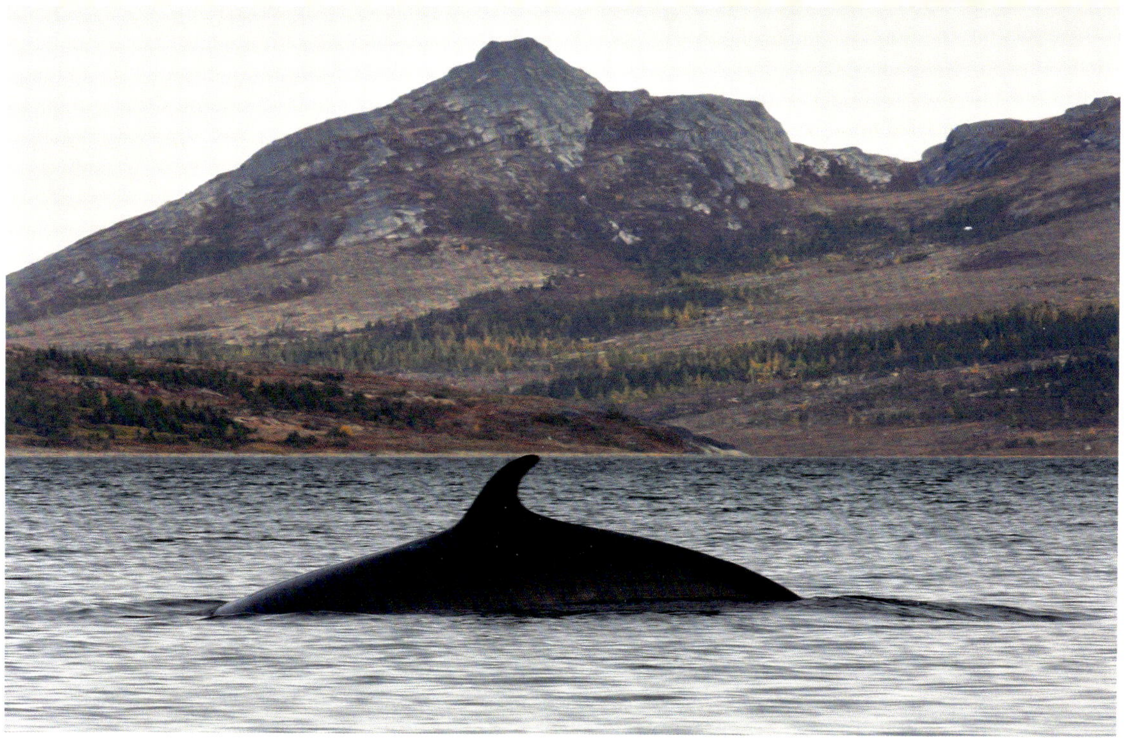

Minke whale welcoming us to Nain. Cole Johnson

FACING: Willow ptarmigan on the Kogaluk. *Wing Light*, oil, Rob Mullen

One incongruously sparkling exception was a star: the endangered freshwater seals of Lac des Loups Marins ("lake of the wolves of the sea"). A goal of ours was to document their existence for the Canadian Species at Risk Act (SARA) database. No worries. They appeared before our canoes touched water at Lac des Loups Marins, posing like professionals.

We also found possible mink frogs in a warm pool, but otherwise, silence.

The forest was typical of the northeastern transition zone: a mosaic of dense spruce, open-canopy stands, and thick caribou moss, and caribou trails everywhere. But no caribou.

Crossing into the Ungava Bay watershed, the silence broke. Near Lac Natuak, we interrupted a black bear's lunch.

CLOCKWISE FROM TOP LEFT:
Inuit children in Umiujaq, Nunavik, Quebec, August 2018. Robin Tapley

Stephen Loring inspecting an old hearth and tent ring, Noonish River.

Caribou boiling out of the woods. Stephen Loring and Stephen Gorman photographing two bulls, August 30, 2018.

Mystery frog (possibly mink frog), Lac des Loups Marins, August 15, 2018.

Landlocked harbor seal, Lac des Loups Marins, Ungava, August 21, 2018. Stephen Gorman

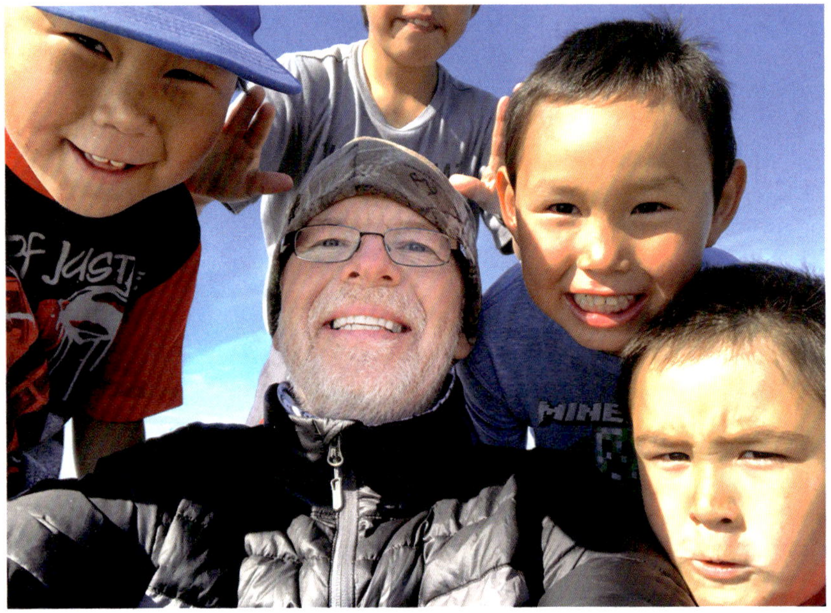

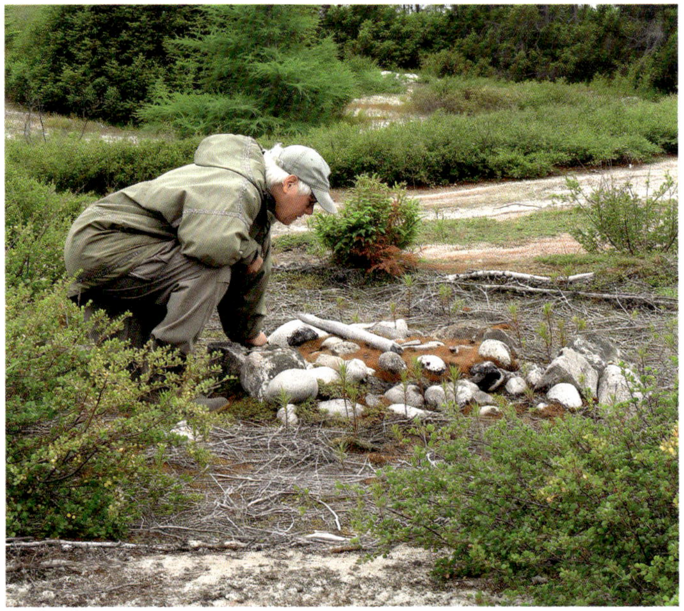

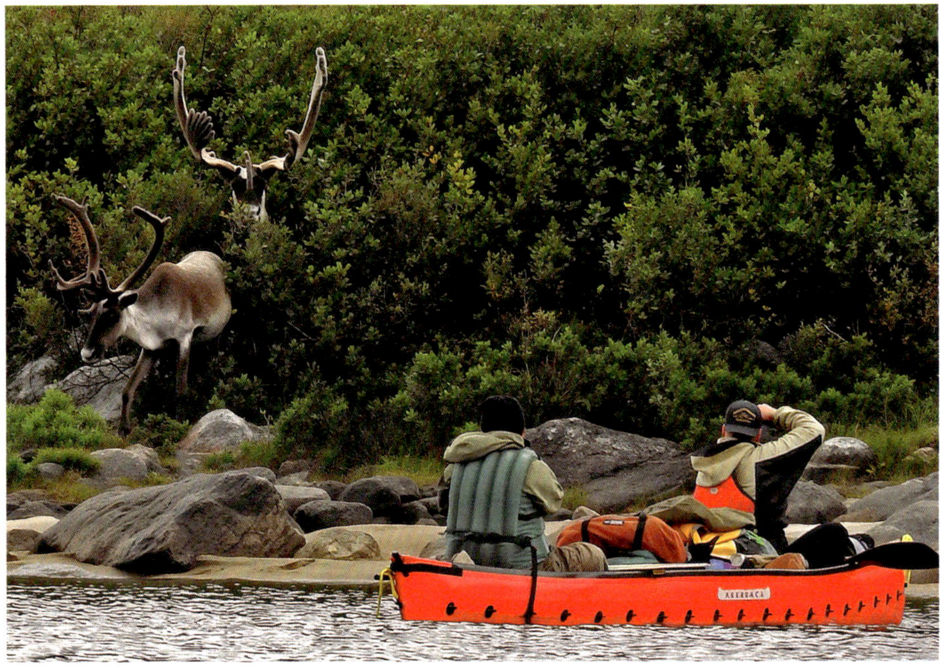

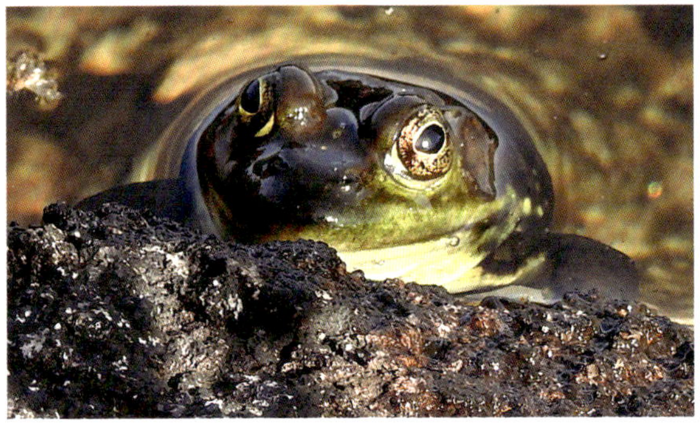

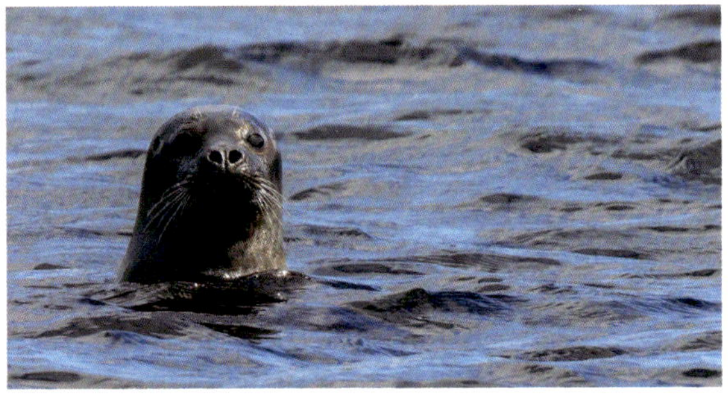

Then, in the still waters above Lac Natuak, a lone young caribou swam in front of us. That night, growing gangs of caribou appeared, swimming on each side of our camp. By the next evening, as we camped above a reportedly "unrunnable" two-mile-long rapid at the end of Lac Natuak, they were, in Stephen's words, "boiling out of the woods." At dawn, a crowd of cows and calves blundered into camp: It was the Leaf River herd on migration.

We searched for a portage trail around the fabled rapid (sometimes caribou make one) but found none. However, to our surprised relief, the rapid looked runnable, at least what we could see of it. I paddled straight through the middle of a gang of caribou. (I doubt any canoeing manual recommends this.)

To our continuing delight, the rapid continued runnable until ending with a steep drop that would have been a Class IV rapid if it was longer. We hadn't tied on our spray decks, expecting to portage, but we didn't stop. After a half hour of fun rapids rather than two days of arduous hauling, we were through, and in high spirits.

And the caribou kept coming, until, 50 miles later, we landed for lunch and locked eyes with a wolf on the beach (see p. 131). We never saw another caribou.

July 12–August 6, 2023: *Skidbladnir*'s Journey Home

Crew: Commodore Jerry Mullen (my dad, aged 54 to 60 during the trips mentioned in the prologue; 91 at the time of sailing), Captain Frank Mullen (my brother; also starring in the prologue; retired from the U.S. Coast Guard as a captain), and Skipper Rob Mullen.

This expedition took the forest's transition to the sea one step further. It was the forest as seen from the sea on a 1,200-mile journey from Chester, Nova Scotia, to Lake Champlain in Vermont, aboard the 32-foot wooden ketch *Skidbladnir* that, at 60, was the youngest of us.

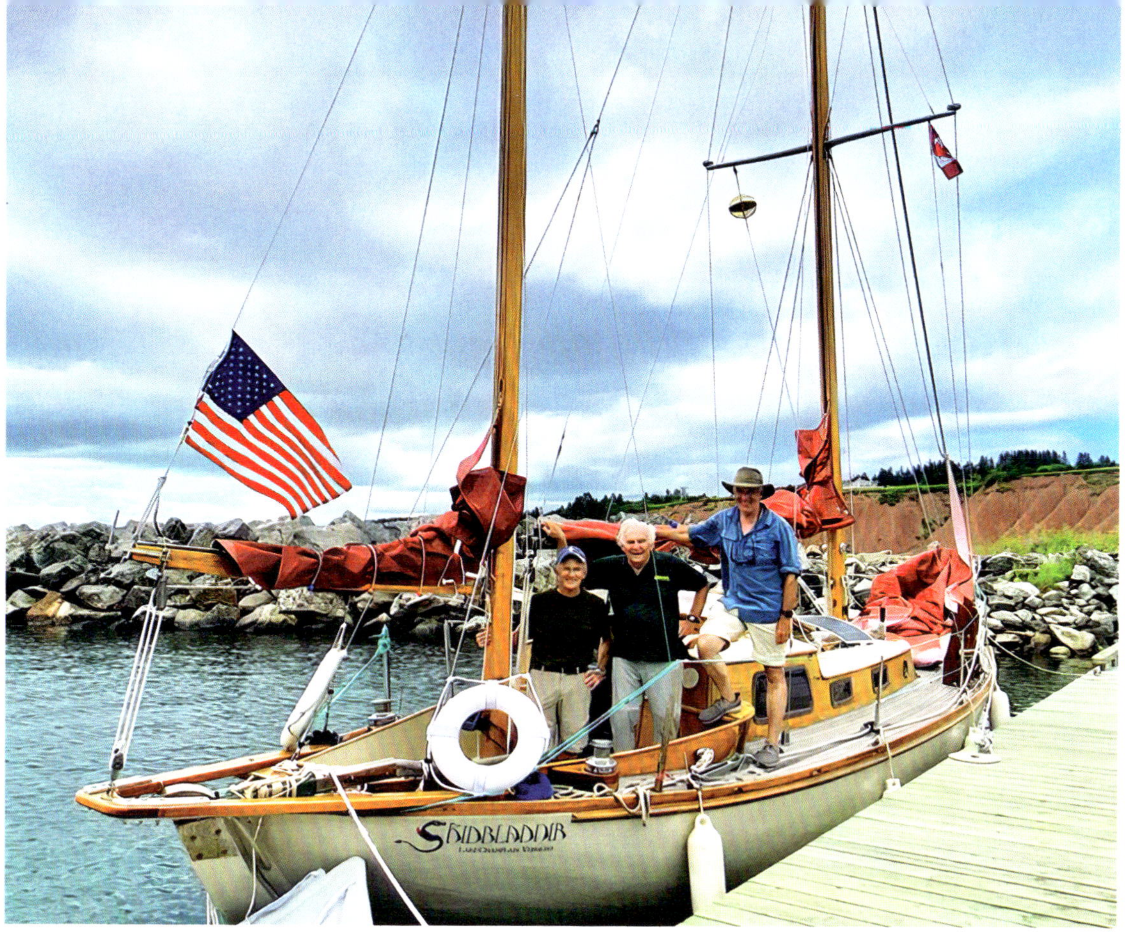

Skidbladnir and crew, Ballantyne's Cove Marina, Nova Scotia. Kailin Howard/Harbour Authority of Ballantyne's Cove

July 22, 2023, 8 bells middle watch (4:00 a.m.): Bearing NNW, nearing the southern tip of Chaleur Bay from the Gulf of St. Lawrence

The pool of blood on *Skidbladnir*'s cabin sole ("floor" to us landlubbers) was a surprise. I'd gone below for the first time since the storm struck the previous evening. Fortunately, it wasn't in the dark, chaotic wee hours when our 91-year-old father had been thrown across the cabin and halfway back again; it was pre-dawn, and I already knew he was fine. Head wounds just bleed a lot.

Twenty hours earlier, the marine forecast had promised 10-to-12-knot SE winds with 1-to-2-foot seas. Perfect for an overnight run—except the forecast was all wet.

The day was easy; my evening (aka "first") watch (8:00 p.m. to midnight), uneventful—at first. But as the sun sank, the wind rose. By nightfall, even with the mainsail furled, *Skidbladnir* flew under jib and jigger (foresail and mizzen) through mounting waves. By 10 p.m., 20-to-25-knot SE winds, gusting over 30, drove 3-to-4-foot storm waves into 4-to-6-foot easterly swells (Frank's expert estimation).

To save words, you can see the painting below. Two nights later, in the Marina de Chandler in Chandler, Quebec, Dad reveled in being the scarred old salt among a mostly young, stylish crowd. We celebrated with a seafood dinner and Pit Caribou, a strong beer brewed in nearby L'Anse-à-Beaufils.

Red Sails at Night—Fulmars Take Flight, acrylic, Rob Mullen

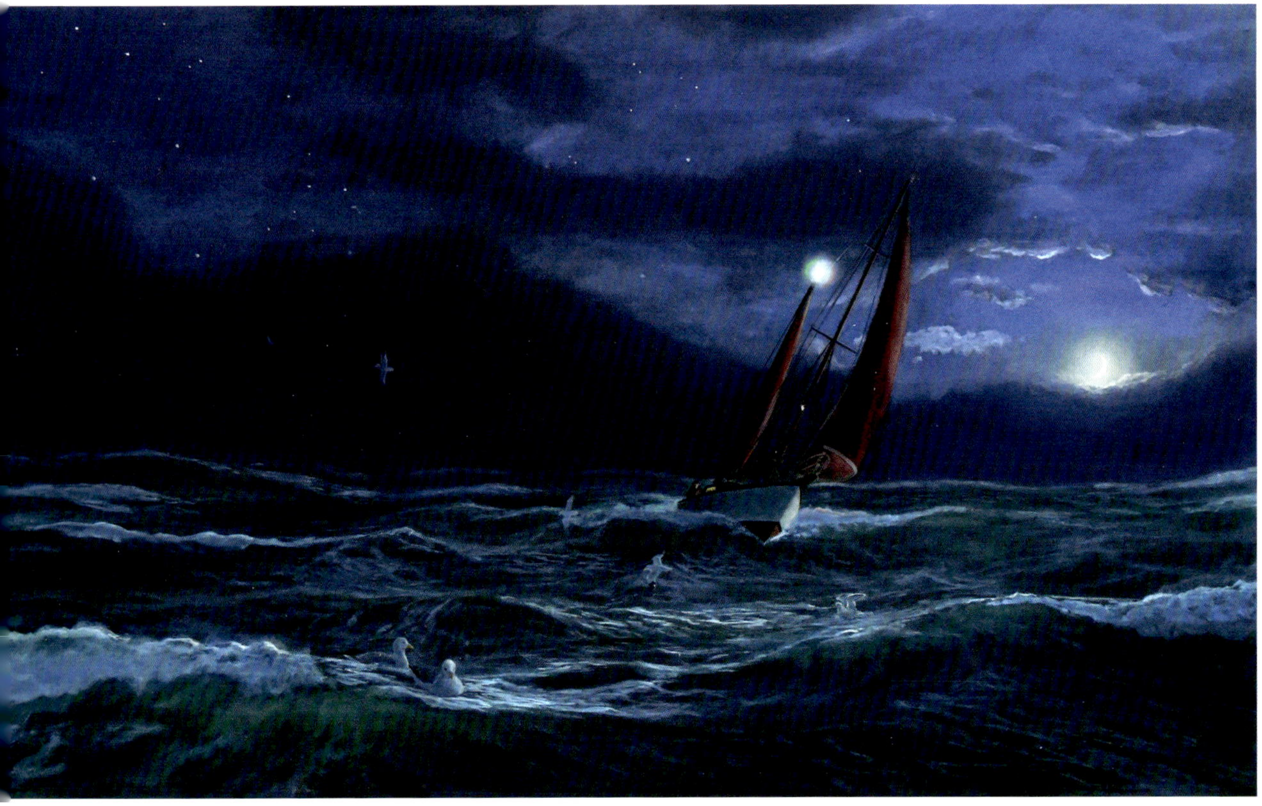

According to some maps, we reached the southern boreal forest the next day, rounding the Gaspé Peninsula (Gaspésie to locals) past Le Rocher-Percé and into the St. Lawrence estuary. According to others, we'd been offshore of the boreal forest nearly the whole way from Nova Scotia: Its spruce-lined hills looked plenty boreal to me, when we could see through the fog.

From deck, our biggest impression wasn't the forested mountains but the sea life. Nutrients from numerous boreal rivers enrich the marine habitat, dramatically at the eastern tip of Gaspé, across from Le Rocher-Percé. Just off the point, Bonaventure Island, now one of Quebec's national parks, hosts North America's largest northern gannet rookery—about a hundred thousand birds. Approaching, we saw lines of ever more numerous gannets cruising low over the waves. Even more remarkable is Tadoussac.

July 31, 2023: Off L'Isle Verte, opposite Tadoussac, Quebec

Draining a vast boreal watershed, the Saguenay River meets the St. Lawrence at Tadoussac. One of the only true fjords south of Labrador, it is very deep, and its already heavy nutrient load stirs more nutrients up from the bottom of the St. Lawrence. Together, they create a buffet for krill, fish, seabirds, and whales—humpback, fin, minke, blue, and a resident beluga population.

Earlier that day, between scattered squalls, a pod of belugas shadowed us past Trois-Pistoles's triple cathedral spires, a rainbow framing the scene. Fighting the ebb tide, we made headway until the Saguenay's outflow and the tide's push stalled our 18 horses off L'Isle Verte at 9 p.m.

In the evening glow and with a rising moon, we ducked behind a rocky islet, rounded up, and anchored in its eddy to await slack water. Dad volunteered for anchor watch. Though he is hard of hearing, it turned out that he can hear whales, and he enjoyed two hours of whale song under a full moon that night.

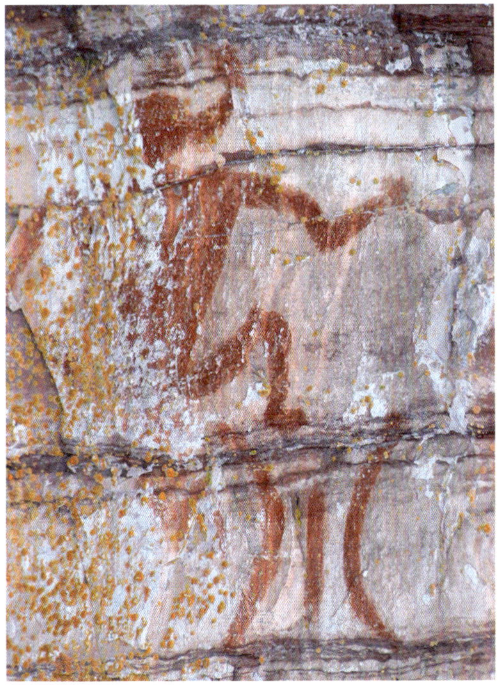

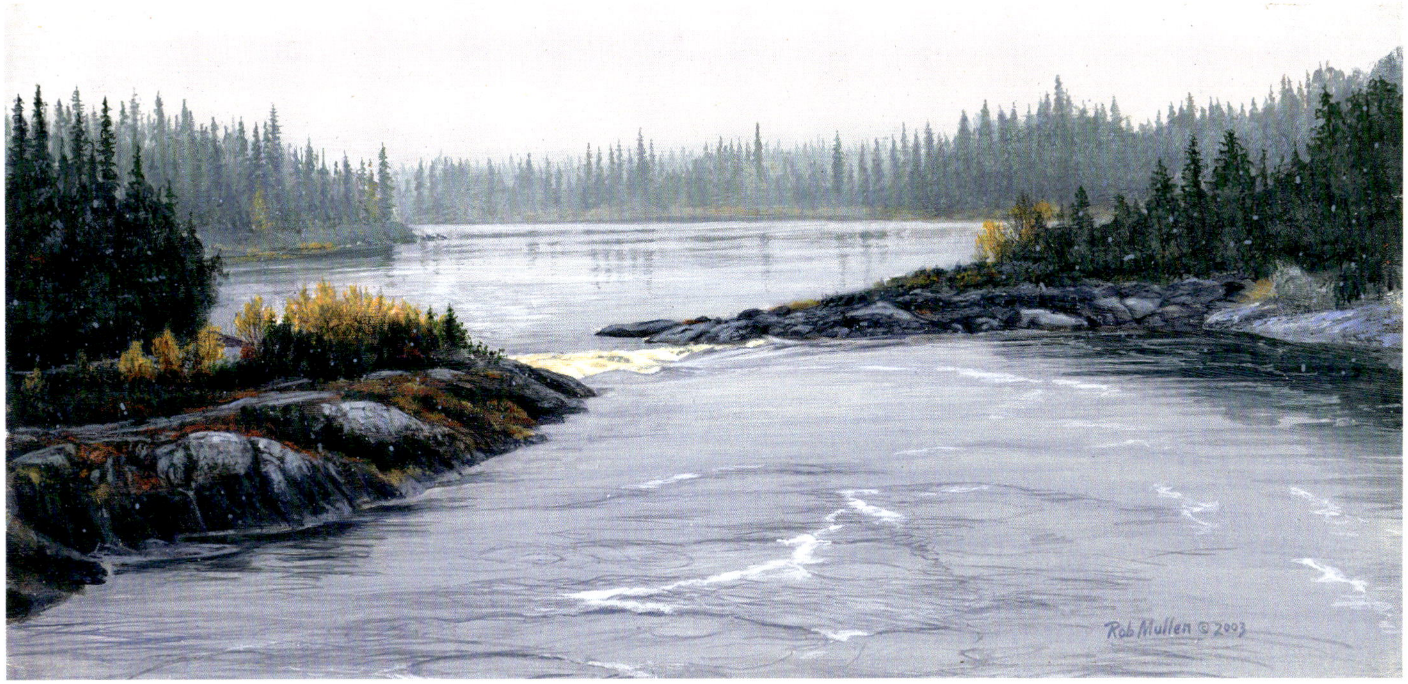

LEFT: Pictographs in Nipigon Bay, Lake Superior, February 2008.

RIGHT: Harricana River, northern Quebec. *Flowing North*, acrylic, Rob Mullen

Heart of the boreal forest—the full canopy (mostly) Southern Quebec, Ontario, and points west*

The distinction between the "heart" and the "Far East" of the North American boreal forest is fuzzy. Both lie on the Canadian Shield, but in this central region, closed-canopy forest is more common. "Closed canopy" doesn't necessarily mean bigger trees but sometimes just more densely packed ones, with moss dominating the shadier forest floor.

Mostly due to the more solid forest cover, the Shield's rock tends to appear more as ledges and cliffs rather than wide outcrops. Those cliffs can stop you cold though: not only because they're 2.5 to 3 billion years old, but because they sometimes bear pictographs left by another artist and canoeist centuries, even millennia, before you.

Forests here are still dominated by black and white spruce, with some jack pine, which isn't a climax species but comes in after fires.

August 24–September 12, 2001: Missinaibi River, Ontario—Barclay Bay on Lake Missinaibi to Moose River Crossing (Moosonee intended)

The first of these art expeditions, and it has gotten plenty of ink already.

The Missinaibi is a "pool and drop" river—rapids and falls separated by calm stretches—typical of the Canadian Shield. Many of the 80 named rapids are runnable, including this one, but we portaged and had lunch during a brief but intense rain. Glacial scarring, also typical of the Shield, is clear on the rocks. (See overleaf spread.)

* Unfortunately, our expeditions beyond Ontario were turned back or canceled by forest fires and COVID-19.

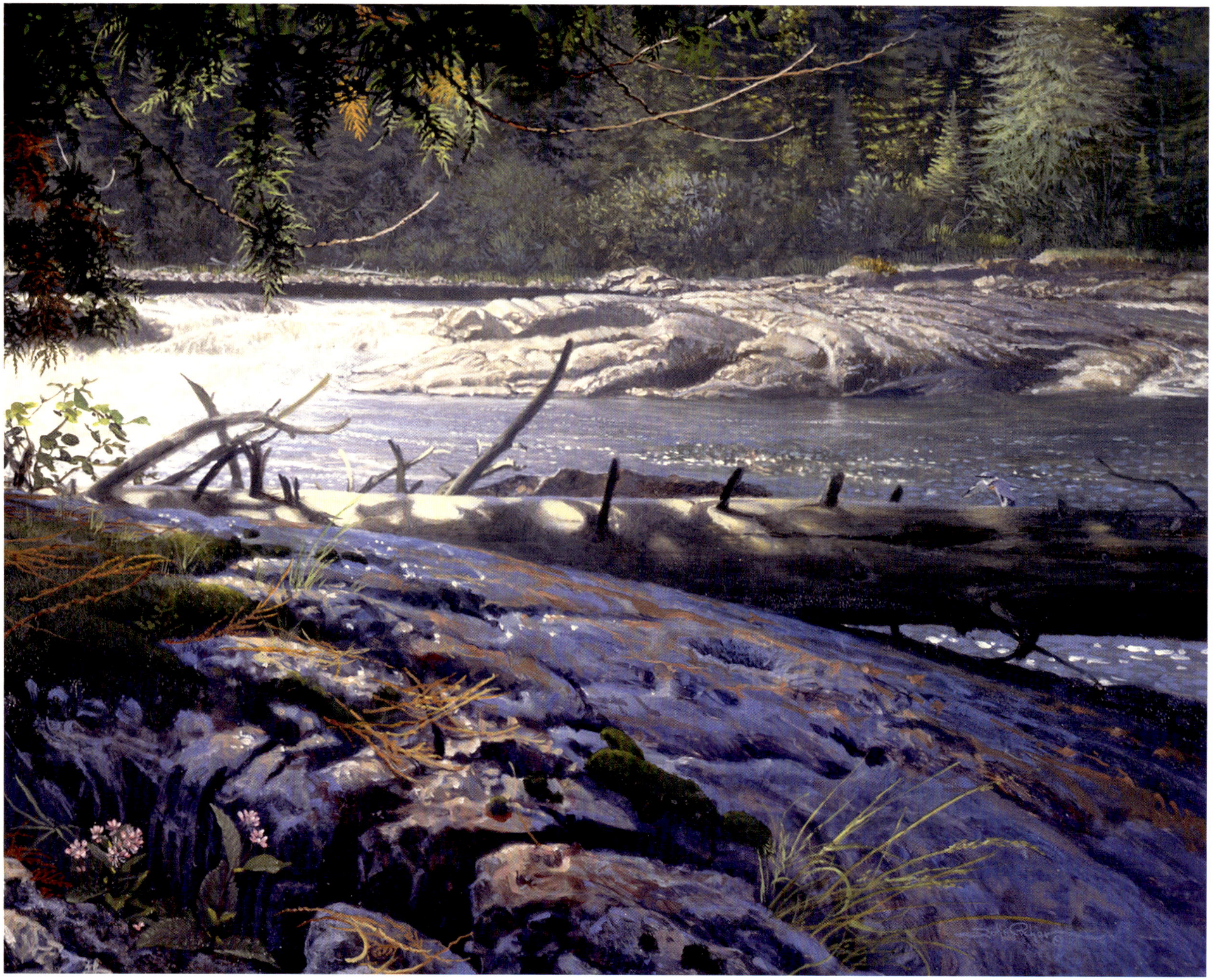

Missinaibi Blues, acrylic,
John C. Pitcher

FACING: Devil Shoepack Falls,
Missinaibi River, 2003.
Acrylic, Rob Mullen

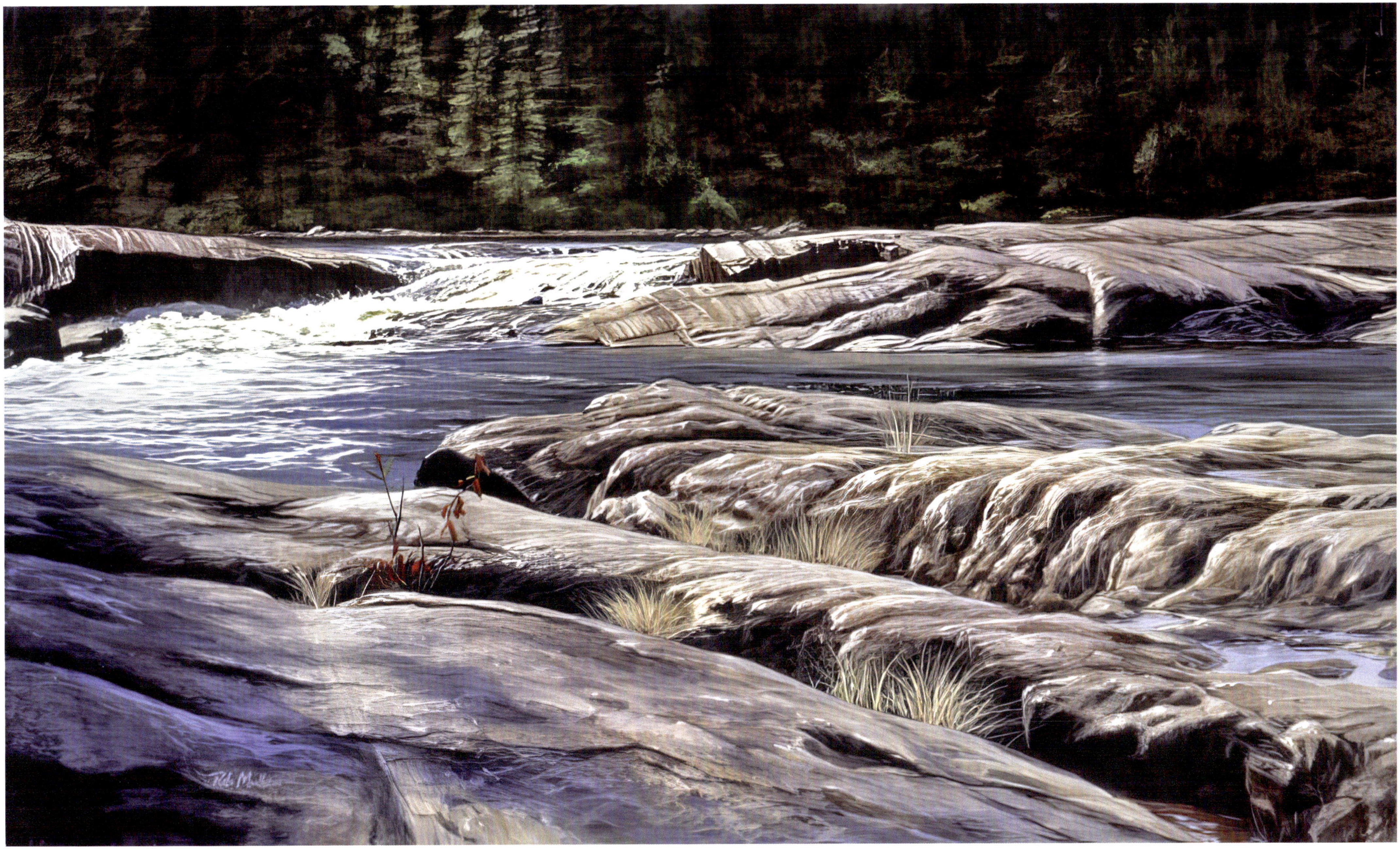

John and I clearly shared the same inspiration, but it was only months later that we discovered we'd both painted the falls when work from the trip was exhibited at a Montana gallery.

October 3–12, 2002: Harricana River, Quebec

An ill-advised solo attempt to James Bay—though since I was the one doing the advising, there was nobody to complain to.

Flowing North began as a field sketch after I abandoned the trip. High water had turned a normally easy Class II rapid into a raging Class IV or worse (*Homeward Bound*, on pp. 6–7), so I holed up to wait for safer levels. On the third morning, I awoke to ten inches of wet snow—that sealed it—along with a possibly weird (some might suggest mystic; I wouldn't, but being alone can be strange) encounter with a lynx (see chapter 4).

The return trip was pure Northwoods serendipity. Reaching the bridge where I'd been dropped off (staying near shore where the current is weaker and using a pole allowed me to work upstream), I cached my canoe and gear, packed a few days' supplies, and started walking. Within an hour, I was picked up by two trucks of *scieurs* (French-Canadian woodcutters). They spoke no English except "big fish" which the driver uttered as he helpfully pointed to a lake flashing past while driving at terrifying speeds along the abandoned mining road. There were no seat belts and every surface inside the cab—including the roof I kept hitting on bumps—was papered with cutouts from adult magazines. When I came back for my canoe a few days later, it was snowing again, so I sat on the bridge in the light snowfall and started a painting (*Flowing North*, on p. 77).

September 26–October 3, 2003: Missinaibi River, Ontario—Mattice to Moose River Crossing

Another intended run to Moosonee. The late date was because Belgian artist Carl Brenders wanted to go on a trip and late September was his preferred season, so we went to test conditions.

After a week paddling north from Mattice (with a send-off gift of top-sirloin moose steak from Bob Britton of the Empire Inn), the rain turned to snow. Lindsey Foggett and I were still 100 miles from Moose River Crossing with the train due in 48 hours. Being English, she never complained, but I suspected that she was about done. I was. "Can you paddle 50 miles a day for two days?" I asked. "Yes!" Lindsey replied.

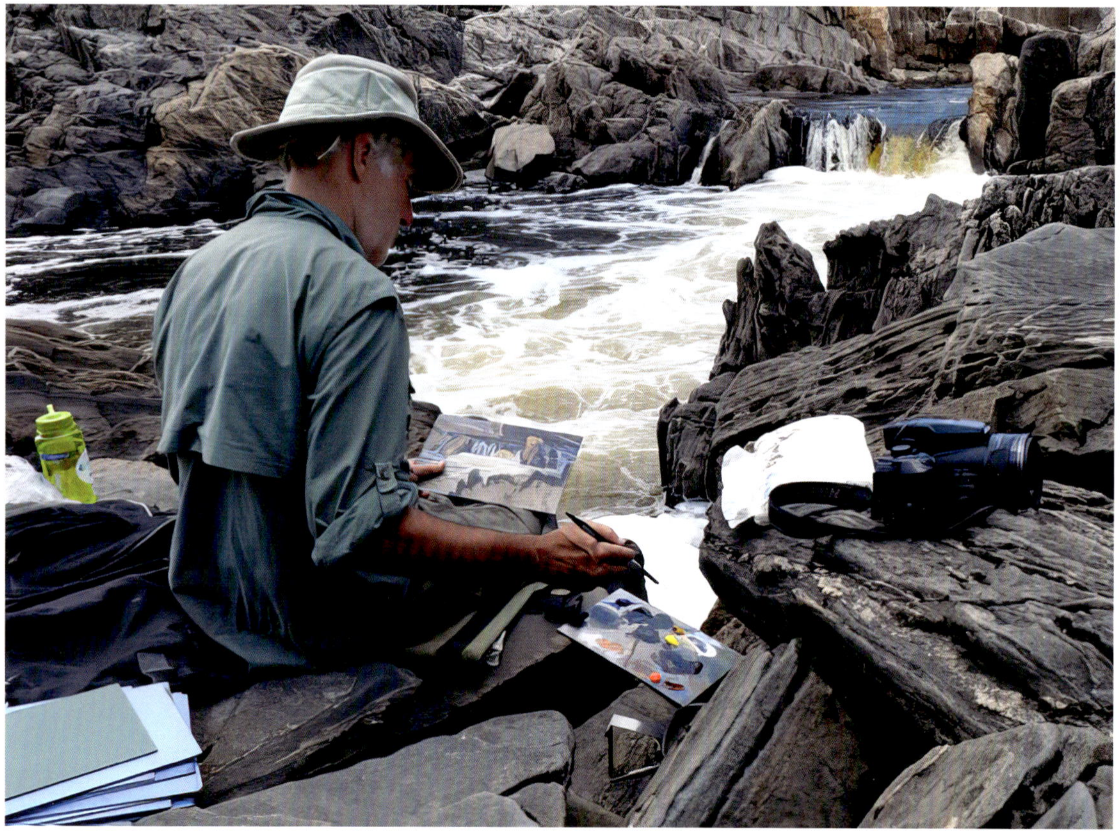

Painting in Thunderhouse Gorge.
Julia Hargreaves

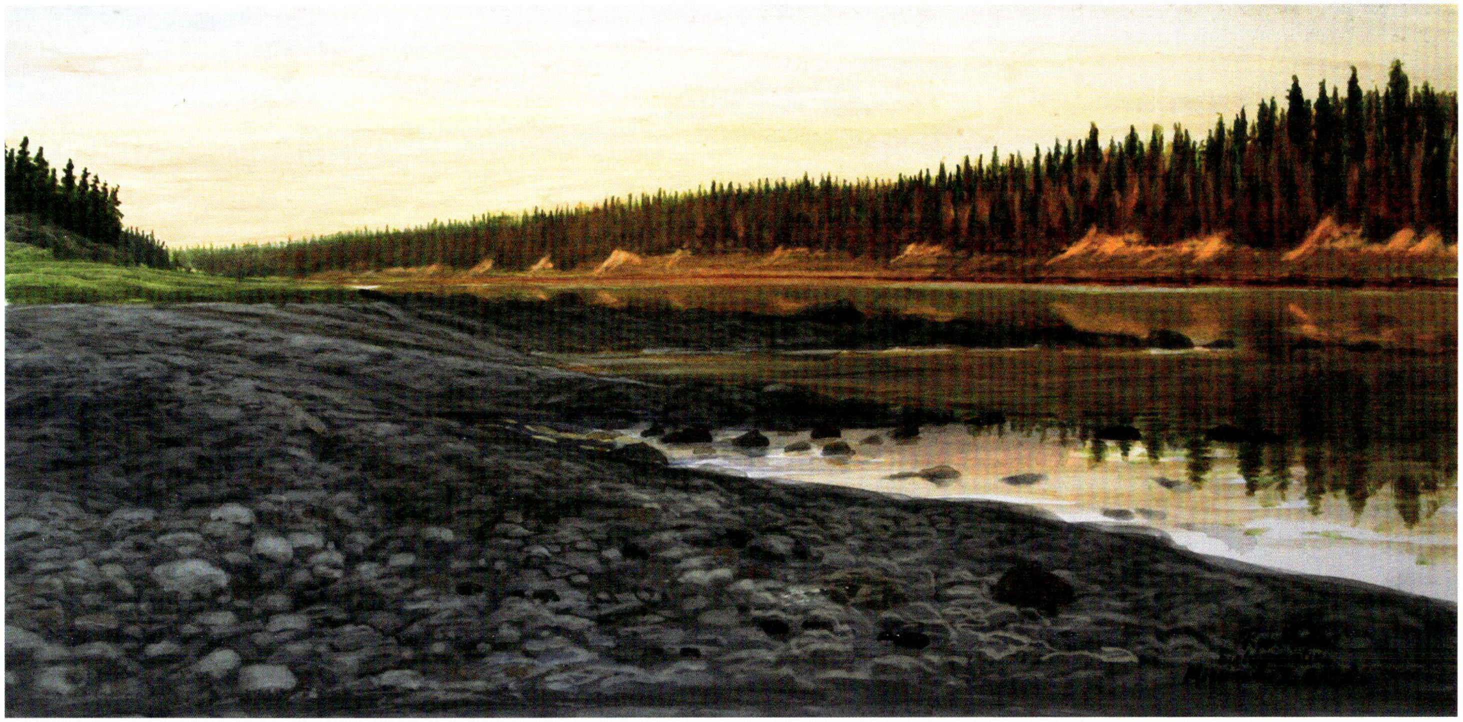

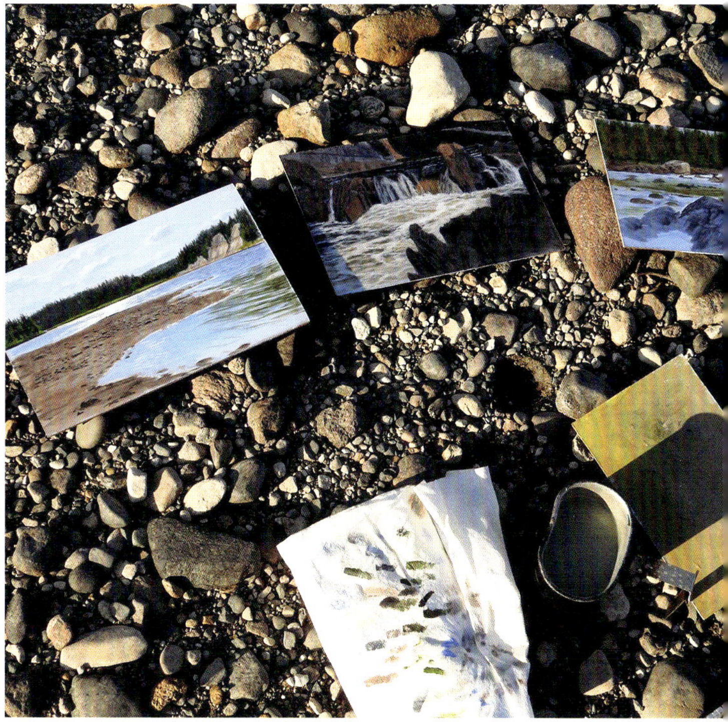

LEFT: *Missinaibi Evening*, acrylic field painting, Rob Mullen

RIGHT: Field studio, acrylic field studies, Missinaibi River.

We made it—and were warmly hosted by the Elder who had told me about 9/11 two years earlier. His young wolf greeted us like family, and two visiting hunters from Moosonee served up "moose pancakes" (from butchery leavings) and tall tales. It was a fine end to a cold, wet trip with a sprint finish. It was also the end of any notions about late-September trips on the Missinaibi.

July 29–August 14, 2016: Missinaibi River—Mattice, Ontario, to Moose Factory and Moosonee on James Bay (!)

After completing the Alaska trips (next section), I needed a break. It seemed the perfect time to finally finish what I had started in 2001.

This trip stood out from the other Missinaibi expeditions in two ways: We coordinated with the Cree from the outset, and we finally reached James Bay.

With a relaxed schedule and mostly easy weather, there was plenty of time for field painting. The Moose Cree First Nation hosted us at Moose Factory in the 1850s Hudson's Bay Company Staff House, a perfect cultural and historic bookend to our "wilderness" trip.

July 29, 2016: Setting off from the Fred Neegan Landing in Mattice, we were honored by a visit from Fred Neegan himself, a Cree riverman. He suggested a good campsite downstream: Like Iñupiaq camps in Alaska, it was back in the bush, not on the beach—unobtrusive and sheltered.

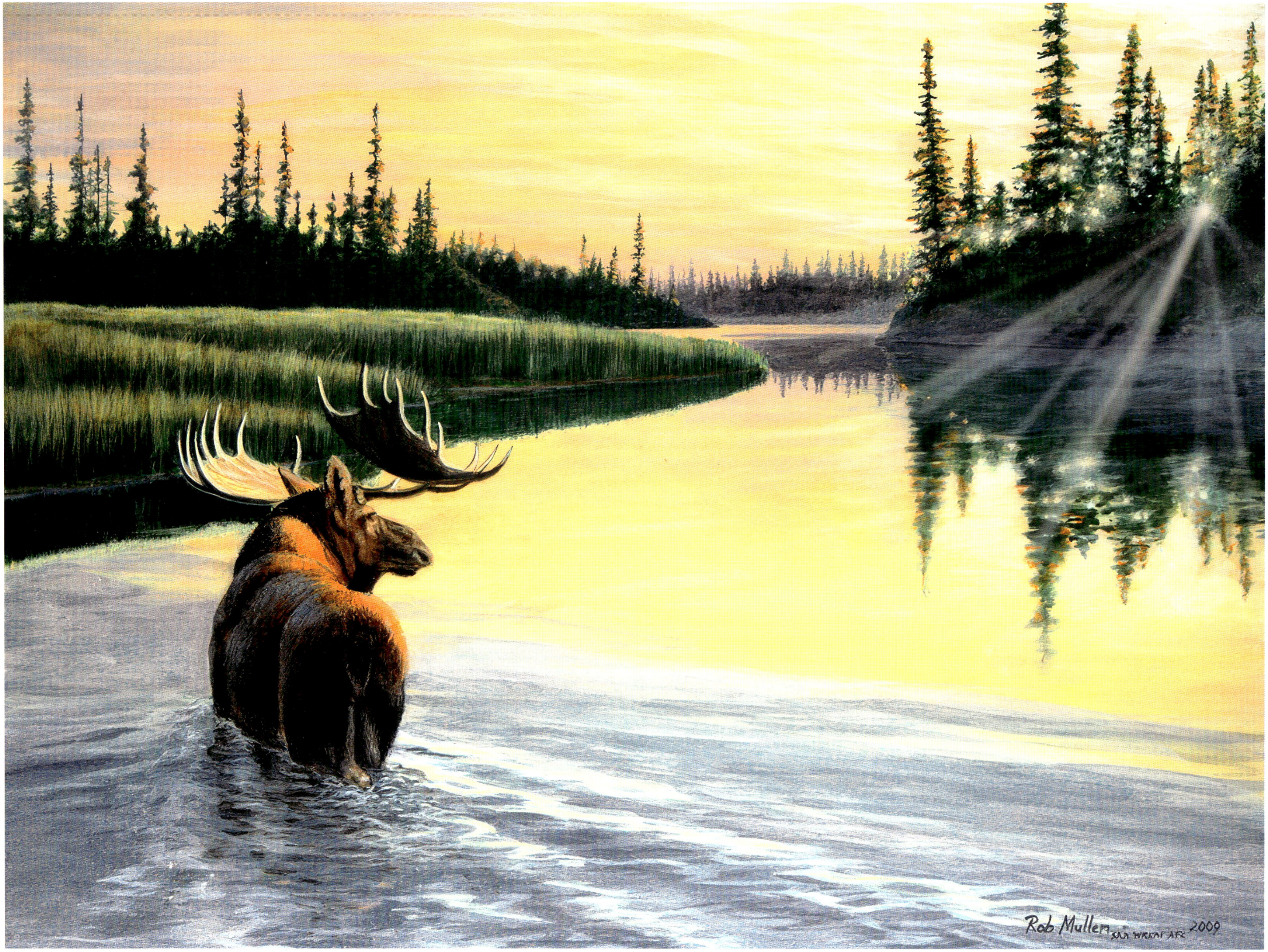
Rob Mullen
2009

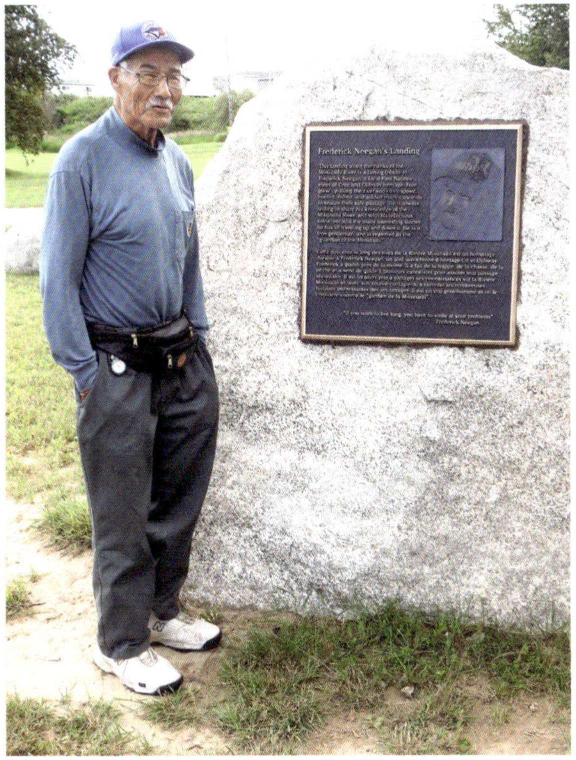

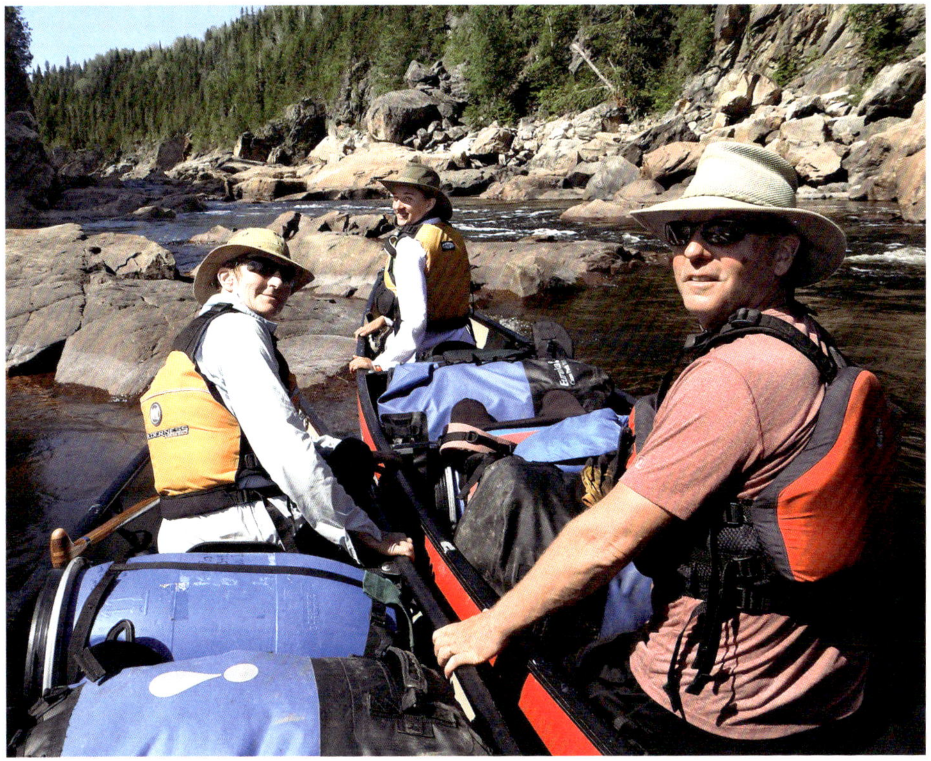

LEFT: Fred Neegan at his plaque.

RIGHT: Missinaibi 2016 crew: Julia Hargreaves, JJ Daniels, and Stephen Gorman below Hell's Gate, Missinaibi River.

FACING: Canadian bull moose. *Chapleau River Dawn*, oil, Rob Mullen

For two wild, scenic weeks on the river, we ran rapids, watched eagles and moose, and listened to wolves and loons. Passing under the railroad bridge at Moose River Crossing was a poignant milestone; I'd never been north of it before.

On August 12, in a scene reminiscent of 150 years earlier, we ground the canoes ashore below the Hudson's Bay Staff House at Moose Factory ("factory" meaning the seat of a Hudson's Bay Company factor, not a moose production facility). The two days with our Moose Cree hosts were a rich cultural exchange, followed by a plunge into James Bay itself. Well, for the youngster in the crew.

The video of the High Ridge Singers and the Longhouse can be found in the online appendix.

September 4–14, 2004: Chapleau River, Ontario

A short, easy trip put together after a larger plan fell through. Entirely within the Chapleau Crown Game Preserve (recently the world's largest after an African preserve was split for geopolitical reasons), it offered regular moose and bear sightings and excellent fishing.

July 15–31, 2007: Lake Superior North Shore—Sleeping Giant Provincial Park to Rainbow Falls Provincial Park

The trip was organized by Destination Ontario to promote the proposed Lake Superior National Marine Conservation Area (Lake Superior NMCA), later established as the largest expanse of protected freshwater in the world. Spanning 87 miles of the north shore from Thunder Point to Bottle Point and offshore to the US border, the Lake Superior NMCA covers 3,861 square miles of lake.

February 2–13, 2008: Lake Superior National Marine Conservation Area Winter Walk—Nipigon Bay, St. Ignace Island, Honeymoon Bay

A follow-up to the summer trip, focusing again on the new marine conservation area. The four-person crew was led by Jon Farshmin.

August 2–29, 2011: Woodland Caribou Provincial Park—circle route

Planned as a Bloodvein River to Lake Winnipeg run, this trip was diverted into a circuit of the southern park by the Force 5 forest fire "Red 124" (see chapter 2). The expedition had been requested by the Canadian Boreal Initiative (CBI—now shut down) Executive Director Larry Innes to help promote the then-proposed UNESCO World Heritage site Pimachiowin Aki ("the land that gives life" in Ojibwe). It was an exciting project, coordinated with the Bloodvein First Nation of the Anishinaabe and supported by CBI, Goldseekers Canoe Outfitting and Wilderness Expeditions, Sabourin Lake Lodge, and a local air service. Pimachiowin Aki went on to succeed and was officially designated a World Heritage site in 2018. It includes the ancestral lands of four Anishinaabe First Nations, over 11,000 square miles of boreal forest, including Woodland Caribou Provincial Park.

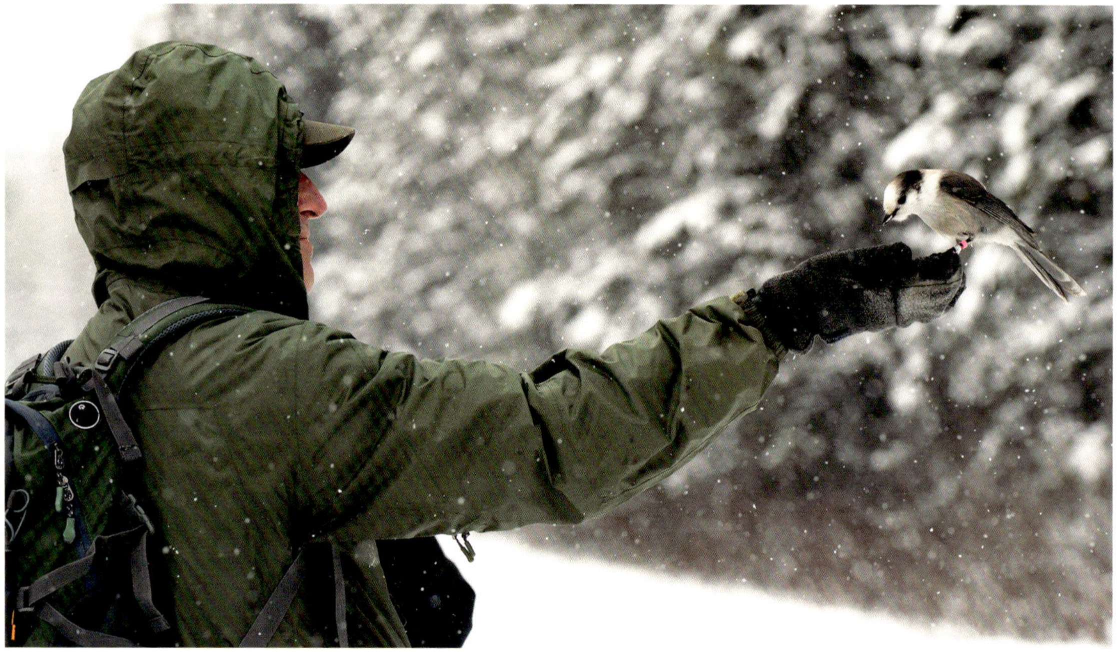

With a Canada jay. Bonnie Rowell

FACING: Alaska. *CariBOO!*, oil, Dustin Van Wechel

March 5–12, 2018: Winter on the edge of the boreal forest—Algonquin Wildlife Research Station

A genteel base-camp trip at the research station arranged by Destination Ontario. Highlights included otters and martens, but the real stars were our hosts' research subjects: Canada jays.

Also known as whisky jacks, camp robbers, gray jays, and other colorful monikers, these friendly, intelligent corvids had been companions on many other journeys.

The northern rim (in North America): Alaska

At a 2011 National Museum of Natural History conference on our boreal forest project, a senior VP from one of our sponsors asked if we'd be doing any expeditions in the US. "Of course!" Alaska was the only option. Two rivers rose to the top of the list: the Noatak on the north slope of the Brooks Range and the Kobuk on the south.

The Noatak flows mostly through arctic tundra before entering boreal forest near the Chukchi Sea. The Kobuk, as Dr. Steven Young pointed out, is a classic boreal forest river

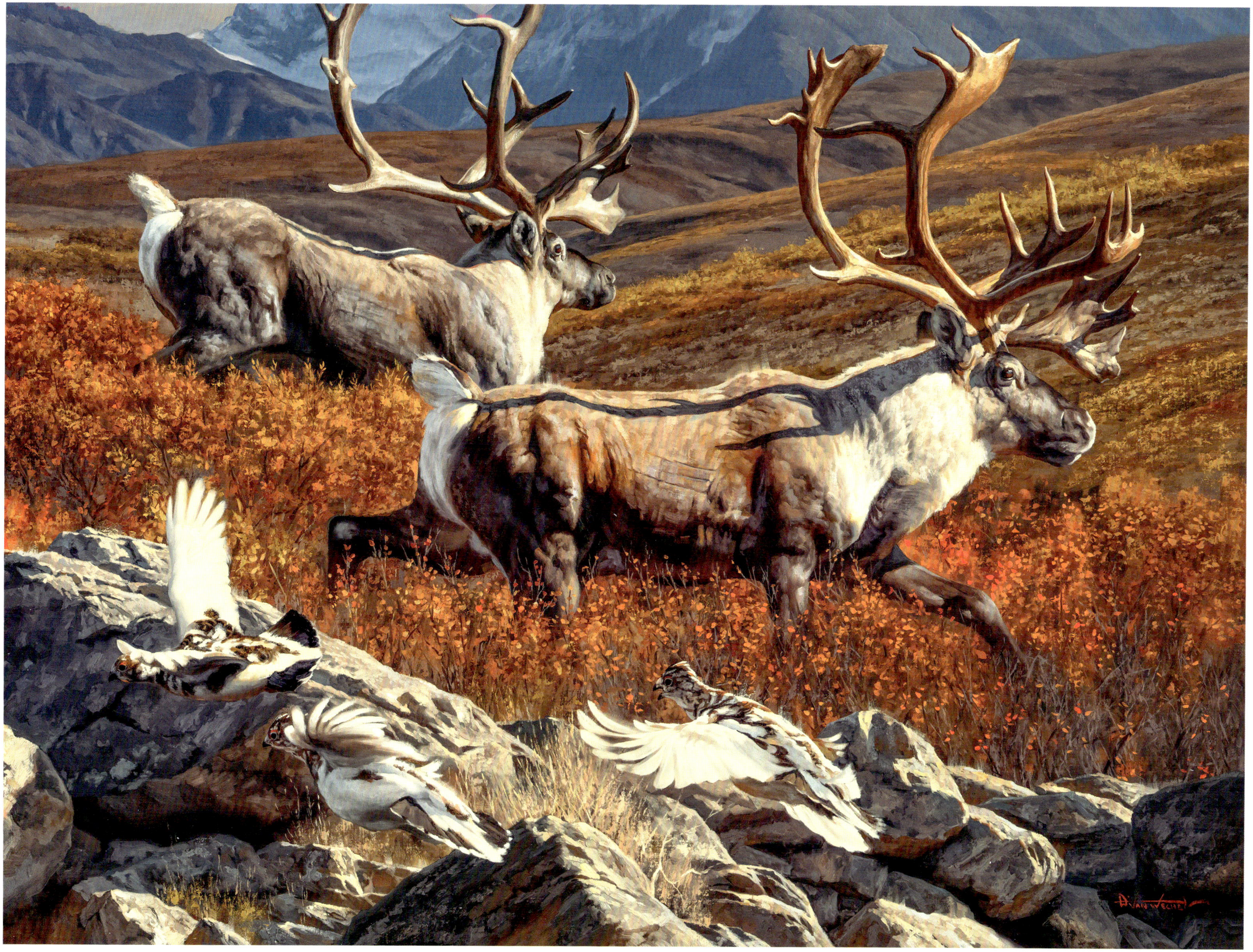

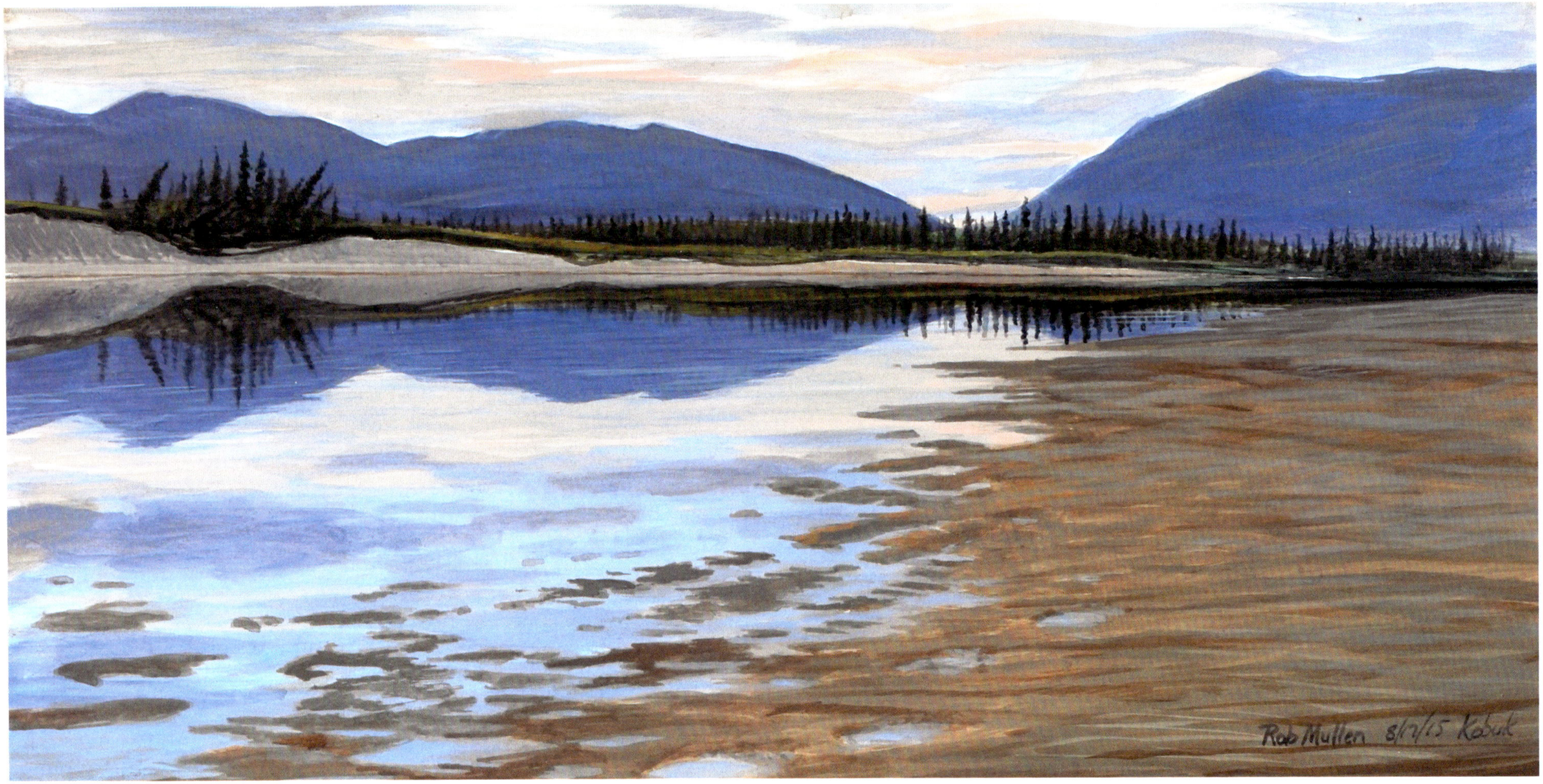

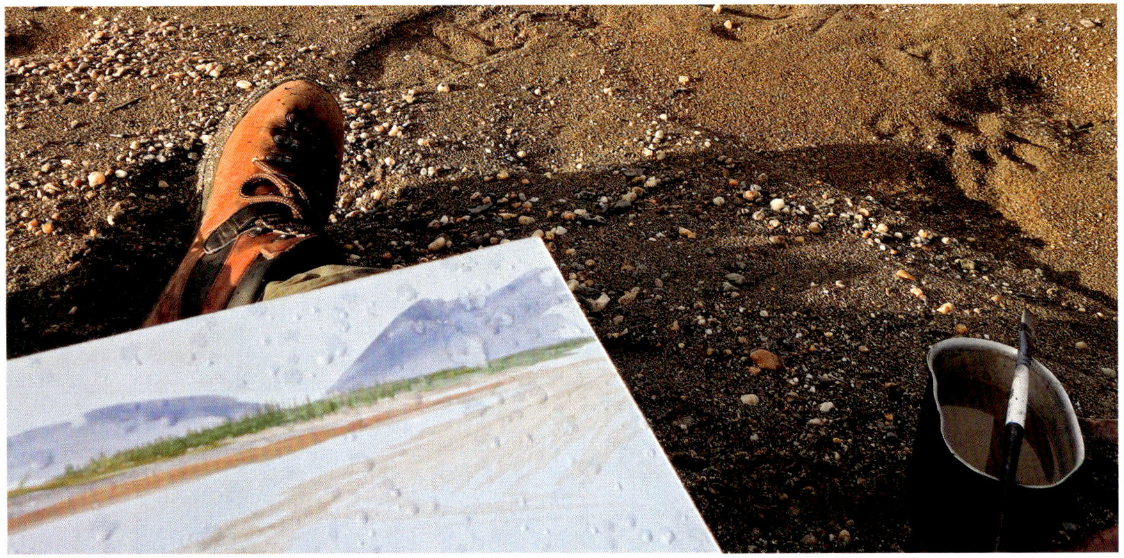

ABOVE: Field painting from camp, Kobuk River, August 17, 2015. Acrylic, Rob Mullen

RIGHT: Painting in passing showers with grizzly and wolf tracks at my feet, August 18, 2015.

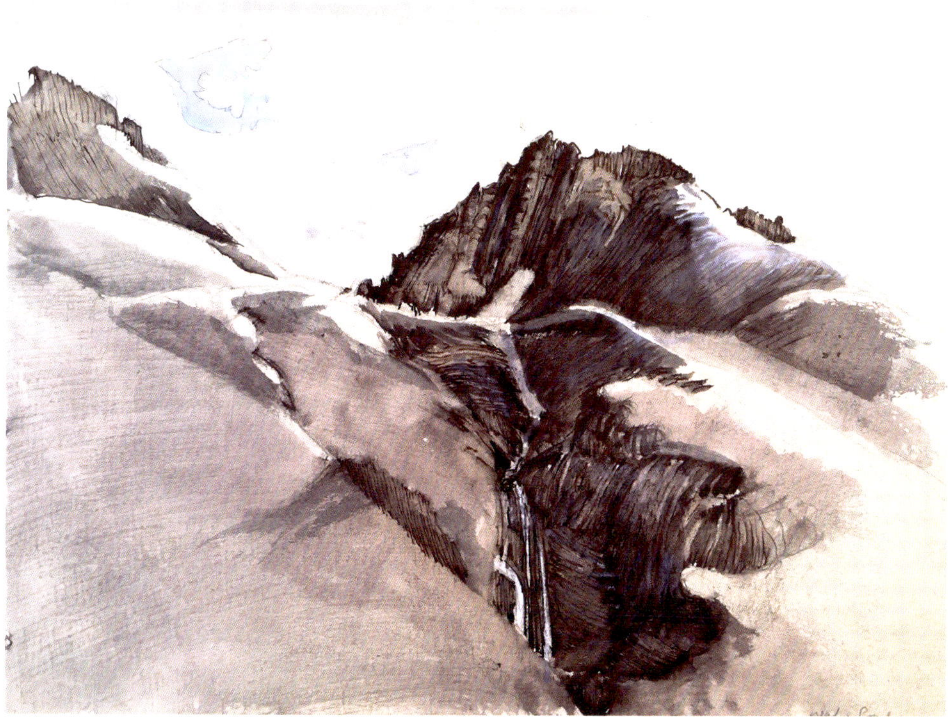

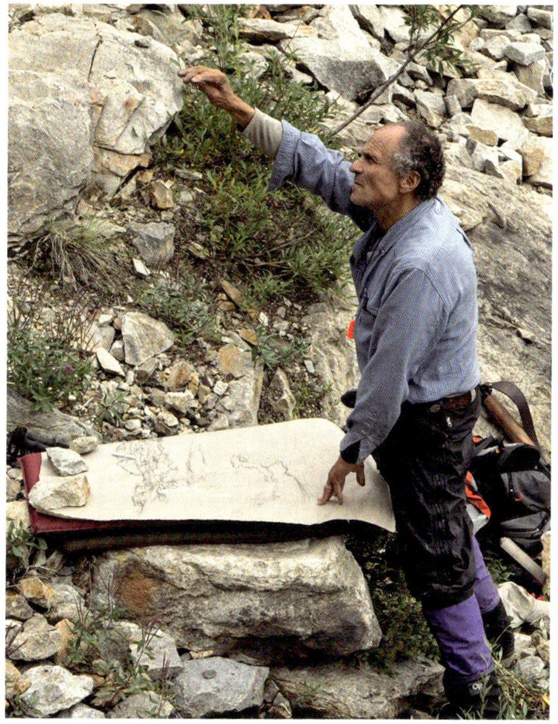

LEFT: Waterfall sketch, ink wash, Bill Brody

RIGHT: Bill Brody sketching the waterfall. Dustin Van Wechel

for almost its entire length. Looking at the map, I realized both rivers rise from glaciers on Mount Igikpak, the highest peak in the Gates of the Arctic National Park and Preserve; run parallel on opposite sides of the mountains; and empty into the Chukchi Sea near Kotzebue. "Why not do both?" I asked. And the "circumnavigation of the western Brooks Range" was born, following the full transition from alpine and arctic tundra into boreal forest.

The Brooks Range, northern tip of the Rockies, arcs west to form a crown south of Alaska's arctic coastal plain. Here, the forest is a transitional mix dominated by white spruce and the northwestern limit for trees in North America. Unlike Labrador's Torngat Mountains, which barely brush the forest edge, the boreal forest washes up over the feet of the Brooks Range on its south flank along the entire Kobuk River, and an adventurous sliver holds on at the northwest extreme of the range along the last one hundred miles of the Noatak. And unlike much of Canada, this forest was never glaciated. It was still too cold and dry for trees during the Ice Age, but its soils remained intact, allowing human habitation thousands of years longer than in Labrador or Nunavik.

Over three expeditions—about a thousand river miles—we traveled through multiple US National Park System units: Gates of the Arctic National Park and Preserve, Noatak National Preserve, Kobuk Valley National Park, Noatak and Kobuk Wild and Scenic Rivers, and the Selawik Wilderness in the U.S. Fish and Wildlife Service's Selawik National Wildlife Refuge.

Field painting, Noatak River. *Gravel Bar Blues*, acrylic, Rob Mullen

FACING: Arrigetch Valley, Brooks Range, Alaska. *Maidens in Cloud*, oil, Bill Brody

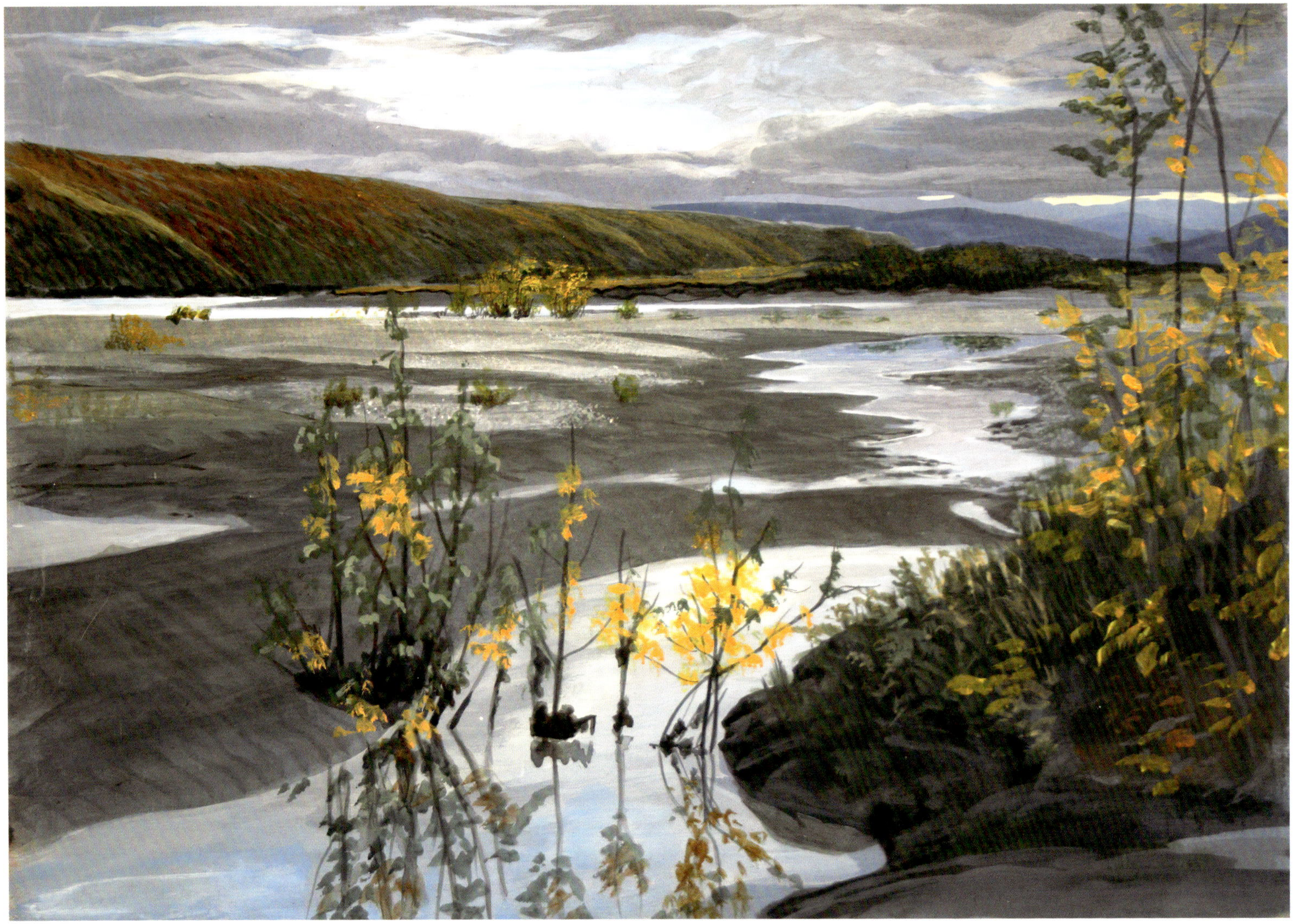

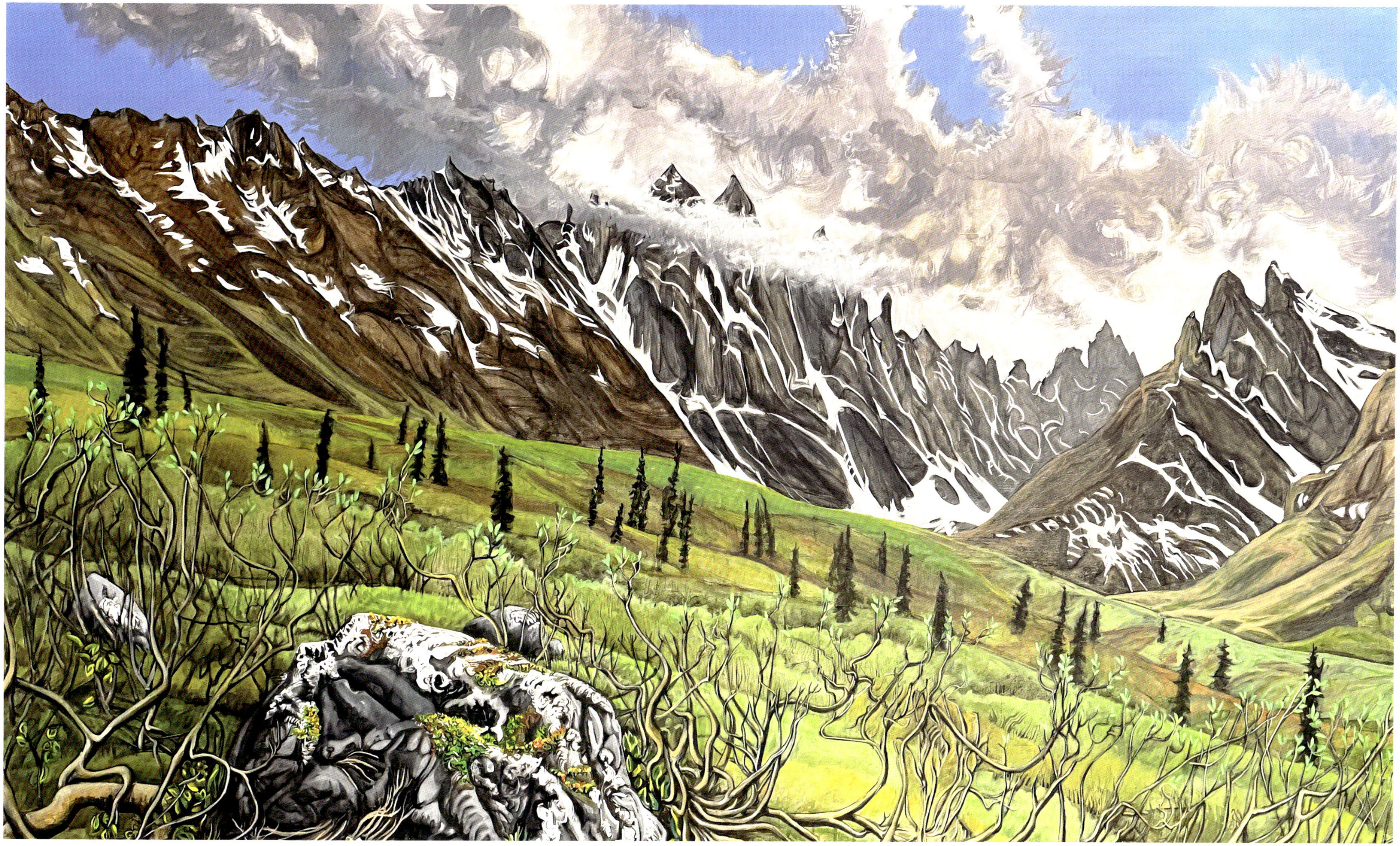

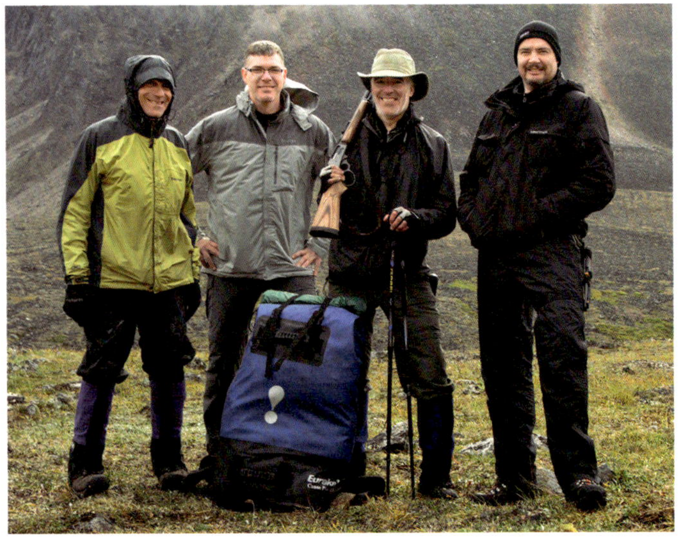

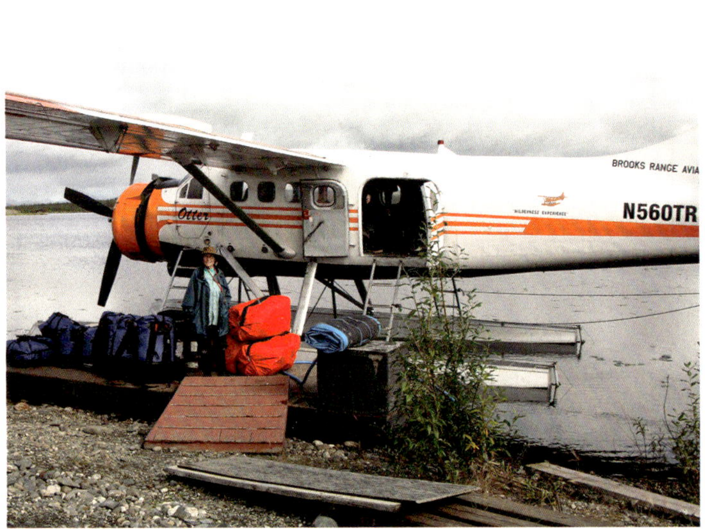
BROOKS RANGE AVIA
N560TR

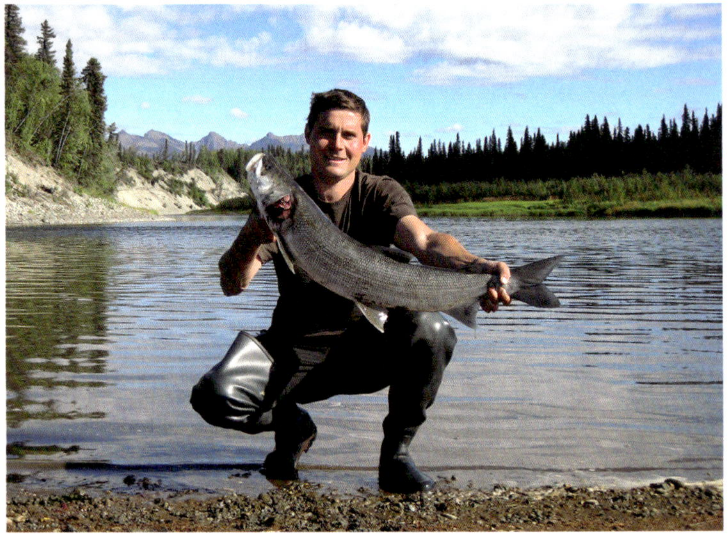

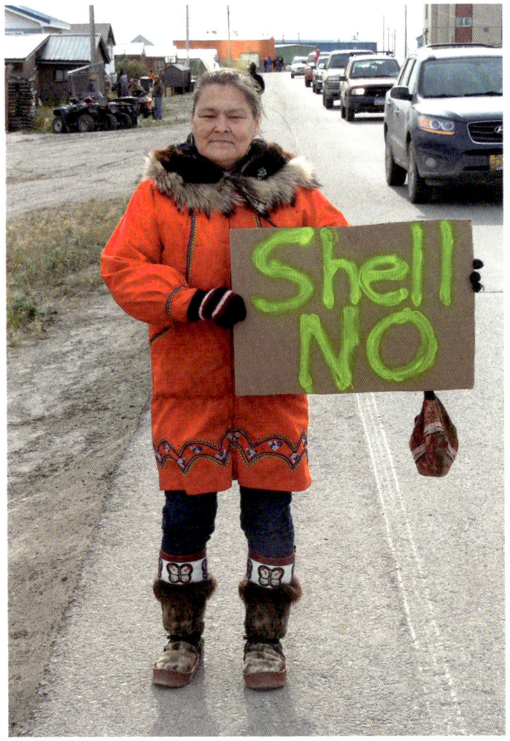
Shell
NO

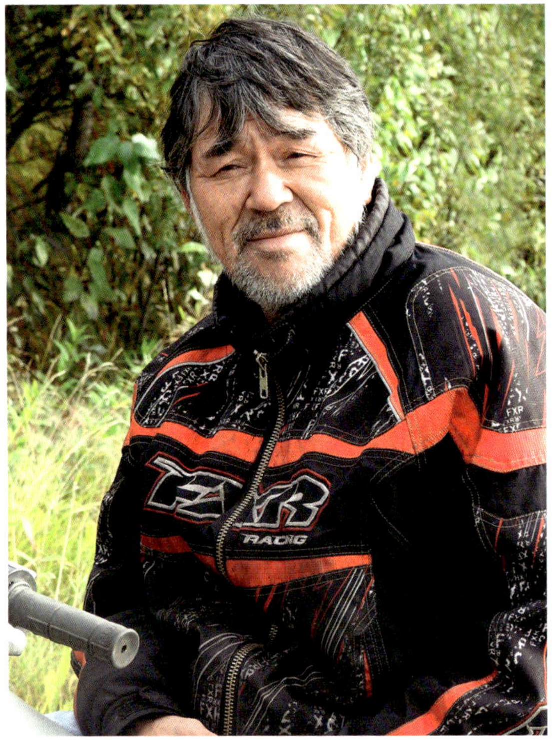
FXR
RACING

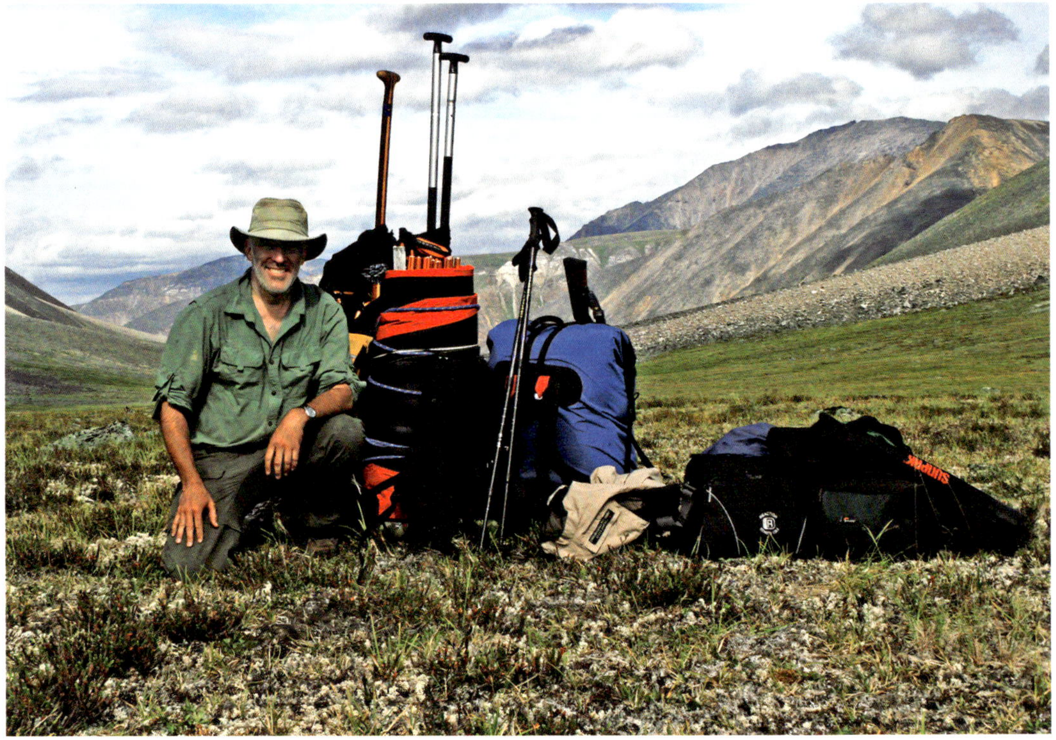

July 29–September 2, 2012: Northern Rim of the Boreal: Part 1—mountain section: Angiaak Pass recon

Arriving in Coldfoot, Alaska, early on July 29, we were supposed to fly to the Noatak headwaters the next day. But with weather closing in, we tossed our gear straight from the car into Dirk Nickisch's de Havilland Canada DHC-2 Beaver at Coyote Air and were airborne within minutes.

Clouds built over jagged peaks, weaving rain and rainbows around them. At the Noatak, high muddy water had flooded Dirk's usual gravel bar, so we zigzagged the narrow valley until he found one just big enough to land.

Crossing the swollen Noatak was tough. We ascended to Angiaak Pass along Tupik Creek under a mountain I dubbed Mons Innominatum (No Name Mountain in Latin—I think) because Bill Brody told us the park didn't allow unnamed peaks to be named. When we topped the pass, fog and rain kept us from descending the south side.

River section: 458.6 miles, Twelve Mile Slough–Kotzebue, Alaska, on the Chukchi Sea

On August 7, Brooks Range Aviation's de Havilland Canada DHC-3 Otter dropped five new crew members and over half a ton of gear at Twelve Mile Slough. The mountain team—except me—boarded for home.

We spent the first day sorting gear, reviewing safety, and assembling canoes. I sketched Mons Innominatum that evening in pencil, then returned the next night for a color version in acrylic (finished around 1 a.m.), which later became the setting for musk oxen in *Echoes of the Ice Age* (see pp. 96–97).

With light until midnight, I painted as much as cold and fatigue allowed. Linda Besse, who doesn't work in the field, generously took over my cooking shifts so I could keep going.

On August 25, we spotted our first white spruce—the boreal forest's very edge. Below the village of Noatak, where we had a very friendly reception, riverbanks exposed massive permafrost blocks, and harbor seals curiously observed us.

September 1 began not with rain but with a grizzly snorting in my left ear. A grizzly had never snorted in my ear before, yet I instantly knew what it was. I suspect there is something deep in human DNA with data filed on the subject. I also heard it start to run away (the rest of that encounter is in the online appendix).

On September 2, with thunderheads massing to the north, we left the last spit of the Noatak's delta—seals sunning behind us—and crossed to Kotzebue. Staring toward the open sea, I joked about seeing Russia.

See the online appendix for the full 2012 trip report.

July 25–August 24, 2013: Northern Rim of the Boreal: Part 2

Two dramatic events from this trip, plus scattered others, appear elsewhere in this book. This journey edged out even Trans-Labrador for harrowing adventure. Doing it justice here would take more space than we have, and a quick recap won't cut it, so I'll give you the beginning and the end, hint at the middle with some photos, and point you to the full trip report for the gory details in the online appendix. (Note: As with all WREAF expeditions, no bears were harmed.)

I was forewarned. On the Noatak in 2012 we met the superintendent and the head ranger of Gates of the Arctic National Park and Preserve, guiding a famous filmmaker and an NPR personality on a private canoe trip. Hearing our plan to carry canoes over the mountains, the superintendent remarked he'd "never heard of any other fools doing such a thing"—let alone one.

The main goal was to link the Noatak to the Kobuk over the mountains; reaching the Chukchi Sea was optional, with possible pull-out points at Iñupiaq villages along the Kobuk. By spring, I had a lean, intrepid team of four. By July, the roster was down to me, myself, and I.

FACING, CLOCKWISE FROM TOP LEFT:
Mountain crew in Angiaak Pass, left to right: Bill Brody, Dustin Van Wechel, Rob Mullen, and Derek Wicks. Dustin is 6′6″ and Derek is 6′4″; Bill and I tricked Dustin into a small declivity for the photo. Dirk Nickisch had a hard time fitting him into the plane. When he finally had Dustin shoehorned into the copilot's seat, he grimaced and said, "Just don't bump anything." Dustin Van Wechel

Loading the Otter in Bettles. Linda Besse

Hendrick Luckey with a sheefish (see section on whitefish in chapter 6). Hendrick Luckey

This grinning "fool" thinks the hard part is over.

Dr. Chris Stein, Kiana, Alaska. (Dr. Stein, if you read this, we lost your email address, but as you can infer from this book, we all came out of it just fine. Thanks!)

The only protestor I saw, presidential visit, Kotzebue, Alaska, September 2, 2015.

July 25: Into the mountains Mount Igikpak was in full glory as Dirk Nickisch set down on the same gravel bar we'd used the year before. With my gear on the gravel, he took my picture and left. As his plane shrank to a dot and the engine faded, the vast silence settled in. It was a weird mix of excitement and trepidation.

The weather held bright and blue, and paddling the next day was wondrous, so it was with mixed feelings that I reached the point at which I figured it was best to start hauling over the mountains. My disassembled 14-foot PakCanoe and gear totaled 160 pounds. I opted for three lighter loads instead of two back-breakers (one was not in the cards; I was a distance runner in school, not a shot-putter). The next days were "three pushes up, two trips down" for each leg of progress. Twenty miles to the Reed River's headwaters would mean about a hundred on foot. With scouting, it ended up closer to 120 miles of rough, sometimes arduous "walking."

And then it started to rain, Angiaak Pass.

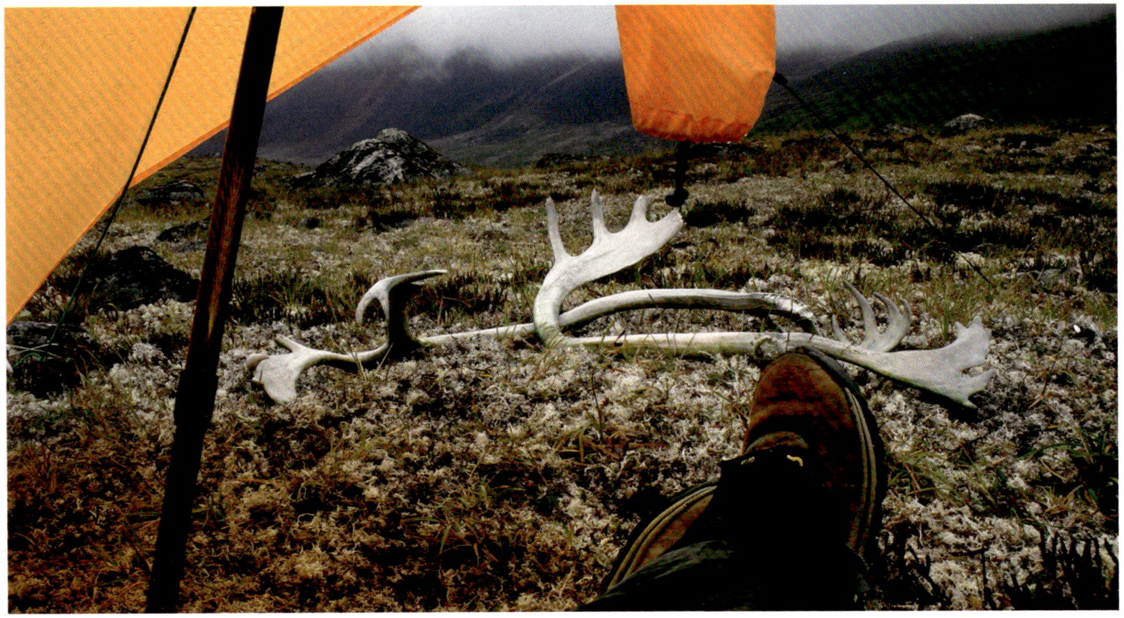

August 1: Over the pass At 1 p.m. I stood on Angiaak Pass's southern brink, where Dustin Van Wechel and I had ended our 2012 recon. Mountain rills quickly coalesced into the nascent Reed River as it began its tumbling descent toward the Kobuk.

A storm rolling up the slope convinced me to camp high. I figured it was "all downhill from here."

Insert two weeks of...

...relentless rain; a flash flood; "Alder Hell" (or as I took to calling it, "Mirkwood," with no trail, magical or otherwise); glacial torrents; black bears and gunfire; grizzlies and more gunfire; stupid mistakes; good calls; a blessed food drop (thank you, Coyote Air); and—minutes after my lowest low—the victorious euphoria of finding the Caribou Road. Then finally, on August 17, having to run a gauntlet of six grizzlies, four of whom were variously aggressive (there was gunfire). The first (later) was shockingly followed by a mom teaching her large cubs how to be mean, which together created a state of suppressed terror that only slowly subsided to watchful wariness—and on the most beautiful river I'd ever paddled. Oh, the irony.

August 18: Down the Kobuk I reached the Kobuk and, moments later, mistook a German solo canoeist, Hendrick Luckey, for a bear (I was still a bit jumpy at that point). He had heard my gunfire the day before and helped stretch my dwindling food until I pushed off again on my birthday, August 21. That night, a father-son pair who were pulling out at the village of Kobuk a few miles downstream shared their last-night steaks with me.

Three windy but pleasant days later, I pulled into Loren Gentemann's Kobuk River Lodge in Ambler. After weeks of stress, I called it. I shipped my canoe home, donated my supplies to a local family, and flew out—sole passenger—to Kotzebue. After one night at Bibber's B&B, I was homeward bound.

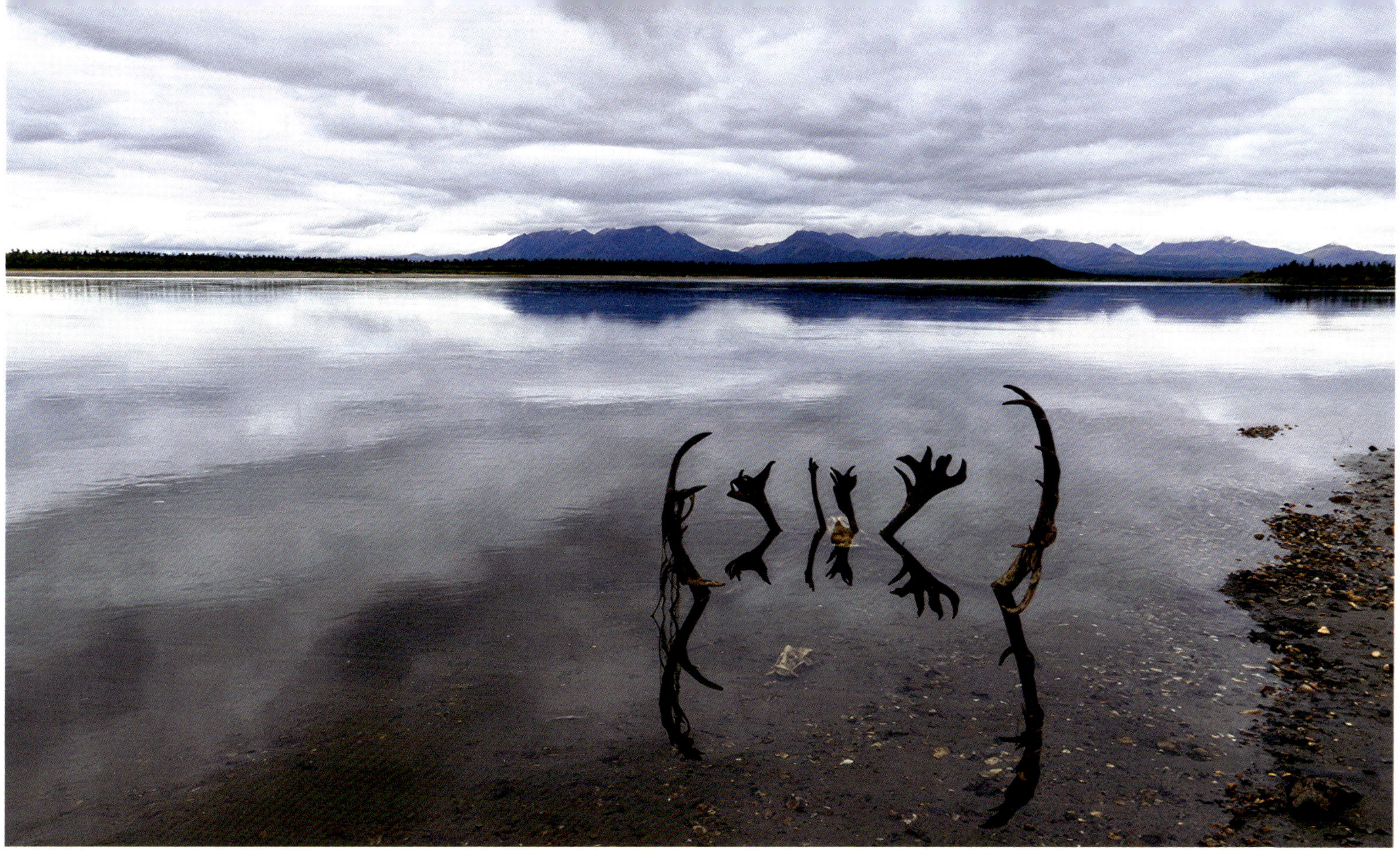

Onion Portage, Kobuk River, August 2015. Stephen Gorman

August 12–30, 2015: Northern Rim of the Boreal: Part 3

This expedition completed the 195-mile gap left when I ended my 2013 solo trip in Ambler. Rather than start there, I wanted the crew to see the upper 110 miles I'd paddled before—some of the most beautiful on the Kobuk. We flew into Lake Minakokosa, which joins the Kobuk River via Beaver Creek just downstream of the Reed River.

Two of the crew, Jim (Dr. James Vogelmann) and Dan Berns, had never seen a grizzly and were keen to do so. They weren't disappointed: We spotted half a dozen before hitting the main river. By trip's end, the tally was 22 or 24, depending on who was counting. All were either frightened or unconcerned with our presence. With the salmon running, wherever we came to shore, herds of bears had preceded us, leaving their tracks everywhere. Avoiding camping near them, as some experts advise, was not practical. They were even difficult to avoid when painting (see p. 86).

August 21–23: Ambler to the Jade Mountains

We reached Ambler on my birthday, where Bonnie had conspired with Shirley Gentemann at the Kobuk River Lodge to surprise me with a homemade cake. Jim had to return due to work and Julia Hargreaves was close to going with him. Steve Gorman, Dan Berns (both expert canoeists), and I assured her that any decision was fine, though we very gently encouraged her to continue. She bravely did. Downstream

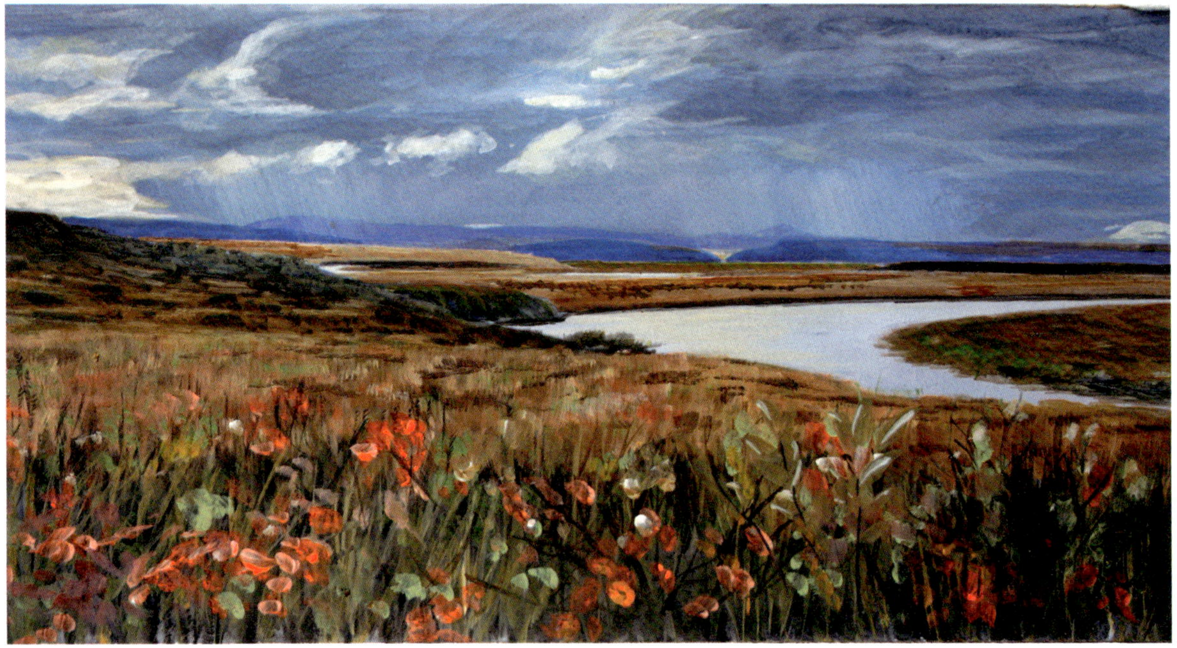

Noatak River, August 19, 2012. *Approaching Rain Squall*, acrylic field painting, Rob Mullen

FACING: Juvenile bald eagle, Kobuk River. *Into Every Life* (I was going to title it "Angry Bird"), acrylic, Rob Mullen

were the Jade Mountains: a 40-square-mile massif and ancient Iñupiaq tool material source. We camped with a perfect view on August 23.

Dawn was announced by a grizzly cub bawling unapologetically for breakfast. It kept the racket up until its harried mother found a half-dead salmon. The sunrise was a "stop and stare" slow eruption of crimson. But as the old saw goes, "Red sky in the morning, sailors [and canoeists'] take warning."

August 22–26: Onion Portage to the sand dunes

Past the Jade Mountains lay Onion Portage, a caribou hunting and habitation site dating back 12,000 years. It has been studied extensively by archaeologists and was clearly still in use when we passed by. The "portage" cuts across a meander's neck. Presumably practical once, meanders (loops in the river's course) move, and this one is now just historical.

We'd looked forward to the Great Kobuk Sand Dunes: 10 miles long, up to 4 miles wide, and the largest dunes in the world above the Arctic Circle. But by then the weather had turned, and the thought of bushwhacking two miles each way in cold rain had lost its appeal. My highlight instead was a soaked juvenile bald eagle who looked a bit like I felt.

August 27–29: Kiana to the sea

At Kiana, our last village stop, we camped upstream to avoid the nighttime chorus of village dogs. In the morning, locals offered advice and concern about river conditions.

A returning fishing boat warned of six-foot standing waves in the main channel from 30 mph westerlies.

We stuck to our plan—following the Melvin Channel along the delta's north edge—and avoided the worst of it. On August 29, we reached the Chukchi Sea and were greeted by Chucky, a playful harbor seal who bombed around the canoe hard enough to make me consider the water temperature.

August 30–September 2: Kotzebue and a presidential visit

With the wind an issue, the four-mile crossing to the Baldwin Peninsula was perilous. So we called Jean Bibber, who arranged a boat pickup at Ekichuk Lake—known as "York Bay" to our Iñupiaq captain. The rendezvous cabin was indeed unmissable: the only one in sight, tipped by ice, and halfway in the sea already.

Two days later, Kotzebue hosted then US President Barack Obama: the first sitting US president to visit Arctic Alaska. Jean Bibber's B&B was full of US Marines, but she'd told Fox News to find other lodging "because Rob Mullen and his canoe crew are coming." I hope I'm not on some Fox enemies list.

Jean and her son were invited to meet US President Obama. I saw the SUV (there were two; no idea which he was in) and one lone protestor—who, judging by later events, seems to have won her case.

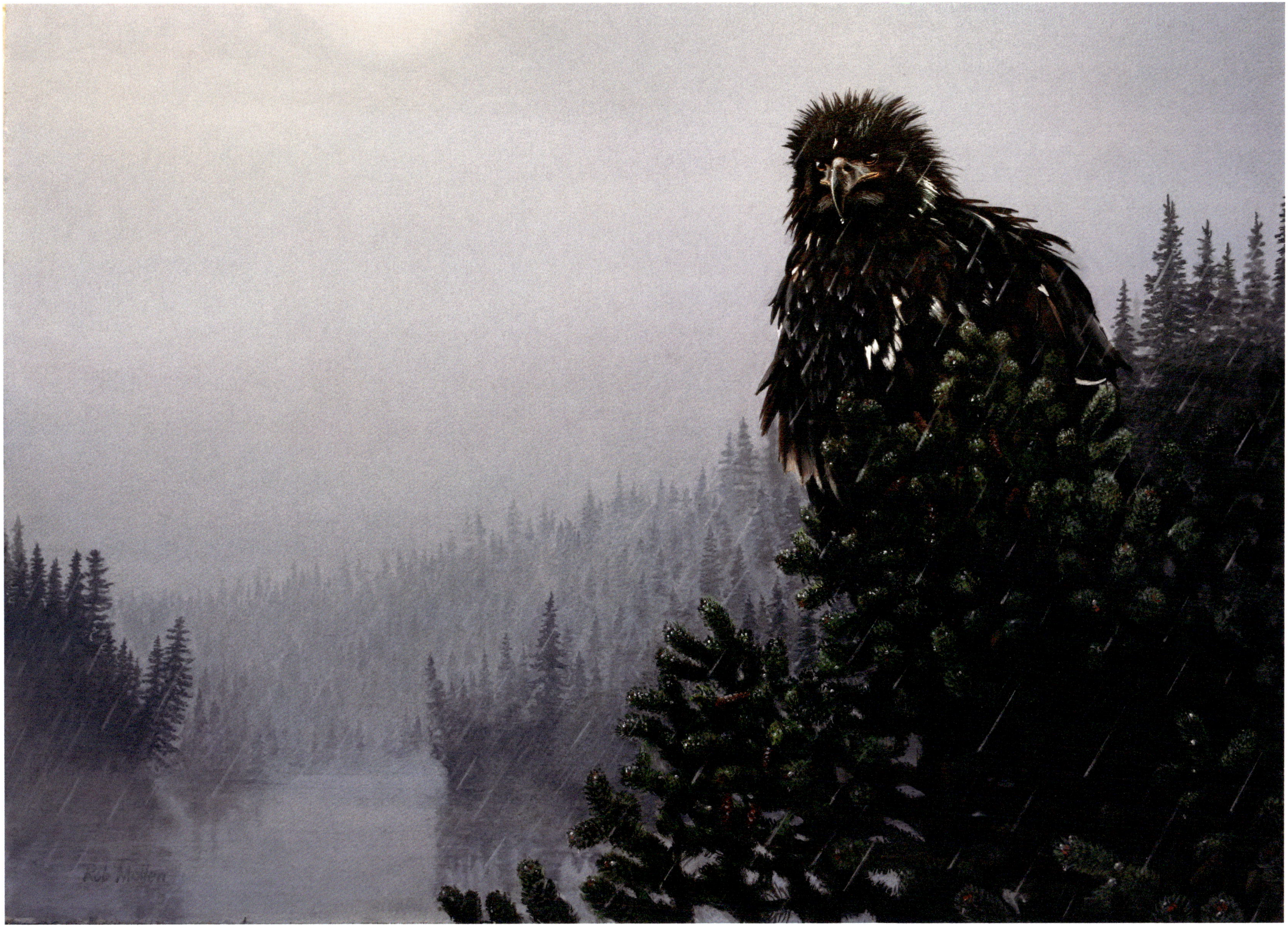

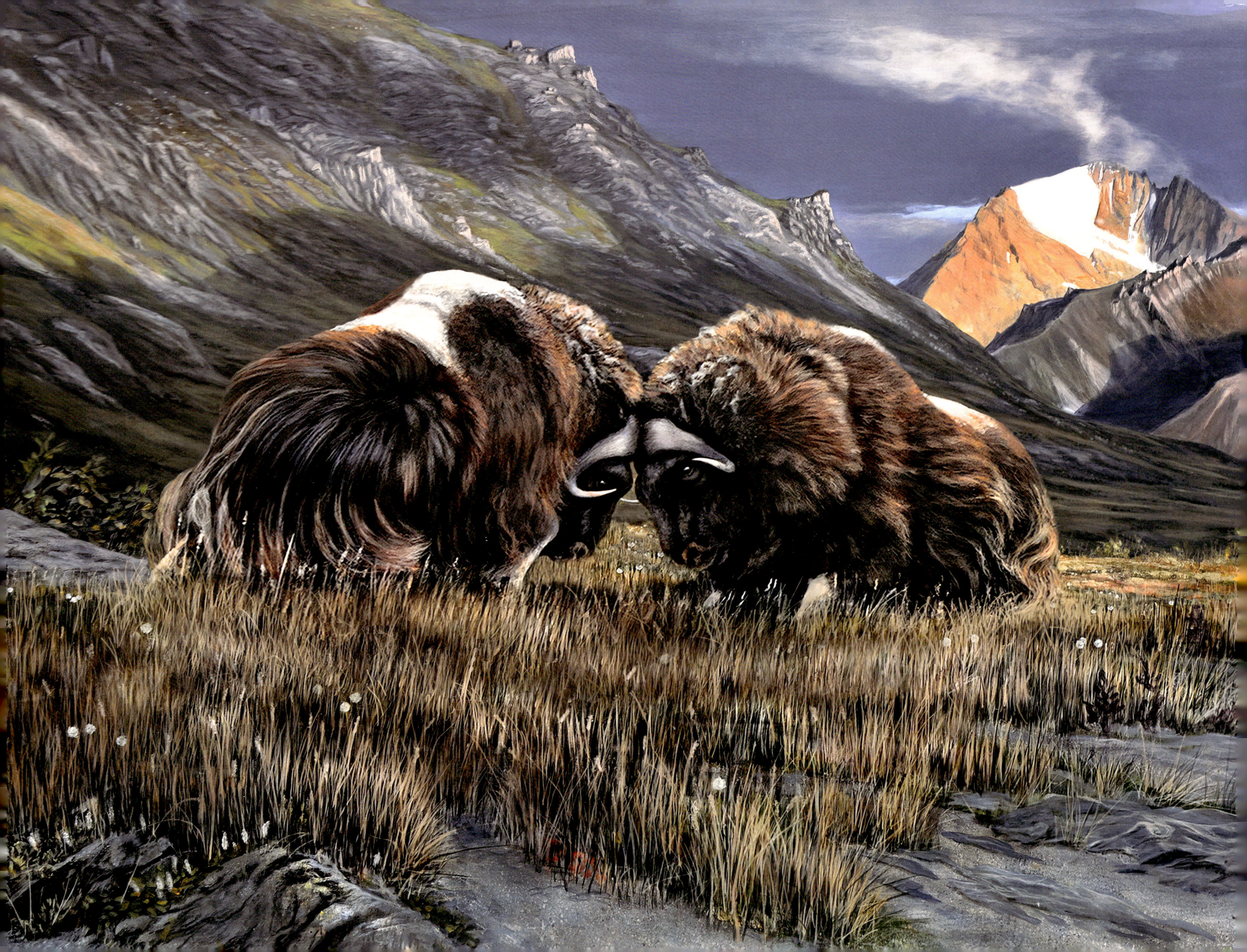

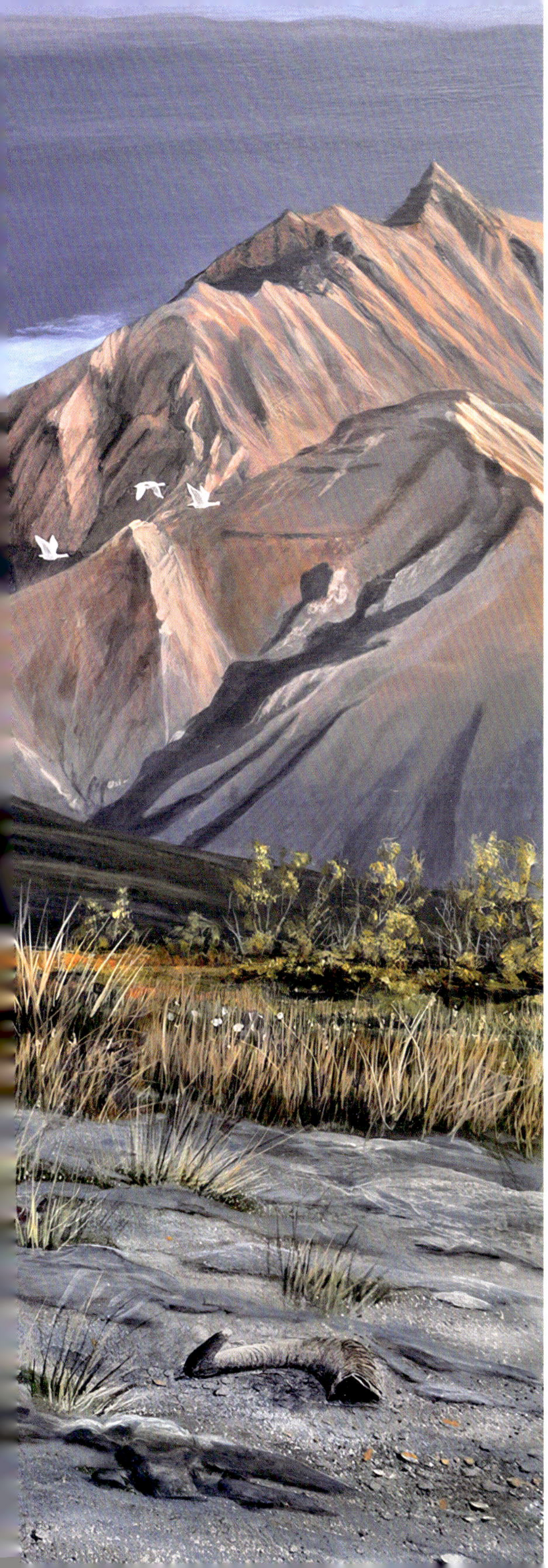

Musk oxen, Dall sheep, tundra swans, woolly lousewort, and cottongrass, Noatak River, Alaska. *Echoes of the Ice Age*, oil, Rob Mullen

PART TWO

LIONS AND TIGERS AND BEARS (AND MORE!)

FACING: Peekaboo bull above Kamestastin Lake, north-central Labrador Barrens.

October 10, 2008: The hills above the southeast shore of Kamestastin Lake, north-central Labrador Barrens

It was a funny dance, each of us maneuvering as the caribou bull tried to move downwind of me to get some idea of what I was while ineptly "hiding" behind the rock dome (I could see his antlers). Curiosity won. He popped his head up. I reflexively triggered the shutter a half moment before he snorted and pranced away. We were about to return to the canoe when the real show started (see *Migration Moon*, page 100).

In remote country, far past the final ruts of the last dirt road, a matter-of-fact air of wildness suffuses everything. It is expansive, quiet, and timelessly elemental with a faint echo of deep memory. It is usually amorphous and subtle—until you meet a wild animal.

Wild animals focus your attention. They are fascinating in their similarities to us and exhilarating in their differences. They have their own sense of things. You are meeting a profound stranger who exists in the same space and time but perceives a vastly different world. Such musings aside, there are more immediate reasons to value, indeed, to cherish wildlife; our burgeoning populations of a few species of domesticated animals notwithstanding, we can't live without them (see chapter 8).

Some bookkeeping

Because part 3 will get into the importance of biodiversity and the threat that extinctions pose to it, the IUCN conservation status for species at any elevated level of risk will be listed in part 2. The IUCN is the International Union for the Conservation of Nature, founded in 1948, and a world authority on the conservation status of species and the threats they face. The ratings are Least Concern (LC), Near Threatened (NT), Vulnerable (V), Endangered (EN), Critically Endangered (CR), Extinct in the Wild (EW), and Extinct (EX). The Red List status will be given for most species and a specific note made if the species has not been assessed. Where, due to space or clarity considerations, no status is given, the species is of Least Concern.

Note: Even if we focus on the vertebrates (again, apologies to invertebrates), there are too many to list, so this is far from an exhaustive bestiary (and probably has a small bias toward animals artists like to paint).

Furthermore, to save space, descriptions of an animal's range will often use the following terms:

Nearctic: North America above the Tropic of Capricorn (roughly the northern tip of Cuba): that is, the northern New World

Palearctic: Eurasia above the Tropic of Capricorn: that is, the northern Old World

Holarctic: Both Nearctic and Palearctic: that is, the northern globe, worldwide

All nesting percentage data for species in chapter 5 come from research by Dr. Jeff Wells and colleagues.

And finally, for those who have forgotten their high school biology or haven't taken it yet, here is a quick review of the basic taxonomic classification levels. They are not essential to know but they provide an idea of who is related to whom. We will only deal with the last few, but for context, they run like this from big group to small: Kingdom, Phylum, Class, Order, Family, Genus, and Species. A classic mnemonic for them is "Kathy Put Chocolate On Fred's Good Shoes."

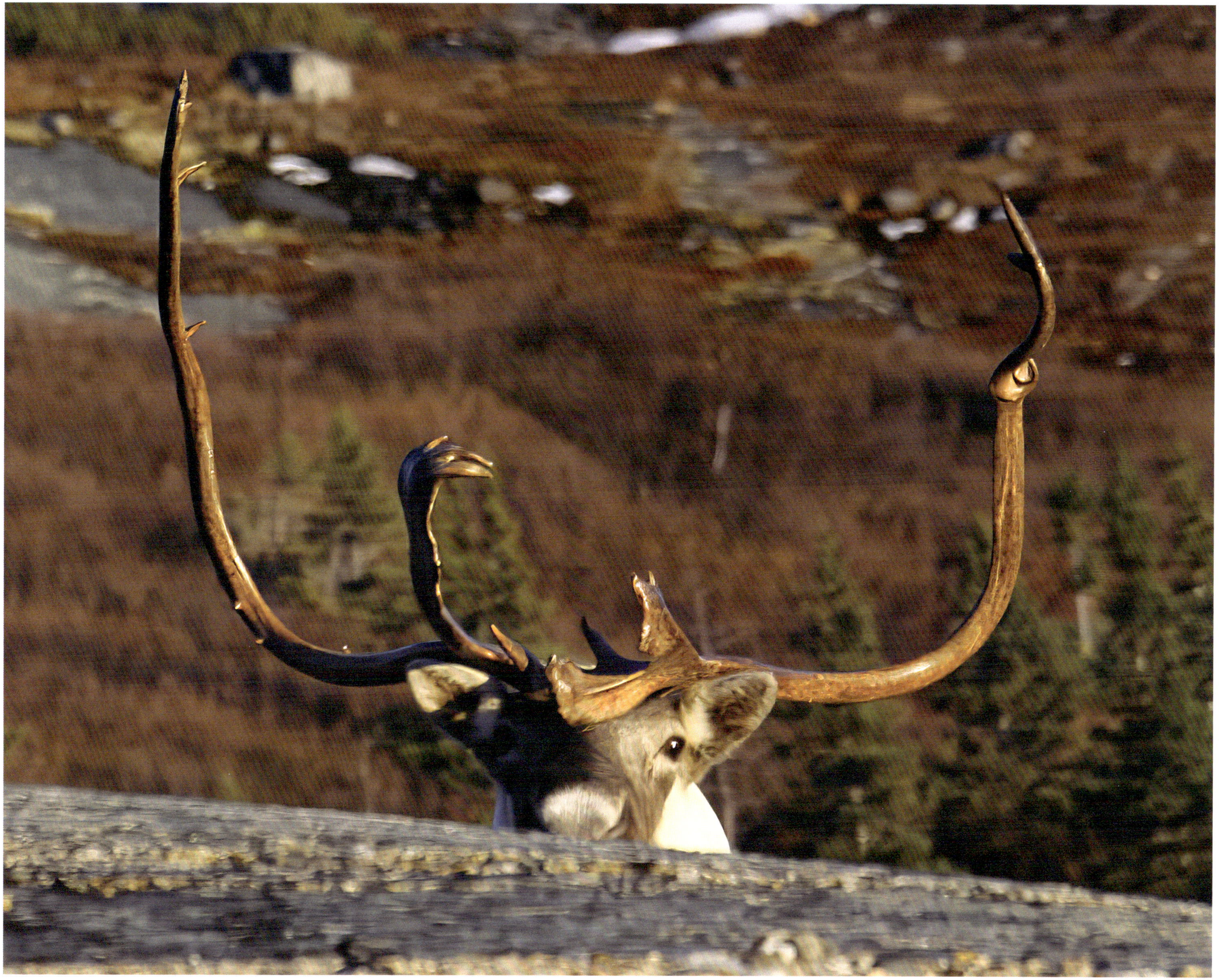

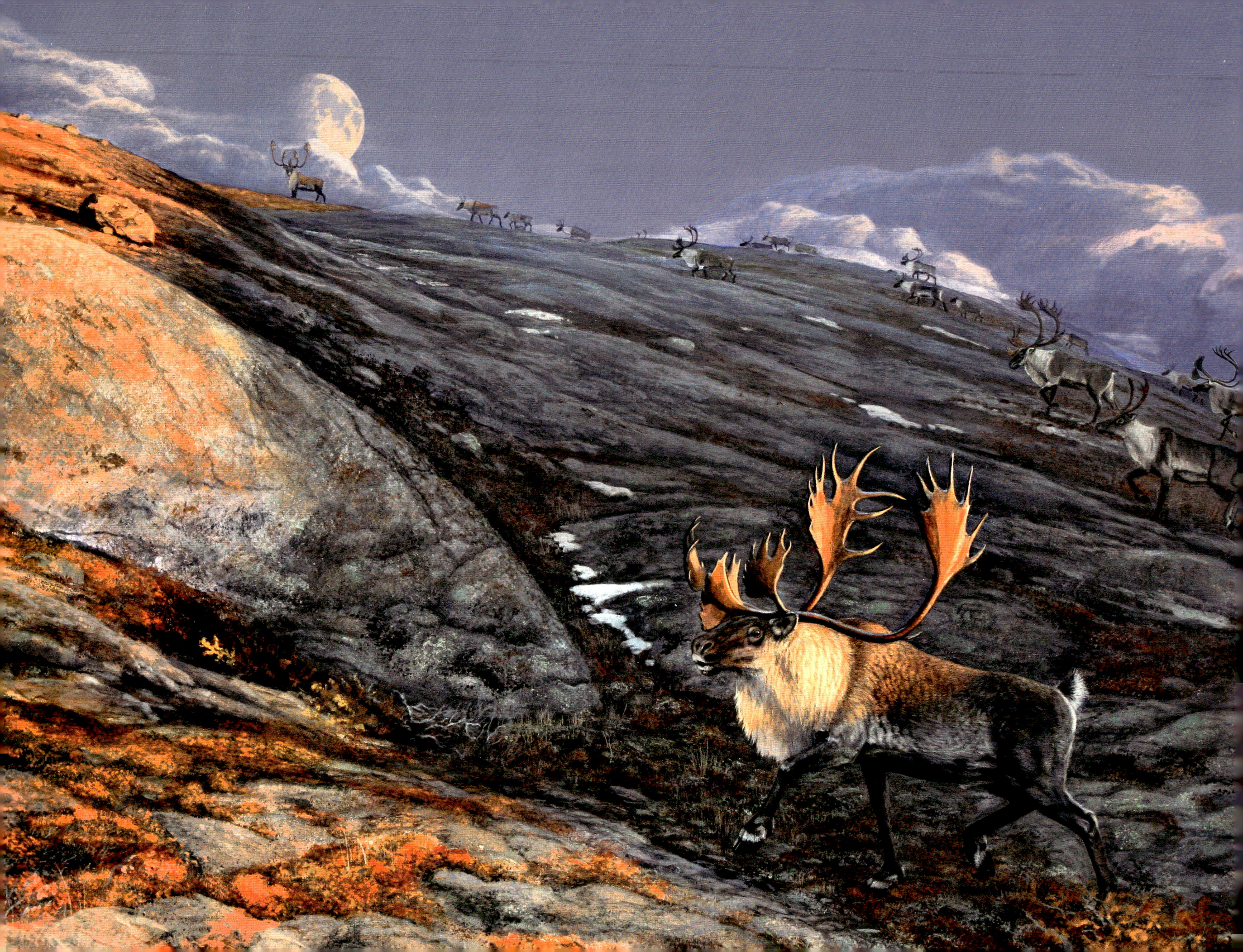

FACING: George River caribou cresting a tundra ridge above Kamestastin Lake under the rising moon. *Migration Moon*, acrylic, Rob Mullen

4

Non-Human Mammals

Order Artiodactyla (Even-Toed Ungulates)

Family Cervidae (the deer family)

Caribou (*Rangifer tarandus*), aka reindeer: IUCN Red List: Vulnerable (as of 2016, from Least Concern in 2008). In 2022, a paper proposed splitting *Rangifer tarandus* into three species, but for now the one will do; taxonomy, like all sciences, is a dynamic field. Caribou are Holarctic in distribution.

Caribou are special. They live farther north than any other member of the deer family and are an iconic (or the iconic) species of the boreal forest and tundra. Of more practical import, they are an indicator species for ecosystem health. They require vast expanses of suitable habitat, so if they are doing well (and they aren't), then the habitat is too. They are also a keystone species (benefiting many other species in the ecosystem). Their wide-ranging browsing and "digestive leavings" are major determinants of ground cover and enrich soils throughout the landscape. Their ubiquitous trails provide easy passage through brush and undergrowth for all other medium-to-large animals, including artists and anthropologists.

Moreover, caribou are a central pillar of Indigenous cultures around the northern globe (see chapter 7).

Two WREAF expeditions in September and October 2008 were specifically designed to observe contrasting herds of woodland caribou (as detailed in "Caribou 'n' Craters" under Papers & Essays in the online appendix): the insular Slate Islands herd on the north shore of Lake Superior, one of the most southerly herds and the smallest woodland caribou in stature, and the large, robust members of the George River herd at Kamestastin Lake in Labrador.

Despite their differences, both herds experience dramatic population boom-and-bust cycles, likely related to being too good at predator avoidance combined with relying on periodically abundant, but slow-growing lichens for food. Those natural cycles, especially the busts, are putting them at great risk as human activities pose additional threats.

October 6, 2008: Kamestastin Lake, north-central Labrador
Hunkered behind a lichen-encrusted boulder to shelter from the wind, I intently watched for movement that would betray an approaching caribou. The crew had been very easygoing about not having seen any so far. Over the previous four days, we had looked and looked.

With such thoughts spurring my search of the tundra heights, it occurred to me that I was probably not the first

Slate Island bull (left) versus a George River bull (right). The photos are scaled so that the cannon bones are the same size in each bull for comparison. The cannon bones of ungulates are analogous to fused human hand and foot bones in the fore- and hind limbs, respectively. This elongates their legs for greater leverage and speed (they run on their toenails).

FACING: Caribou of the George River herd, Kamestastin Lake. *Labrador Bull*, charcoal, Cole Johnson

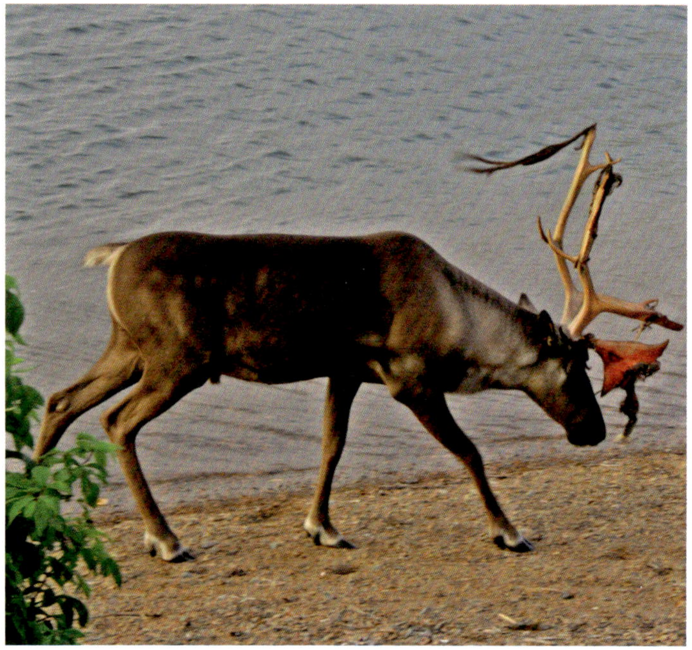

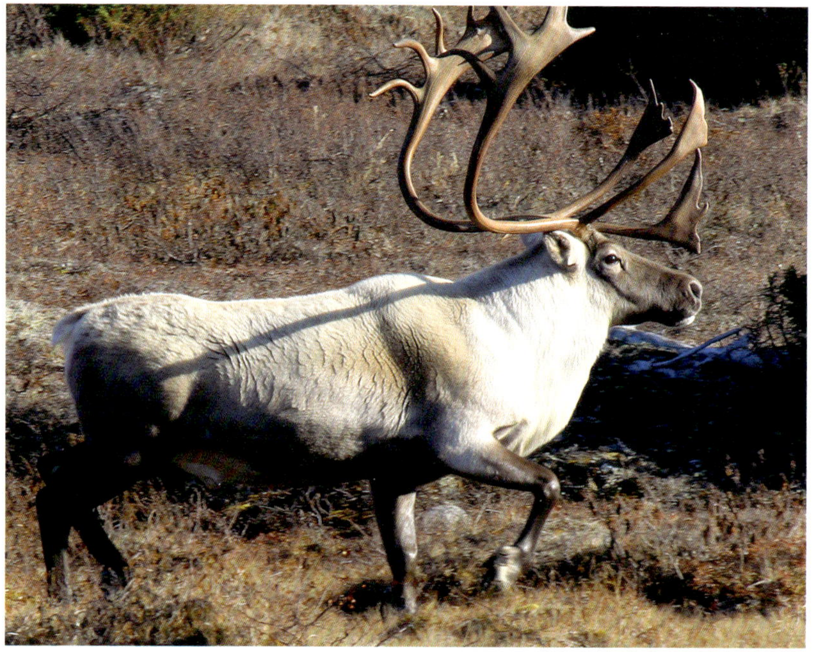

person to sit on this billion-year-old outcrop looking for caribou. My vantage point would have been just as attractive to an Innu hunter 500 or 5,000 years ago. Dr. Stephen Loring's research, with Tony Jenkinson, the Tshikapisk Foundation, and the Innu, firmly established that the Innu have continuously used Kamestastin during the spring and fall migrations for about 7,500 years. The glacier only departed from Kamestastin 8,000 years ago.

Then, motion. It was the first. The next day, there were a few, then hundreds, eventually thousands, but . . .

When we arrived in Kamestastin on October 2, 2008, the IUCN had recently assessed caribou as being of "least concern." That was at odds with what we saw, impressive as it was, compared to what some of our party had witnessed in past years: caribou streaming through the camp and covering the land as far as one could see.

In the 80s and 90s, the George River caribou herd numbered nearly a million animals. The largest ungulate herd in the world, they traversed the longest land migration of any land animal on Earth.

By 2001, the herd had already decreased by about 50 percent. When we visited Kamestastin in 2008, it was probably reduced by 75 percent. Concerned, in 2010, the Quebec, Newfoundland, and Nunatsiavut (Labrador Inuit) governments initiated regular aerial surveys. By 2014, the herd was down to 14,000 and continued to decline to about 5,000, a loss of over 99 percent.

In 2020, the herd recorded its first increase in population, recovering to 8,100 animals as estimated by the aerial survey program. As of July 2024, according to these surveys, there were 8,600 caribou in the George River herd and the proportion of young was increasing. While encouraging, it is too early to tell if the George River herd is on course for a recovery, and even if so, unaccustomed, new challenges caused by humans face them. The George River herd has a history of previous dramatic declines, and there is evidence that at least some of this boom-and-bust cycle, as mentioned above, is a natural consequence of the herd's migratory pattern that reduces wolf predation and relies on a periodically profuse, though slow growing food supply (lichens). Whatever the cause, as the 21st century got underway, it wasn't just the George River herd; caribou herds were declining worldwide. The IUCN relisted caribou as "vulnerable" in 2016. Canadian federal and provincial governments listed virtually all caribou populations in Canada as "threatened" or "endangered." The Beverly herd in the Northwest Territories was declared likely extinct in 2012.

Natural cycles notwithstanding, Human activity is exacerbating the declines, possibly causing some, and likely delaying or preventing natural recovery. Directly, forest clear-cuts and hydro dam impoundments the size of small countries; mining; the vast roadbuilding done to support them; and illegal hunting all pose challenges. Climate change adds more. Winter rain, previously unheard of in the far North, now freezes into solid crust, sealing forage under ice thick enough to starve whole herds. Warming also pushes other deer species north, bringing more wolves. Easier to catch than moose and

> "Caribou roam the whole of this land, seeking summer and winter range, safe calving grounds, nutritious forage, and relief from insects and predators . . . over thousands of miles and thousands of years. *Rangifer tarandus* is both from Latin and Greek and means 'to range' and 'wild and untamed'. . . indeed the very essence of this great animal."
>
> SUSAN C. MORSE

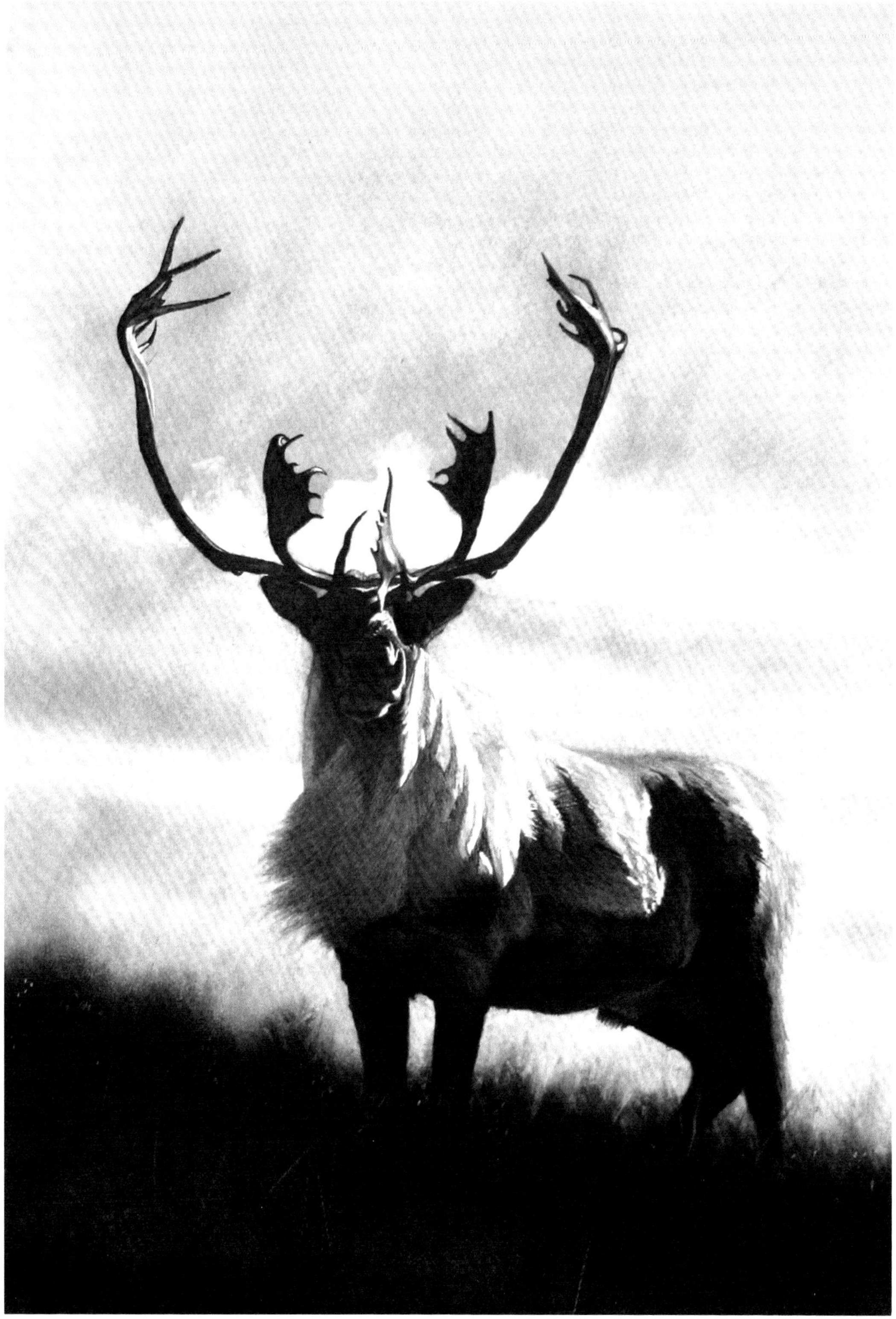

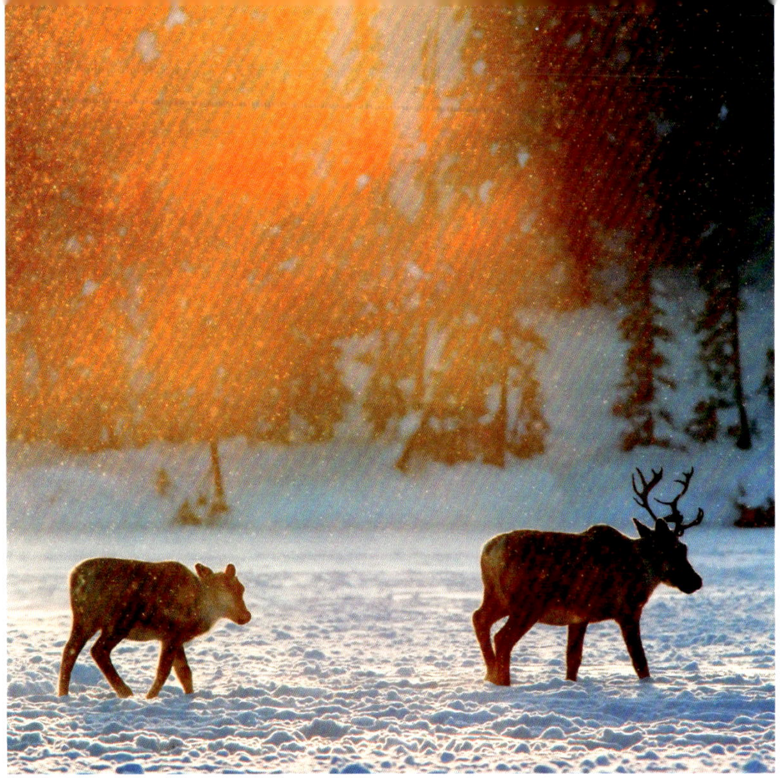

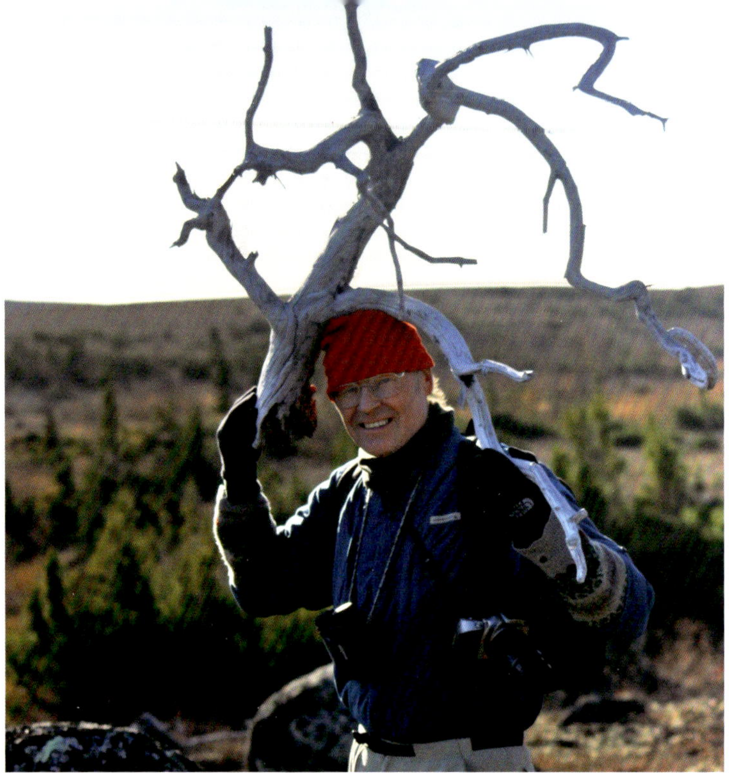

LEFT: Sunset cow and calf, Ungava. Susan C. Morse

RIGHT: John Pitcher and his antler trying to look like a caribou (it worked), Kamestastin.

slower to rebound than white-tail deer (caribou have one of the lowest reproductive rates among deer), caribou lose out. In 2009, Cole Johnson and I found moose droppings along the Kogaluk in northern Labrador. Inuit hunters had started to report seeing them occasionally only shortly before.

For a species adapted to the severe cold of the boreal forest and arctic tundra, there is no "north" left to go to as our world warms.

Moose (*Alces alces*), aka "elk" in Europe: Some taxonomists split the New World moose off as a separate species: *Alces americanus*. For now, we will stick to *Alces alces*.

The largest deer in the world, an adult male Alaska moose (*Alces alces gigas*) can top seven feet at the shoulder and weigh more than 1,600 pounds. Canadian moose are not far behind. Their range overlaps most of the boreal forest worldwide.

The Shiras subspecies (*Alces alces shirasi*) in the American West and into British Columbia is the "smallest," but small is relative: All moose range from huge to ginormous, and up close, they're unforgettable.

Wildlife artists and photographers spend hours—days—waiting, often for nothing. Susan C. Morse tells it well:

Hiding in the thick cover of spruce, I watched for migrating caribou. I took several photos and enjoyed the experience, but there had been nothing special. Then, hours later, the setting sun made the whole scene glow, and a shower of backlit snow crystals was stunning. But, no caribou. I fashioned a makeshift tube from a foam mat so that I could shoot directly into the sun, and suddenly, it all came together. These two lovely animals walked into the scene!

Woodland caribou on the move. *Near Kamestastin*, oil, Jay J. Johnson

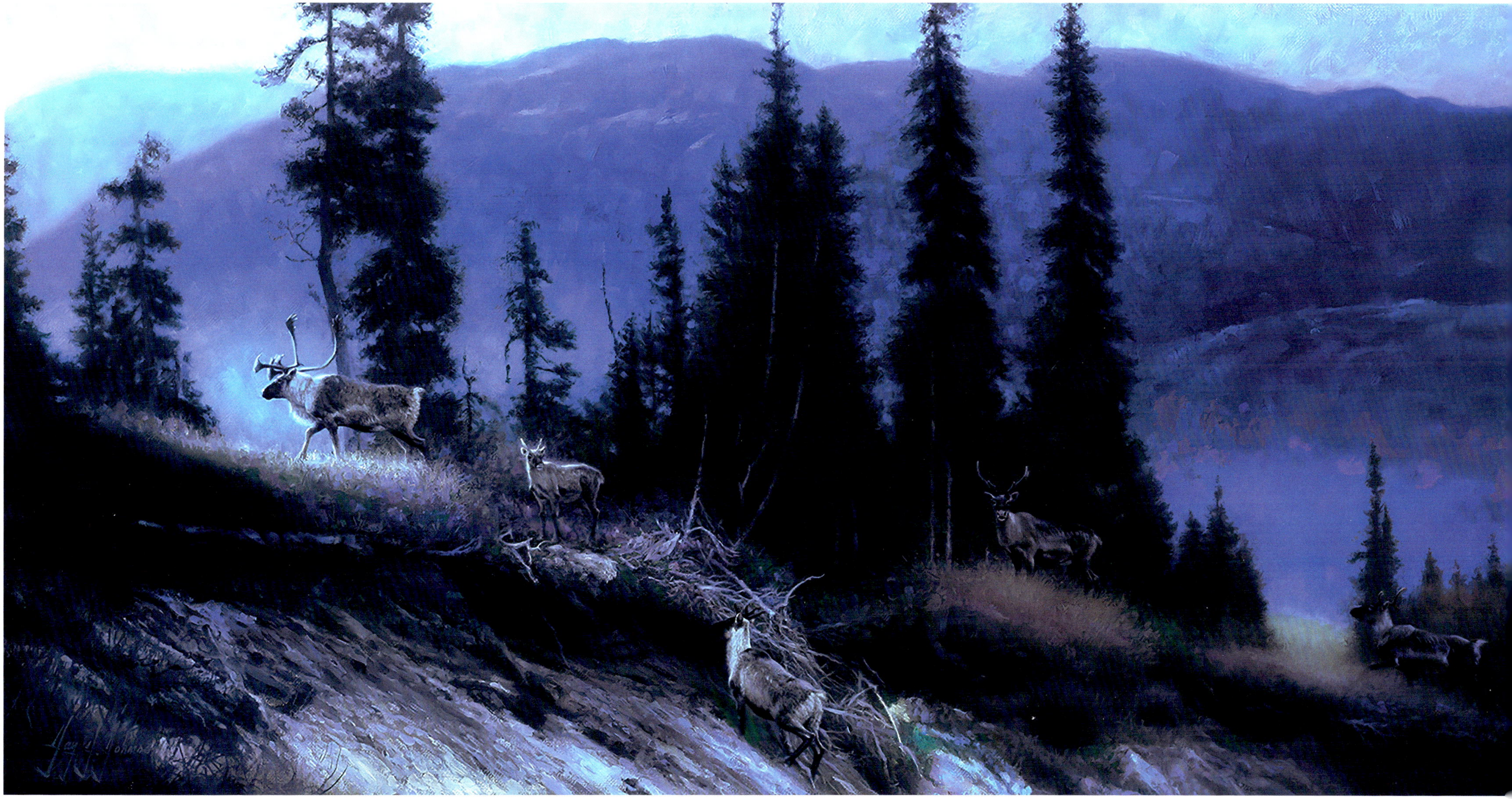

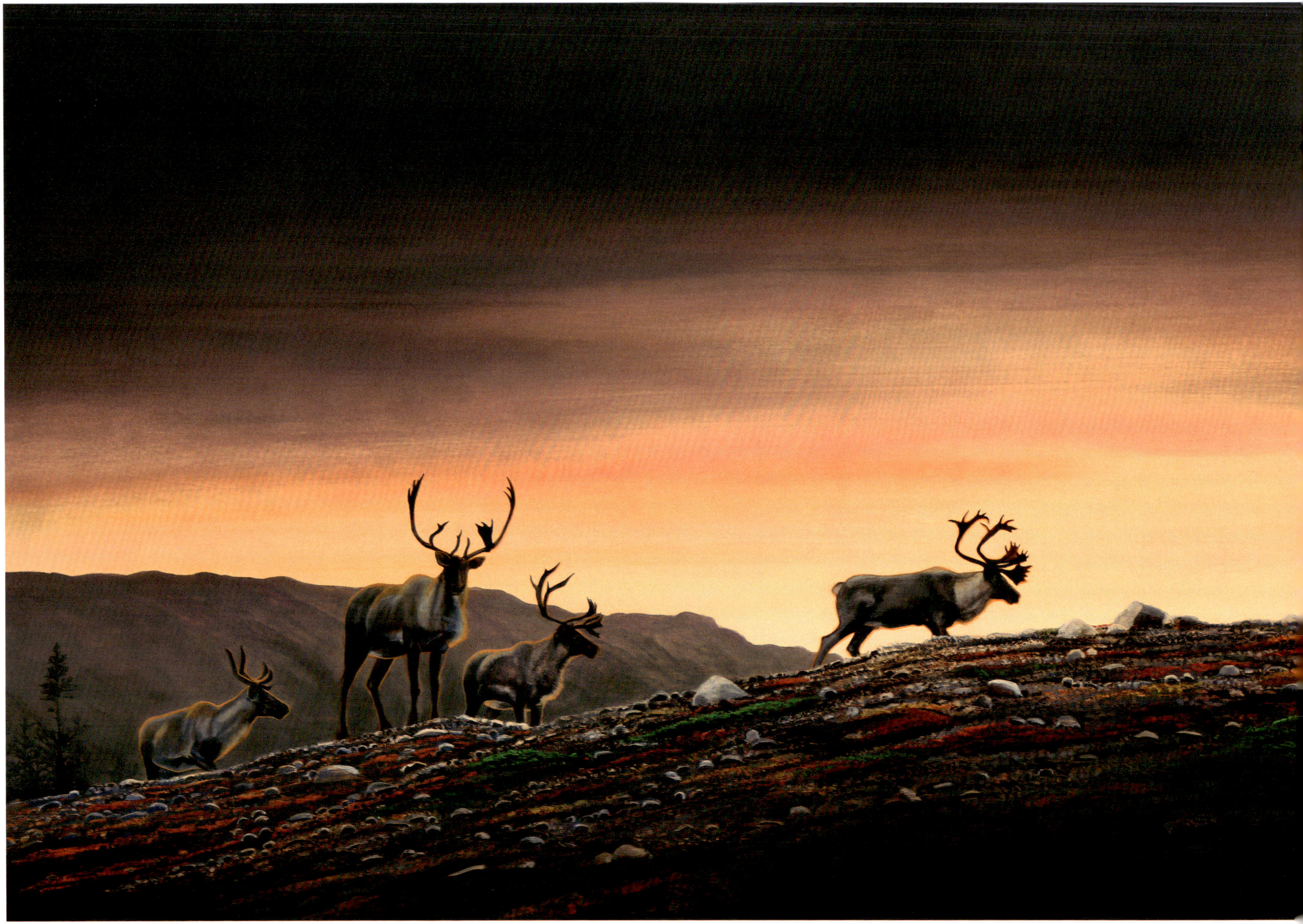

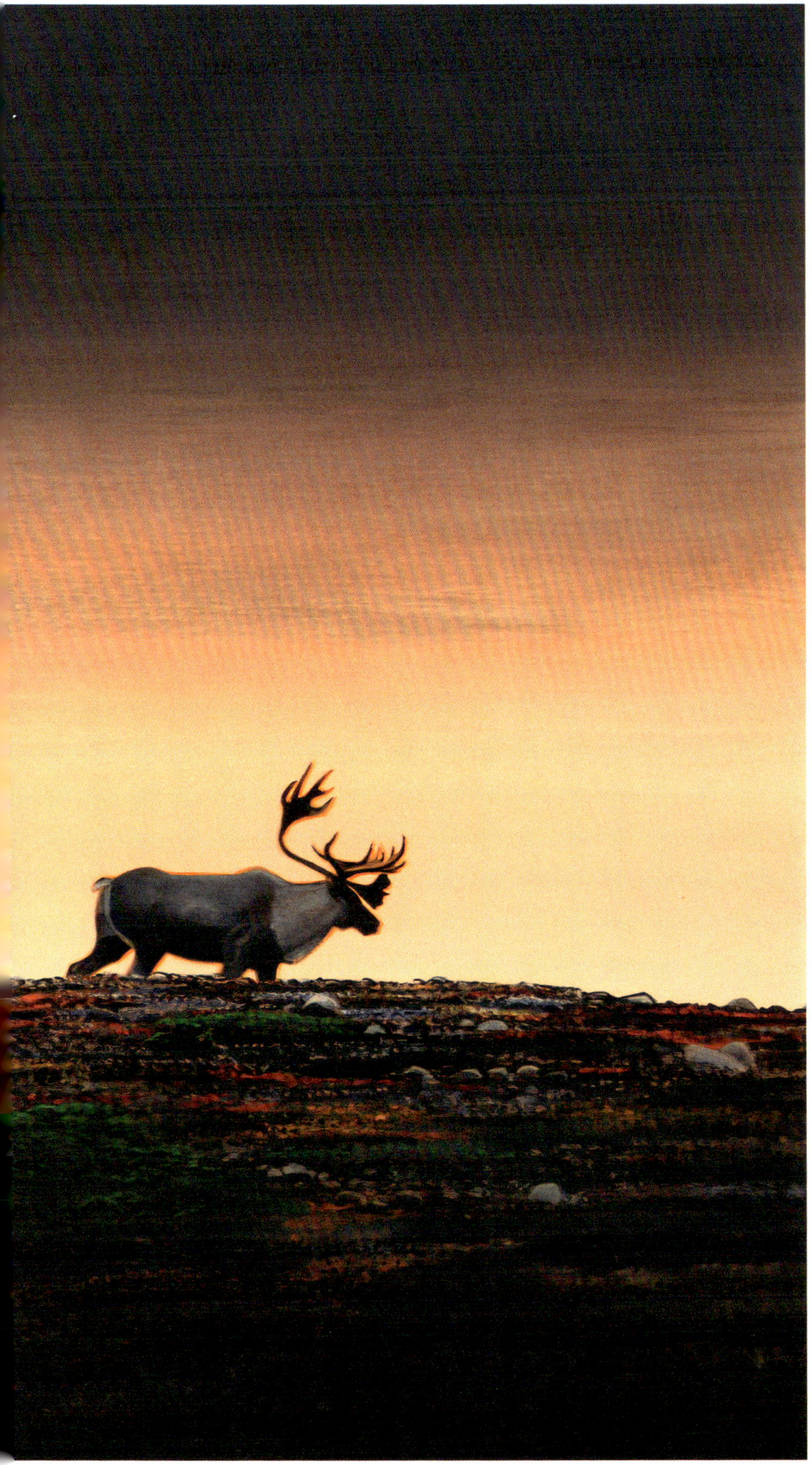

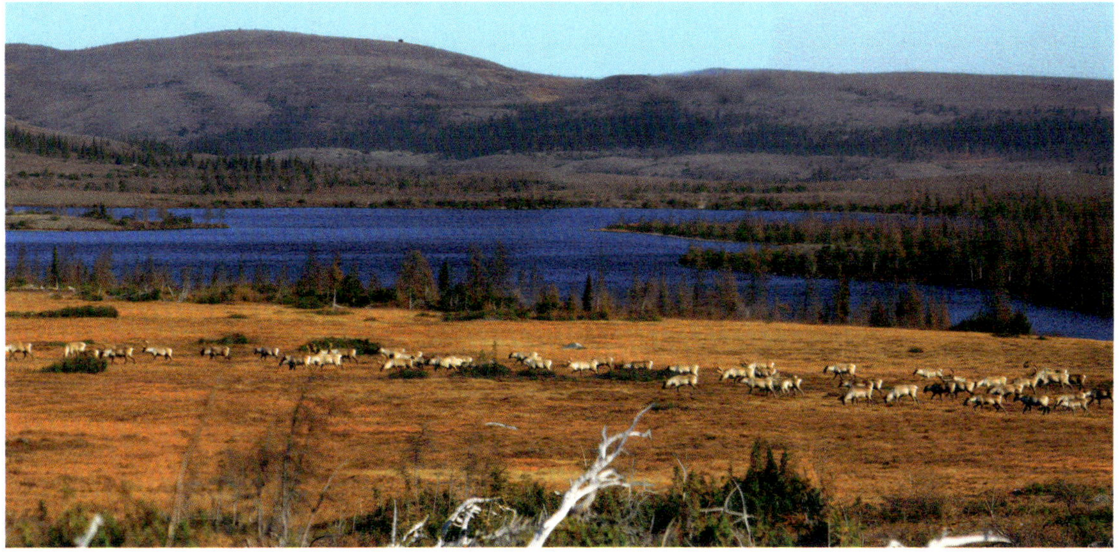

The first large gangs of caribou arriving at Kamestastin, October 8, 2008.

FACING: Caribou at Kamestastin, 2008 Kamestastin expedition. *On the Move*, oil, Steve Oliver

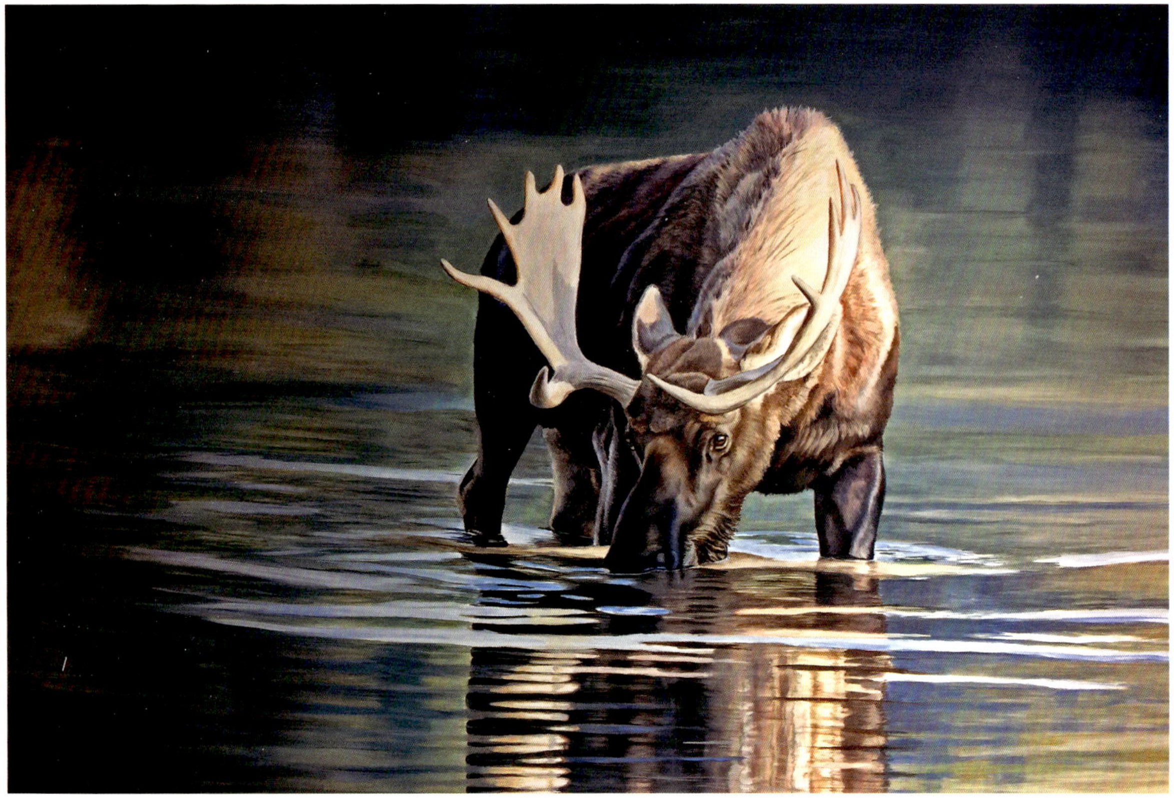

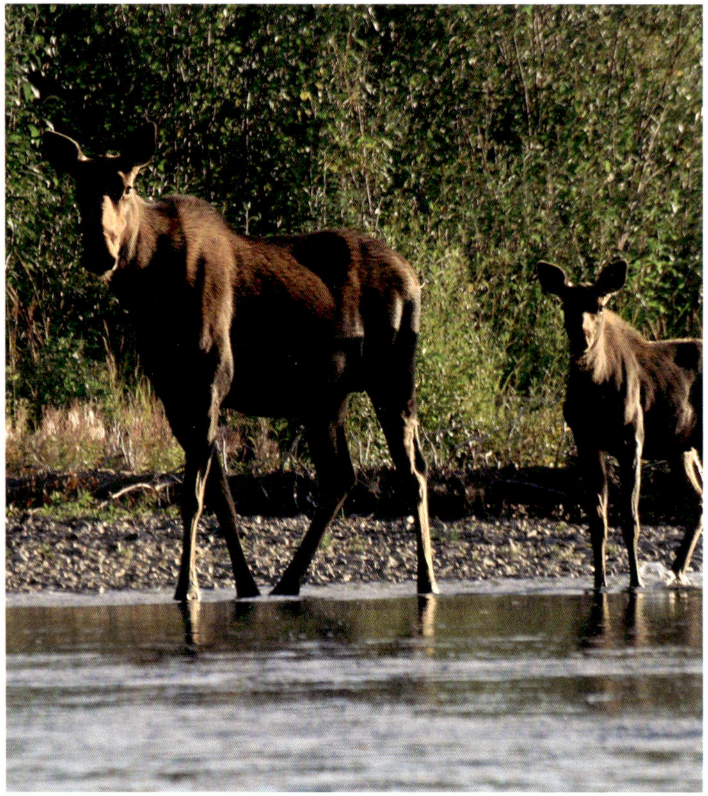

LEFT: Shiras moose, Yellowstone National Park. *Peaceful Solitude*, acrylic, Lindsey Foggett

RIGHT: Intensely curious cow and calf moose on the lower Reed River nearing the Kobuk River.

September 7–8, 2001: Alone on the edge of the Hudson Bay Lowlands below Hell's Gate, Missinaibi River

I had seen dessert plate-sized moose tracks in the sand where I was setting my tent, but there wasn't room to avoid them. I had a pleasant evening and enjoyed a fresh-caught walleye for dinner under my tarp in a warm, light rain.

At dawn, I woke to a slow, deliberate series of earth-shuddering thuds: The author of those monster moose tracks was approaching. Warn him off? No, a startled bull might decide to stomp first and ask questions later. I froze. I relaxed as he veered to avoid the tent—until I remembered the guy lines. Sure enough, he hit one. The tent shook, but the stake popped out cleanly, no entanglement. He snorted, took a couple of quick steps, then stopped—probably to glare back in annoyance—before moving on. As he continued, I snuck my head out the door and watched him. He was a beast!

Their enormous size helps moose withstand cold. Their long legs allow them to move through deep snow and reach high into the trees to browse and, in the summer, to forage on summer water plants while wading in wetlands, making them ideally suited to the boreal forest.

The name "moose" is almost certainly a European effort at the various Algonquin-speaking peoples' names for this echo from the Pleistocene Age of megafauna: *mush* (Innu), *moz* (Abenaki), *moos* (Narragansett), *mooz* (Ojibwe), generally meaning "twig eater," "bark eater," or "stripper of bark." Moose were known in Scandinavia, though called "elk," and wapiti were common in the Americas, so possibly, early Europeans in eastern North America, maybe having heard of Scandinavian "elk" but never having seen them, mistook wapiti for them and named the wapiti "elk." Then, with the name elk already in use, they may have simply adopted the Indigenous name for moose when they ran into the real deal. Twig/bark eater fits in any case. Moose have room to move north as the climate warms (and are). At least for now.

Elk (wapiti) (*Cervus canadensis*)

The second-largest deer in the world after the moose. Bulls may reach five feet at the shoulder and weigh more than 750 pounds. Until the late 1990s, elk were thought to be a subspecies of the red deer in Europe, but genetic research established that they are a separate species. Prehistorically, elk were a widely distributed Holarctic species. Historically, elk were common in the boreal forest and throughout eastern North America, though today, they occur only in scattered areas along the southern margins of the ecosystem in both North America and northern and eastern Asia. They are still common in the North American mountain west.

Exclusively Eurasian boreal forest cervids (Palearctic)

Red deer (*Cervus elaphus*)

Similar in appearance to elk, red deer are the fourth largest deer species in the world.

Siberian roe deer (*Capreolus pygargus*)

The Siberian roe deer is closely related to and hybridized with the European roe deer (*Capreolus capreolus*), whose range extends into Scandinavia. IUCN Red List for both species is Least Concern. Holarctic in distribution.

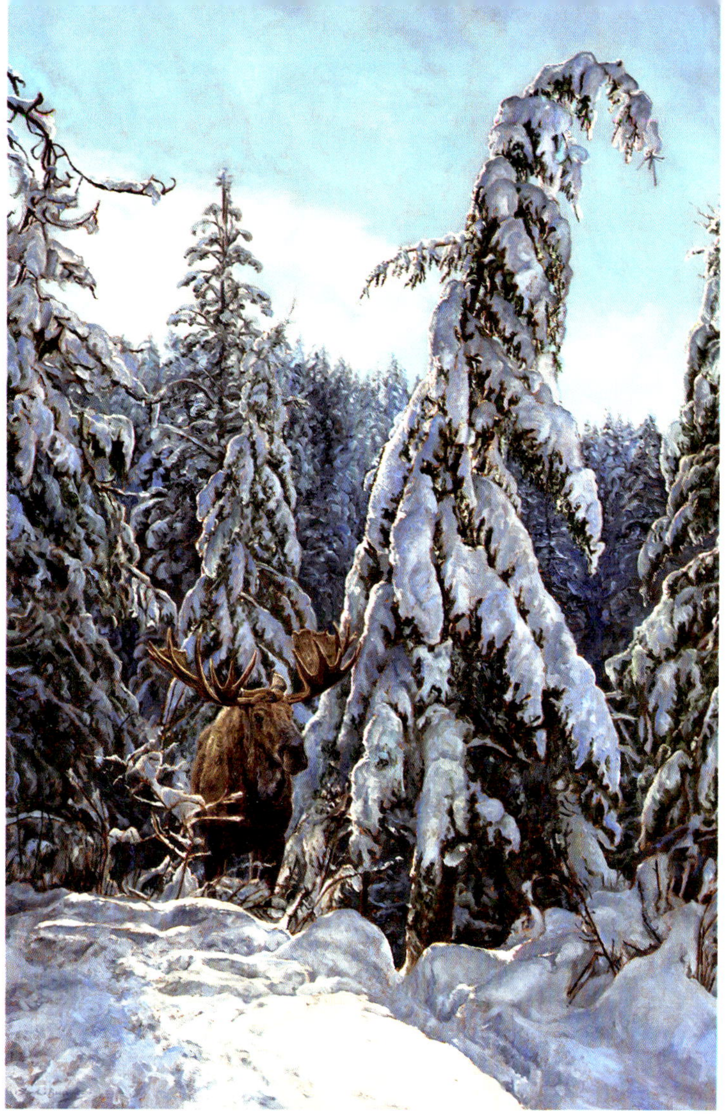

Moose in boreal forest.
Winter Solstice, oil, Linda Besse

Family Moschidae

Siberian musk deer (*Moschus moschiferus*)

IUCN Red List status: Vulnerable, Palearctic

Once classified with the deer, musk deer now have their own family. At a glance, they look like small deer—until you notice the fangs (no photo, unfortunately). Favored prey of Siberian tigers.

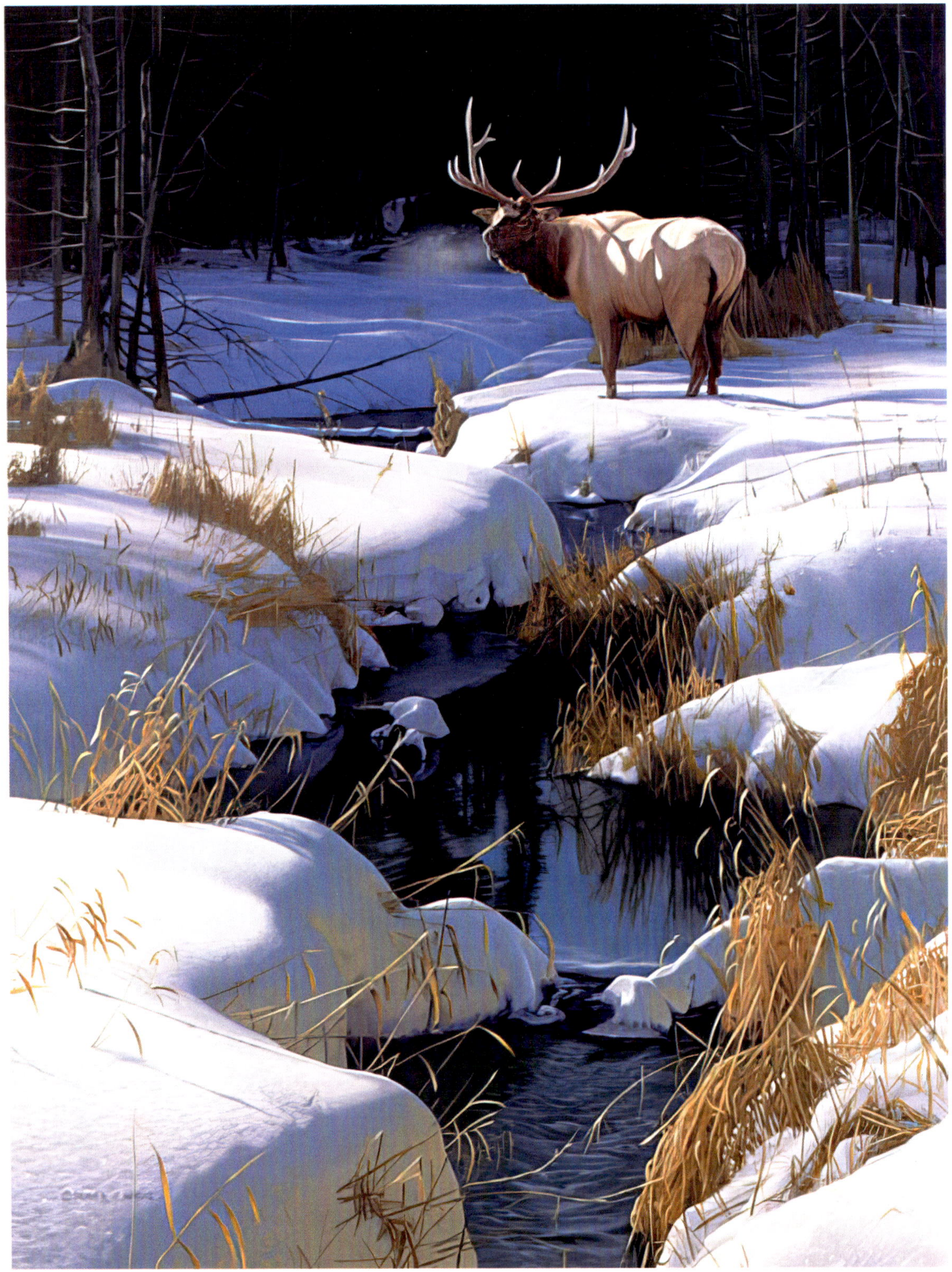

Elk. *Winter Sonata*, oil, Derek Wicks

FACING: Bull moose (Canadian) in typical habitat. *Boreal Morning*, oil, Patricia Pépin

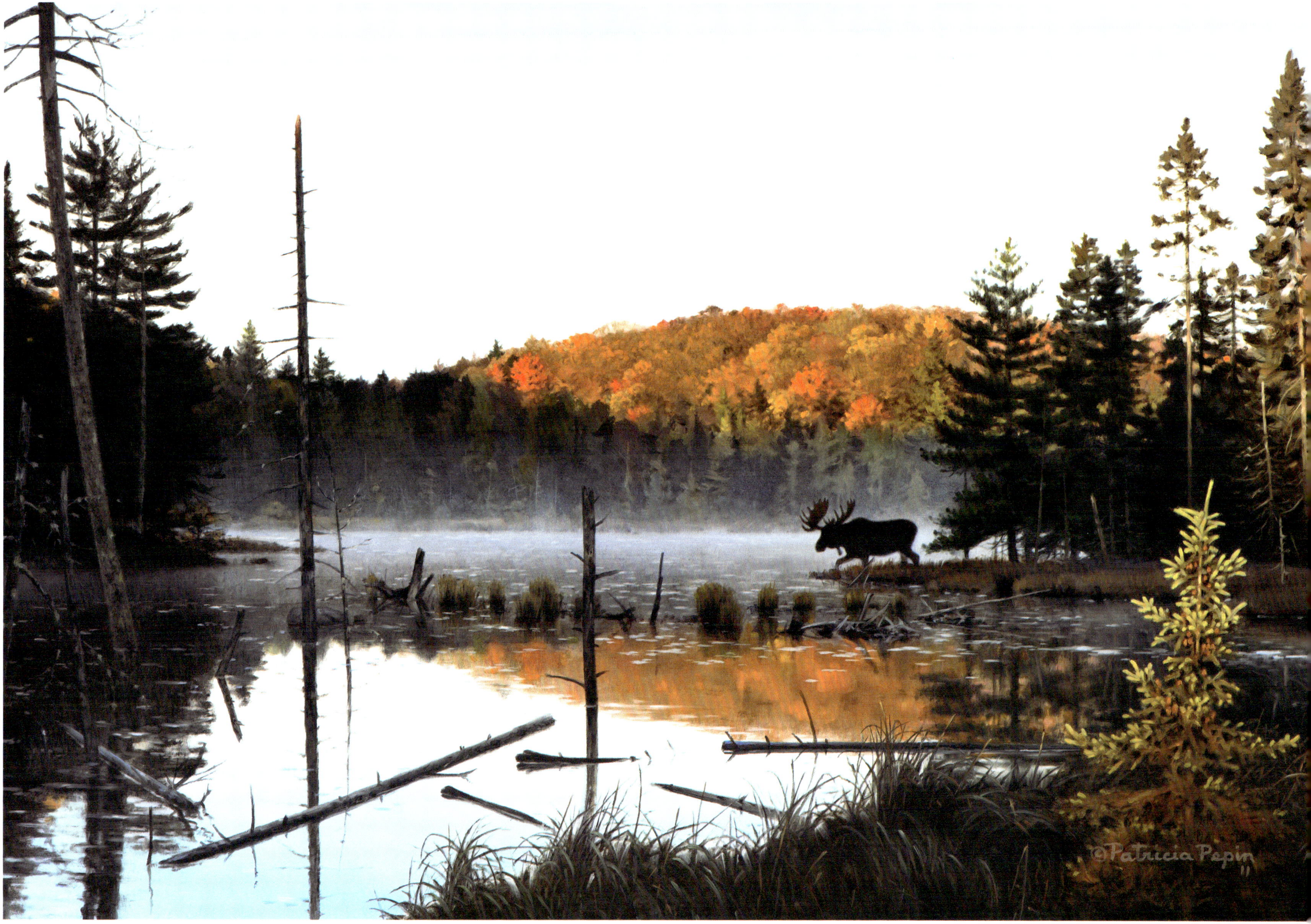
©Patricia Pepin

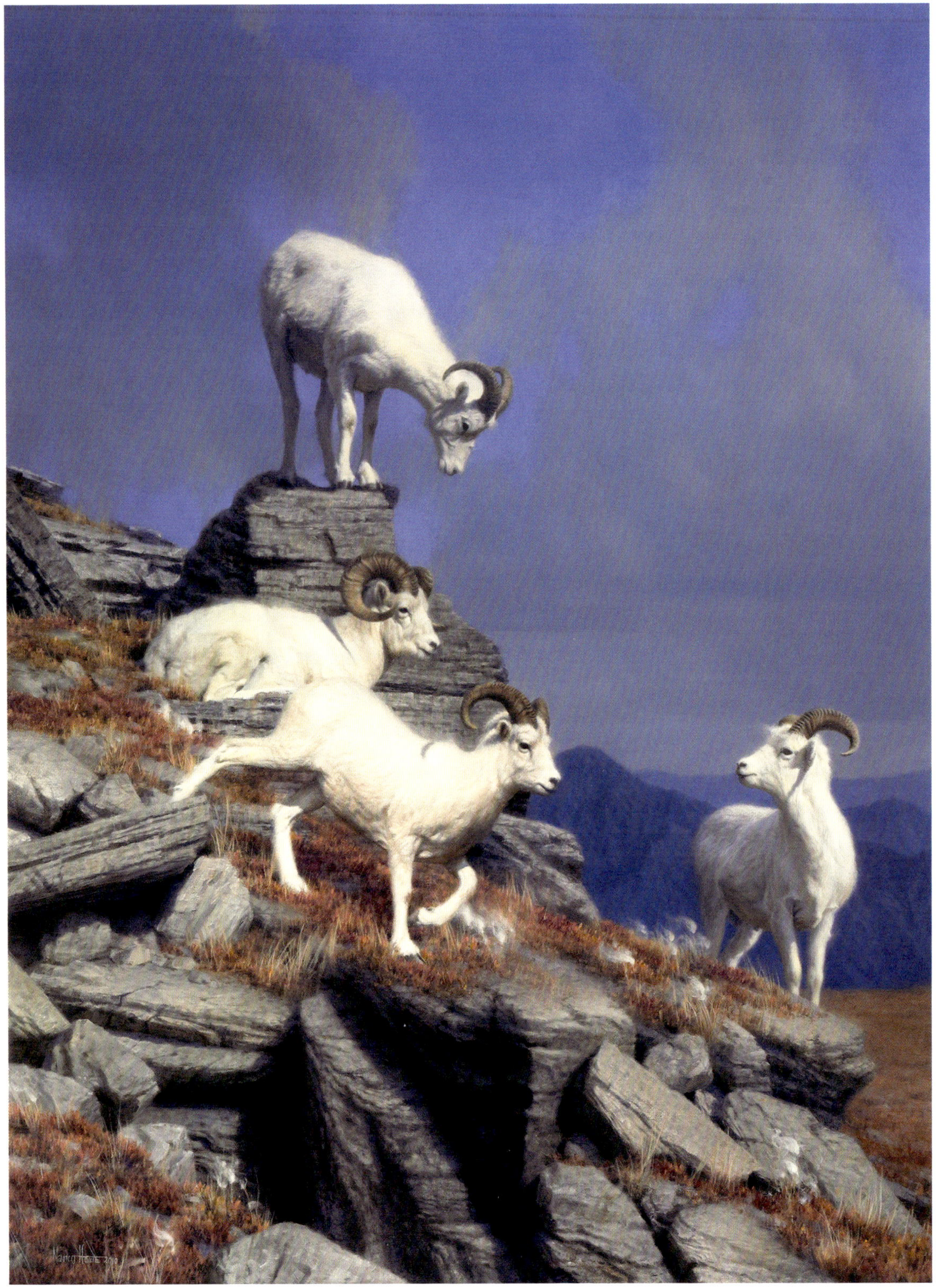

Family Bovidae

Musk oxen (*Ovibos moschatus*)

August 26, 2012: Camped upstream of the entrance to the Grand Canyon of the Noatak, Arctic Alaska: We were spending a wet, chilly day ashore. I'd been sketching the entrance to the canyon, when I realized that I'd been hearing an odd *thunk*-ing sound for some time. I found two musk ox bulls peacefully grazing in the willow brush, until one slowly swung his head around and squared up as his companion rounded to receive the looming blow. Then . . . they gently clunked heads. To be fair, the rut (June and July) was largely over. Nonetheless, they picked it up considerably a few times. To paint them later, I moved them (my artistic license) from the willows to the mountains 300 miles upstream, where I'd seen my first musk ox three weeks earlier (*Echoes of the Ice Age*, pp. 96–97).

Musk oxen, *oomingmak* (bearded one) in Inuktitut and *Mathi-mos* (ugly moose—an amusing distinction) in Cree, are primarily tundra habitués, but use the forest. An introduced population on the Koksoak River near Kuujjuaq, Quebec, is in forested habitat and doing well.

Thinhorn sheep (*Ovis dalli*), aka Dall sheep

Dall sheep range in the mountains of western Canada and Alaska. We've only seen them high on mountain slopes. They do come down—we found the horn in *Echoes of the Ice Age* on the riverbank—and their trails through alder thickets were vital for our climb up Tupik Creek to Angiaak Pass. Like musk oxen, they are not primarily a boreal forest species.

Wood bison (*Bison bison athabascae*)

IUCN Red List status: Not listed separately, but American bison status is Near Threatened, Nearctic

The largest living bovid in North America: about a third heavier and taller than plains bison. Mature bulls stand six

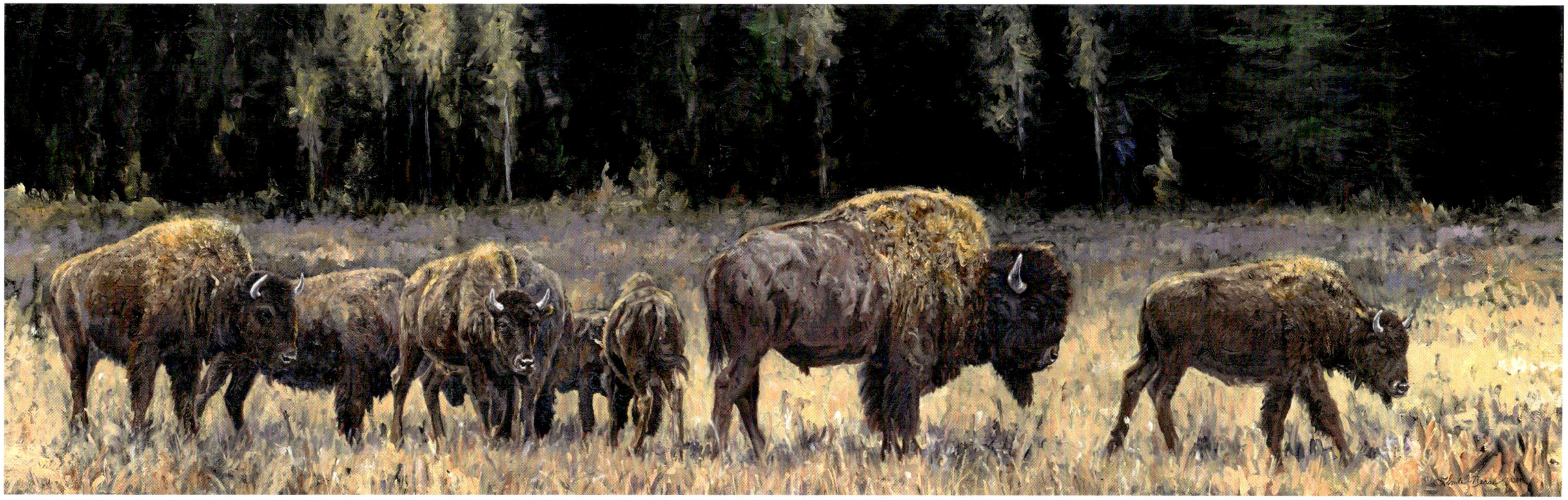

Wood bison. *A Reintroduction*, oil, Linda Besse

FACING: Dall sheep. *Above Us Only Sky*, oil, Nancy Howe

feet at the shoulder and weigh well over a ton. Nearly extinct by 1900, several herds now roam Canada, most notably in Wood Buffalo National Park in Alberta, and in 2015 a reintroduced herd took hold in Alaska near the Innoko and Yukon rivers.

Order Rodentia (rodents)

Rodents are defined by their paired upper and lower incisors: constantly growing, self-sharpening tools for gnawing. They are the most diverse and numerous mammals on Earth.

September 13, 2006: Confluence of the Nutillilik and George rivers, Nunavik (Northern Quebec)
I'd just finished a sketch of the last waterfall before the Nutillilik joined the George when an insistent red squirrel (*Tamiasciurus hudsonicus*) intercepted me. Not unusual—except this one wouldn't quit. He circled me, sniffed my ankles, then dashed off only to return for another round. Amusing at first, then almost annoying. I finally threatened to paint him if he didn't leave me alone. He didn't. So, I did.

Back at camp, I found the crew with their cameras aimed into a spruce tree. Seven or eight feet up sat the cutest—and by then the most photographed—baby porcupine in Quebec.

Rodents are primary consumers like moose and caribou, but instead of fearing only apex predators, they're on nearly everyone's menu: martens, weasels, mink, wolverines, lynx, foxes, coyotes, birds of prey, even wolves and bears when larger prey is scarce. In Alaska, we saw ground ripped apart by grizzlies digging for Arctic ground squirrels (*sik-sik* in Iñupiaq, a language closely related to Inuktitut).

Though mostly herbivorous, many are opportunistic omnivores. Mice will eat injured small animals, including

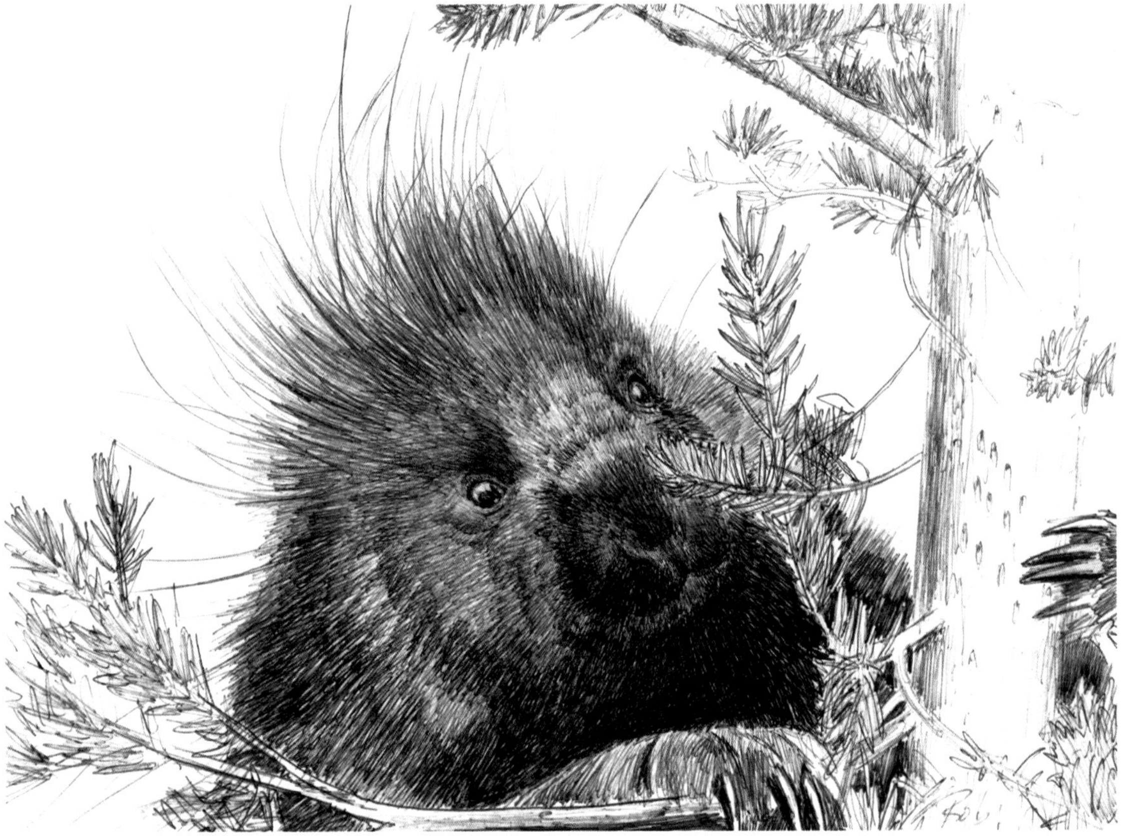

Cute young porcupine, Nutillilik and George rivers. Ballpoint pen, Rob Mullen

FACING: Nutillilik River red squirrel. *Curious George I & II*, acrylic, Rob Mullen

other mice. Chipmunks, those cute little ground squirrels of suburban backyards and the boreal forest, will happily eat an injured songbird or stranded nestling.

Breaking the mold: Big rodents of the boreal

While most rodent populations persist by producing huge numbers of small, short-lived, early reproducing individuals to offset heavy losses to predation, two species thrive with a different strategy: Grow big, make yourself difficult and/or unpleasant to eat, and live long. And then, there is the pretender.

Porcupine (*Erethizon dorsatum*)

The North American porcupine is the second-largest rodent in North America (adult males often weigh 20–30 pounds; females, about 15 pounds) and can live 20 years or more. Slow-moving but well defended, they carry roughly 30,000 stiff, barbed, hollow quills everywhere except their face, feet, and underside. Add a strong scent and tree-climbing ability, and few predators succeed against a healthy adult. Even for those who do, it can be a Pyrrhic victory as they often are gifted a mortal load of quills in the process. Consequently, even though females aren't sexually mature until over age two and usually only have one pup a year, they thrive in the boreal forest and beyond.

One notable exception among the would-be diners: Male fishers, twice the size of females, are the porcupine's most effective regular predator.

For Indigenous people, porcupines have long been valued as a provider of quills for decoration and meat for food. At Kamestastin, Labrador, we learned they're an Innu delicacy.

North American beaver (*Castor canadensis*) and Eurasian beaver (*Castor fiber*)

IUCN Red List status for the Nearctic North American beaver and the Palearctic Eurasian beaver is Least Concern; both species are coming back from being virtually extirpated in large portions of their range.

> Squirrels can be heavy "predators" on seeds in conifer cones. Feeding on immature cones has led to some trees evolving smaller, tougher cones with fewer seeds to make squirrel foraging less profitable. If they haven't already, the squirrels will likely adapt their tactics to counter the trees' defenses.

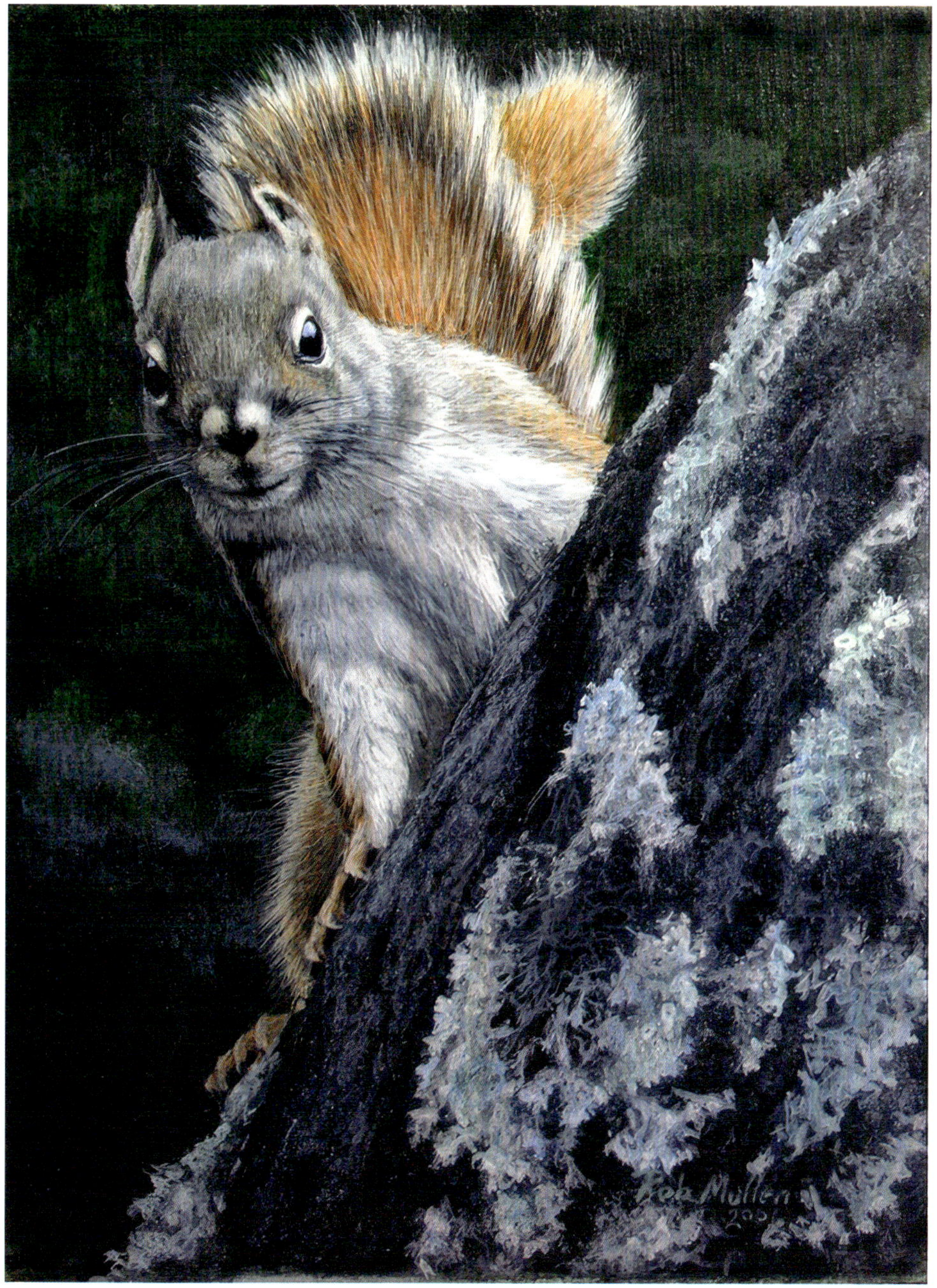
Rob Mullen

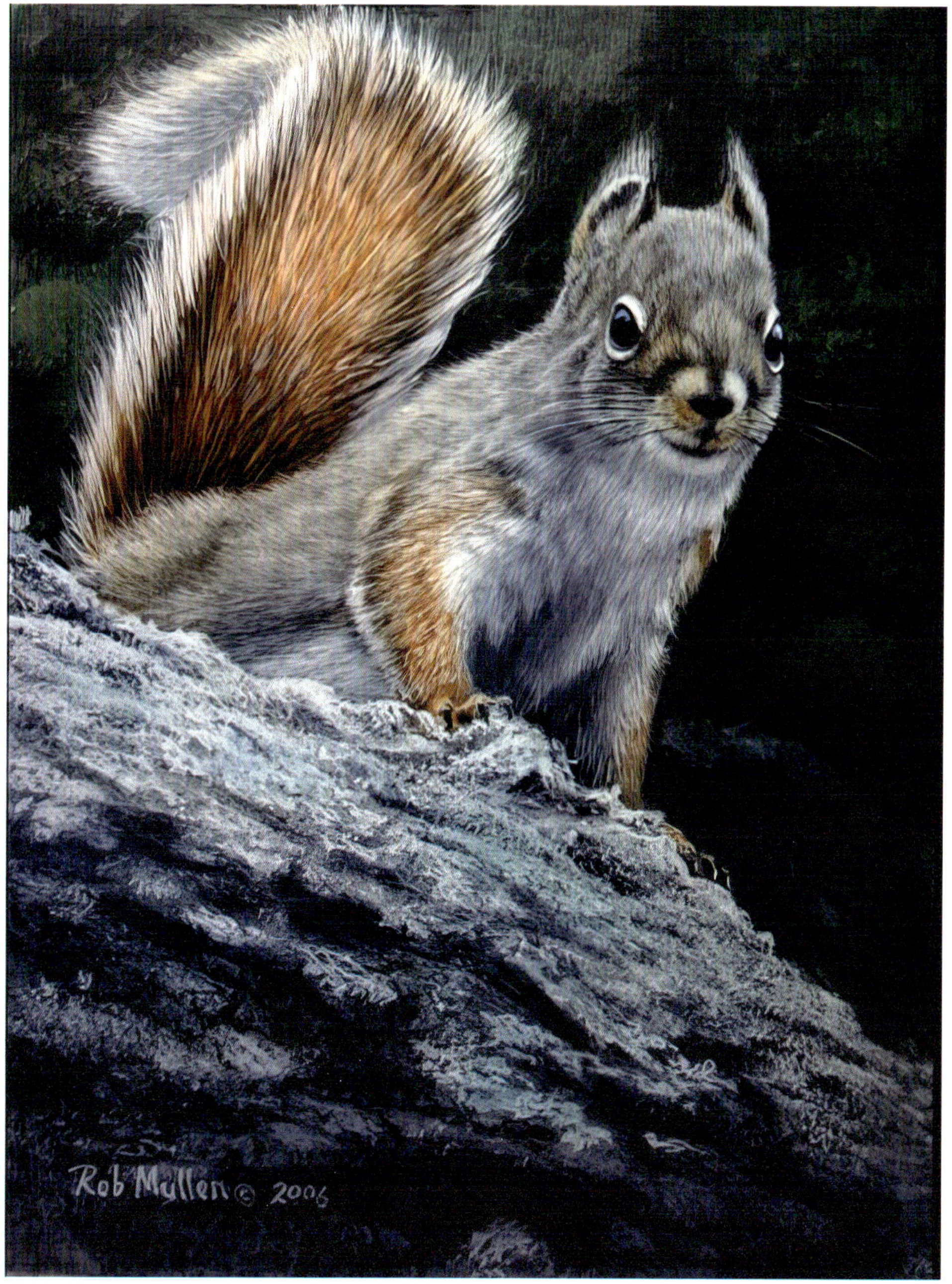
Rob Mullen © 2006

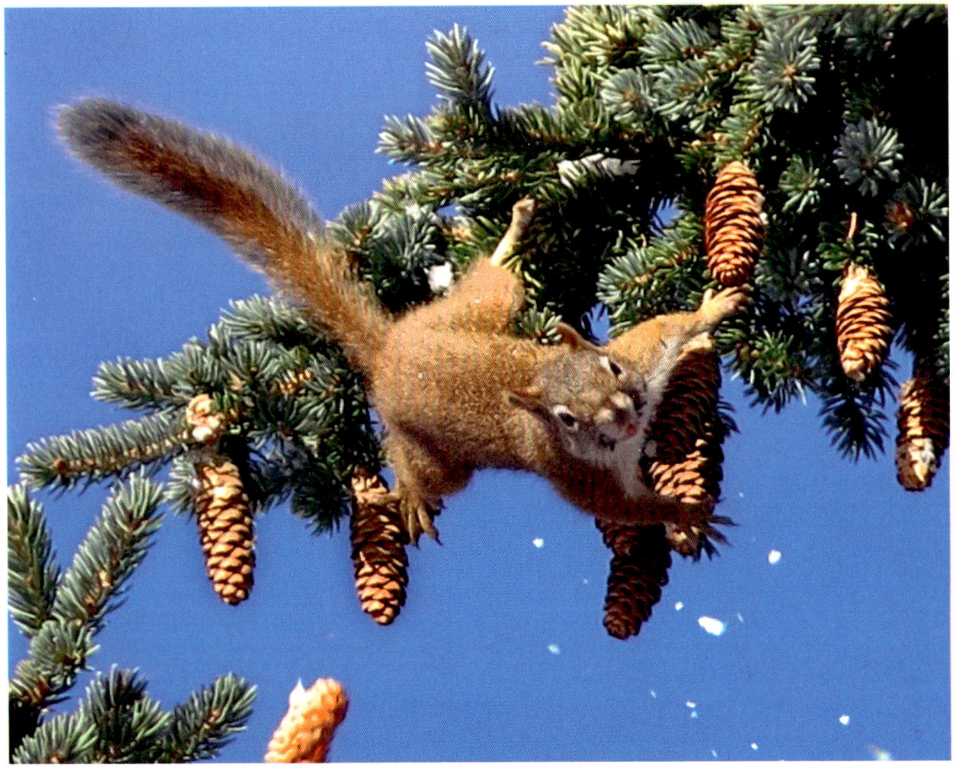
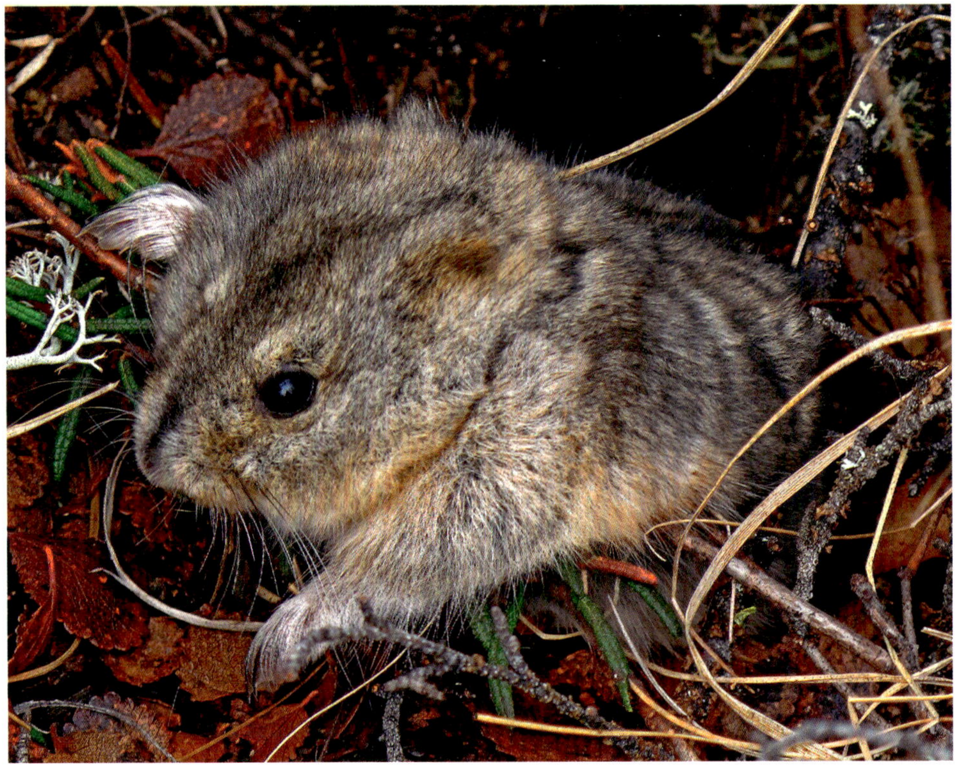
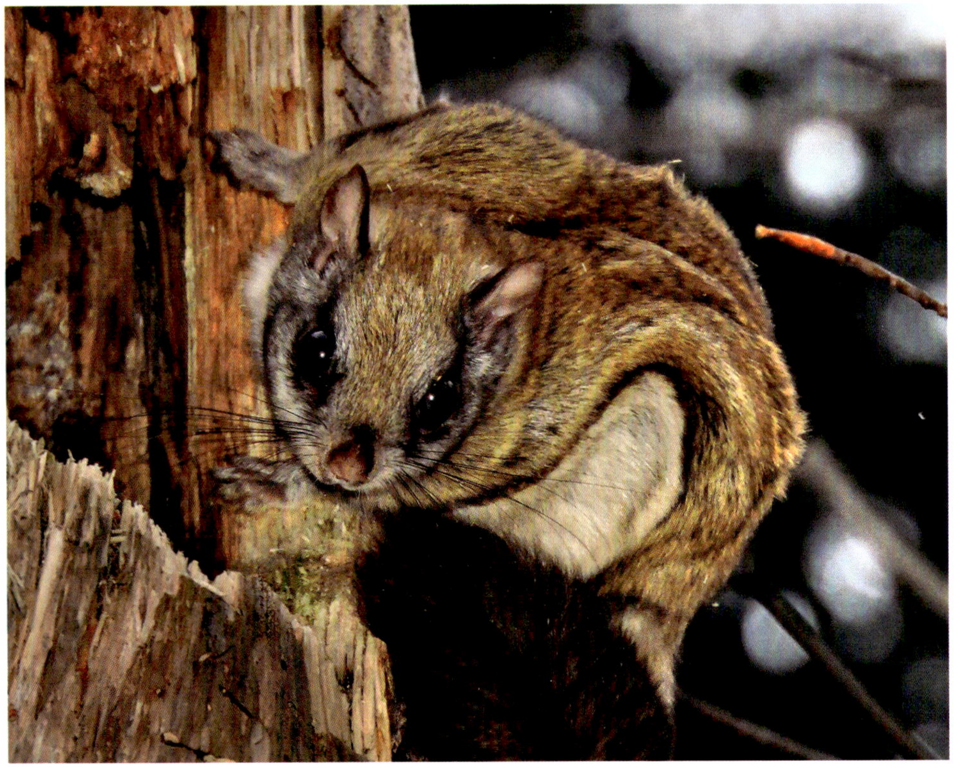
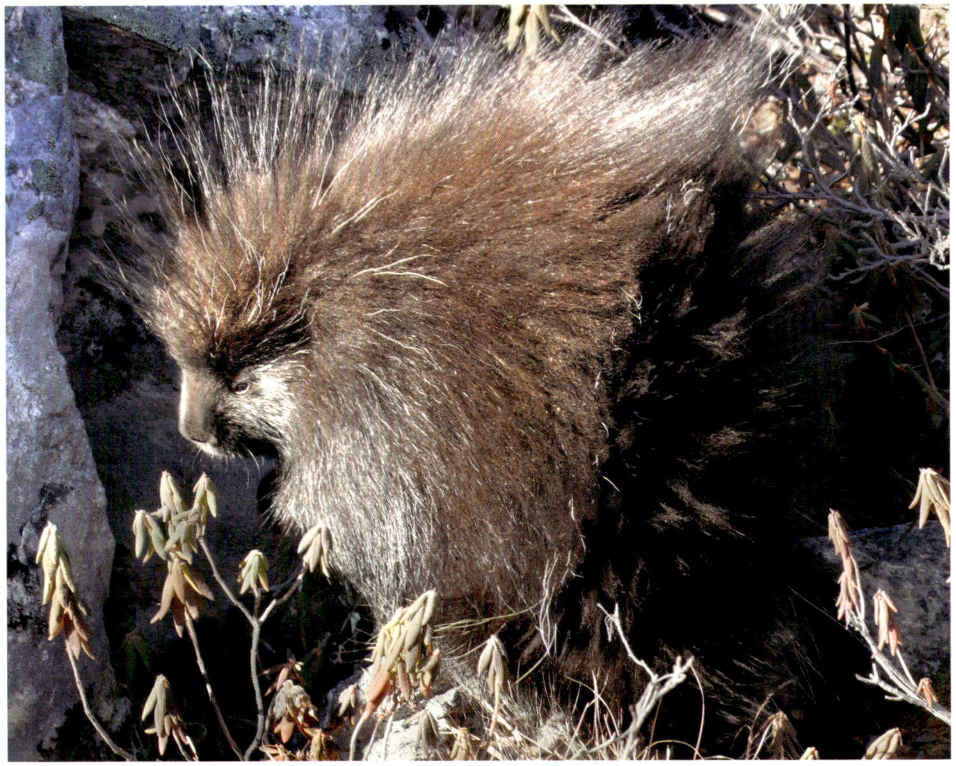

Beavers are one of the most important keystone species in the boreal forest and the largest rodents in the northern hemisphere, topped globally only by South America's capybara. Typical adults weigh 40–50 pounds, with records over 100 pounds. They show little difference in size between sexes, and, unlike most rodents, grow slowly but steadily over a long life.

Beavers are slow and ungainly on land like porcupines, and to make things worse, they lack quills and are unable to climb trees. However, their even beefier builds, remarkable engineering ability, and behavior result in very low adult mortality and high survival rates for kits. Not only are beavers some of the larger animals in their habitats, but they build virtually impregnable lodges of interwoven sticks, logs, and dried mud, with doorway entries to the snug inner chamber submerged in ponds that they create with dams made of the same materials.

As seen in the painting to the right, lodges can be, and often are, surrounded by water, which is deep enough that there is room under the ice for entry and exit with quick access to their stored food: cut branches, which they anchor in the mud on the bottom next to the lodge.

Behaviorally, they usually don't venture more than 100 yards from shore when foraging and unlike most rodents, to protect their food supply, they are aggressively territorial. Even their own children are sent packing in their third summer to establish their own ponds. This is one notable way that their populations are more self-regulative than most herbivores, let alone rodents.

Beavers are second only to humans in the scale of their constructions. The largest recorded beaver dam in the world is over half a mile long. It is in Wood Buffalo National Park in Alberta, Canada, and it is visible from space. The wetlands beavers create almost always benefit the ecosystem they build in, storing water and mitigating droughts, floods, and erosion, all the while creating some of the most varied and biologically diverse habitats in North America and Eurasia.

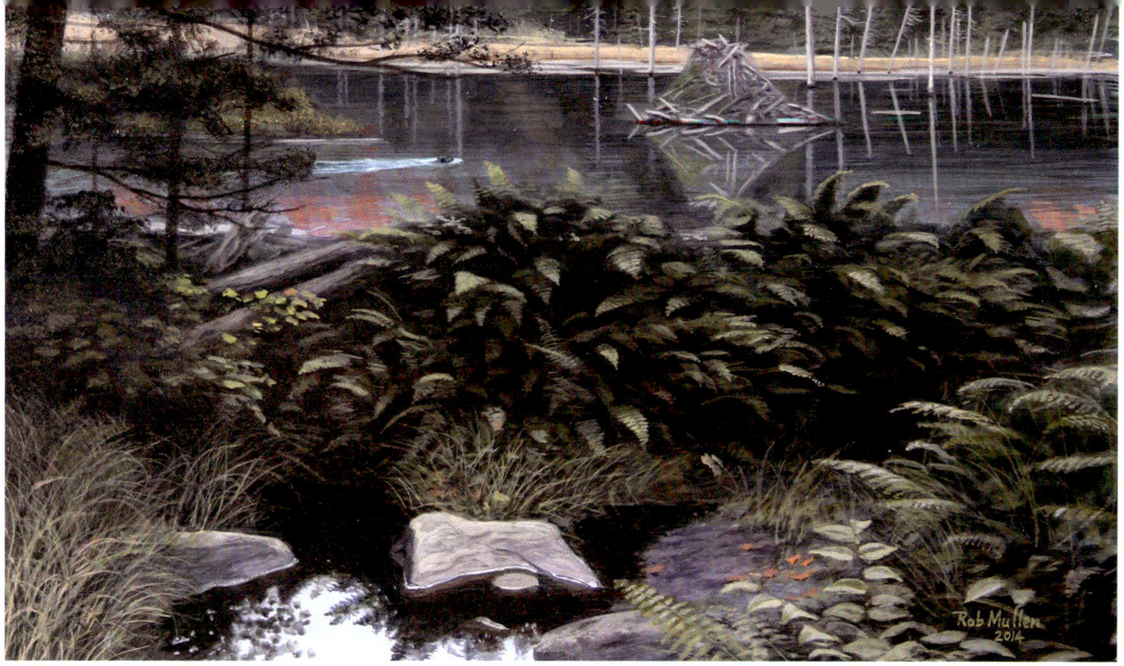

Beaver lodge in boreal forest habitat along the Long Trail in Vermont. *Condo on the Water*, acrylic, Rob Mullen

FACING, CLOCKWISE FROM TOP LEFT:
Red squirrel cutting white spruce cones. Susan C. Morse

Collared lemming, Ungava. Susan C. Morse

Mature porcupine enjoying the Labrador sun, Kamestastin. Jay J. Johnson

Northern flying squirrel. Susan C. Morse

(For a fascinating book on the subject, I strongly recommend Ben Goldfarb's *Eager*.)

Yet, for centuries (circa 1550–1850), humans perversely measured the value of beaver only in hats (felt made from beaver fur was all the rage). The resultant fur trade pushed beavers to virtual extinction in Europe in 100 years and was on the verge of repeating the feat in North America in 200 years, when the fickle finger of fashion and economics turned to silk hats. Saved by a moth.

Today, beavers are doing well and are reestablishing themselves in their former ranges, especially in North America. Their reception, however, is mixed. A collaboration of state, provincial, Indigenous, and local governments and environmental NGOs in the more arid and heavily grazed western US and Canada are introducing beavers to restore degraded habitats. For instance, British Columbia's 10,000 Wetlands project aims to restore, unsurprisingly, 10,000 wetlands using beavers and BDAs (beaver dam analogs) throughout the province. Jennifer Lovett's book *Beavers Away!* recounts how a game warden reintroduced beavers into a

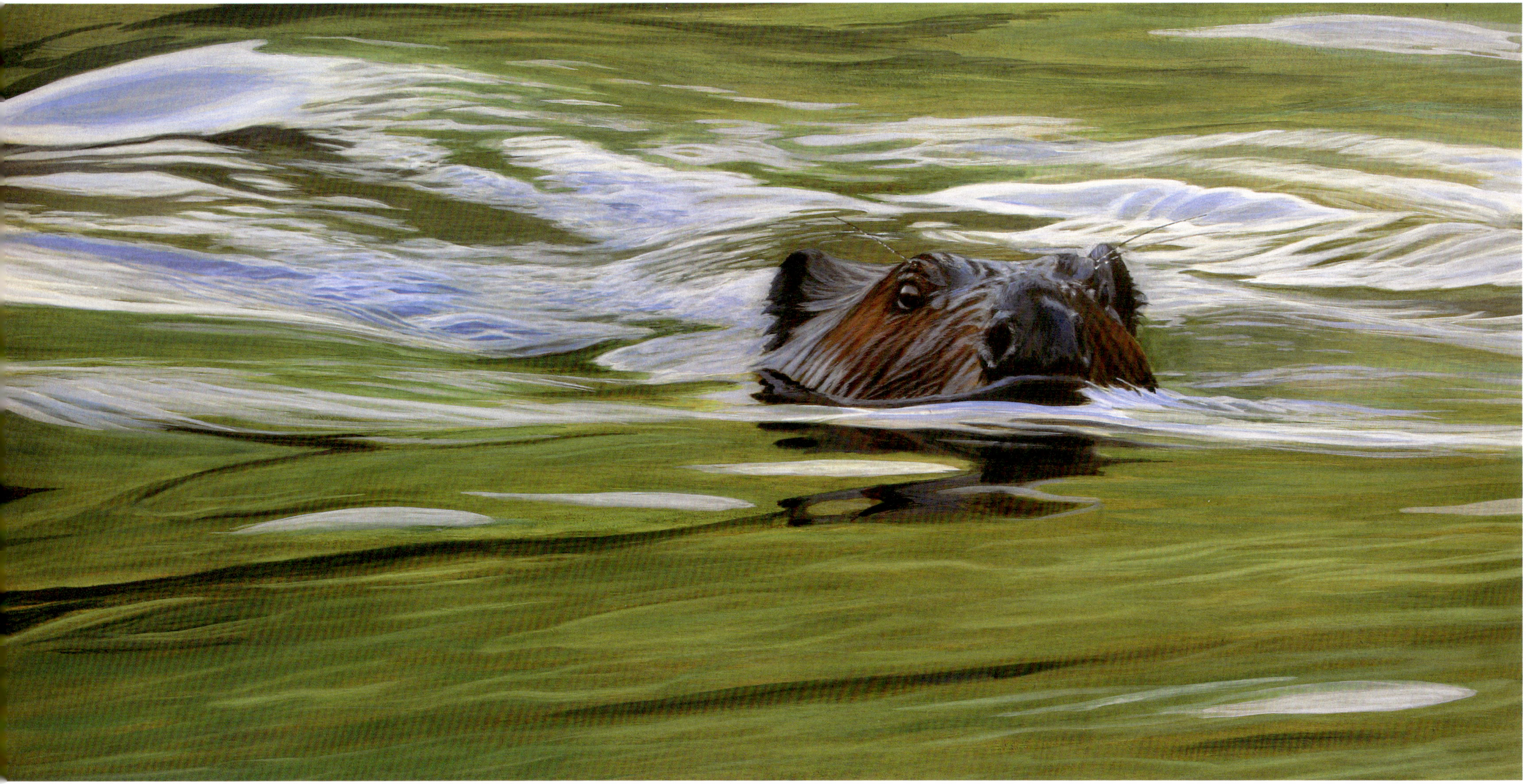

North American beaver. *Neck Deep*,
acrylic, Rob Mullen

remote Idaho location by parachute where, led by Geronimo (the adult male and possibly the first beaver pushed out of the plane), they restored the eroded landscape into lush habitat for many more species of animals. On the other hand, some landowners and highway departments, concerned with flooding, are locked into an unending cycle of natural recolonization and trapping. Flow control devices can break that cruel and expensive merry-go-round, maintaining the benefits of a healthy wetland while avoiding unwanted inundation. Skip Lisle pioneered and leads in the development of flow control devices with his company Beaver Deceivers in Grafton, Vermont.

The pretender: Muskrat (*Ondatra zibethicus*)

IUCN Red List: LC, Holarctic, though recently, there have been alarming declines in some muskrat populations in North America.

Naturally, muskrats don't pretend to be anything: They are just muskrats. However, considering that they are closely related to lemmings and voles and only distantly to beavers, it is interesting that they are so large and reminiscent of beavers in lifestyle. To the untrained eye they can be mistaken for beavers, especially for young ones. A keener eye will notice the narrow, keeled tail, and the fact that they are a fraction of the size of a mature beaver, weighing about 4 pounds to a beaver's 40–50 pounds. They don't build dams, but they do build lodges after a fashion. As one might guess from their name, these "push-ups" are far flimsier than beaver lodges and are often destroyed by storms or spring floods.

Despite their similarities to beavers, muskrats face constant heavy predation—from nearly every sizeable forest or aquatic predator. Fortunately for the species, a trait they share with their lemming kin is high reproductive rates to offset the losses. Females mature in their first year and can raise several litters per season.

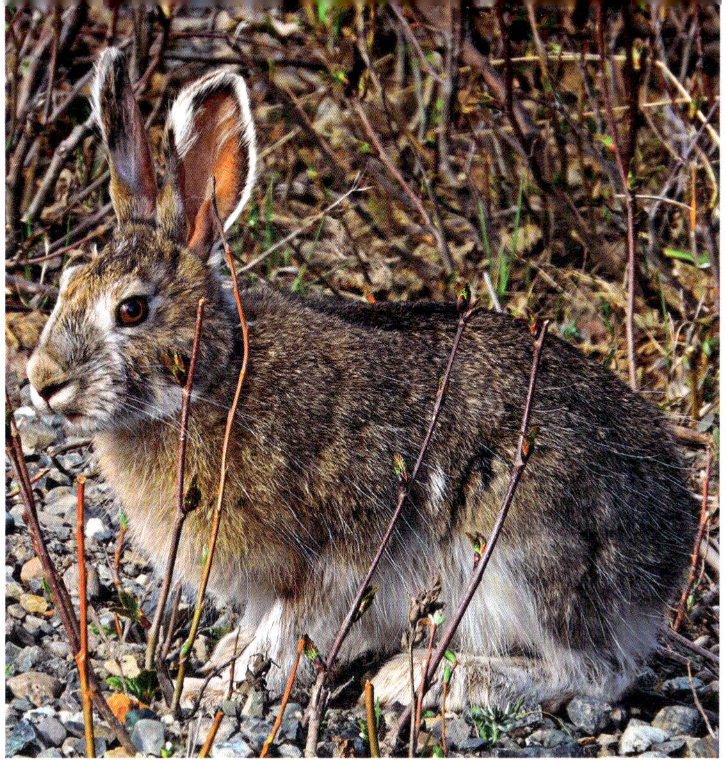

Snowshoe hare molting to summer coat. Susan C. Morse

When their populations are high, their feeding can dramatically shape plant communities, and their travel corridors and feeding platforms open channels in dense vegetation. These edges attract fish, birds, and other wildlife, making muskrats key—if less celebrated—players in wetland ecology.

Order Lagomorpha (hares, rabbits, and pikas)

Once considered rodents, lagomorphs were reclassified in the early 20th century when scientists noted their extra set of small upper incisors and their unique digestive system that produces two types of droppings: dry, hard pellets which are waste, and a soft, nutrient-rich paste which is re-eaten. Each to their own.

Hares are larger, with longer ears and legs, and unlike rabbits that are famous for sitting quietly when a predator threatens, hares prefer to RUN! They are very good at it.

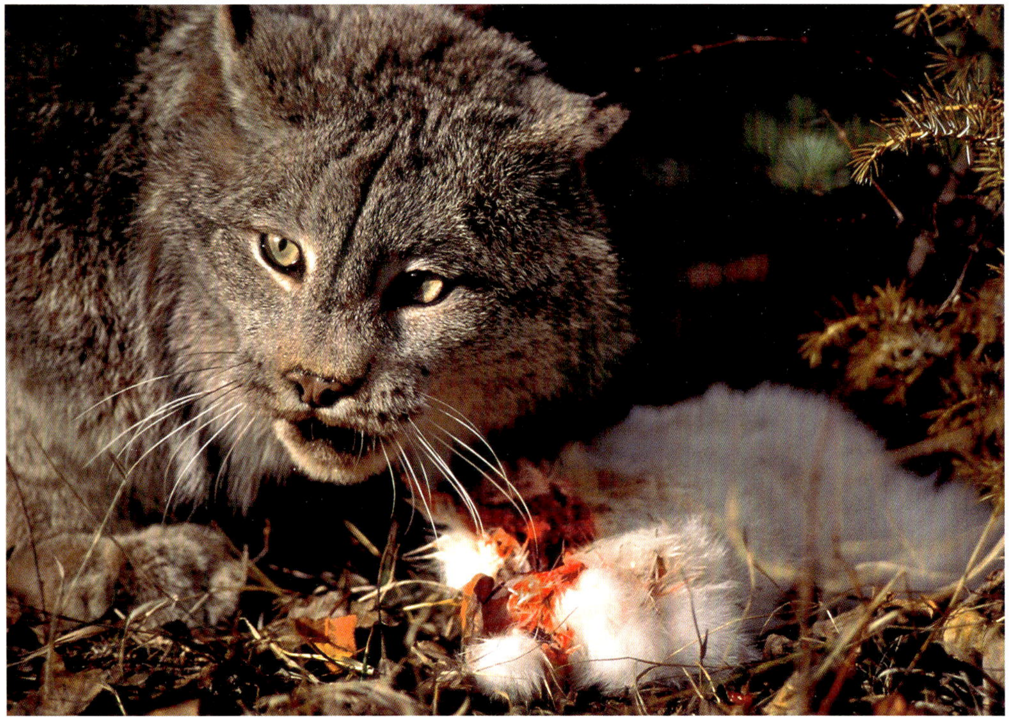

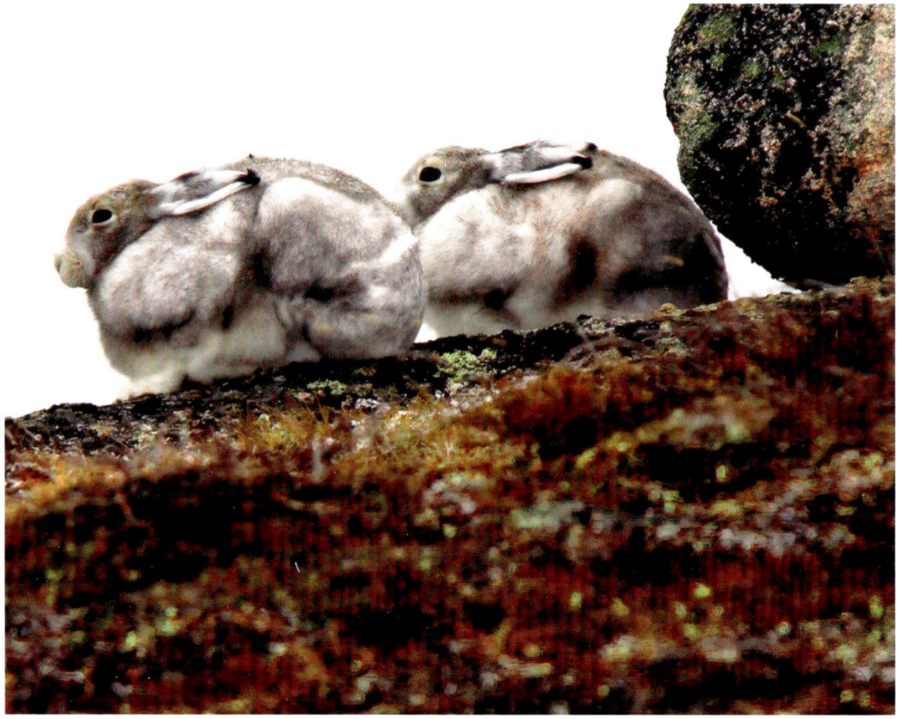

LEFT: Lynx tom eating a snowshoe hare. Susan C. Morse

RIGHT: Arctic hares in Ungava. Susan C. Morse

Snowshoe hare (*Lepus americanus*)

Named for their oversized, furry feet, snowshoe hares swap brown summer coats for white in winter, though rapidly shifting snowfall patterns from climate change may be disrupting that strategy. In summer, they browse greenery; in winter, buds, needles, and bark, favoring birch and aspen where available and, like many small herbivores, meat when they can get it—including hare (especially the young). They're prey for nearly everything from weasels to wolves, and especially lynx, which rely on them when hares are abundant.

As in a mysterious pattern for many species throughout the boreal forest, snowshoe hare populations swing from high to low in dramatic, roughly 10-year cycles documented since the 18th century in Hudson's Bay Company fur records. In peak years, they may be the most numerous vertebrates in the boreal forest; in low years, scarcely one per acre. Lynx populations track these boom-and-busts closely, a textbook predator–prey example that's far more complex in real life than the simple model taught in biology classes.

Arctic hare (*Lepus arcticus*)

Well-insulated for the far north, and adhering to Bergmann's and Allen's rules, Arctic hares are among the largest lagomorphs, up to 15 pounds and over two feet long, with small ears and noses and shorter legs than their southern kin. Most stay white year-round, but those in southern ranges, including parts of Ungava we visited, turn brown in summer.

Order Chiroptera, family Vespertilionidae (bats)

May 11, 2017: A stolen boreal river in bandit country—Upper Missouri River Breaks

After a long day on the river, I usually have no problem sleeping. Moreover, Bonnie and I had settled into our sleeping bags in unaccustomed luxury: a stoutly constructed wooden lean-to at a national monument camping ground. And yet, in the pitch dark, there were odd clicking sounds. Worse yet, they were right between Bonnie and me. They came in series. Then silence. Then, more clicks. What the hay? I turned on a flashlight and checked.

It looked like . . . well, something rude (yes, there was liquid too). It could be coming only from one direction. With some trepidation, I swung the flashlight up.

Bats, with about 925 species, are the second most diverse mammals, behind rodents. Yet only about six species live in the North American boreal forest. Bats are most common in the tropics and absent from the Arctic, so it fits a pattern. Many bats hibernate, and it simply may be difficult to find sufficient shelter for the severe boreal forest winters.

Bats are important insectivores, though the few we've seen seem overwhelmed in the boreal forest. Two boreal forest bat species are at risk.

Little brown bat (*Myotis lucifugus*)

IUCN Red List status: Endangered, Nearctic

Little brown bats were abundant in Vermont and in the boreal forest of North America through most of my life, but recently, their populations have plunged due to white nose syndrome, a fatal fungal infection that was first detected in New York in 2006. In 2008, the IUCN still assessed them as of Least Concern on the Red List; now, they are Endangered. The U.S. Fish and Wildlife Service is "reassessing."

Conversely, the northern long-eared bat (*Myotis septentrionalis*) is listed as Near Threatened on the IUCN Red List but is federally endangered in the US. It ranges through much of the Midwest and eastern US and the southeastern Canada. Possibly the boreal forest portion of its range is buffering the stresses it may be dealing with in more developed areas.

Other boreal forest bats

Silver-haired bat (*Lasionycteris noctivagans*)
Big brown bat (*Eptesicus fuscus*)
Hoary bat (*Lasiurus cinereus*)
Eastern red bat (*Lasiurus borealis*)

There are bats in the Eurasian boreal forest too. In a study conducted in Finland, six species of Palearctic bats were identified. There was evidence that their diversity varied with the degree of human alteration of the habitat and that diversity was most dependent on mixed, coniferous and deciduous, old-growth forests.

Bats in the belfry, Missouri River Breaks lean-to. I swear they were laughing.

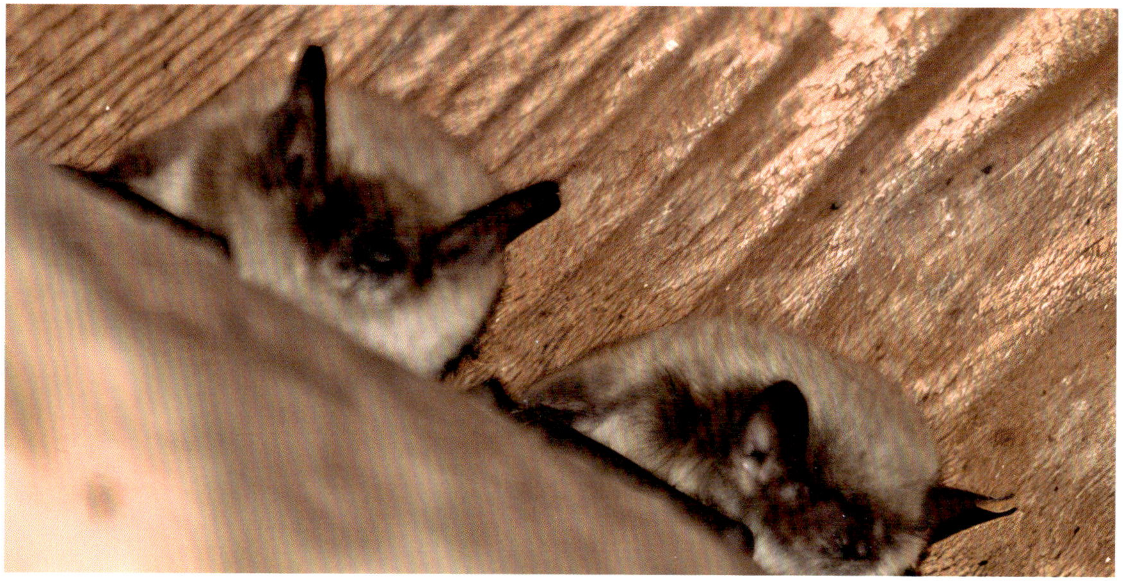

LEFT: Ghost cat in spruce cover—Canada lynx. Susan C. Morse

RIGHT: Cougar tom with spruce hat (he was a collared research cat treed by hounds). Susan C. Morse

FACING: Lynx in summer coat. *Sun Catcher*, acrylic, Lindsey Foggett

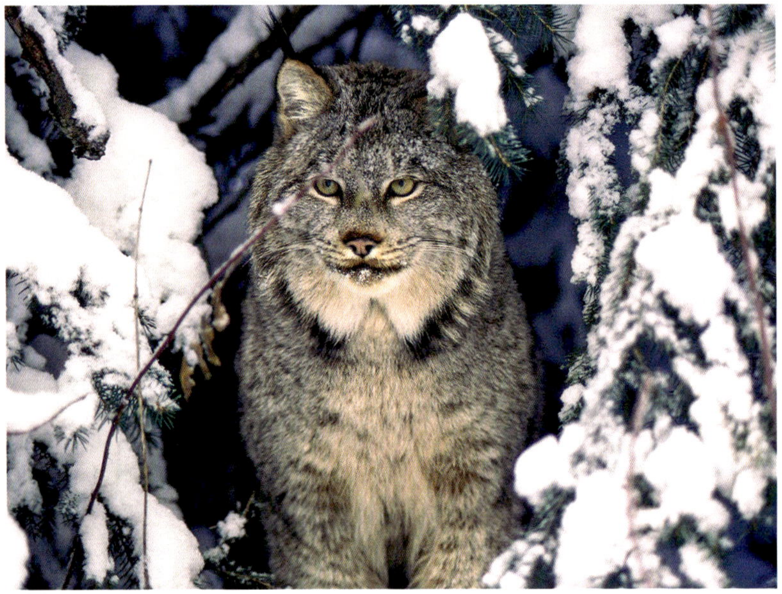

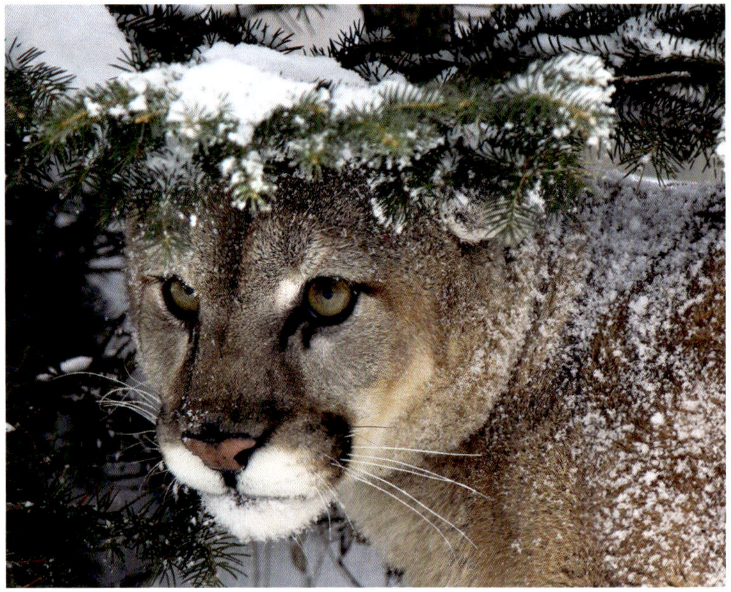

Order Carnivora (carnivores)

Somebody has to eat all those squirrels, lemmings, hares, and caribou. Carnivores are distinguished largely by their teeth: six incisors flanked by large, sharp canines (fangs) and premolars, often modified into self-sharpening shearing teeth (carnassial) in both the upper and lower jaws (maxilla and mandible). All the better to eat you with.

Family Felidae (cats)

Eurasian lynx (*Lynx lynx*): IUCN Red List: LC, Palearctic

Canada lynx (*Lynx canadensis*): IUCN Red List: Least Concern, though listed as endangered in its former range in the US (except for Alaska where they are common and Hawaii where there never were any), Nearctic

Bobcat (*Lynx rufus*): IUCN Red List: LC, Nearctic range only just into the southern boreal forest

Mountain lion or cougar (*Puma concolor*): IUCN Red List status: Least Concern, though extirpated throughout much of their original range Nearctic and Neotropical

Siberian tiger (*Panthera tigris tigris*): IUCN Red List status: Endangered, Palearctic range: extreme eastern Siberia, northern China, and possibly North Korea

Amur leopard (*Panthera pardus orientalis*): IUCN Red List status: Critically Endangered, Palearctic range: extreme eastern Siberia, even more restricted than the Siberian Tiger

CONSIDERING THAT a housecat can be invisible until dinner time, it should be no surprise that seeing cats in the wild is a relative rarity. It takes an expert tracker like Susan C. Morse to do it with any regularity. Of the boreal forest cats, the only

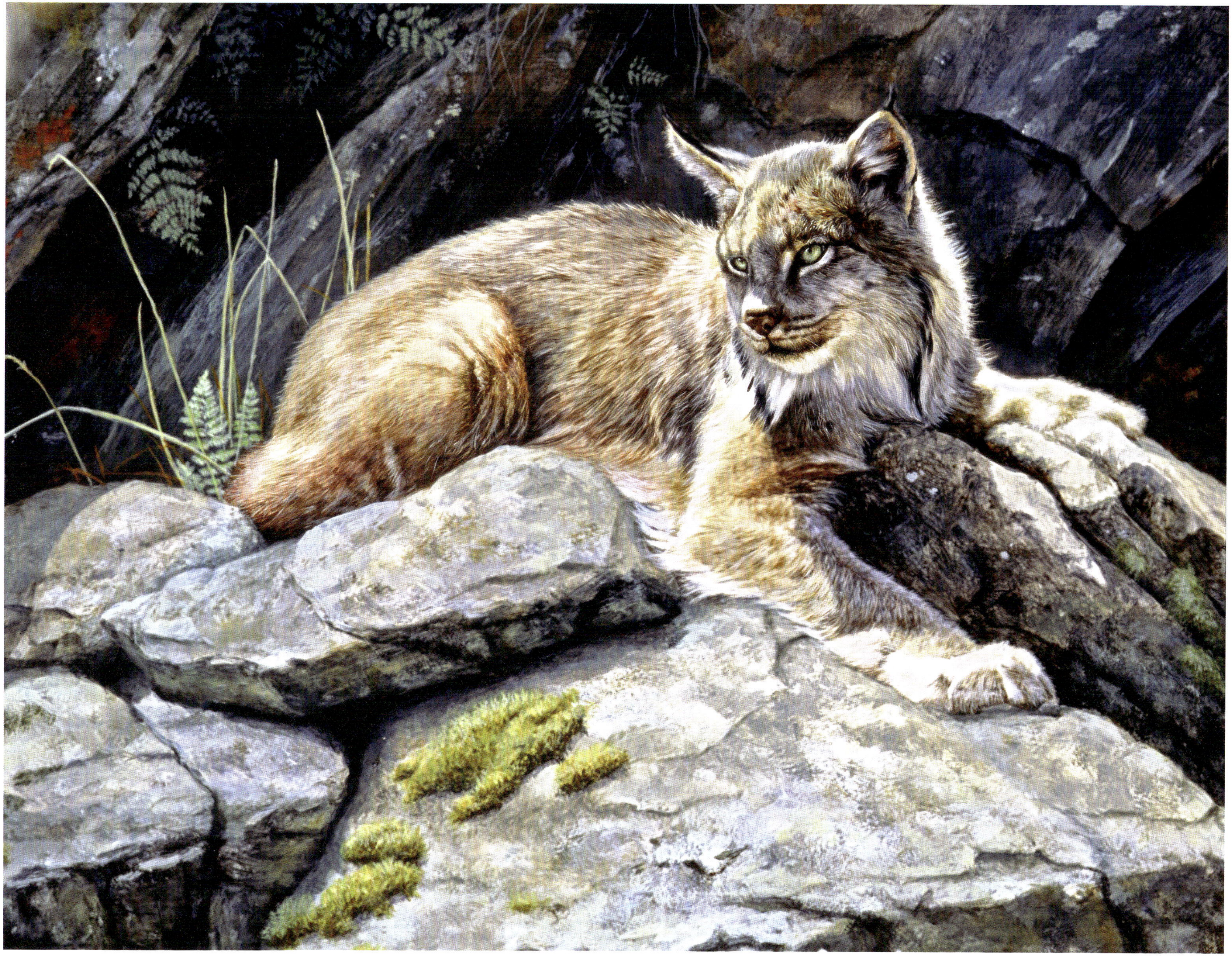

Bobcat. *Of a Secret Nature*, oil, Nancy Howe

FACING: *Lynx in Snow*, acrylic study, Robert Bateman

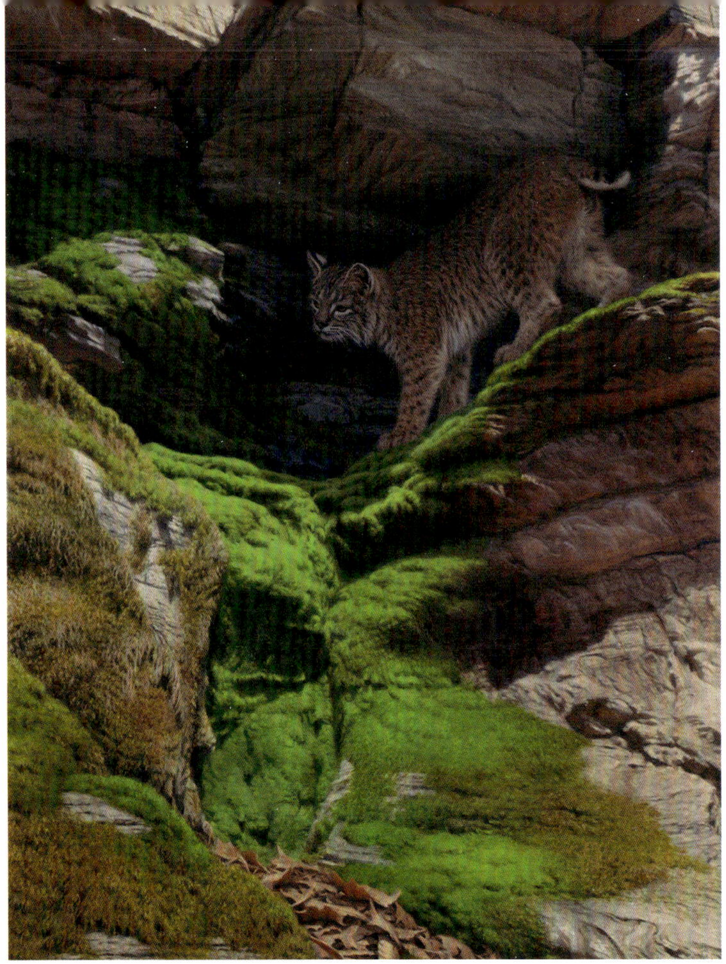

species we've seen on any expedition is the Canada lynx. As previously mentioned, lynx prey preferentially on snowshoe hares, and their respective populations are linked in a marked boom-and-bust, predator–prey cycle. The year 2002 must have been a "boom" year because I saw four lynx on one of the shortest journeys in this project: the aborted Harricana River trip to James Bay.

October 7, 2002: Harricana River, Northern Quebec, stranded by dangerously high water in a gorge

The next morning was strangely dark, and the usual drumming rain on the tarps was silent. Hopeful, I looked out, only to discover a thick blanket of snow even under the forest canopy, and it was still coming down heavily. That was the last straw. I had been debating the wisdom of continuing, and the snow seemed to be a sign.

Leaving my camera safely in the tent, I went to fill my canteens for breakfast. As I knelt by the rock pools on the ledge, meditatively watching the peaceful snowfall over the raging rapid, a lynx materialized among the trees on the opposite side of the gorge. He looked briefly toward me and then back out over the river for a few seconds before turning back, vanishing into the snow-laden boughs. As I cursed not having the camera, something suddenly struck me: This gorge was a worthy destination. My turning back now was okay too. That was final.

There is a good reason that lynx are the only cats we've seen on these trips. They are the only cats to see in most of the boreal forest. Mountain lions, tigers, and leopards are rare and found only in a few small regions of the biome. Bobcats are very common in their range, but overlap only slightly with the southern boreal forest. The Canada lynx and Eurasian lynx, on the other hand, occupy the entire ecosystem, so even if they are not technically the reigning monarchs of the forest, they are the effective stewards of the boreal forest feline kingdom while the Siberian tiger hopefully recovers.

The Eurasian lynx is the largest of the four lynx species, followed closely by the Canada lynx. While there are reports of very large bobcats, they are generally notably smaller than lynx (the Iberian lynx is the smallest, is very rare, and restricted to, unsurprisingly, Iberia).

Sadly, the Siberian (aka Amur) tiger is endangered, and the Amur (also aka Siberian) leopard is critically endangered. There are strong efforts at conserving them. Hopefully, it is not "too little, too late." Their populations seem to have stabilized or even grown over the last 20 years due to concerted work between the Russian and Chinese governments and conservation NGOs.

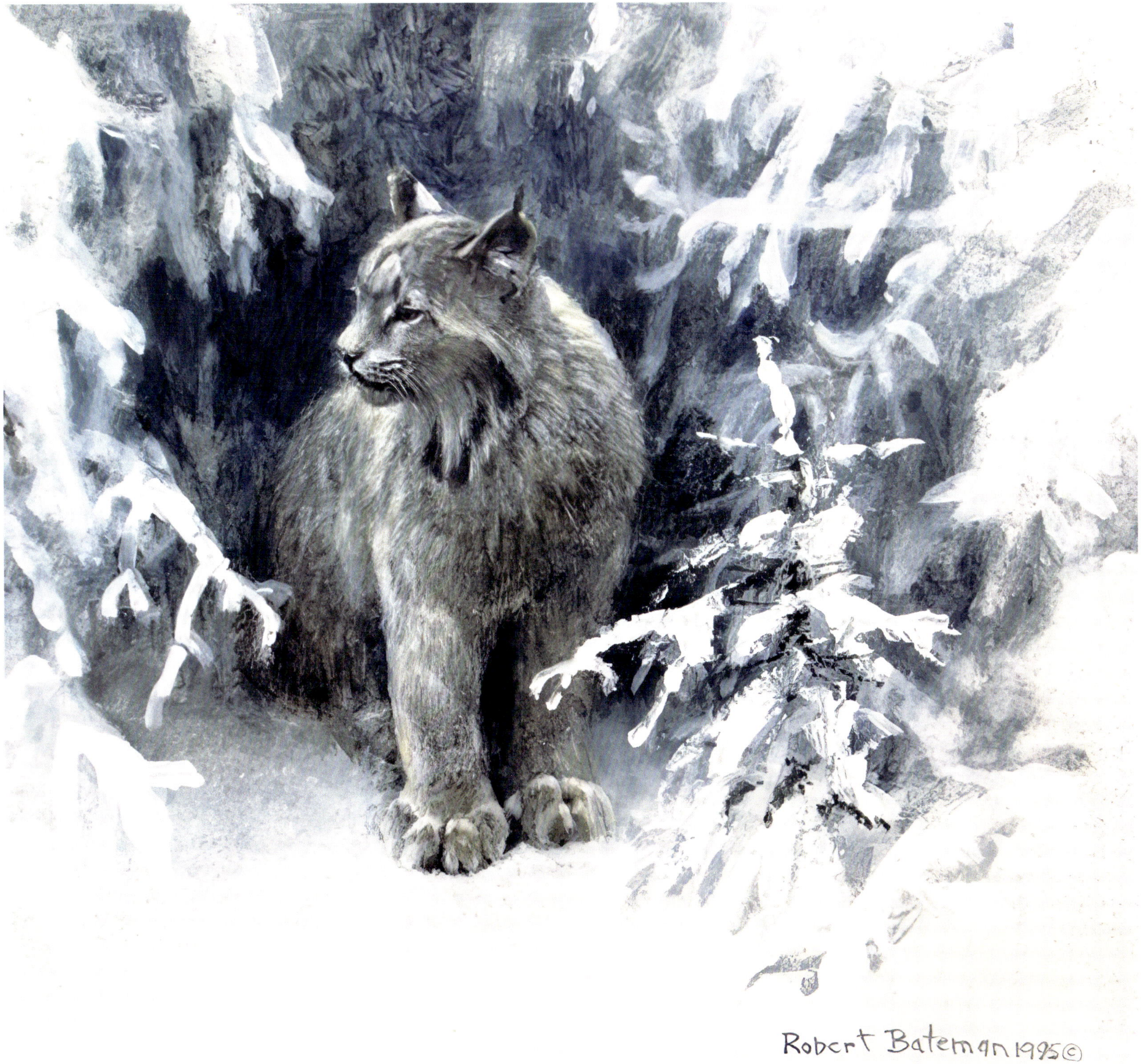
Robert Bateman 1995©

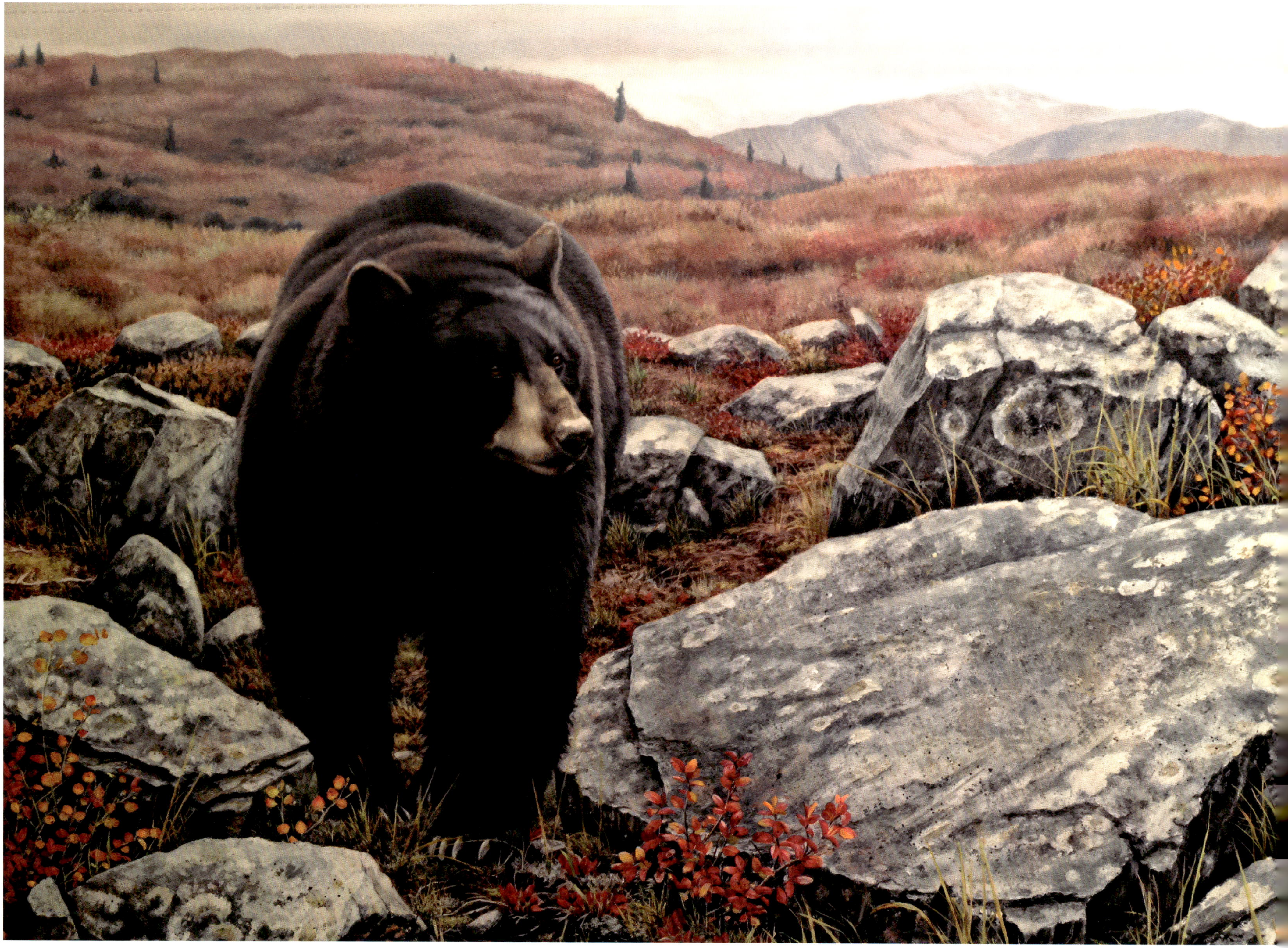

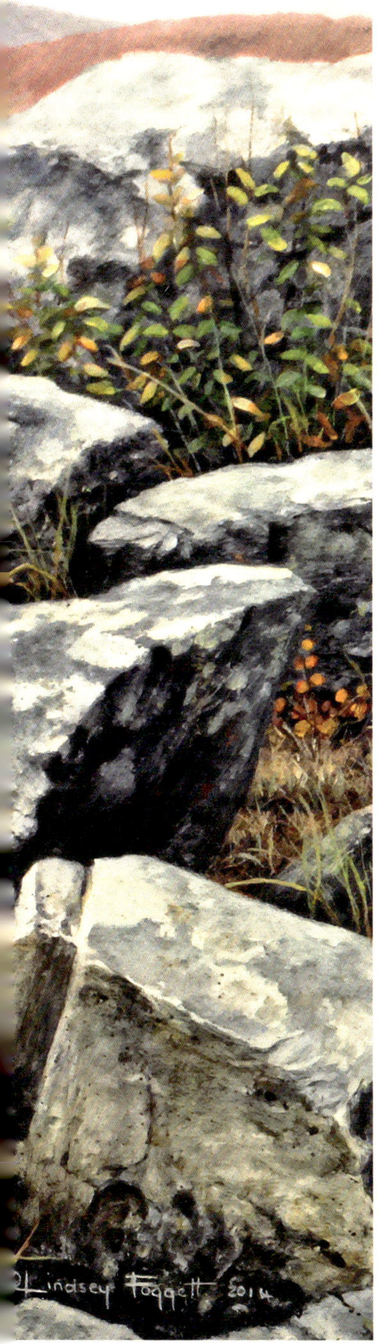

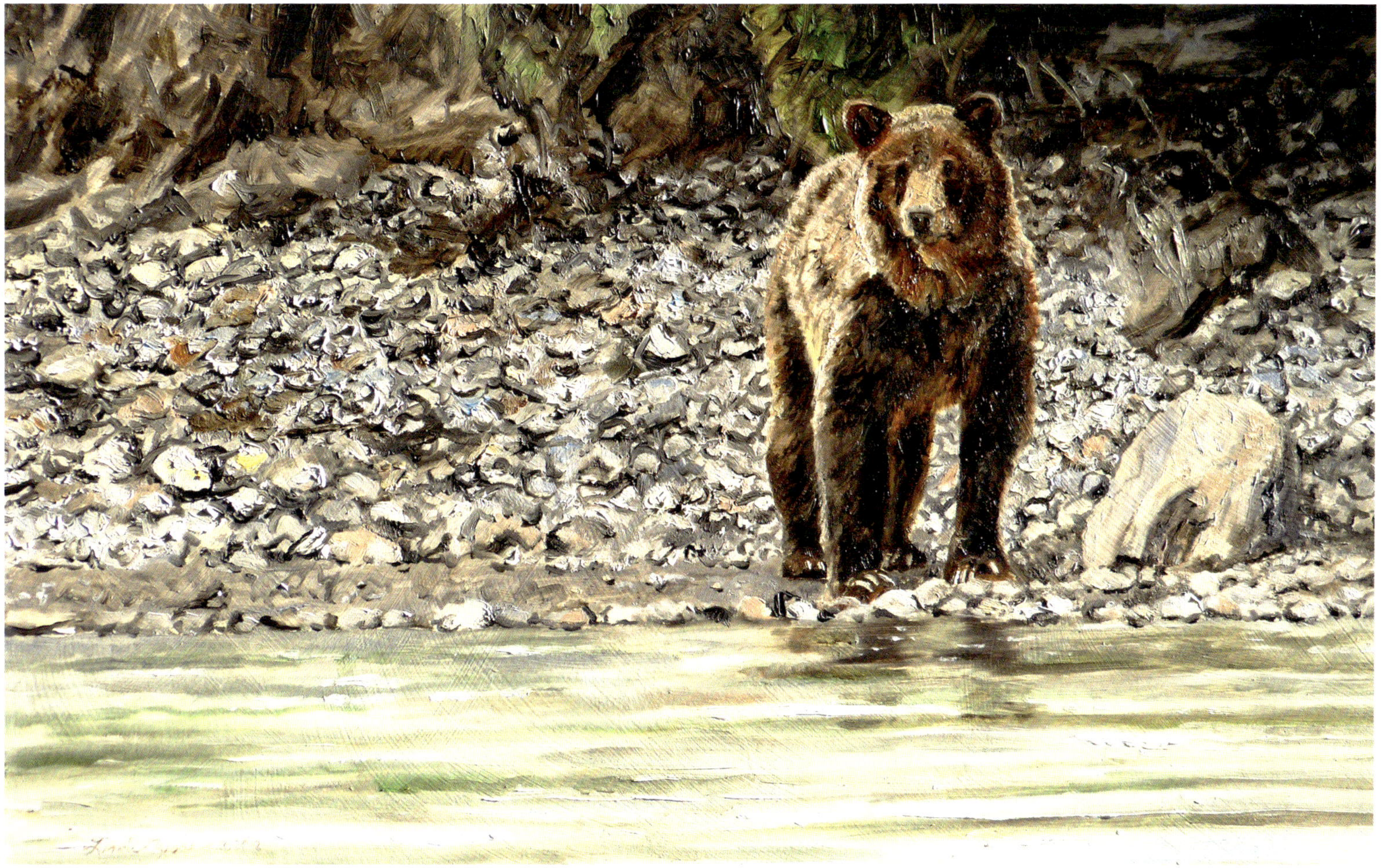

Young grizzly on the Noatak River. *Onlooker*, oil, Linda Besse

FACING: Black bear above the George River. *Autumn Trail*, acrylic, Lindsey Foggett

Missinaibi River black bear.
Trouble Bruin, acrylic, Rob Mullen

FACING: Black bear study. Graphite, Cole Johnson

Family Ursidae (bears)

American back bear (*Ursus americanus*); Asiatic black bear (*Ursus thibetanus*), aka moon bear: IUCN Red List status: Vulnerable, Palearctic

Brown bear (*Ursus arctos*): Eurasian brown bears (*Ursus arctos arctos*) and grizzly bears (*Ursus arctos horribilis*) are respective Palearctic and Nearctic subspecies of brown bears. They are not separately assessed by the IUCN. Grizzlies have suffered a massive reduction in their former range in North America including regional extinctions in California and Nunavik/Labrador. They are protected to varying degrees under the Endangered Species Act in the US.

Polar bear (*Ursus maritimus*): IUCN Red List status: Vulnerable, Holarctic

September 8, 2001, late morning: Lower Missinaibi River—the beginning of the Hudson Bay Lowlands below Thunderhouse Gorge, Hell's Gate, and Long Rapids: Working against the chill wind and rain, I had gotten into a paddling trance. Or it could have been something more insidious—likely an early sign of the hypothermia that struck unmistakably at the end of the day. (Hypothermia is a leading cause of death in outdoor sports and is discussed in "Being There" in the online appendix.)

Lining up for the open *V* (a funneling of the current that is usually a safe route) of a small rapid, I noted a shoreside rock "river right" moving along a gravel spit and headed for my passage. For several seconds, I considered how to avoid it. But rocks don't move! Fully awake, I realized that the large bear and I would likely meet in the *V*, so I spun into an eddy "river left" near shore. I had done a 180° eddy turn, so I had to crane around to watch the bear.

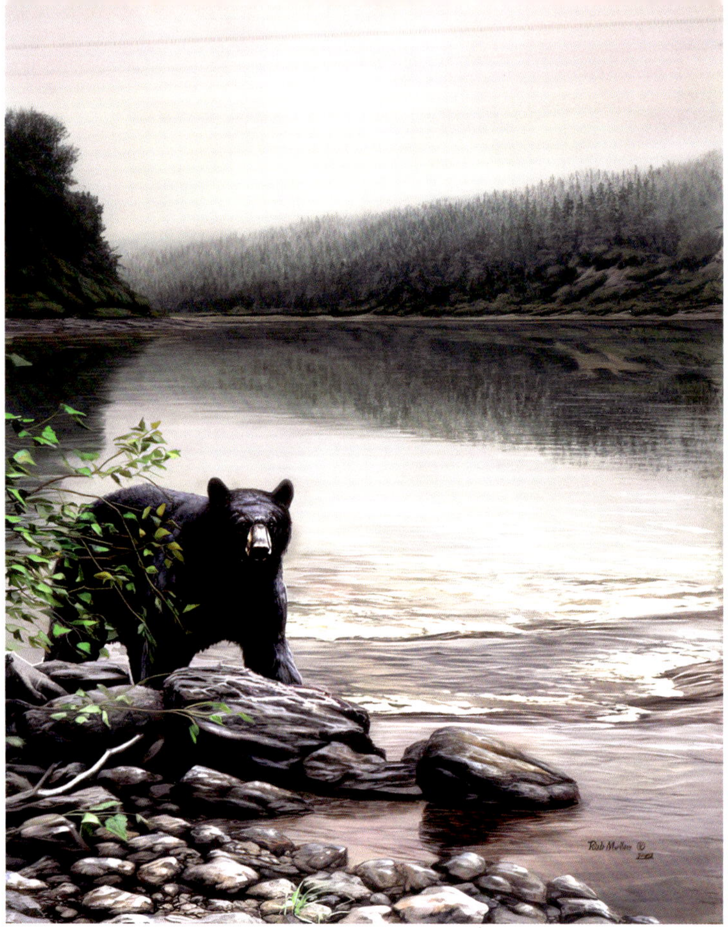

He was large and walking across the river without swimming. Despite the rain, I dug out my camera. Contorted as I was and with the canoe jostled by the current, it was difficult to compose a Pulitzer Prize–winning photo, and while I fussed, the bear reached the left bank. I expected he would flee once on shore, so, absorbed in steadying the camera, for an instant, I was encouraged when he didn't. He looked right at me. Then he headed up the bank toward me. With a growing distrust of his intentions, suddenly, the photo didn't seem important. Stowing the camera and regretting not having a weapon, I peeled out of the eddy and headed for the *V*. That did it. To my considerable relief, he wheeled and lumbered up into the brush.

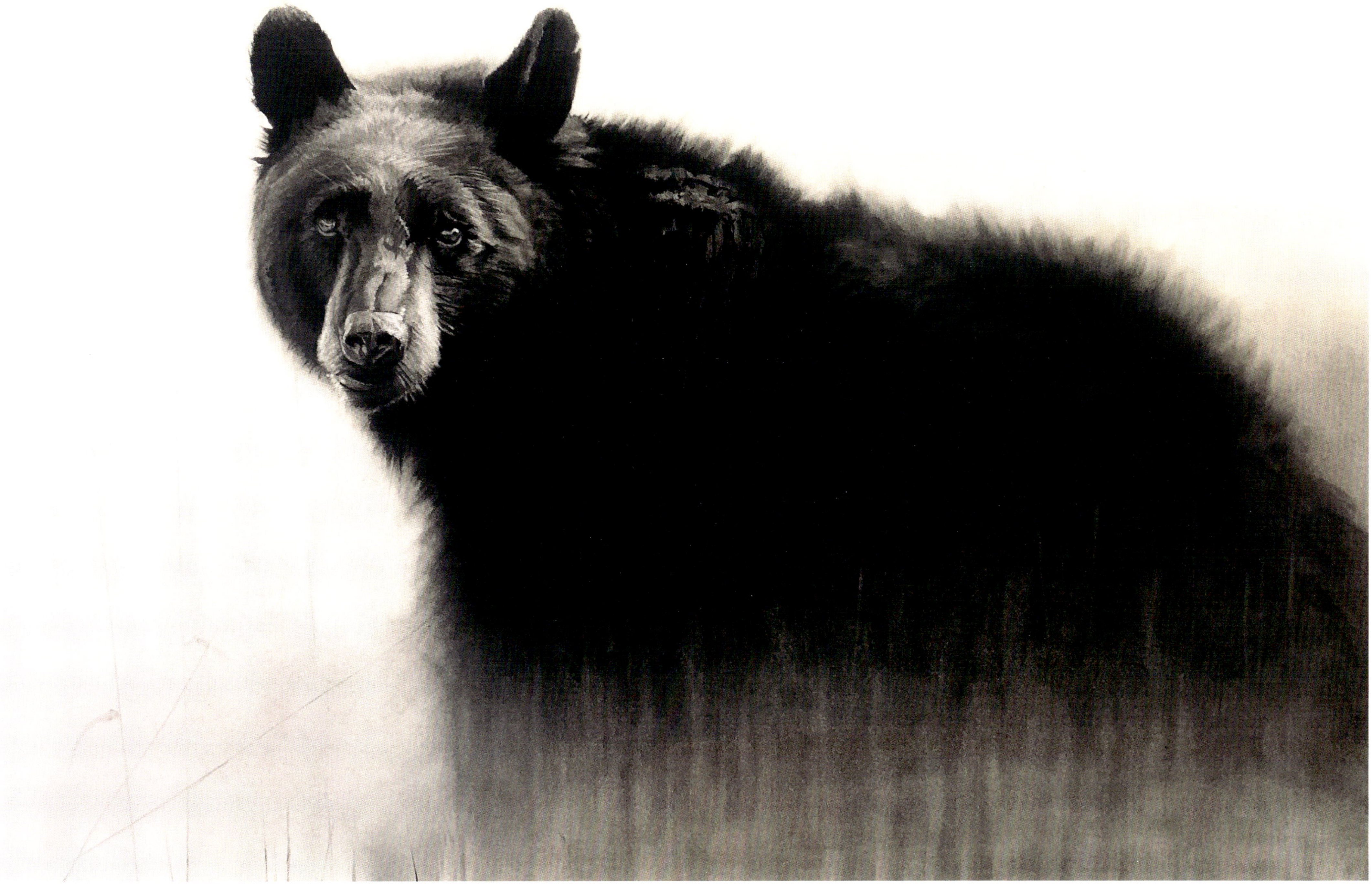

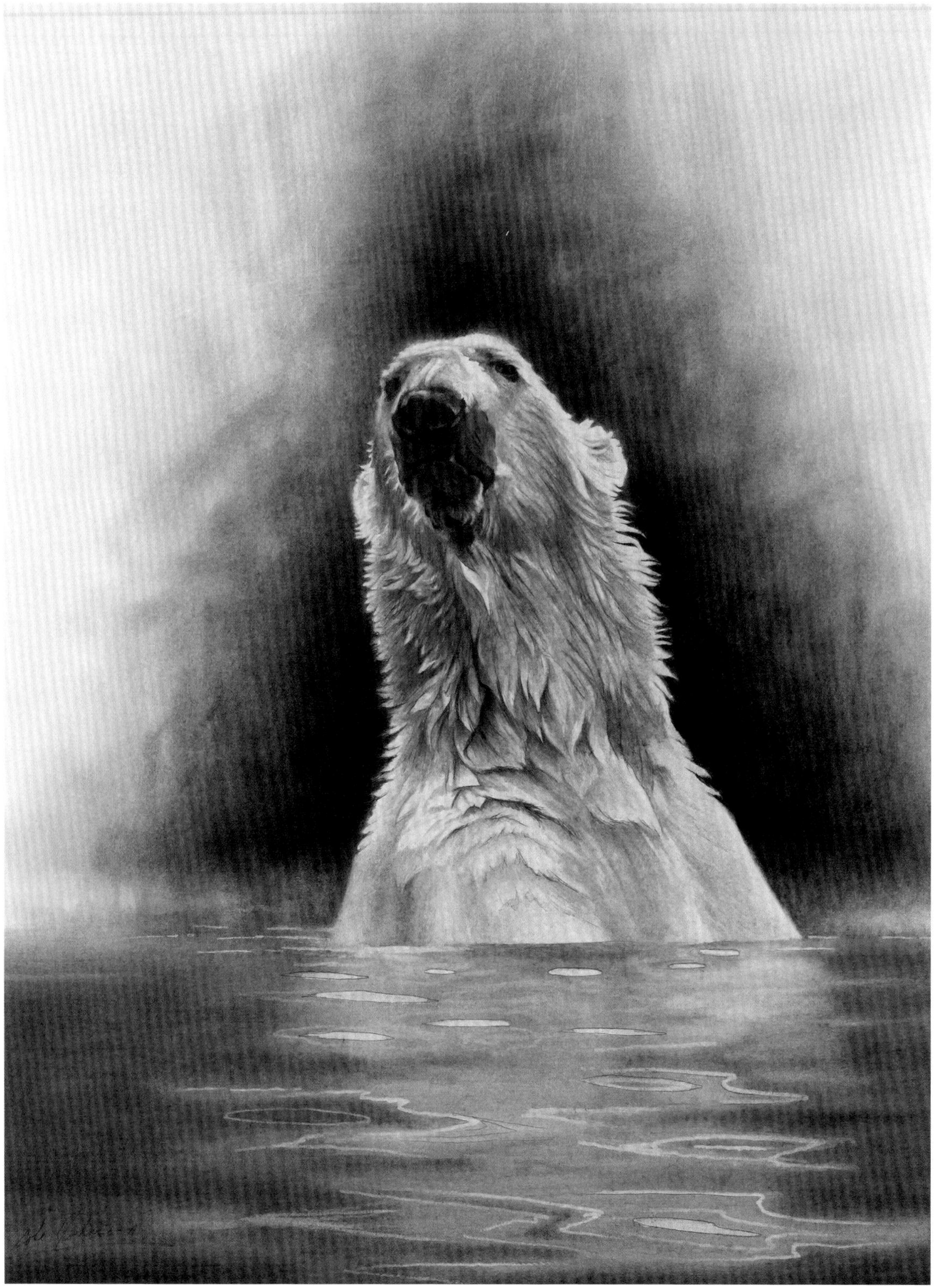

September 13, 2001, late afternoon: Cochrane, Ontario

At the Ontario Provincial Police office, an officer thanked me for letting them know I was off the river. I was the last person they knew about, and they would close the book on the season. An Ontario Ministry of Natural Resources biologist was there having coffee, and the two of them asked about my trip. The biologist perked up when he heard about my bear encounter. In a Canadian accent, he said, "Oooh, given where you were, that probably was a wild bear."

I thought he was pulling my leg. "What other kind of bear would it be?"

They laughed. "Wild bears are what we call bears that have never seen a human before. They can be dangerous." He speculated that it was probably peeling out of the eddy that discouraged him: "You suddenly went from being a three-foot thing [the canoe with me kneeling in it, seen end-on] that was interesting, to a fifteen-foot thing [the whole canoe] that was scary."

There is something about bears. Bears are intelligent, have distinct personalities and moods, eat like us, walk on their flat feet like us, can stand up like us, can be adorably funny and cute or deadly, and are threaded into our cultures. And I seem to bump into them a lot. I've had 54 grizzly and uncounted black bear encounters, several of which have been "up close and personal." By contrast, I've seen maybe a dozen wolves and half a dozen lynx, all brief meetings. Interestingly, back home, our town's sobriquet is The Land of Boulders and Bears, and true to the title, we often have bears in the yard (especially since we live in the gone-wild remains of an apple orchard), and the glacier does seem to have burped out a special load of boulders here.

Even though the gulf between us is smaller than we may have been taught to imagine and regardless of their sentience, when in their country, it is critical to remember that they are bears, and at the top of the food chain if so inclined.

Vegetation being scarce on ice floes, polar bears are generally so inclined. However, other species' frequently omnivorous habits can make them quirky apex predators. Fearsome grizzlies will graze on grass and salivate over white bark pine (*Pinus albicaulis*) seeds, and we've watched big, glossy black bears on the tundra hillsides along the George River gorging on blueberries while lying down and lazily rolling to a fresh patch when they had stripped all within reach. Yet they have large, stabbing canine teeth set in powerful jaws, sharp claws, and muscled legs that can rip through a bear barrel, all designed to capture living prey, and the instinctive predator drive to use them.

This bear had been feeding on berries the entire time it took us to break our frozen camp on the opposite shore of a tundra pond. He stood up to watch us as we crossed to our next portage and later tried to take our lunch—while we were eating it. Cole grabbed his 12-gauge, and I grabbed my 300 mm (camera lens; he had gotten the earlier shots while I steadied the canoe, so it only seemed fair).

While several of our expeditions have traversed polar bear territory, and the Inuit in Nain told Cole and me that there were six polar bears on an offshore island (awaiting passing canoeists), when we arrived in Nain in October, we never encountered one.

Family Canidae (wolves, coyotes, and foxes, among others)

Gray wolf (*Canis lupus*)

September 13, 2004: Trip end at Elsas, Ontario (Kapuskasing Lake): Coming to the warped dock at Ron and Jean Quigley's bush bed-and-breakfast (B&B) from where we were catching the train the next day, we heard strange, yelping cries. Ron, who was on the dock, temptingly waving a cup of coffee at us, nonchalantly said, "Oh, that's just the mother fox, she's upset."

That got our attention. "Upset about what?"

"Oh, the wolves."

As it turned out, Jean had rescued and nursed a female wolf back to health after it had been hit by a train the year before. Released, the wolf had gone its way but then returned with her new family. The female was quite tolerant of Jean, as were her pups, who were gangly teenagers when we met them. But her mate, a big white male, would have nothing to do with people.

After dinner, while the others settled in for drinks, I returned outside in the drizzling dark to finish prepping the bags for their possible rough handling on the train in the morning. I had a headlamp, and a light on the building spread a glowing halo through the fine Scottish mist (precipitation that can't decide if it is very light rain or heavy fog). I had two canoe packs open and was absorbed in impact-proofing the contents of one when I smelled something oddly familiar... wet dog! With a touch of apprehension, I glanced up. The big male with his dripping wet, dirty white coat was not more than three feet away—with his head deep in the other bag. I must

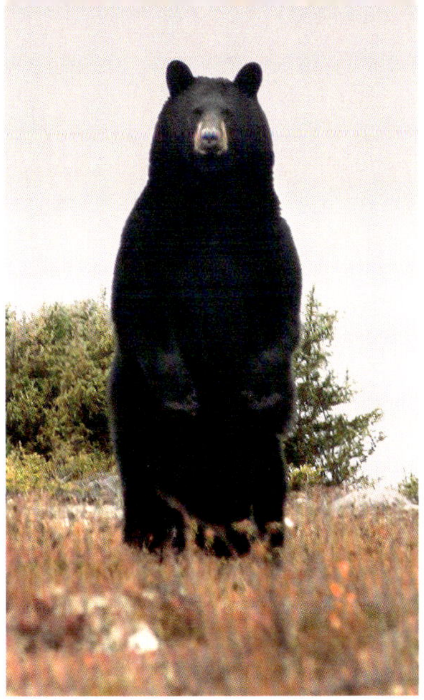

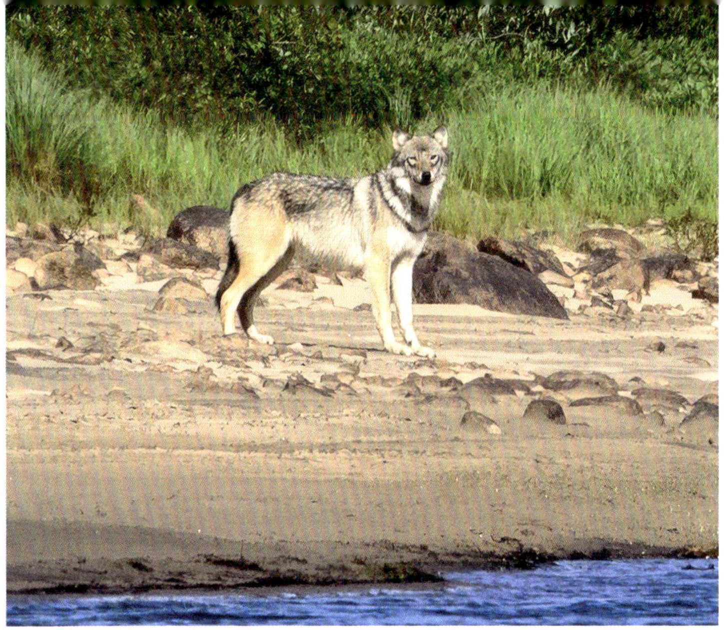

LEFT: Labrador Barrens black bear. Cole Johnson

RIGHT: Wolf on "our" lunch beach, Rivière aux Mélèzes, Nunavik, Trans-Ungava Expedition. Stephen Gorman

FACING: *Ice Bear*, charcoal, Cole Johnson

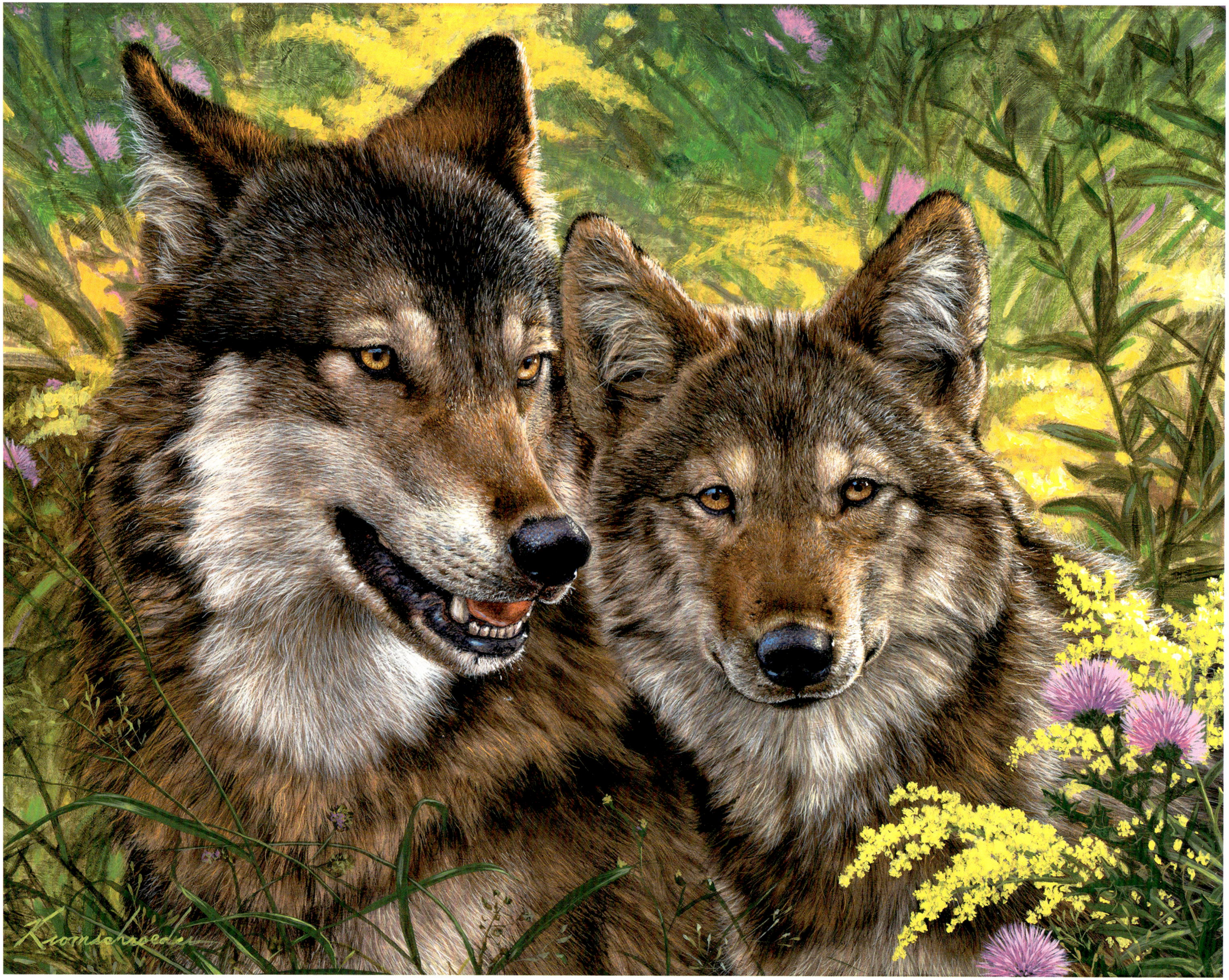

have gasped because the wolf whipped his head up. We stared at each other for an intense moment, before he wheeled about and exploded away into the soggy night before I could even start breathing again.

Over the years, dozens of wolf subspecies have come and gone with new DNA techniques and different philosophies of wolf researchers and taxonomists. For our purposes, we will go with a 2022 listing of five subspecies for North America, two of which inhabit the boreal forest: the northwestern wolf in the west (*Canis lupus occidentalis*) and the Great Plains wolf (*Canis lupus nubilus*) in the east. When Cole Johnson and I were in Labrador, there was still a Labrador wolf (*Canis lupus labradorius*), though as of 2022, they are now part of *Canis lupus nubilus*. The Great Plains wolf, as unsatisfying as the name may be for a Barrens predator, is widely distributed in eastern North America, though, ironically, not in the Great Plains anymore. It is a large wolf, with big males reaching 150 pounds.

That comports well with the wolves that nosed through our camp on a dried-up lake in the Labrador Barrens while we were scouting our next portage. The largest sunk into the gravel as deeply as I did, and his front feet were nearly the size of Cole's hands.

The guy who checked out our camp was, as Cole respectfully intoned, "a big dog."

The northwestern wolf (*Canis lupus occidentalis*) is the largest in the world, with a top weight of nearly 160 pounds. There are claims of larger wolves, but I suspect that one is inclined to overestimate their size in the field.

Wolves make up for being smaller than bears through strength in numbers. They are famed as pack hunters. They can and regularly do take down moose, even adults. Wolf packs can compete with grizzlies for kills, though such contests can go either way. One species that wolves cannot compete with in eastern Russia is the Siberian tiger. Where

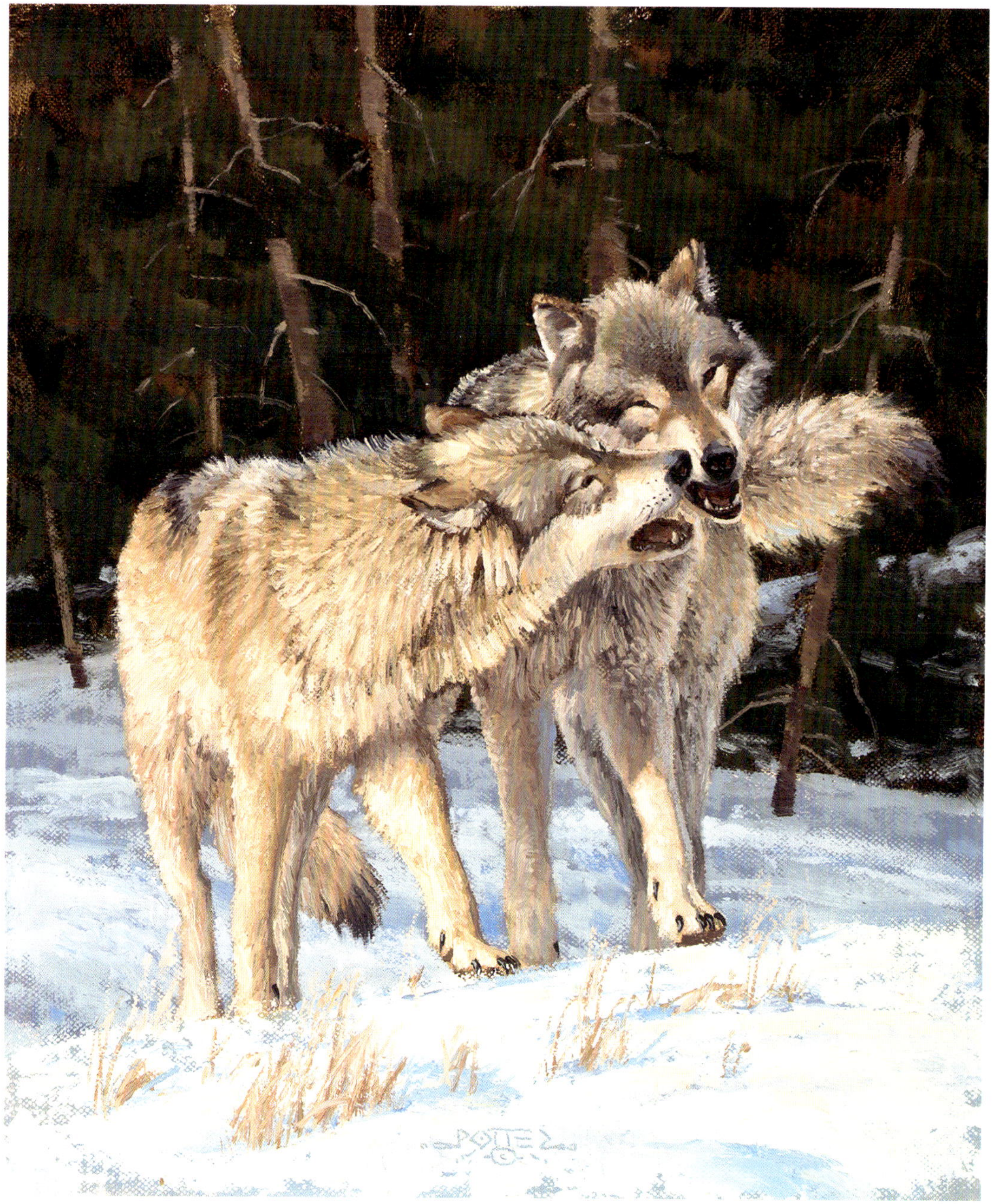

Gray wolves, Montana. *Play Date*, oil, John Potter

FACING: The rescued wolf and one of her sons, Elsas, Ontario. *Mama's Boy*, acrylic, Lee Kromschroeder

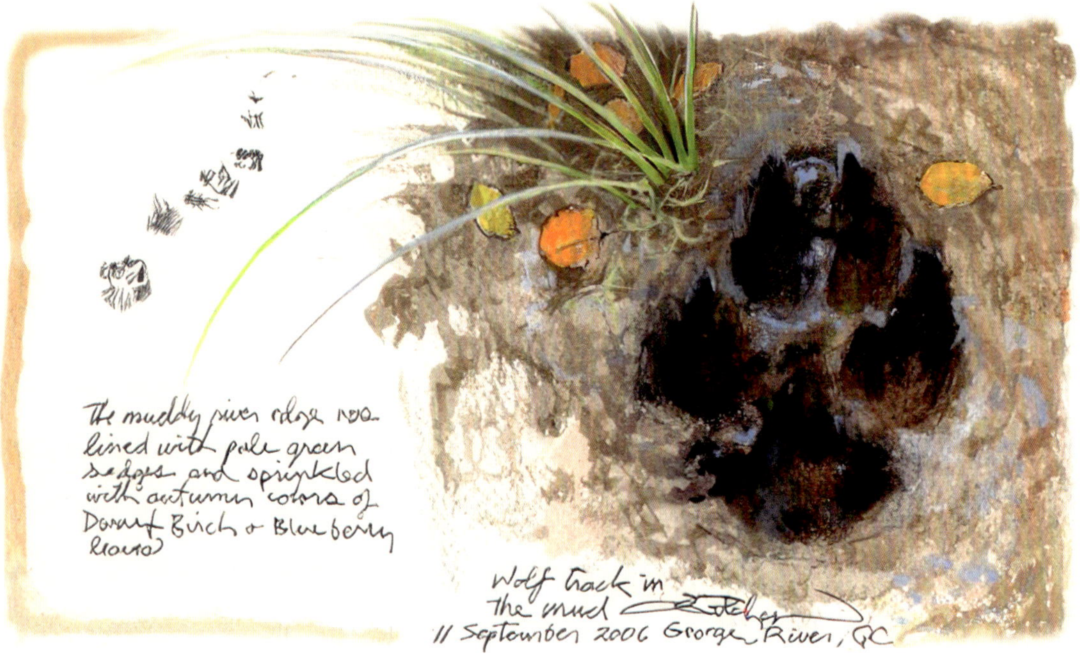

Wolf track in mud, George River. Acrylic field sketch, John C. Pitcher

FACING: Coyotes and field mouse. *Ruh Ro*, graphite, Cole Johnson

there are tigers, there are few, if any, wolves. While wolves will prey on a wide variety of smaller animals, they prefer ungulates when available.

Wolves' contributions to the ecosystems they inhabit are profound, changing the browsing impact herbivores have on plants by controlling their numbers and possibly changing their behavior, a phenomenon referred to as the "landscape of fear."

Cole had his own close encounter with a wolf a bit later on that Labrador trip. However, he was not simply preoccupied like I had been in Elsas, Ontario, but completely unconscious. After reaching the Kogaluk and finding a way to descend the walls of its 1,200-foot canyon (we found a bear trail that threaded around the sheer cliffs), we gratefully set camp on a beach next to the river and slept like logs.

In the morning, large wolf tracks came up to Cole's side of the tent, right next to where his head had been. Knowing dogs, we assume it probably investigated with its nose—with its teeth inches from his ear. Happily, it didn't mark the tent as its territory.

Coyotes (*Canis latrans*)

October 1962: Jericho Center, Vermont: It was a brief sighting, but the impression it made on my six-year-old self is vivid to this day. Up the hill in the pasture behind our house , my brother Frank and I watched in amazed excitement as two large, lean gray wolves loped along the wood line. They were out of a storybook: huge and wild. Utterly agog, we could hardly wait to tell our parents. Dad, who was the entire science department at Jericho High School (replaced by Mount Mansfield Union High School, my alma mater), casually dismissed them as coyotes. "Wolf" or "coyote" didn't matter to me. Nobody could throw cold water on the fire that the sight of such wild beasts ignited in my imagination.

Coyotes have expanded rapidly from their historic western US center in the late 19th century. They are now in the southern boreal forest from Alaska to Quebec. However, most of these coyotes are not the Wile E. Coyote of Road Runner fame. As they have expanded their range, they have interacted (and more) with wolves. What is now commonly called the eastern coyote is part wolf (Frank and I weren't all wrong). Opinions vary on how much wolf DNA has snuck in and from what subspecies: a nebulous consensus seems to be in the 10–25 percent range.

Eastern coyotes are larger and more likely to form packs than their western cousins. They are more like wolves. From the perspective of genes, you could see it as the "return of the wolf" to the eastern United States in a form that can more easily coexist with people after their extirpation over a hundred years ago.

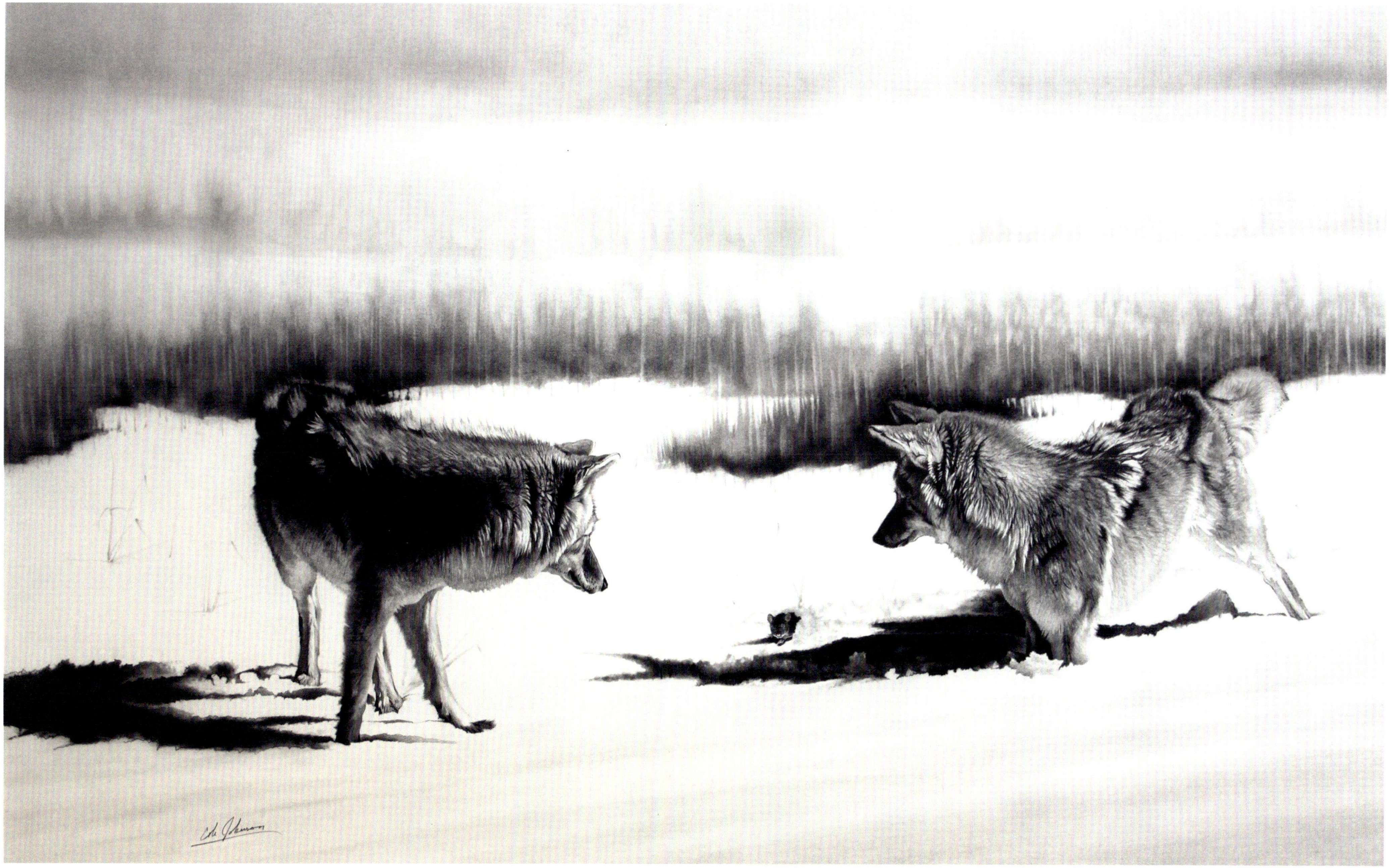

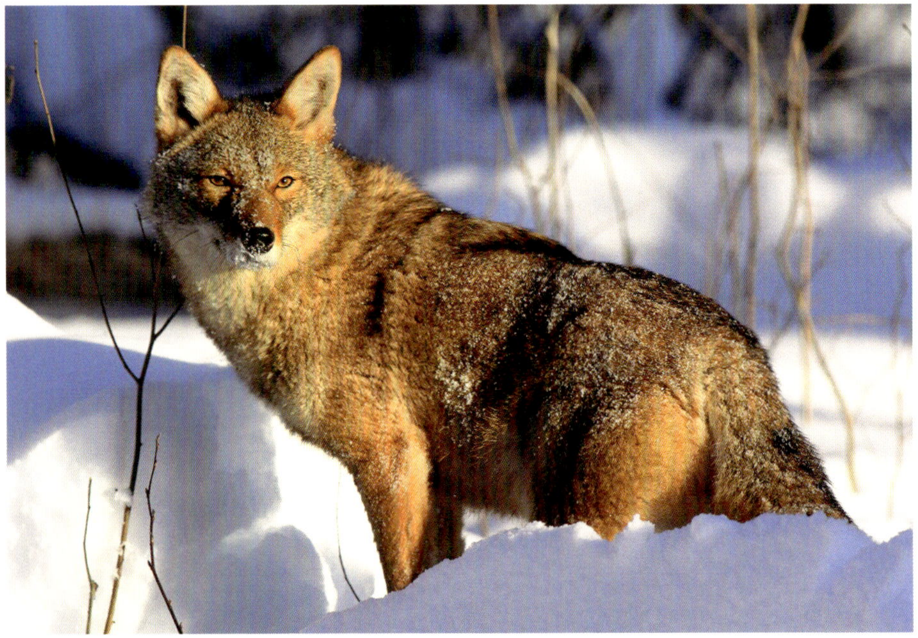

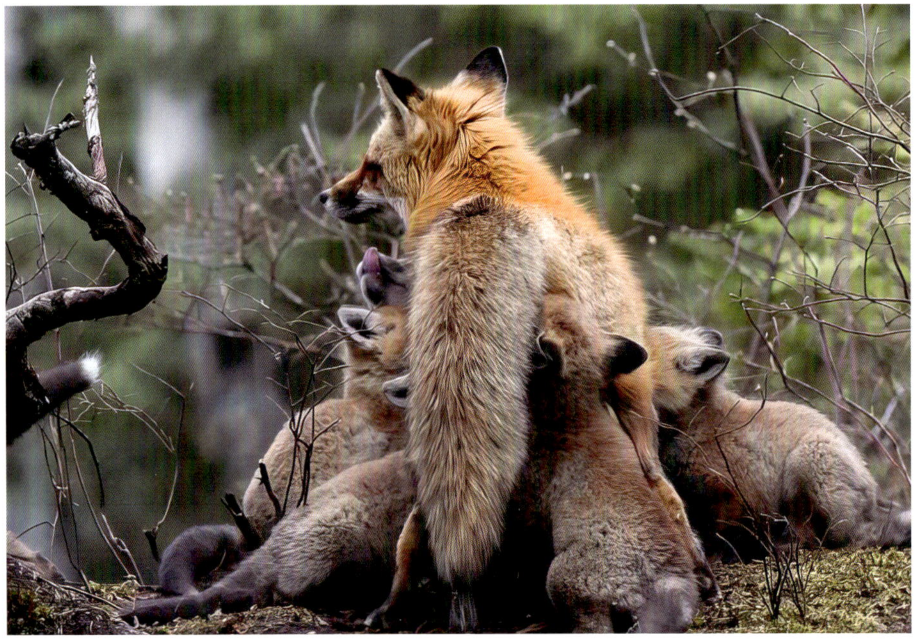

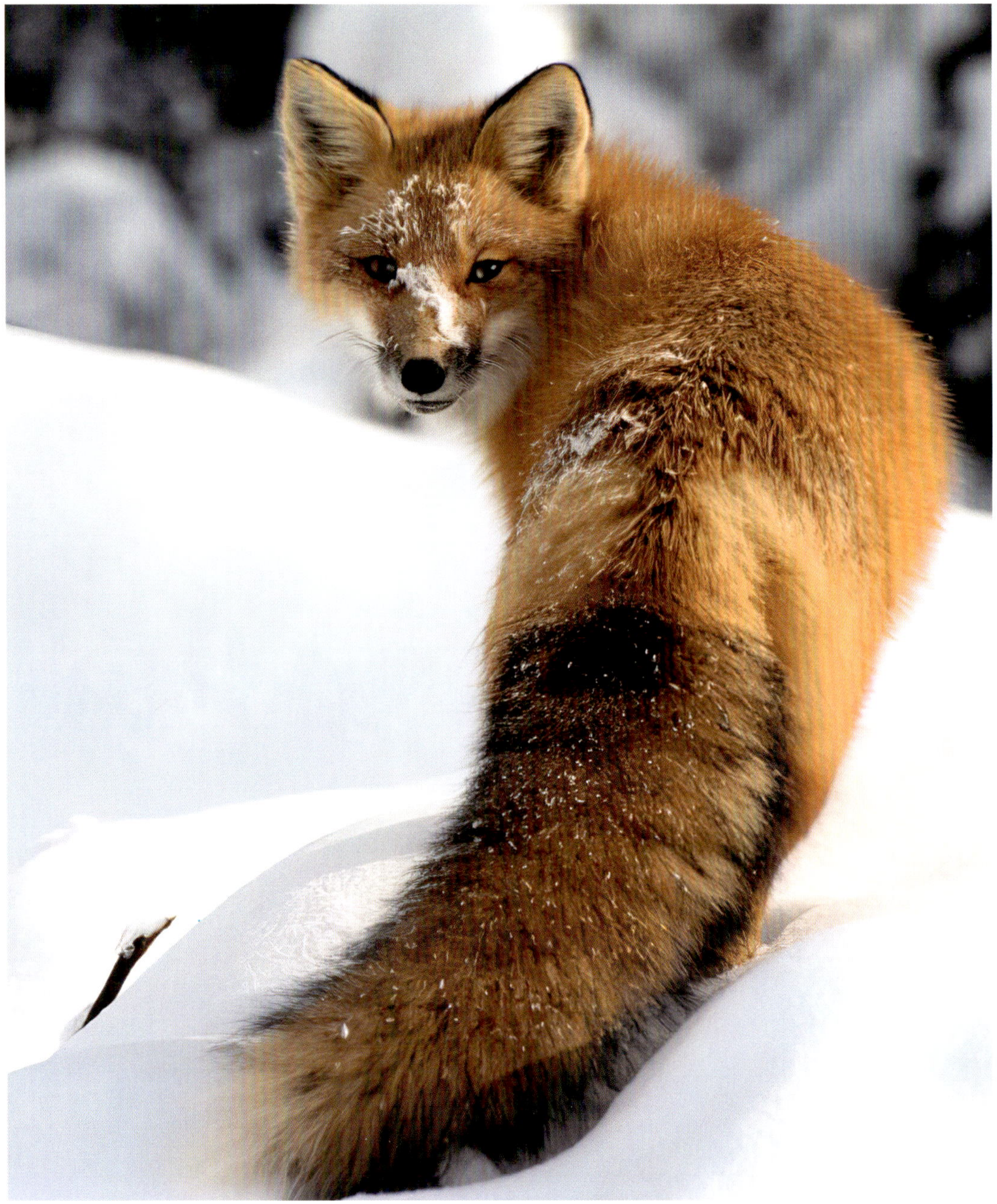

Foxes

Red fox (*Vulpes vulpes*); Arctic fox (*Vulpes lagopus*)

September 9, 2005: George River, Nunavik: It is difficult to avoid setting camp on animal tracks: There always seems to be someone using the beach as a promenade. Four years earlier, I had nearly come to grief on a small beach with big moose tracks, but this was a big beach with little fox tracks. What could go wrong?

As it turned out, not much. It was amusing though. Usually, we go to bed with the sun on river trips, but we were all up around the dying campfire watching the northern lights when, right through the middle of us all and right next to the coals, trotted the fox. There was no hesitation, no acknowledgment that we were there, and not a single glance back as it continued down the beach!

I was too startled to definitively ID it as a red or Arctic fox, especially in the dark. The tip of the tail would have been the easiest summer field mark as red foxes usually have a white tipped tail. However, they don't always, and both species can be gray to dark gray in the summertime (Arctic foxes are white in the winter).

Ranging throughout the entire boreal forest worldwide and topping out at about 10–25 pounds and occasionally 30 (usually the size of a big house cat), the red fox is the largest fox and preys mostly on rodents, rabbits, hares, and almost anything else it can catch. The Arctic fox's range is also circumpolar, but as its name suggests, it is limited to the more northerly reaches of the boreal forest and across the Arctic. One Arctic fox was radio-tagged and all by herself ranged almost worldwide, from the Norwegian island of Spitsbergen to Hudson Bay. Interestingly, foxes share a trait with small felines: They have vertical pupils in their eyes.

Red foxes have undergone a dramatic range expansion in the last hundred years, especially into the tundra and the Arctic, likely due, at least in part, to climate change.

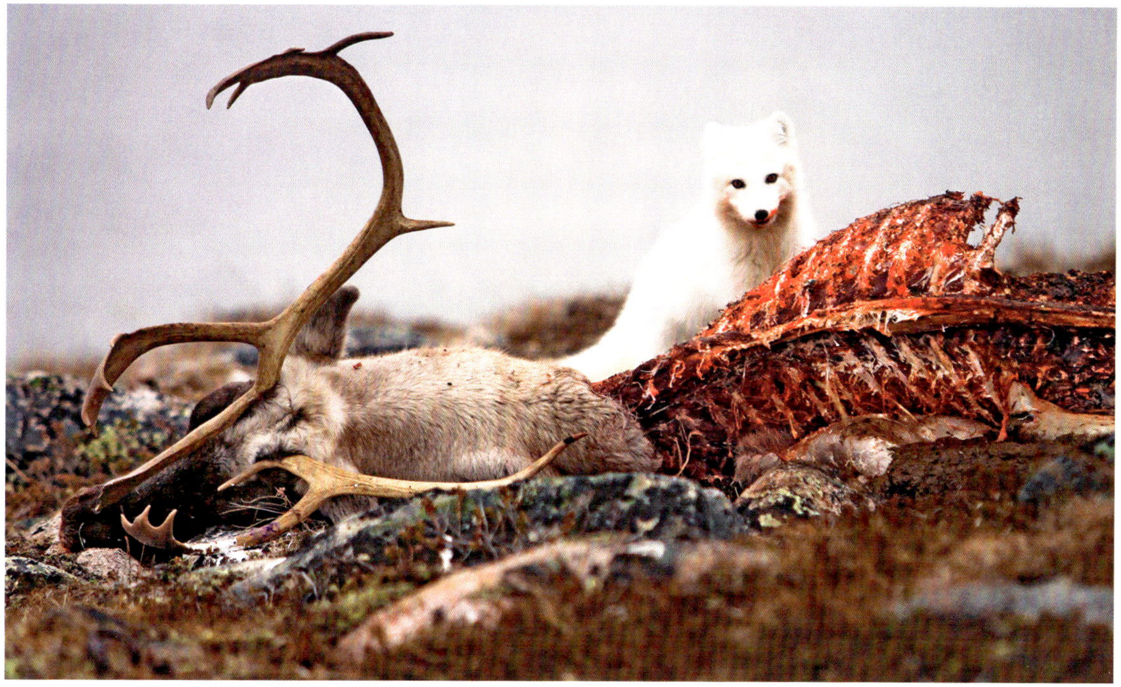

Arctic fox scavenging a caribou. Susan C. Morse

FACING, CLOCKWISE FROM TOP LEFT:
Eastern coyote late light portrait. Susan C. Morse

Red fox in full winter pelage (coat), Labrador. Winters in Labrador inspire beautifully thick winter coats on both wildlife and people. Neil Simmons

Belly up to the bar: Vixen feeding her hungry kits, Algonquin, Ontario. Steve Bruno

They have been observed to outcompete Arctic foxes in areas of Scandinavia and Alaska, but increasing human presence (extra food source for the red fox) was a possible factor. A 2022 paper studying the interactions of Arctic and red foxes where they coexist in Arctic Canada, far from significant human presence, indicated little competition.

There is also debate (to put it mildly) as to whether the expanding red foxes in North America are native, or non-native. This is an important distinction when it comes to conservation policy in a rapidly changing boreal forest and Arctic. A 2002 literature review suggested that all native boreal forest red foxes in North America had been displaced by descendants of English red foxes brought in for sport in colonial times, or escapes from fur farms. Yet a 2012 genetic study, conducted in part by the U.S. Department of

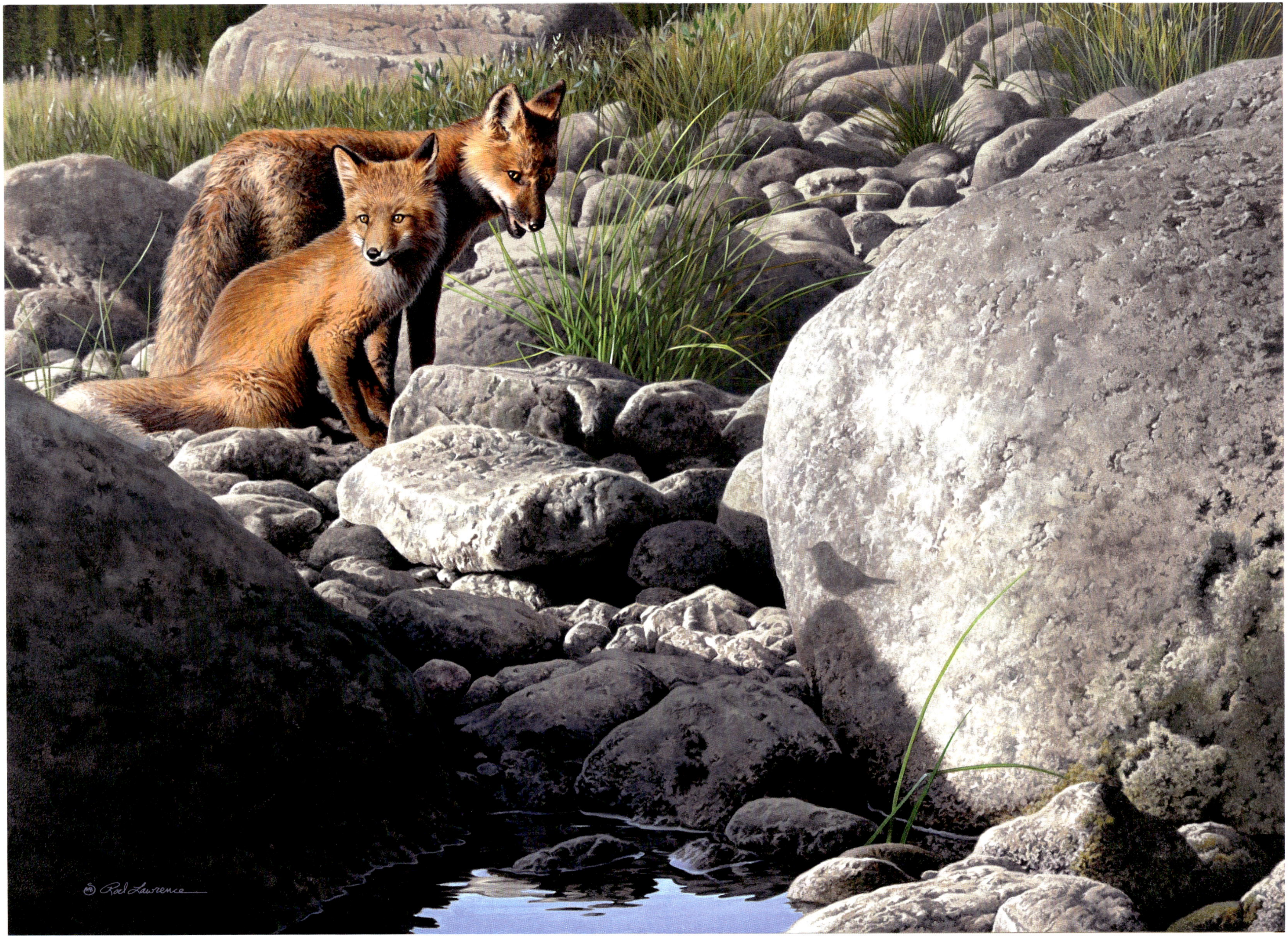
Rod Lawrence

Agriculture Forest Service, found evidence only for native foxes and none for non-native American red foxes in the boreal forest and Arctic, and even in the east, which would have been the origin point for English red foxes.

Family Mustelidae (wolverine, otters, martens, fishers, mink, badgers, polecats, and weasels)

September 17, 2006: Helen Falls, George River, Nunavik: I was waiting out a sleet squall while trying to get a sketch done of Helen Falls in the logbook when I smelled something reminiscent of a skunk. It was faint, though from how it kept changing direction, I could tell it was close and moving quickly and erratically. Not knowing what to expect, I was taken aback and amused when a short-tailed weasel zoomed up to me, darting beneath my legs and sniffing at my boots. There was no taking his picture. The best I got was a blurry six-foot streak snaking through the rocks: They are fast!

Small to medium carnivores with long bodies, short legs, and potent scent glands, mustelids are the world's largest and most diverse mammalian predator family. Of the approximately 63 living species, 21 or more live in the boreal forest: 11 in Eurasia (Palearctic), 7 in North America (Nearctic), and 3 Holarctic species are found around the northern globe. They range from the tiniest carnivore on Earth—the least weasel (*Mustela nivalis*), under one ounce and celebrated as a symbol of courage in Indigenous lore—to the 50-pound wolverine (*Gulo gulo*).

A common name for the mustelids is the "weasel family." The family does include "true" weasels, but other boreal forest mustelids include tree-hunting martens, aquatic otters, and burrowing badgers, so the true weasels are tricky to define. The name covers species also called stoats, ermines, ferrets,

The smell in this incident brings to mind a formerly famous New World boreal forest "mustelid," the striped skunk (*Mephitis mephitis*). Skunks aren't considered mustelids anymore. They were until DNA technology in the 1990s helped reclassify them as their own family: Mephitidae. They are related to mustelids, just not as closely as their smell would suggest. The striped skunk ranges throughout the central North American boreal forest.

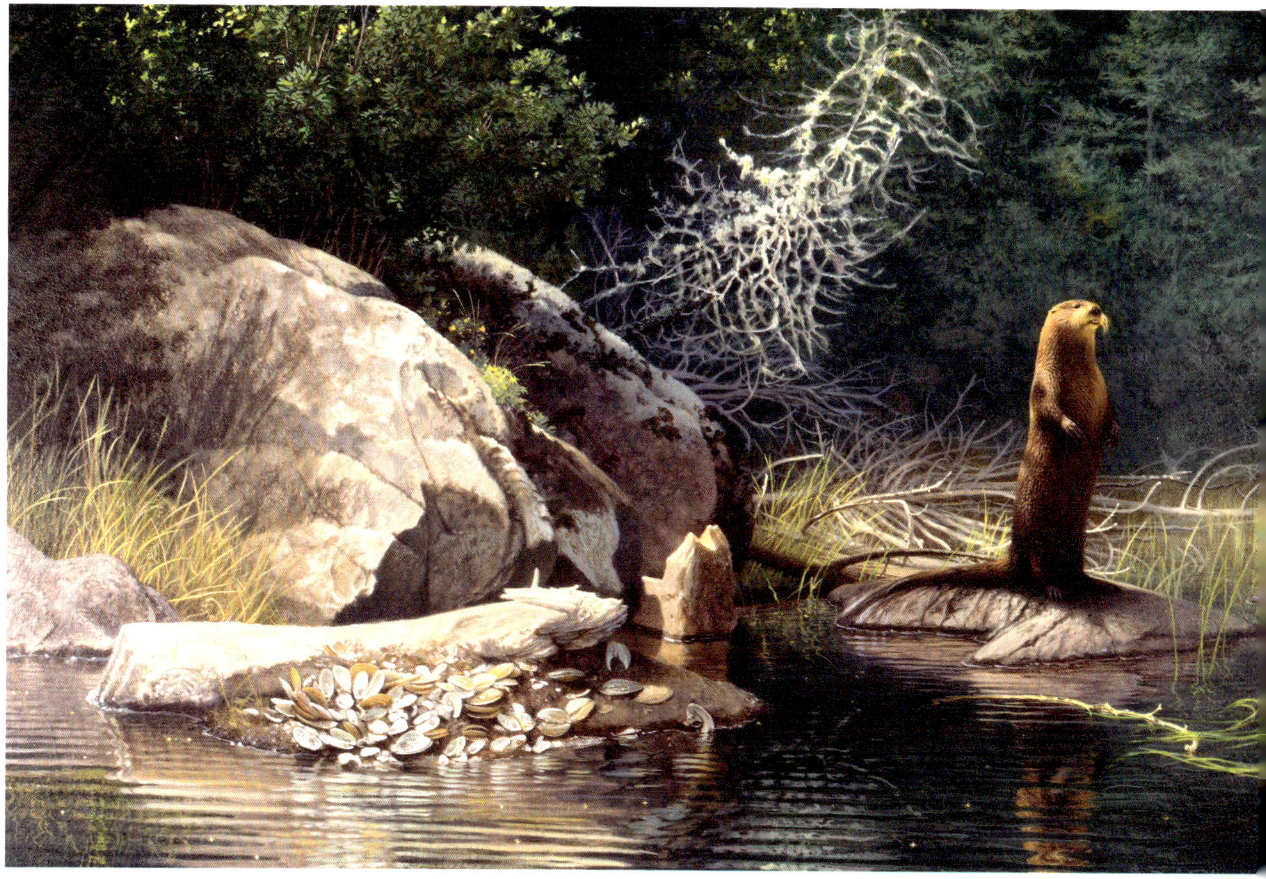

Most mustelids are secretive and infrequently seen. That is not so true of otters, who can be quite curious when feeling secure in the water. Playful young otters cavorting around our canoes were part of the inspiration for this painting. *Sharing the River*, acrylic, John C. Pitcher

FACING: Young red foxes. *In the Shadows*, acrylic, Rod Lawrence and polecats—and even mink. The name "ermine" adds to the confusion as it can be a species common name or simply the white winter coat of any species.

Taxonomy recently added another wrinkle. The short-tailed weasel or ermine (*Mustela erminea*), once treated as a single Holarctic species, was split in 2021 into two species:

- Eurasian ermine (*Mustela erminea*)—IUCN Red List: LC, still Holarctic in range
- American ermine (*Mustela richardsonii*)—Nearctic, overlapping with Mustela erminea in North America; indistinguishable in the field; not yet assessed by IUCN

One more northern species turns white in winter: the long-tailed weasel (*Neogale frenata*), which ranges into the southern boreal forest in North America. Only northern populations change color; others simply grow heavier winter coats, as do the rest of the boreal mustelids.

Fishers (*Pekania pennanti*), the second largest weasel and a close relative of the wolverine, are one of the only predators able to prey on porcupines regularly without serious injury or death. They don't flip the poor porcupine as has been conjectured and posed as fact. Rather, they force the porcupine out of its tree (not too hard since they fall out unassisted fairly often). Then, the fisher will repeatedly attack the porcupine from the front, circling it until it tires.

As tough as they look, this largest weasel scales up his smaller relatives' pugnacious attitude and will famously face off against grizzlies if in the mood.

FACING, CLOCKWISE FROM TOP LEFT:

Male fisher next to porcupine kill, Vermont. Susan C. Morse

Mature buck mink, Vermont. Susan C. Morse

American marten. An arboreal specialist and one of the cutest killers in the north woods. Susan C. Morse

Wolverine. *If Looks Could Kill*, oil, Steve Oliver

Conservation status and habitat comparison

Globally, mustelids face significant pressures: 46 percent are listed by the IUCN as Near Threatened, Vulnerable, Endangered, Critically Endangered, or Extinct. Among the 41 or so species outside the boreal forest, that number rises to 64 percent.

The boreal forest tells a different story. Of the three Holarctic species, all are Least Concern. In Eurasia (Palearctic), most boreal species are stable:

- Western and steppe polecats (*Mustela putorius* and *Mustela eversmanii*)—both LC
- Altai weasel (*Mustela altaica*)—Near Threatened
- Siberian weasel (*Mustela sibirica*)—LC
- European mink (*Mustela lutreola*)—Critically Endangered
- Pine marten (*Martes martes*)—LC
- Sable (*Martes zibellina*)—LC
- Yellow-throated marten (*Martes flavigula*)—LC
- Eurasian otter (*Lutra lutra*)—Near Threatened
- European and Asian badgers (*Meles meles*, *Meles leucurus*)—LC

In North America (Nearctic), all boreal species are evaluated as Least Concern:

- American mink (*Neogale vison*)
- Long-tailed weasel (*Neogale frenata*)
- Fisher (*Pekania pennanti*)
- American marten (*Martes americana*)
- North American river otter (*Lontra canadensis*)
- American badger (*Taxidea taxus*)

By percentage:

- All mustelids: 46 percent at risk
- Non-boreal: 64 percent at risk
- Eurasian boreal: 20 percent at risk
- North American boreal: 0 percent at risk

This pattern of boreal forests as a relative refuge will reappear in the next chapter with birds.

We've not mentioned raccoons (*Procyon lotor*), which are Nearctic to southern Mexico (Neotropic). They are not mustelids but are in the family Procyonidae, and they range into the southern boreal forest.

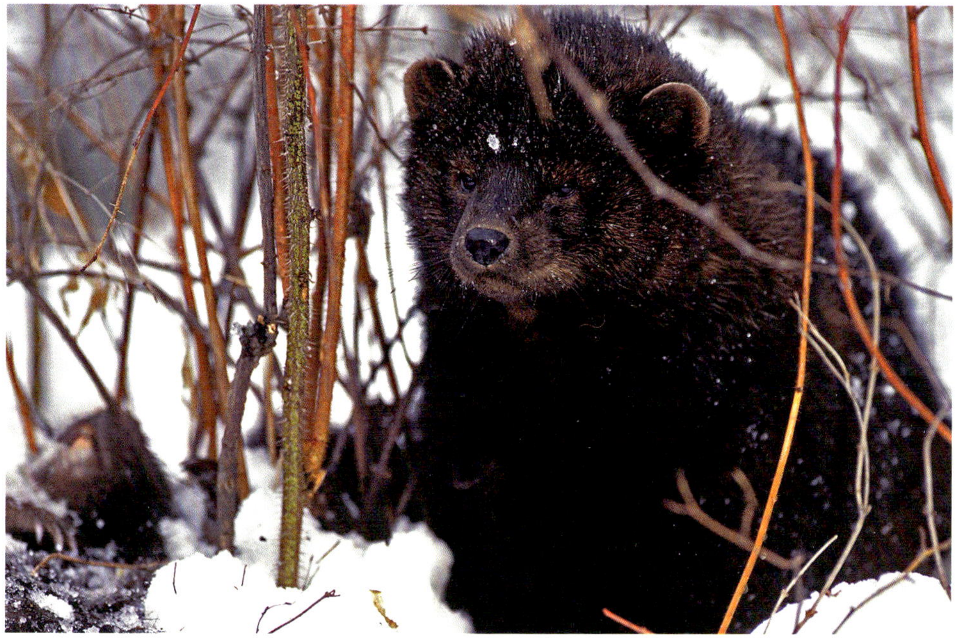
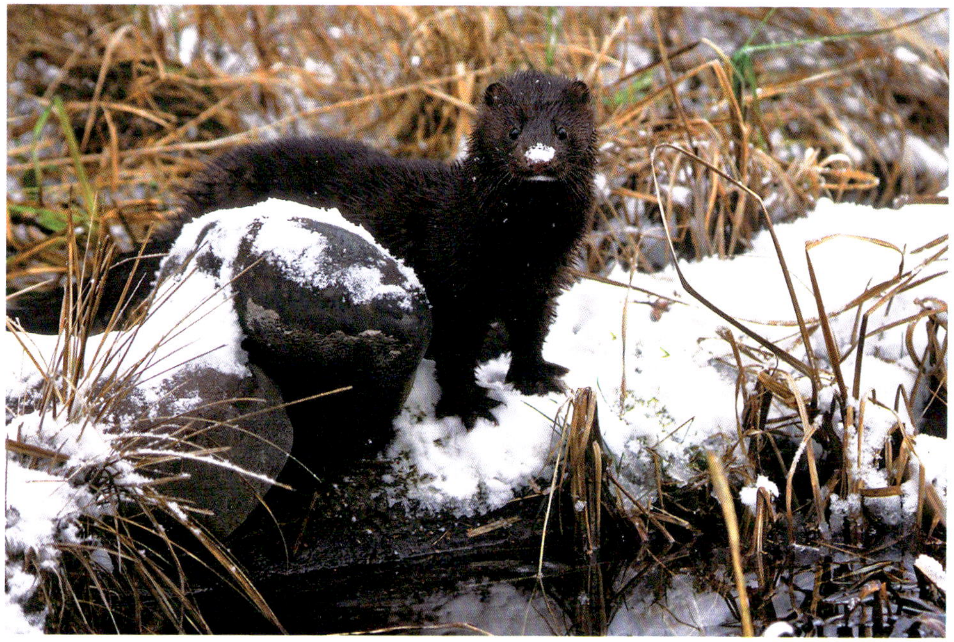
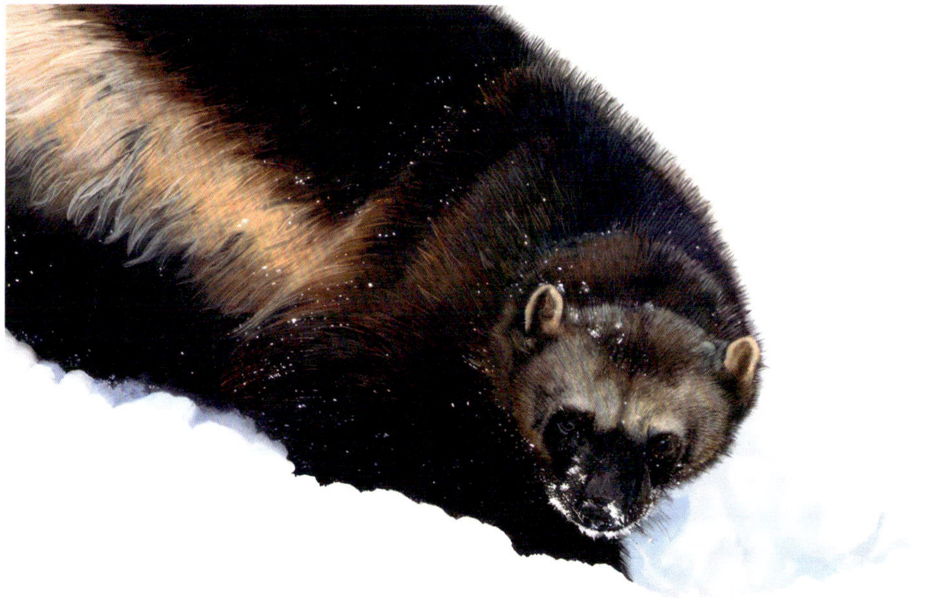
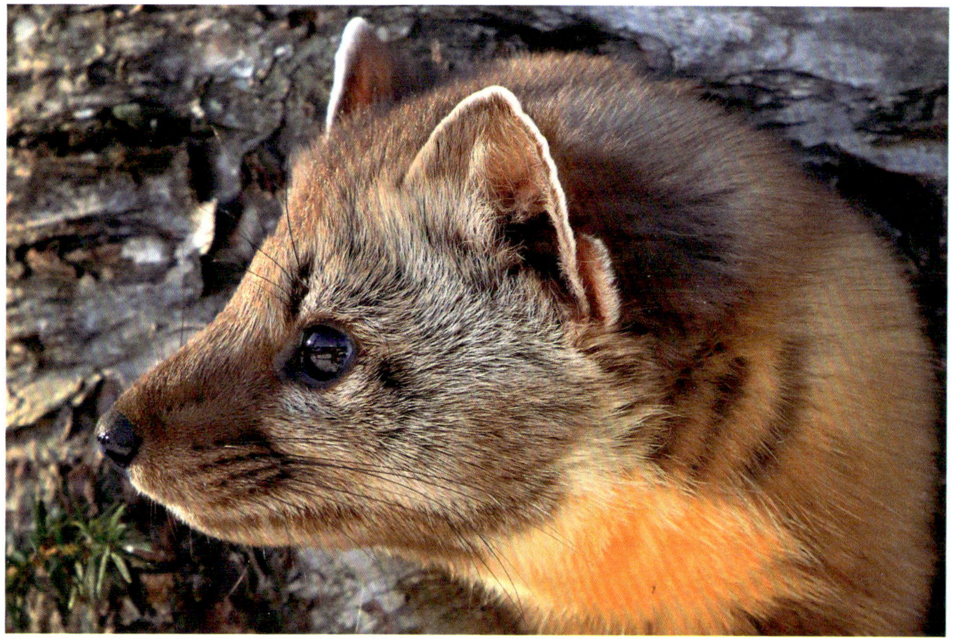

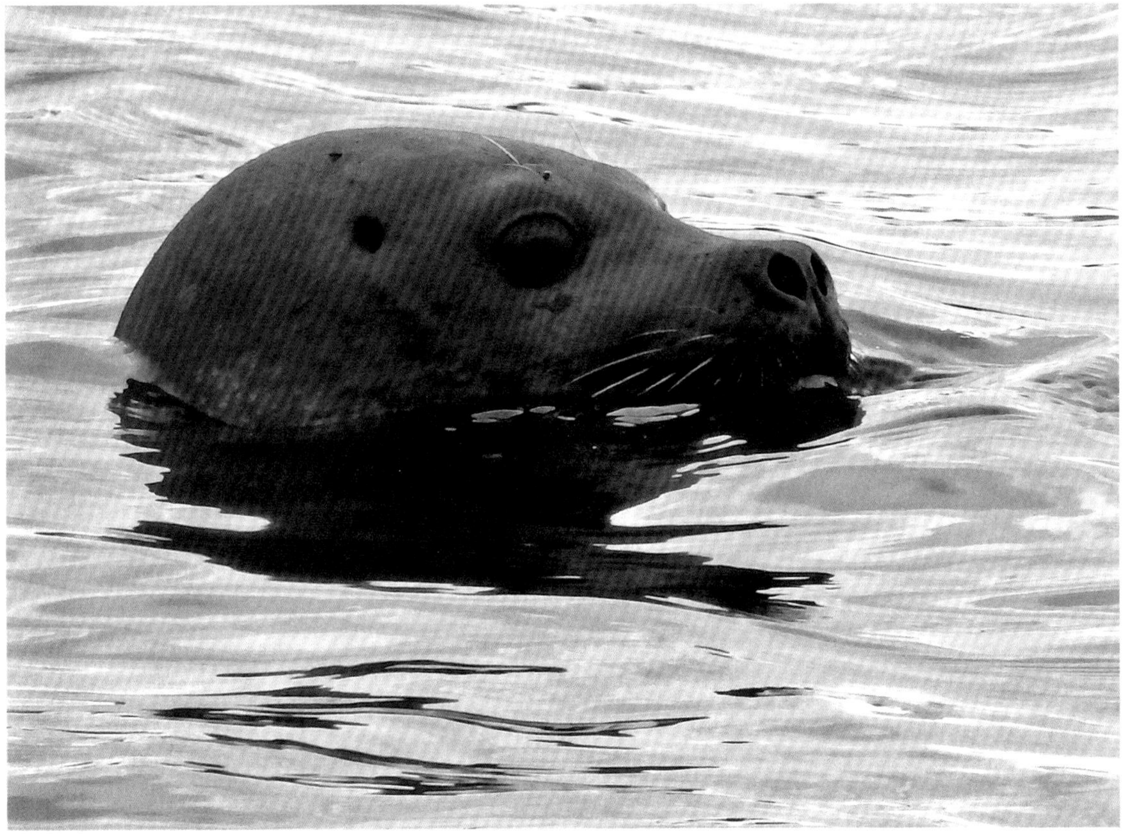

Our first Ungava seal, Lac des Loups Marins, Ungava.

There is one last group of mammals that may not leap to mind when thinking of the world's greatest forest . . .

Family Phocidae (seals)

Baikal seal (*Pusa sibirica*): The Baikal seal is the only species of seal that is exclusively freshwater. They are endemic (found nowhere else) to Lake Baikal in Siberia. One possible factor in this lake having its own unique species of seal is that unlike the Great Lakes (and Lake Saimaa, Lake Ladoga, and the Lac des Loups Marins), which are only about 10,000 years old, having been created at the end of the last glaciation, Lake Baikal is between 25 and 30 million years old.

Saimaa ringed seal (*Pusa hispida saimensis*): IUCN Red List status: Endangered, Palearctic. Found in Lake Saimaa, Finland's largest lake, this population of ringed seals was isolated from the Baltic Sea ringed seals by isostatic rebound.

Ladoga ringed seal (*Pusa hispida ladogensis*): IUCN Red List status: Vulnerable, Palearctic. Lake Ladoga is in Russia, just across the Finnish border not far from Lake Saimaa.

Ungava seal (*Phoca vitulina mellonae*): IUCN Red List status: Endangered, Nearctic. These seals are a subspecies of harbor seal. In the same manner as the Saimaa and Ladoga seals, isostatic rebound trapped the Ungava seals in Lac des Loups Marin. After thousands of years with a mile or two of ice sitting on it, the land had been depressed hundreds of feet so that it was below sea level. As the ice slowly retreated, the ocean crept in. The seals came with it. But then, with the ice gone, the land slowly rebounded, gradually rising above sea level. It kept rising until the seals, who hadn't paid close enough attention, ended up in a freshwater lake 70 miles from Hudson Bay.

August 15, 2018: Searching for a way to Lac des Loups Marins, Quebec, after ascending the Noonish River: There was no portage trail between the Noonish and the lake, so we split up to scout one. Dr. Stephen Loring and Steve Gorman took the direct route, and I looped west to follow the higher, hopefully more open ground route. There was no good way down when I reached the lake, so I headed back east, whistling after a bit to link up with the others. When I was within hailing distance, I heard an urgent call: "Have you got a camera?"

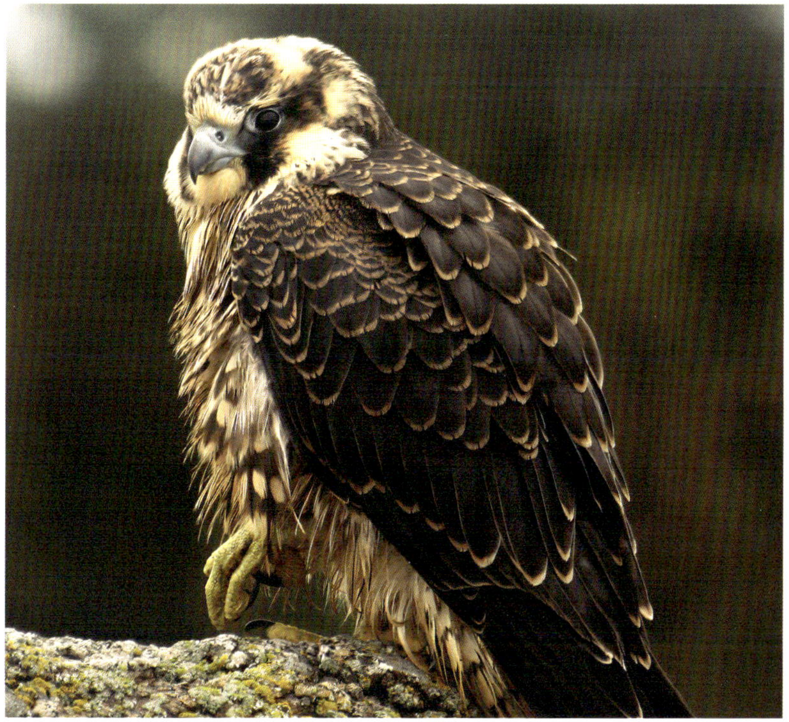

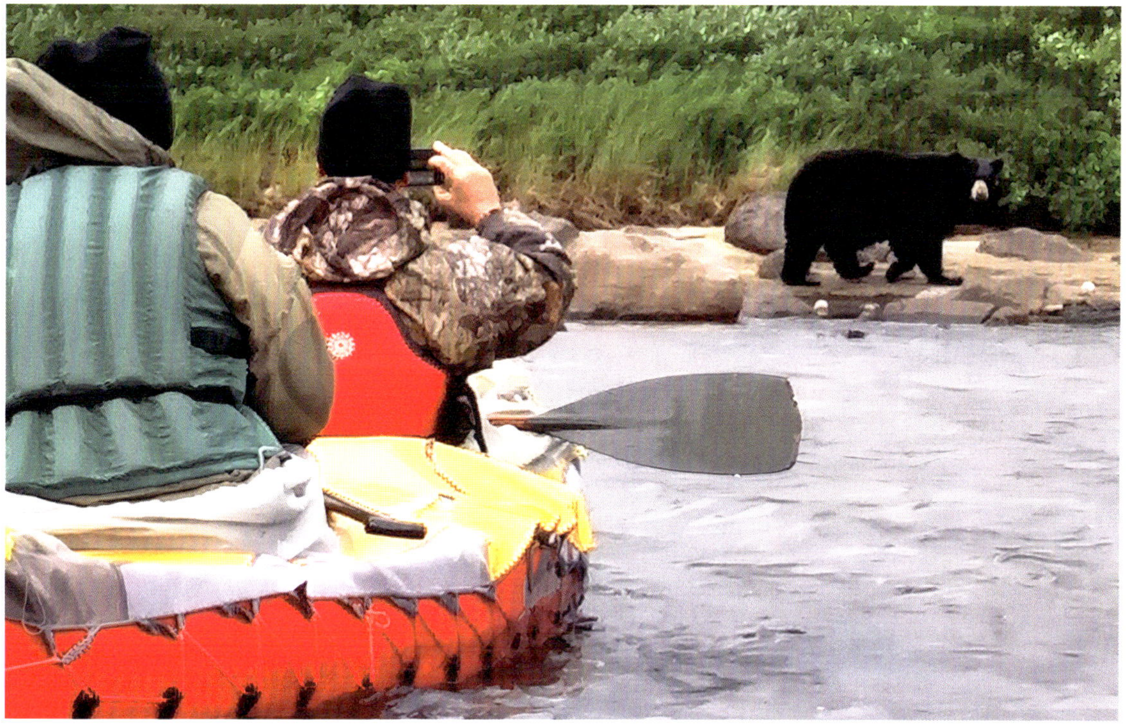

LEFT: Immature peregrine falcon on Voisey's Bay, Labrador Sea.

RIGHT: Unconcerned mama bruin, Rivière aux Mélèzes.

Funny. Steve was our professional photographer. Luckily, I had followed my ABCs of photography: Always Bring Camera.

"Yes," I called back.

"SEAL!"

We were to record seals because of recent threats to dam the Nastapoka. The Ungava seals were endangered and presumably rare, so we were concerned that we might not see any. No worries. When I reached the shore, I could see the seal up where Stephen and Steve were. The moment I arrived, it swam over to me and posed. It was fascinated with us. My camera was equipped with GPS, and we recorded four seals in three days.

The inquisitive reaction of this seal is a perfect lead-in to another phenomenon of wilderness travel . . .

When the locals don't know what you are

August 29, 2015: The mouth of the Ogriveg River (northernmost branch of the Kobuk River Delta) as it entered an inlet of the Chukchi Sea: Dan Berns and I were in the lead canoe when I heard a rush of water behind us, and almost in the same instant, a large animal torpedoed close abeam, then careened under the canoe. A moment's startled alarm turned to amusement as the whiskered face of a very playful seal popped up directly ahead.

Naturally, I grabbed my camera, but it was difficult to get a shot. He (she?) was fast and cavorted around and under us like he was showing off. Every few seconds, he came up to tease a headshot, only to duck beneath the water again.

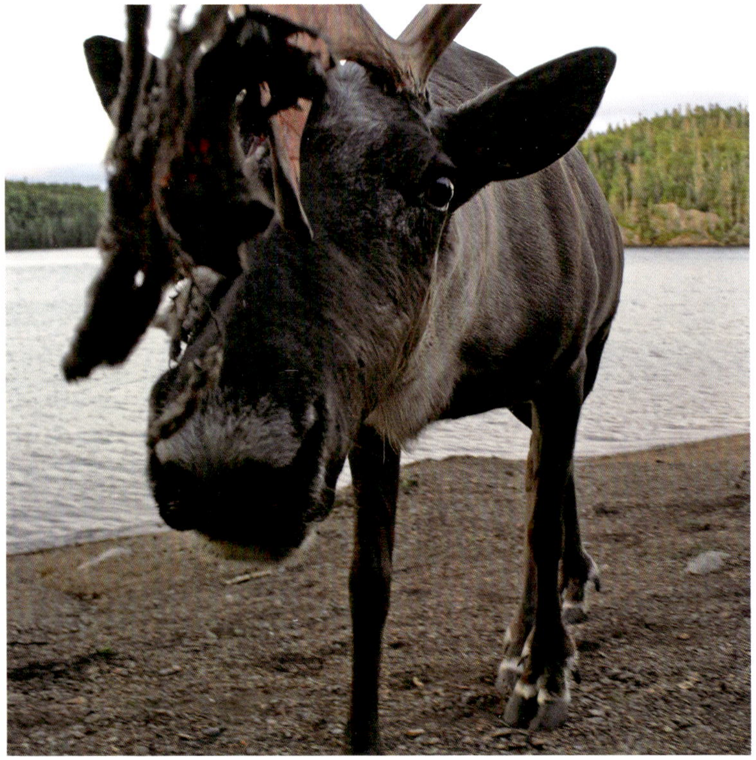

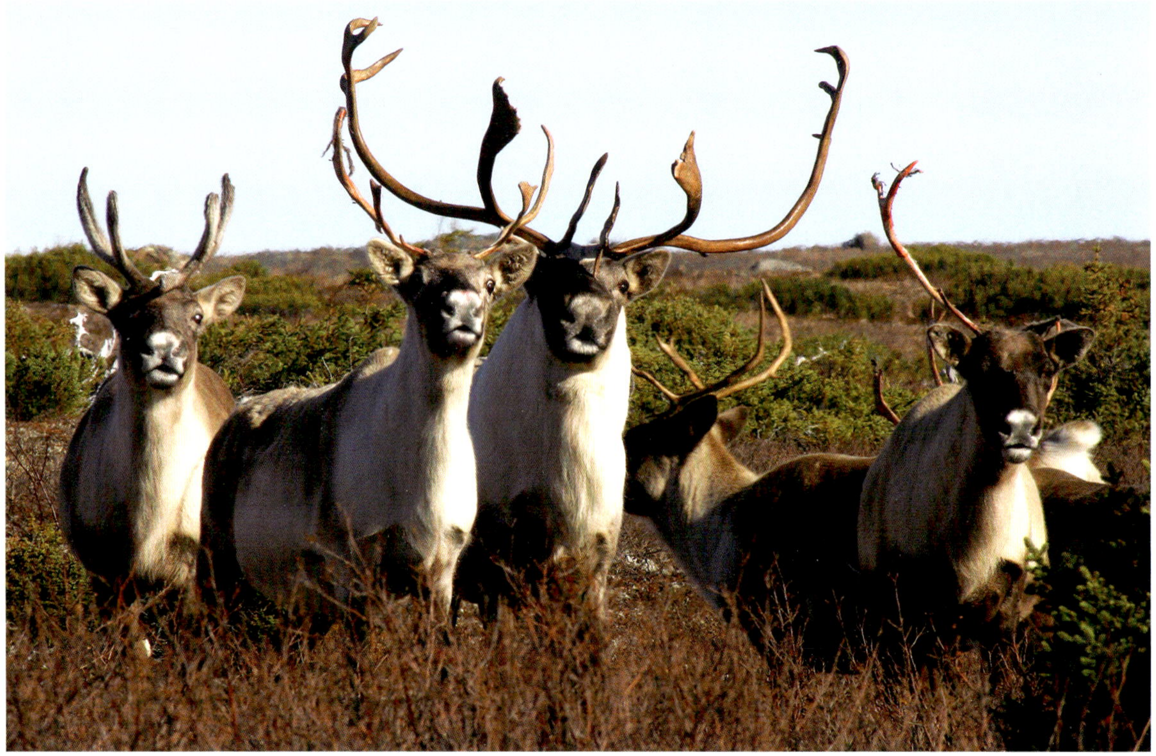

LEFT: "Curious" Woody (that's his name, we didn't give it to him), Patterson Island in the Slate Islands, Lake Superior.

RIGHT: And "curiouser" caribou above Kamestastin.

FACING: Gray wolf on crusted snow under a full moon. *Dangerous Lunatic*, graphite, Cole Johnson

Finally, he relented and posed.

Large black bears were common on the George River: We usually saw several lolling about in the tundra above the river every day. They paid us no mind as they were focused on the bumper crop of various berries, especially blueberries. The bears we met in Ungava on the Rivière aux Mélèzes were universally unfazed by our presence, and a sow and her two large second summer cubs accompanied us along the riverbank for the better part of a mile.

A juvenile peregrine falcon allowed Cole Johnson and me to walk up to and around him on the Labrador coast, almost causing us to lose our canoe. We saw the bird as we were departing our island camp after the stressful crossing of Voisey's Bay. We were so preoccupied with the unexpected encounter that we forgot that we had reached the ocean the day before and that there were tides, and this one was rising. Luckily, we noticed the canoe start to float free in the nick of time and recovered it without a frigid and dangerous swim.

These were all examples of a general pattern: In areas where wild animals have rarely if ever seen people, they tended to range between indifferent to curious (or worse, as will be related later). No species exemplified this phenomenon more entertainingly than caribou. Forget sneaking around. Put your hands in the air and with their poor vision, they mistake them for antlers and think you might be a long-lost cousin.

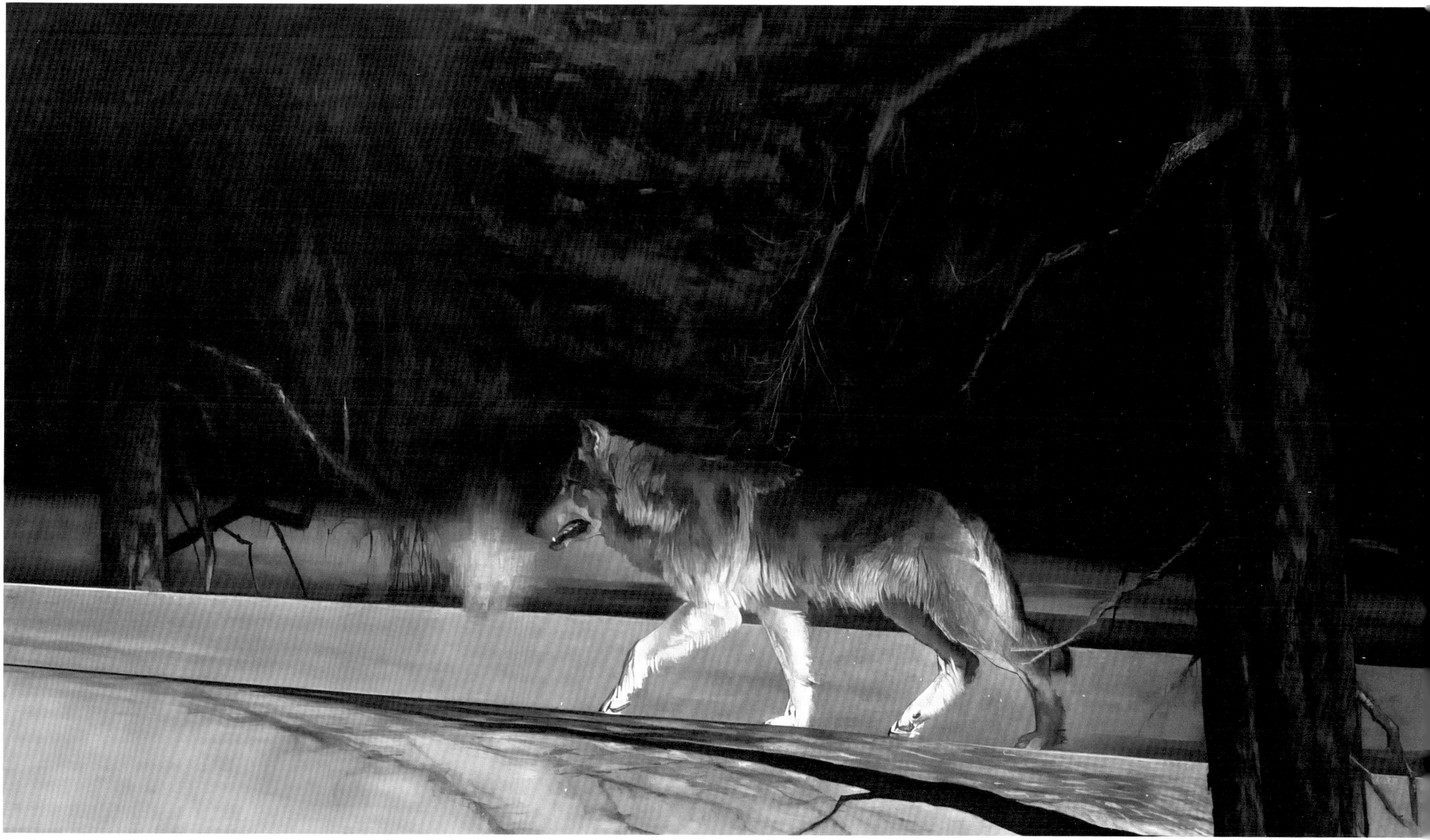

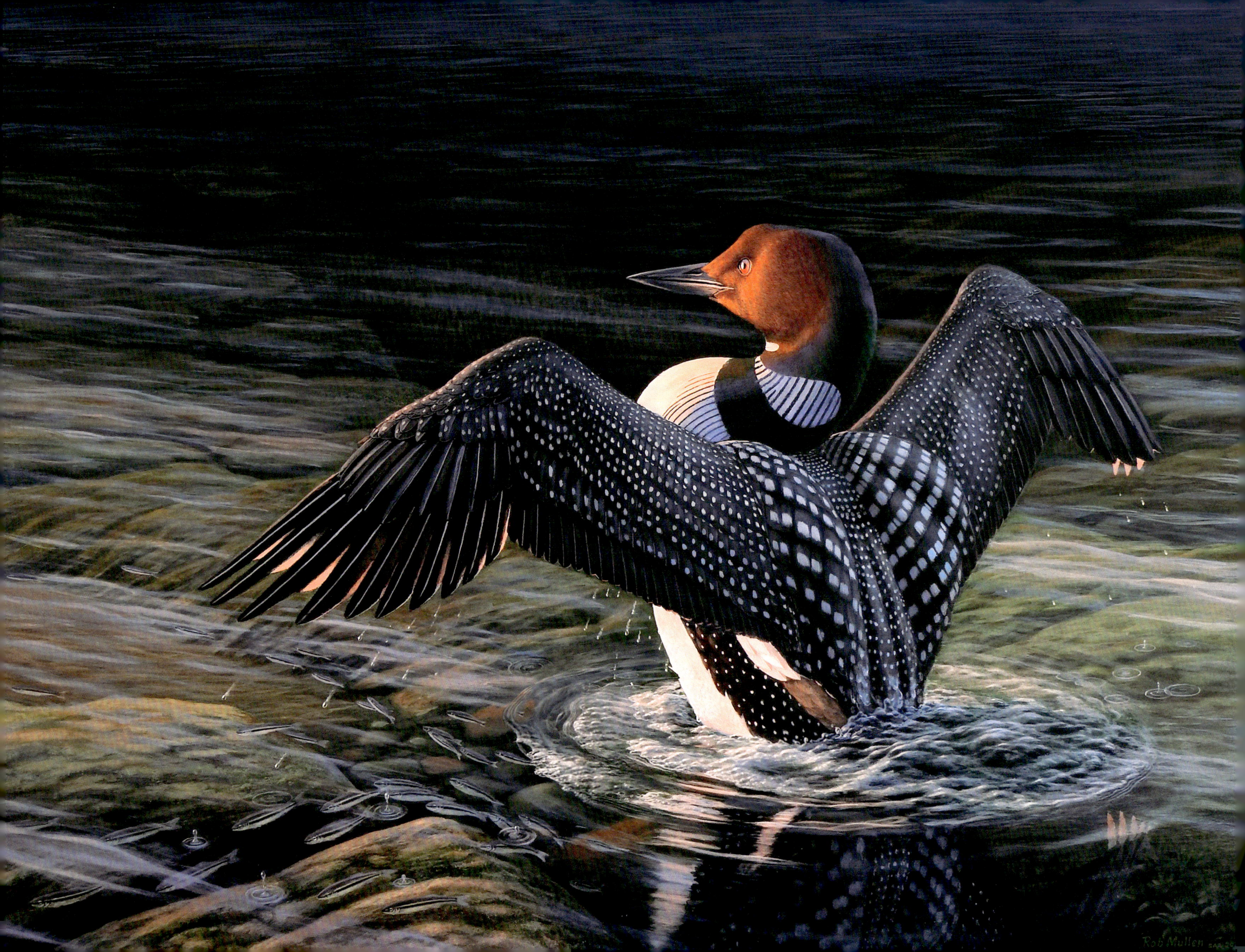
Rob Mullen

FACING: Common loon, black-nose dace, and largemouth bass. *Loon Rise*, acrylic, Rob Mullen

5

Birds

August 17, 2018: Height of land between Hudson Bay and Ungava Bay, Trees and Tundra Expedition

It might have been the overcast sky and lack of any life, but the country had an eerie and slightly foreboding aura as though it were the realm of some awesome and potentially malevolent spirit. I took to thinking of it as the Land of the Windigo. As we paddled through a narrow, twisting slough connecting two lonely lakes, I spotted a pair of Canada geese. It was heartwarming to see the familiar birds (a pair of Canadas nest on the pond behind our house and migrating flocks use it as a stopover from September to November), but I was concerned when I saw their unfledged young. It seemed late in the summer, especially this far north, not to be closer to their adult plumage. I wondered if they were possibly a second clutch after a failed nest. In any event, I wished them well and for winter to hold long enough for the kids to get their flight feathers.

October 9, 2018, around midnight: Preston Pond, West Bolton, Vermont

It was late, and the pond behind our house had shed its comfortable sunlit familiarity for a touch of the strange and eerie: colors sucked into the shadows, cheerful songbirds replaced by self-important owls, and unseen things *splopping* in the dark. Fortunately, I am long past any unease at imagined threats lurking in stygian crevices because an inexorable force of nature compels me into the gloom almost every night. Our dogs last nocturnal necessity before bed.

I had been home from our Trans-Ungava expedition for a month and was working hard on the trip report, so it was late when I gathered Nainette, Shiloh, and Tigger to head over to the pond. It was an oddly wonderful, warm, clear, moonless night, and as the dogs were wrapping up their correspondence, I'd been idly trying to pick out the constellation Draco. Mostly, I just wanted to go to bed. Shiloh, who always must have the last word, had lifted his leg for a final flourish, and we had started back to the house when an evocative sound stopped me.

From far above came the thin, distant honking of geese. Not the strident trombone blasts our nesting residents occasionally direct at the dogs, but the clarion call that signals the change of season, age-old yearnings for freedom and travel, and images of far-off lands (the geese would probably be on a golf course in New Jersey by morning). Some of their earlier

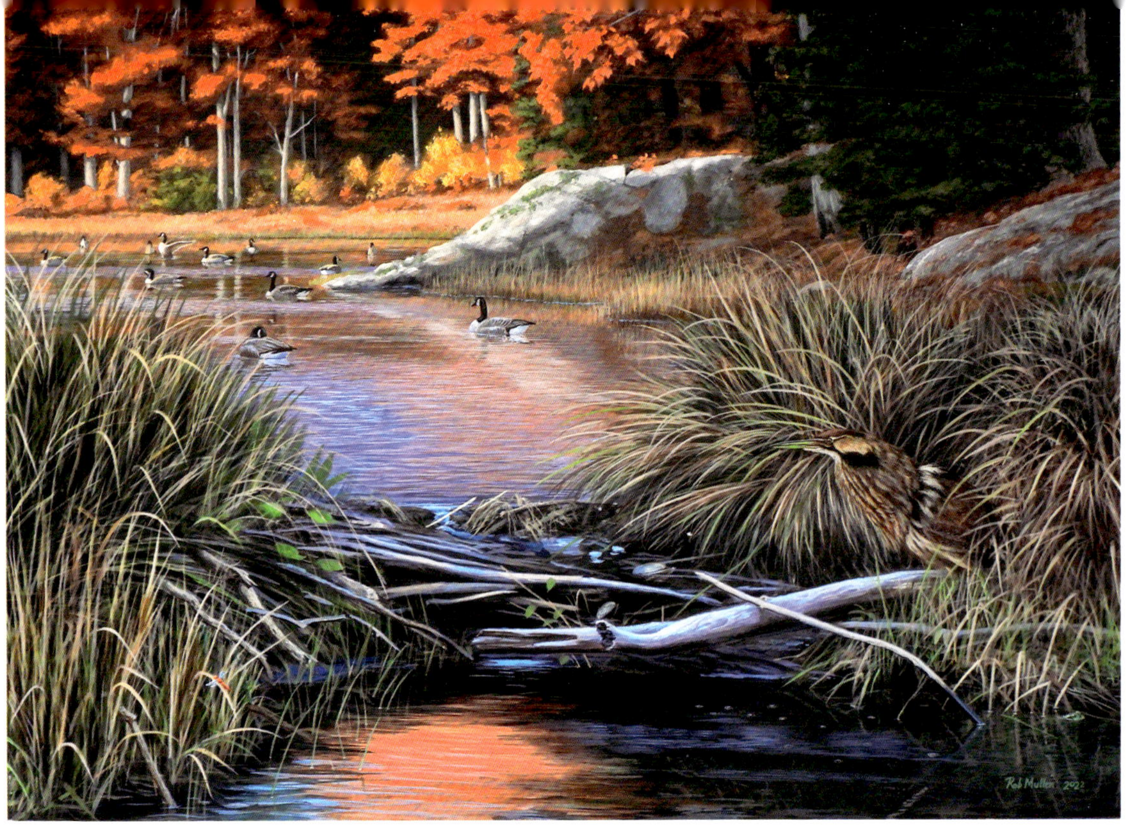

Canada geese and American bittern on Preston Pond. *Birds of a Feather*, acrylic, Rob Mullen

colleagues had rested on the pond two evenings before, but this bunch was making a night run, thousands of feet up against the midnight sky, lost to sight among the stars.

Except that they weren't quite. As I stood gazing up, I thought I saw a star wink out for an instant. I wasn't sure until I saw another and then another. In their vertiginous overflight, the migrating geese were turning the lights of the heavens on and off. And with that small but remarkable pattern of blinking stars, as the flock scudded south, a powerful sense of connection crystallized: the dogs and me standing by the edge of a quiet Vermont pond where migrating geese often sleep, and now, more migrating geese high above, toying with the universe.

Thinking back six weeks and a thousand miles to the north, it pleased me to think that those parents got their kids flying and were now passing far over my head.

This experience and others in this book highlight a value of birds beyond their ecological roles. Being able to fly and ubiquitous worldwide, birds provide billions of people with the potential for frequent and visceral connections to nature wherever they live. Almost anyone can see birds every day. People care about birds. This makes them particularly critical as ambassadors for the boreal forest (and all of nature) to people whether they live in Cape May, New Jersey, or Cape Town, South Africa.

Birds are the second most diverse order of vertebrates on Earth, behind fish. Globally, fish have an almost unfair advantage: Most of the planet is water. In the boreal forest, that edge is dulled. The wetlands are vast, but the brutal winters are hard on fish. Birds, on the other hand, can head to the beach. Migration lets them reap the summer's bounty and then flee the worst of winter. No other animal uses this strategy more successfully.

The result? In the boreal forest, birds overtake fish as the most numerous vertebrates. More than 400 species live in the global boreal forest, and over half of all North American bird species use it for nesting. That includes more than 60 percent of the continent's waterfowl species and nearly 90 species that depend on it for most—or all—of their breeding. In spring, the forest fills with billions returning to nest; by fall, billions more pour south. In 2006, International Migratory Bird Day named the boreal the "Bird Nursery of the North."

The migration wave includes not only boreal nesters but also millions of Arctic breeders, many of whom rely on boreal wetlands as vital refueling stops. Migration connects ecosystems, moves nutrients, and knits together landscapes across thousands of miles, from the Arctic to Africa to the Amazon.

But it's risky business. For millennia, birds balanced the hazards—storms, predators, and exhaustion—against the rewards and it paid off. Then came the human era: Habitat loss, industrial hazards, and climate change added new dangers to the old calculus. For long-distance migrants, it may not be enough to have safe breeding grounds. They depend on a chain of healthy habitats, and a single broken link can put a species at risk. However, a safe breeding ground is at least a port in the storm.

IUCN data show a clear pattern: Many bird families have high numbers of species under threat worldwide—except those that rely on the boreal forest. In some cases, healthy boreal populations have been the reservoirs that allowed species teetering on the brink of extinction to recover.

We'll begin this chapter with three emblematic comeback stories: the whooping crane, the trumpeter swan, and the bald eagle.

Whooping crane (*Grus americana*): IUCN Red List status: Endangered (but population increasing), Nearctic.

Five feet tall with a wingspan of seven-and-a-half feet and weighing up to 15 pounds or more, whooping cranes are the tallest North American bird, with about 70 percent of their population breeding in the boreal forest. They're hard to mistake for anything else in the boreal forest except possibly the white pelican on a foggy day. White pelicans (*Pelecanus erythrorhynchos*), about 20 percent of whom breed in the boreal forest, are also an enormous North American white bird with black wingtips; however, they are easily distinguished by a field mark made famous in an old poem that notes: Oh, that wondrous bird the pelican, his beak can hold more than his bellican.

A growing conservation success and a flagship species for endangered birds worldwide, whooping cranes were on the teetering edge of extinction by the late 1930s, with fewer than 20 birds left in the wild. They are the most intensely monitored birds in the boreal forest, and accurate counts are strictly maintained. The largest population (543 birds as of 2022) is the original, surviving wild population that nests in the boreal in the Northwest Territories and northern Alberta in Wood Buffalo National Park and winters along the Texas Gulf Coast. It is the world's only self-sustaining, migratory population of whooping cranes.

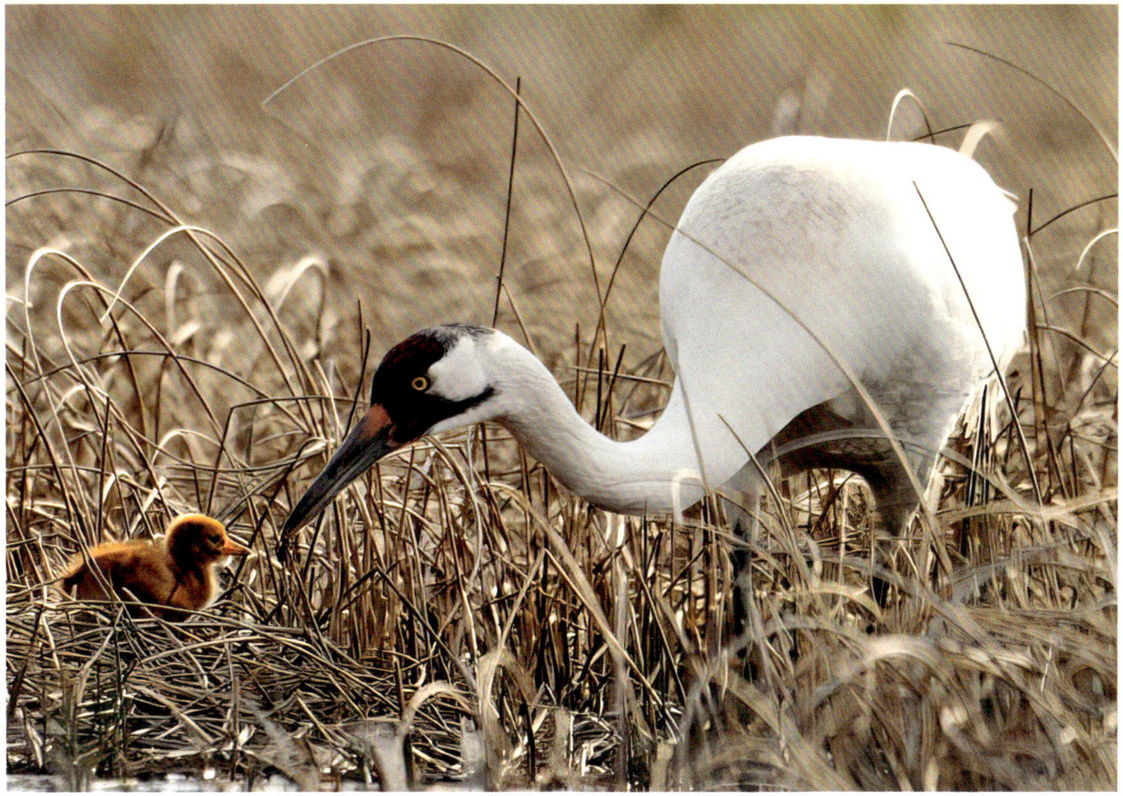

Whooping crane and chick. Klaus Nigge/U.S. Fish and Wildlife Service (USFWS), public domain

Trumpeter swan (*Cygnus buccinator*): IUCN Red List status: LC (!), Nearctic. With a seven-to-ten-foot wingspan and attaining weights of over 30 pounds, trumpeters are the largest waterfowl, and one of the largest birds, in North America. Currently, about 57 percent rely on the boreal forest for nesting. Nearly exterminated by overhunting and habitat loss by the 1930s, reintroductions from a population in Alaska

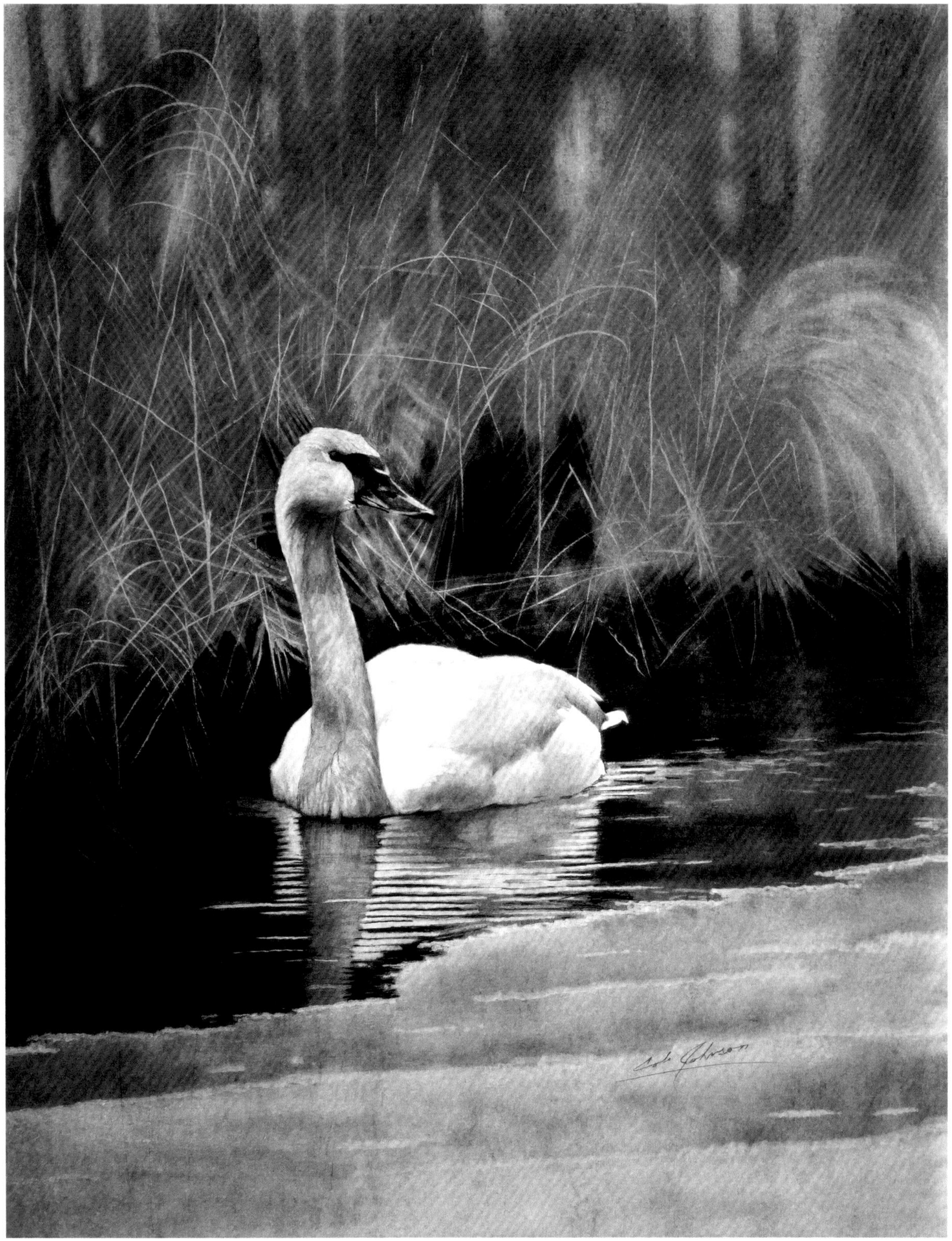

(the boreal forest as wildlife reservoir), strict protections, and ardent conservation work are bringing them back with a current population estimate of 63,000. I saw them on the Kobuk River in Alaska.

Bald eagle (*Haliaeetus leucocephalus*): IUCN Red List: LC (!), Nearctic. National symbol of the United States.

This sea eagle's size varies with latitude, per Bergmann's rule, with Alaskan birds attaining weights and some measurements almost double that of birds from the southern US. Bald eagles have a wingspan of up to eight feet and weigh up to 15 pounds and possibly more (whether they are larger than the golden eagle is a matter of some debate). Their range spans most of North America, with about 16 percent of their breeding habitat in the boreal forest. That said, according to every range map I referenced, the northern limit of the bald eagle's eastern breeding range is 250 miles south of where we commonly saw these birds on the Rivière aux Mélèzes near Ungava Bay in August 2018.

Being the de facto national symbol of the United States did not protect bald eagles, but it may have saved them from extinction, and much of the rest of nature to boot (an example of an umbrella species on steroids, simply because of its mass political appeal). The bald eagle was hailed as the national bird as early as 1782, but the industrial revolution and population growth led to habitat destruction and overhunting. Spurred to action by a precipitous decline in the population, in 1940, the US Congress patriotically stepped in with the Bald and Golden Eagle Protection Act, making it illegal to kill them or possess parts. However, there was no provision for protecting their habitat, nor protecting them from indirect killing, so things did not improve much. Then, they got a lot worse.

The green revolution introduced a chemical for every facet of agriculture, principally fertilizers, herbicides, rodenticides, and pesticides. The most famous of the pesticides was

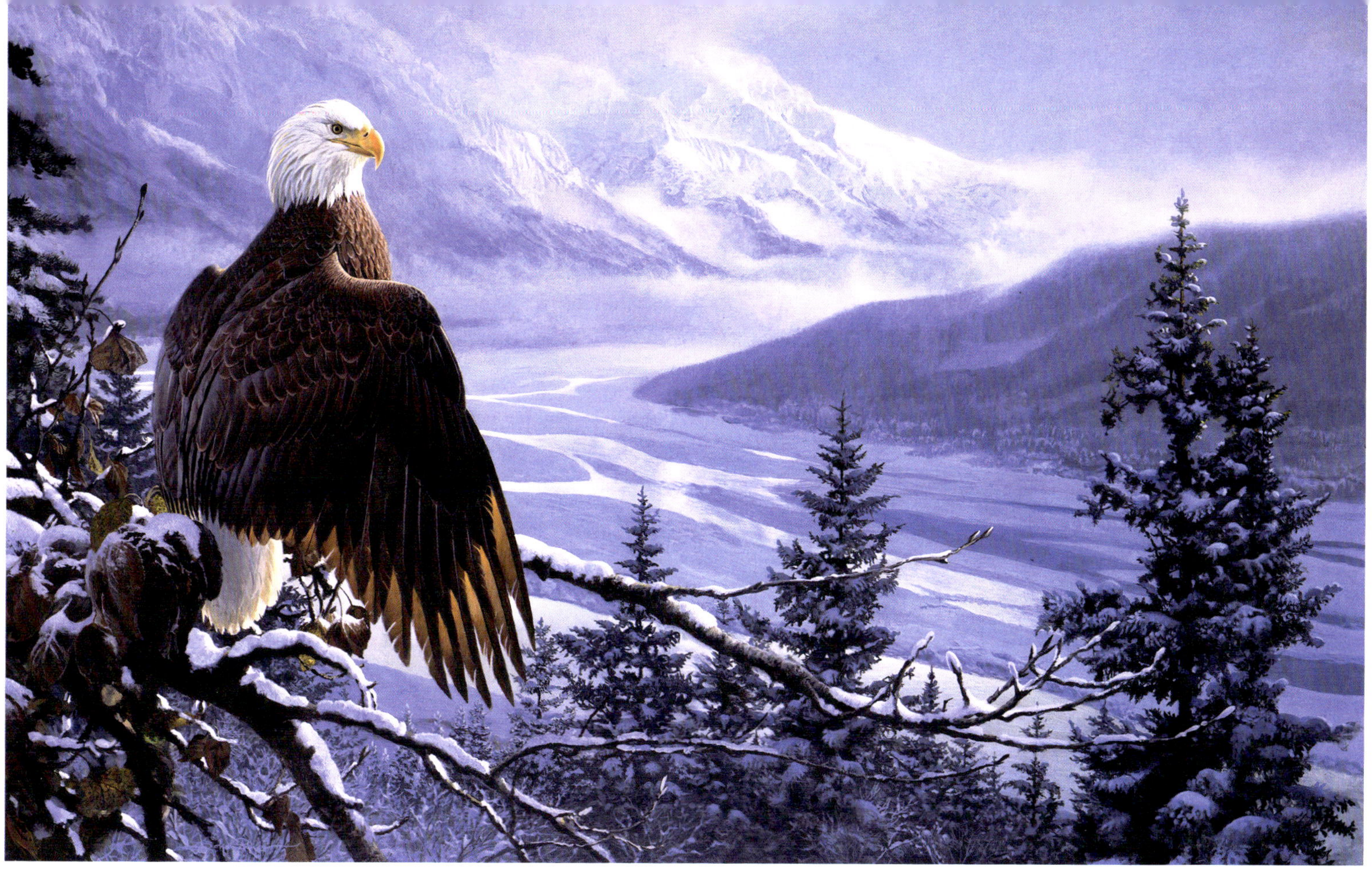

Bald eagle. *Chilkat Sentinel*, acrylic, John C. Pitcher

FACING: *Trumpeter Swan*, charcoal, Cole Johnson

DDT—fabulous for killing mosquitoes. It didn't necessarily kill wildlife directly, at least not quickly; however, it accumulated in predators such as birds of prey as they ate sub-lethally laced rodents, which led to thinning eggshells and declining reproduction. Of such seemingly little things are catastrophes made.

> The first bald eagle I ever saw was on Cape Breton Island in Nova Scotia, Canada, in 1972. I was very excited, but even as a teenager, I noted the sad irony that to see our "national bird," we had to go to Canada.

Nesting bald eagles in the mainland United States plummeted by 99.5 percent by 1963.

Only the boreal forest of Canada and Alaska, harboring population reservoirs of raptors including bald eagles, saved them (and others such as ospreys and peregrine falcons). Rachel Carson's *Silent Spring* was a major catalyst, spurring public concern that led to the banning of DDT and the passage of the Endangered Species Act. The act notably provided protections for threatened species' habitat and against indirect killing (e.g., DDT) rather than just the Bald and Golden Eagle Protection Act's simplistic ban on killing them on purpose (causing direct harm).

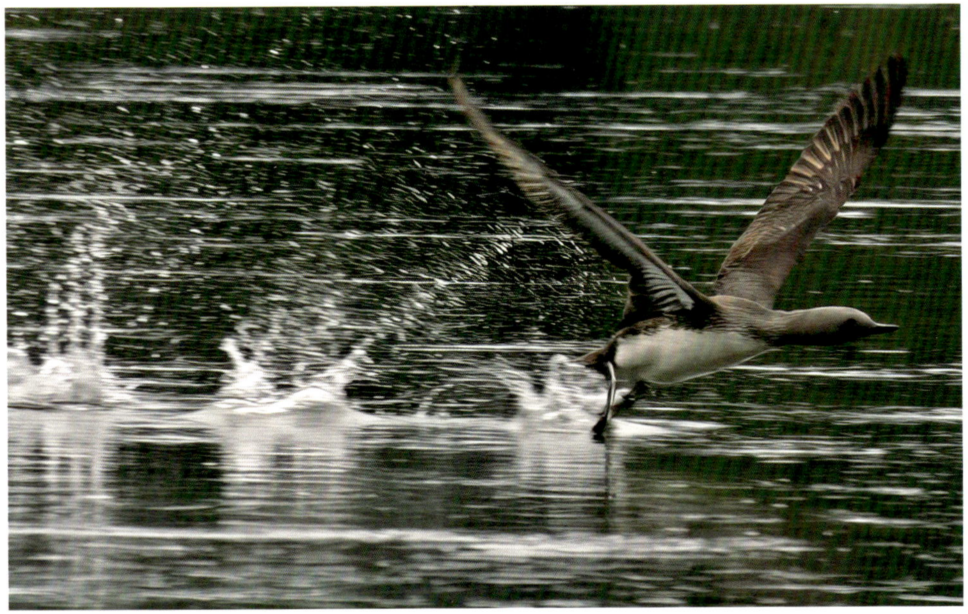

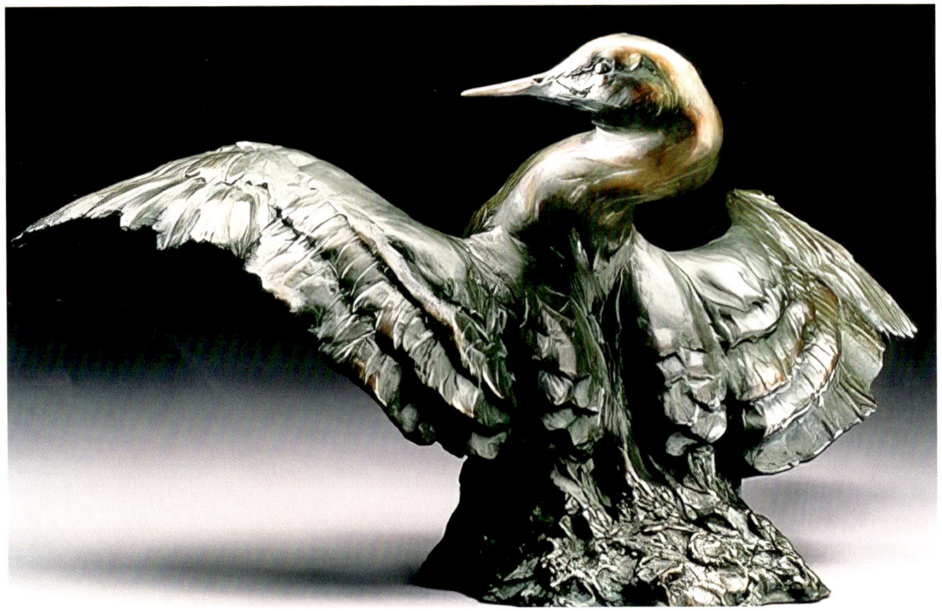

LEFT: Red-throated loon taking off on the Kobuk River.

RIGHT: Common loon. *Height of Land*, bronze sculpture, Sandy Scott

FACING: Diving loons, Quebec. Oil, Jean-Louis Courteau

The boreal forest again proved to be a wildlife refuge—not because it's a cold northern forest, but because it is an intact ecosystem (though that may be because it is a cold northern forest).

For an overview of some of the other birds, since we are talking about wilderness, and mostly Canadian wilderness, there is only one place to start . . .

Family Gaviidae (loons)

The haunting call of a loon is often used to set a mood in movies and TV. And for good reason: It is an aural wash of mystery and wilderness that never gets old, no matter how many times it echoes across a moonlit lake or remote river.

Loons breed around the northerly reaches of the northern hemisphere, including most of the boreal forest, and winter mainly along more southern coasts. As an adaptation for swimming and diving, their legs are so far back on their bodies that they cannot move well on land (there is speculation that their name comes from their awkward lunging about), and all but the red-throated loon require a long run across water to get airborne. But on the water and in the air, they match any bird. Underwater, they reign supreme.

Red-throated loon (*Gavia stellata*): They are the smallest loon, which may be why they are the only one that can reportedly take off from land. (I've only ever seen them take off like any other loon.) Red-throated loons are Holarctic in distribution.

Pacific loon (*Gavia pacifica*): Pacific loons are mostly a North American breeder but ranging into extreme eastern Siberia; 65 percent of the North American population nests in the boreal forest.

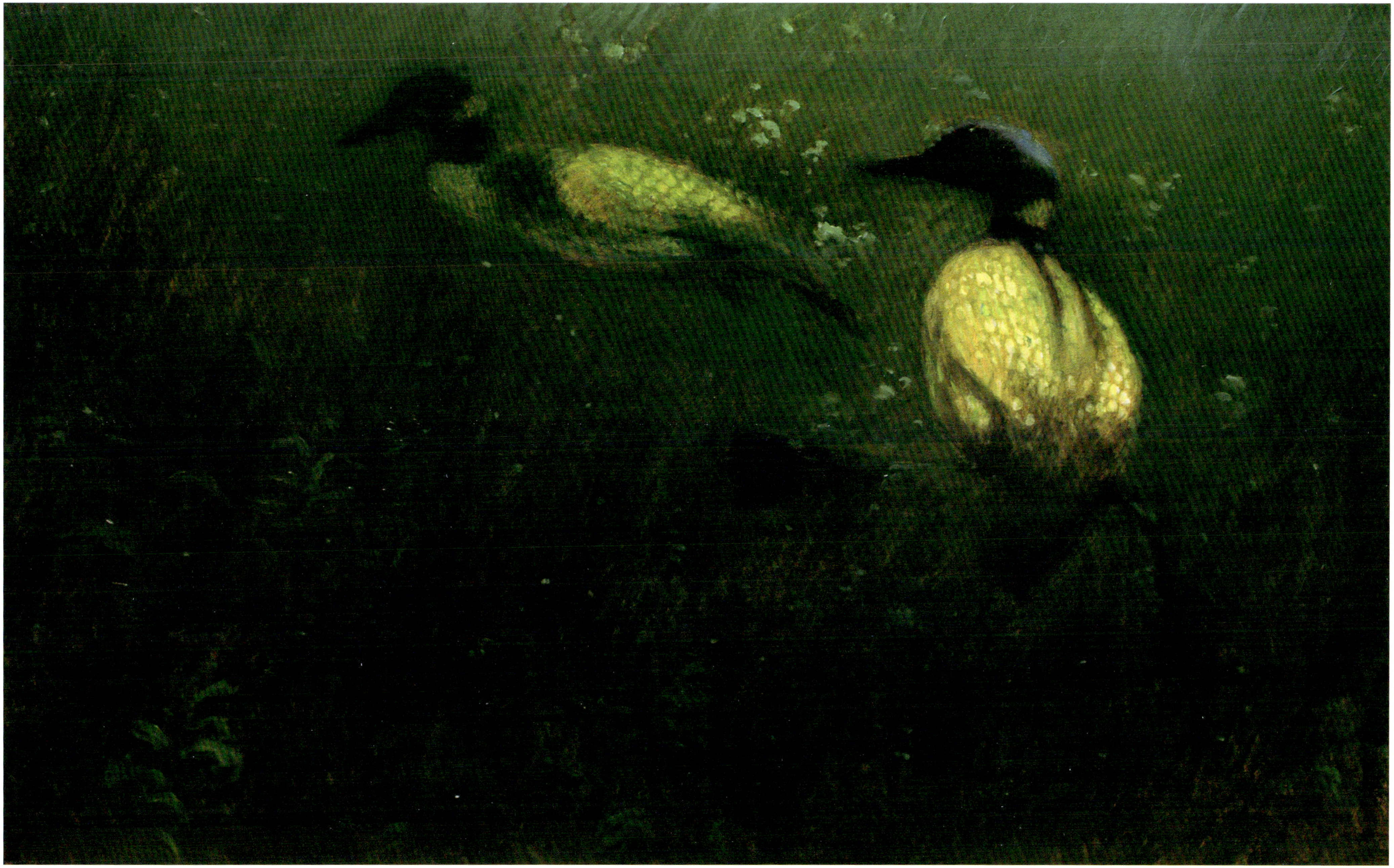

Loon and young, Fletcher Lake, Algonquin Park, Ontario. Steve Bruno

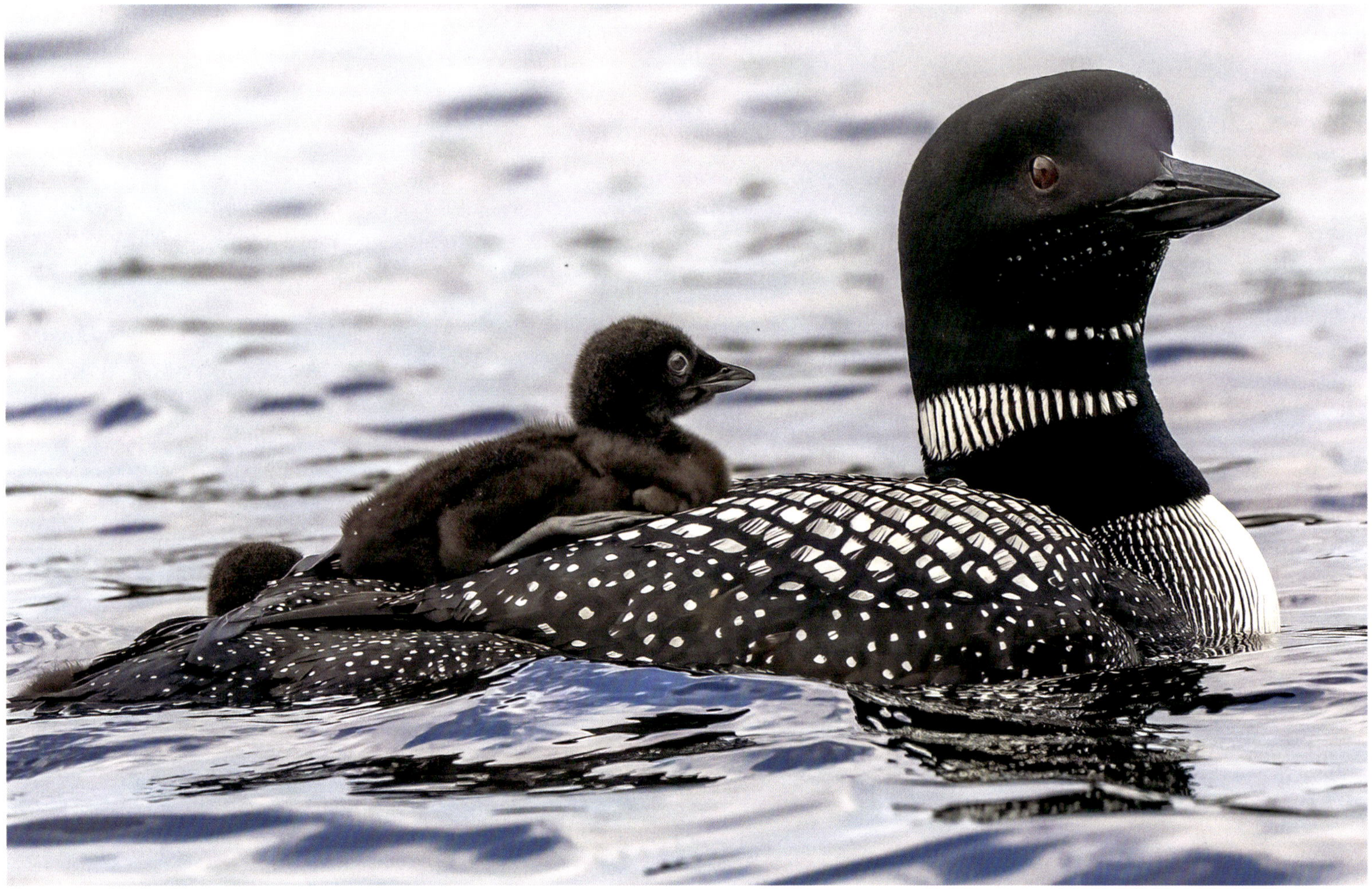

Arctic loon (*Gavia arctica*), aka black-throated loon: Widespread in Eurasia's boreal forest, Arctic loons breed in far western Alaska, including the Kobuk River Delta.

Common loon (*Gavia immer*): About 74 percent of the population breeds in North America's boreal forest with a few nonconformists breeding in southwestern Greenland, Iceland, and the island of Spitsbergen (some winter on the coasts of western Europe and Scandinavia).

Yellow-billed loon (*Gavia adamsii*): IUCN Red List status: Near Threatened, Holarctic. Their decline is suspected from harvest levels in Alaska.

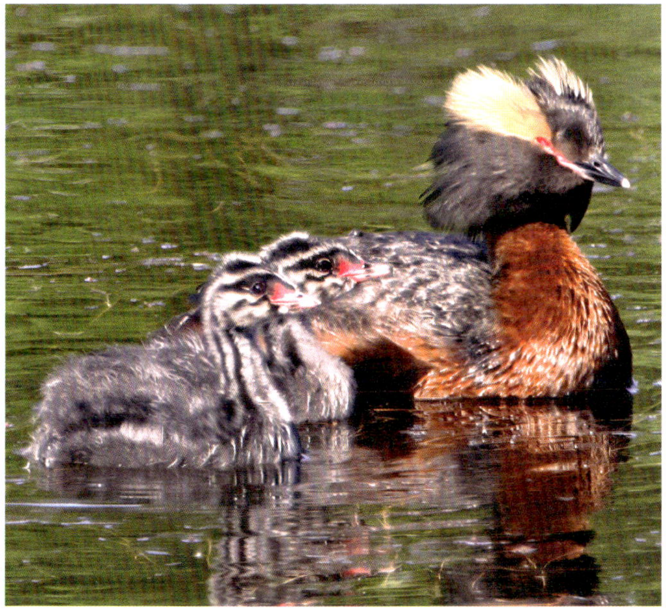 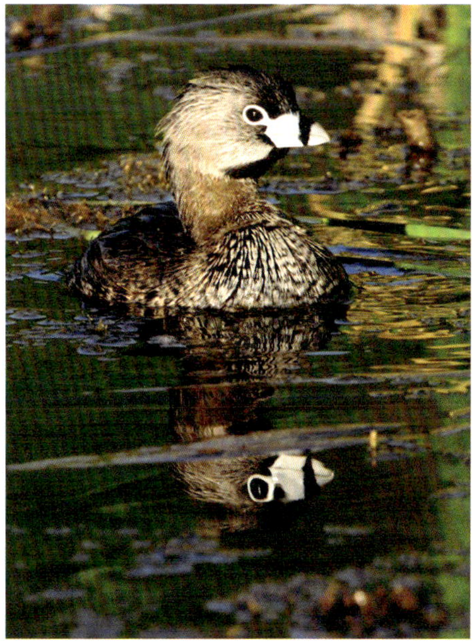 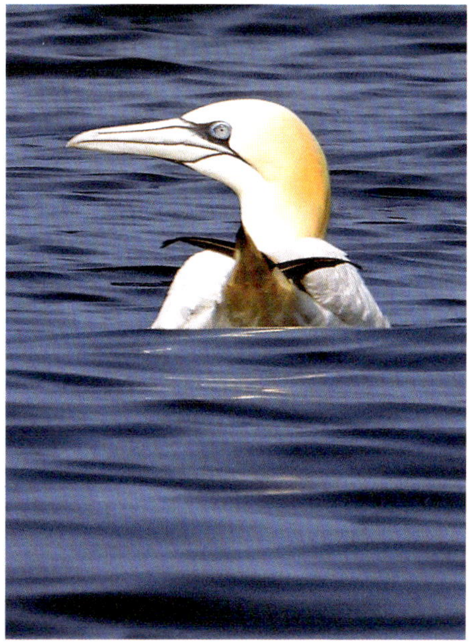 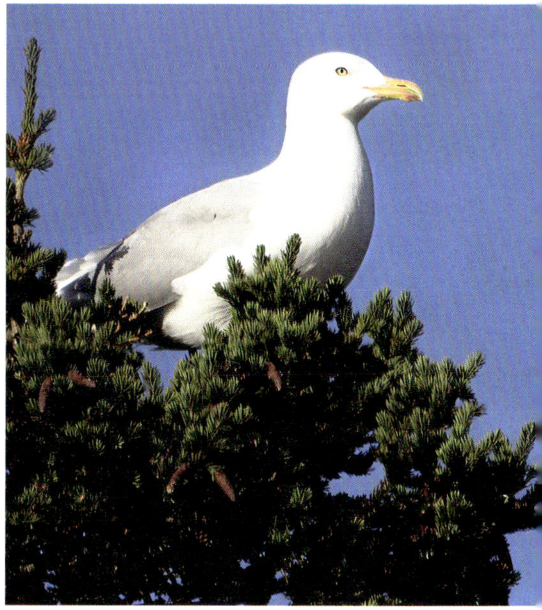

Family Podicipedidae (grebes)

Grebes may resemble small, tailless loons, but the similarity is convergent evolution. Their lobed toes hint at their true kinship (though it was confirmed via DNA) with flamingoes, not loons.

Horned grebe (*Podiceps auritus*): IUCN Red List status: Vulnerable, Holarctic. About 79 percent of the North American population nests in the boreal forest.

Pied-billed grebe (*Podilymbus podiceps*): Pied-billed grebes are widespread in the New World; about 10 percent of the population breeds in the boreal forest.

Red-necked grebe (*Podiceps grisegena*): About 93 percent of the North American population breeds in the boreal forest.

Seabirds: Families Laridae, Alcidae, and Phalacrocoracidae

A surprising number of seabirds depend on the boreal forest, though with its countless lakes, including Baikal, Great Slave, Great Bear, Winnipeg, and the Great Lakes, perhaps it's not so surprising. Some species are "boreal-adjacent," nesting along coasts backed by forest and feeding in marine waters enriched by forest-fed rivers. These include alcids (puffins, razorbills, murres and/or guillemots, dovekies), northern gannets (*Morus bassanus*), and some gulls and terns.

One seabird that penetrates far inland is the double-crested cormorant (*Nannopterum auritum*), a pelican relative. Unlike most birds, its feathers aren't waterproof, so it must dry on shore (or docks and piers) and sits low in the water—useful for telling it from loons, which ride low only when alarmed or hunting.

LEFT TO RIGHT:
Horned grebe and young, Alaska. Donna A. Dewhurst/USFWS

Pied-billed grebe, Île Bizard, Quebec. Bonnie Rowell

Resting gannet, Northumberland Strait, Gulf of St. Lawrence.

Glaucous gull in a white spruce, near Ambler, Alaska, Kobuk River.

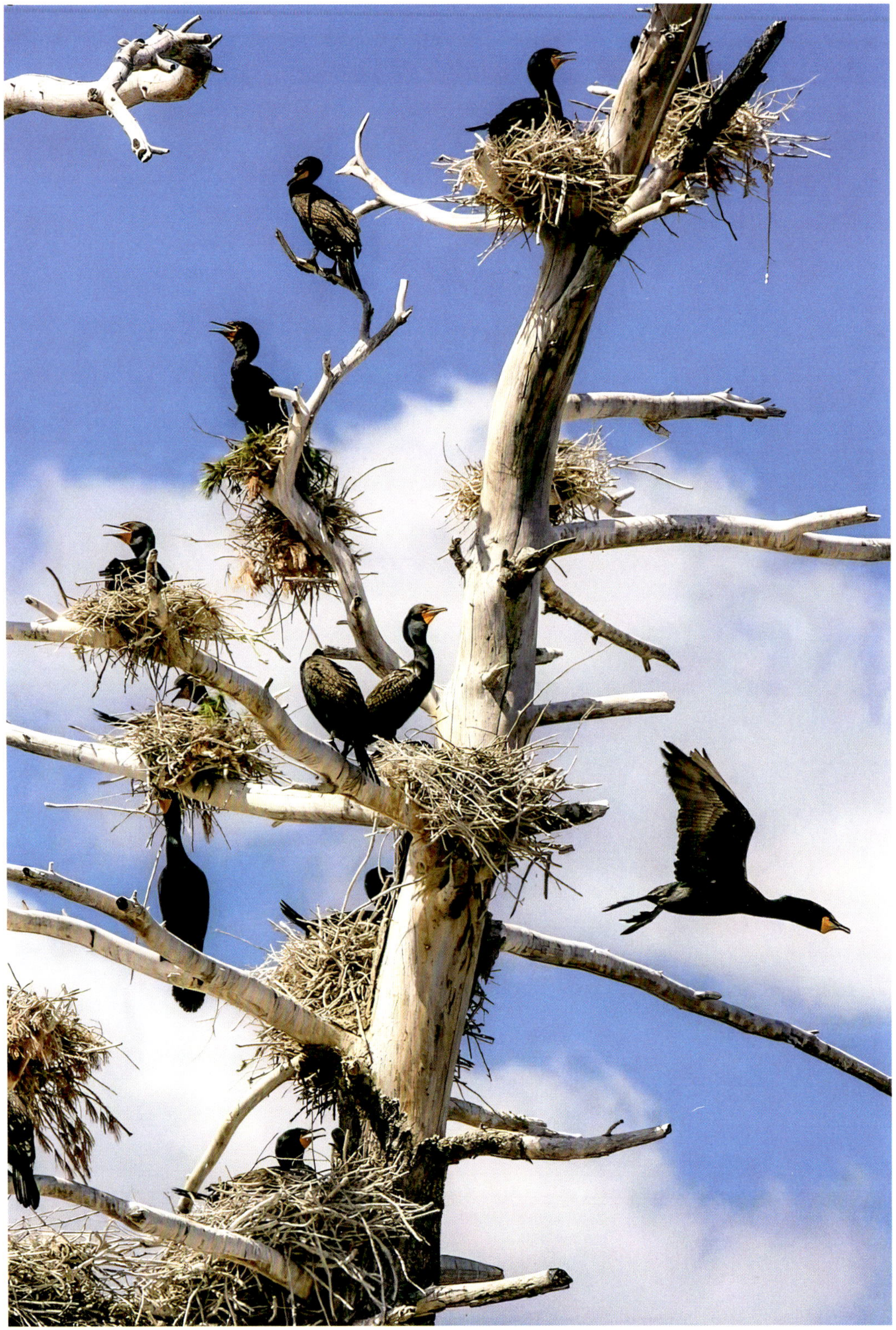

Roughly a quarter of gull and tern species—often icons of the seacoast—nest in the boreal. Many are equally at home on northern lakes, major rivers, or urban landfills, which may help explain why none are IUCN-listed as threatened. Examples include the following:

- Bonaparte's gull (*Chroicocephalus philadelphia*): 95 percent nest in the boreal
- Glaucous gull (*Larus hyperboreus*): a high-arctic and boreal breeder
- Great black-backed gull (*Larus marinus*): the largest gull, 84 percent of whom nest in the boreal
- Ring-billed gull (*Larus delawarensis*): abundant, often urban
- European and American herring gulls (*Larus argentatus* in Eurasia, *Larus smithsonianus* in North America): widespread boreal nesters

On Alaska's Kobuk River near Ambler, I've seen glaucous gulls perched atop white spruce: perfectly logical for the setting but comically at odds with the "gull on a pier" stereotype.

Among terns, both the common tern (*Sterna hirundo*) and the Arctic tern (*Sterna paradisaea*) rely heavily on boreal wetlands worldwide for nesting.

Waterfowl: Family Anatidae (swans, ducks, and geese)

Needing open water, which is in short supply at -40°F (-40°C), waterfowl are motivated migrators, which, with their generally large size, makes them of ecological importance for nutrient and resource transport between ecosystems.

The largest are the swans. In addition to the recovering trumpeter swan, two other species are found in the boreal forest, though only one is a regular nester.

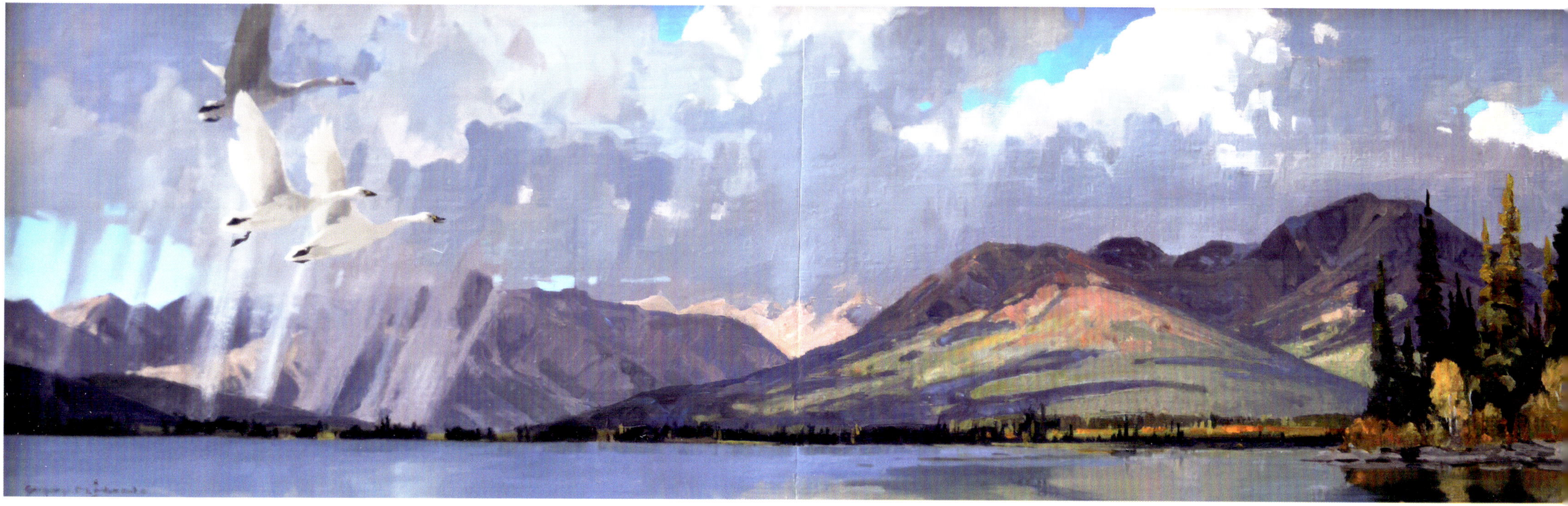

Whooper swan (*Cygnus cygnus*): Nesting throughout the Eurasian boreal forest, whooper swans are closely related to trumpeter swans but with extensive yellow on their bills.

Tundra swan (*Cygnus columbianus*): Smaller (though still over 20 pounds) and with only a yellow spot at the bill base, tundra swans are regular migrants through boreal wetlands.

Swans are devoted, powerful, and aggressively defensive parents (unlike most ducks, they mate for life), and their size makes them potentially hazardous to approach during their nesting season. For that matter, it is not a good idea to tangle with Canada goose parents either, as they can have a wingspan of five to six feet and weigh 10 pounds (our dogs are afraid of them for good reason).

Canada goose (*Branta canadensis*): Nearctic with a population of over a million, Canada geese are ubiquitous both in the wilderness of the boreal forest and on lawns and golf courses across suburbia. This leads many people to take them for granted and even revile them, which, their lawn decorations notwithstanding, they don't deserve. Geese generally also mate for life.

Greater white-fronted goose (*Anser albifrons*): More of an arctic breeder, these geese nest in the boreal forest across Alaska and Western Canada and into the boreal forest's northern margin in Siberia. They are Holarctic in distribution.

Tundra swans, Alaska. *Brooks Range Haven*, oil, Greg McHuron

FACING: Double-crested cormorants, Lake Opeongo, Algonquin Park, Ontario. Steve Bruno

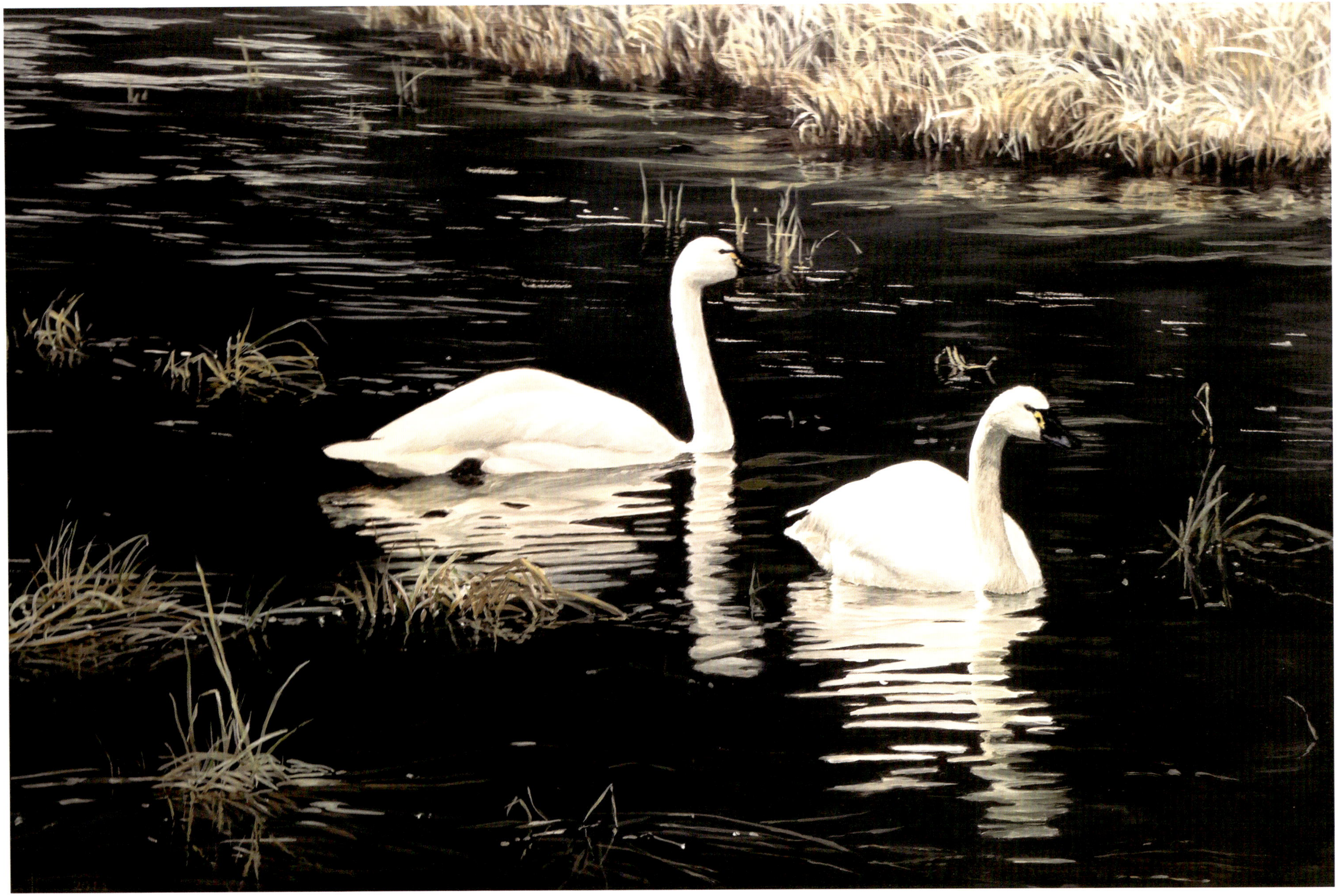

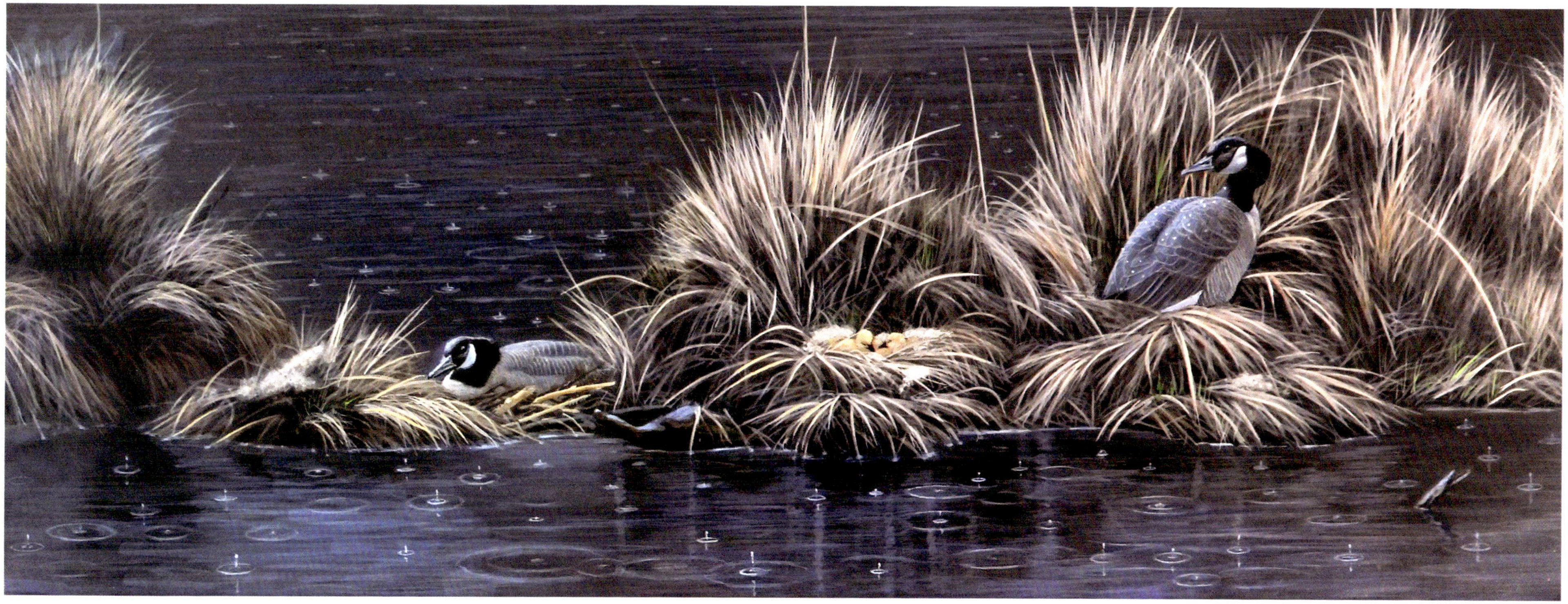

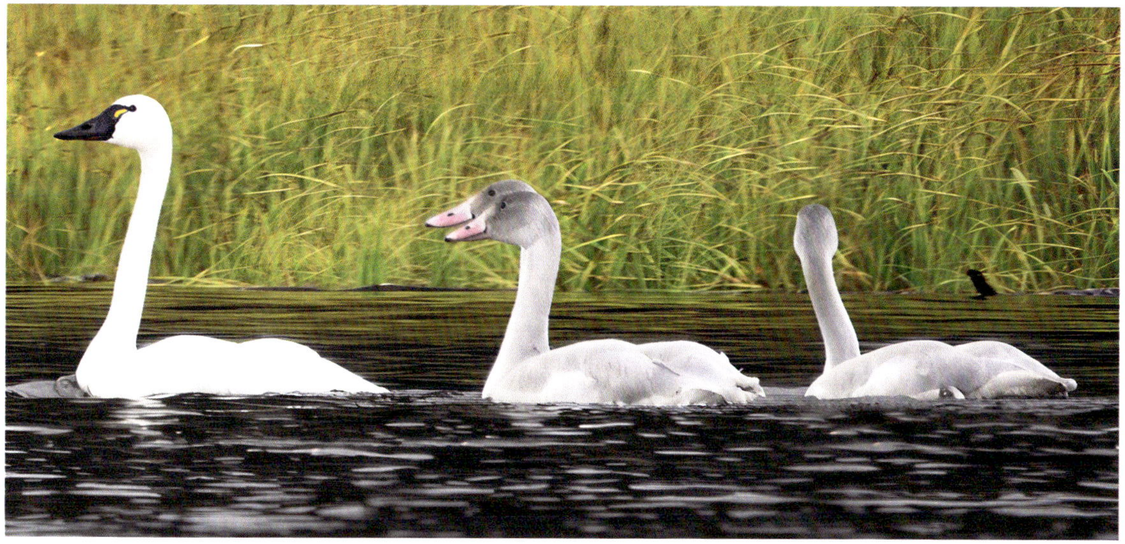

ABOVE: Canada Goose family, Preston Pond, West Bolton, Vermont. *Old Dam Family*, acrylic, Rob Mullen

LEFT: Tundra swan family, Kobuk River, Alaska.

FACING: Tundra swans. *Tundra Beauties*, acrylic, Lindsey Foggett

CLOCKWISE FROM TOP LEFT:
Whooper swans, Iceland. Steve Bruno

Greater white-fronted geese in flight, Kobuk River, Alaska.

Northern pintail. *Flight of Pintails*, charcoal, Cole Johnson

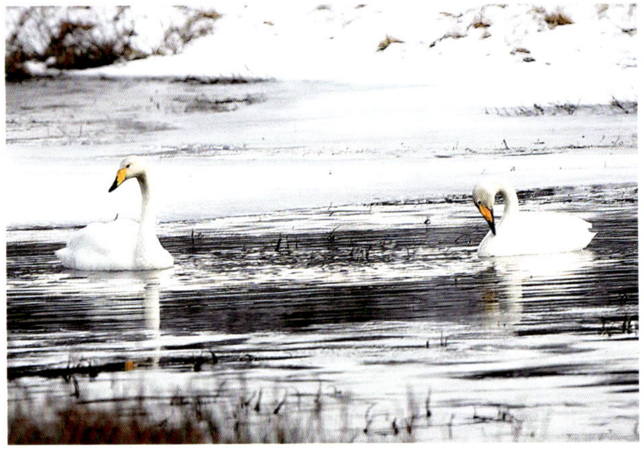

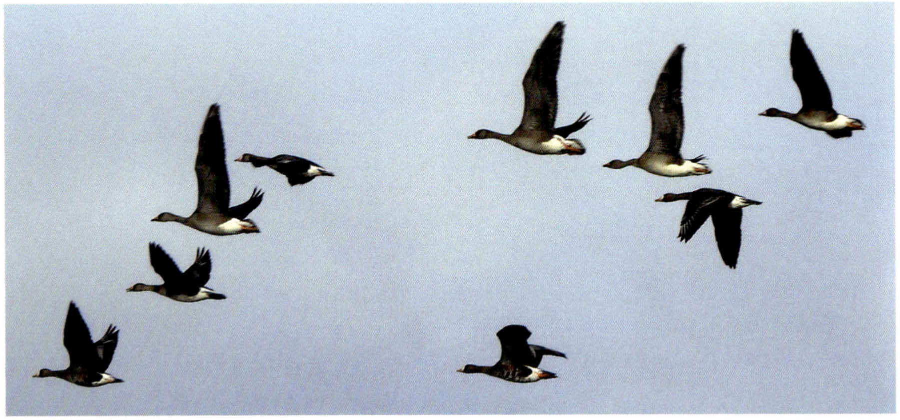

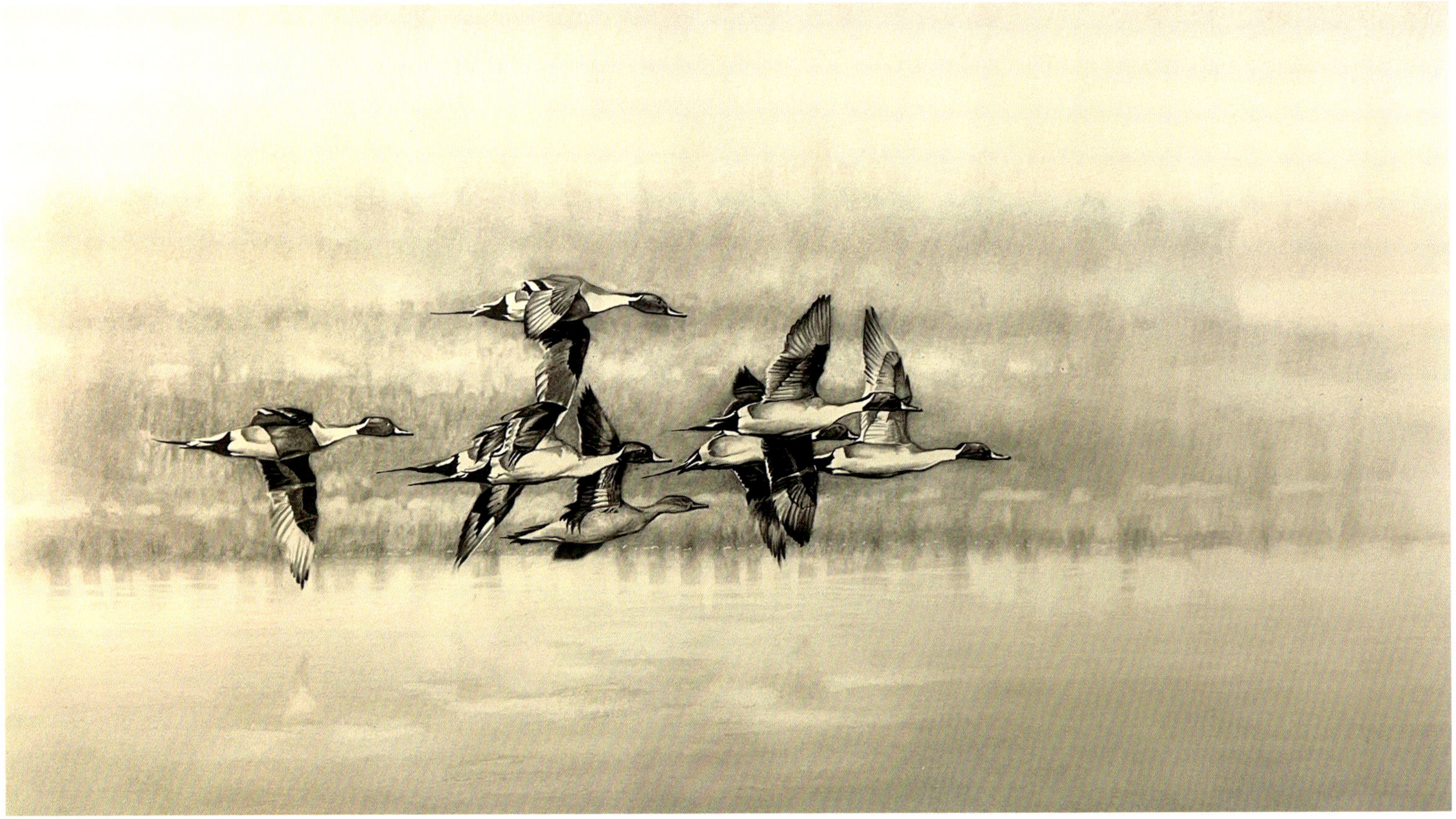

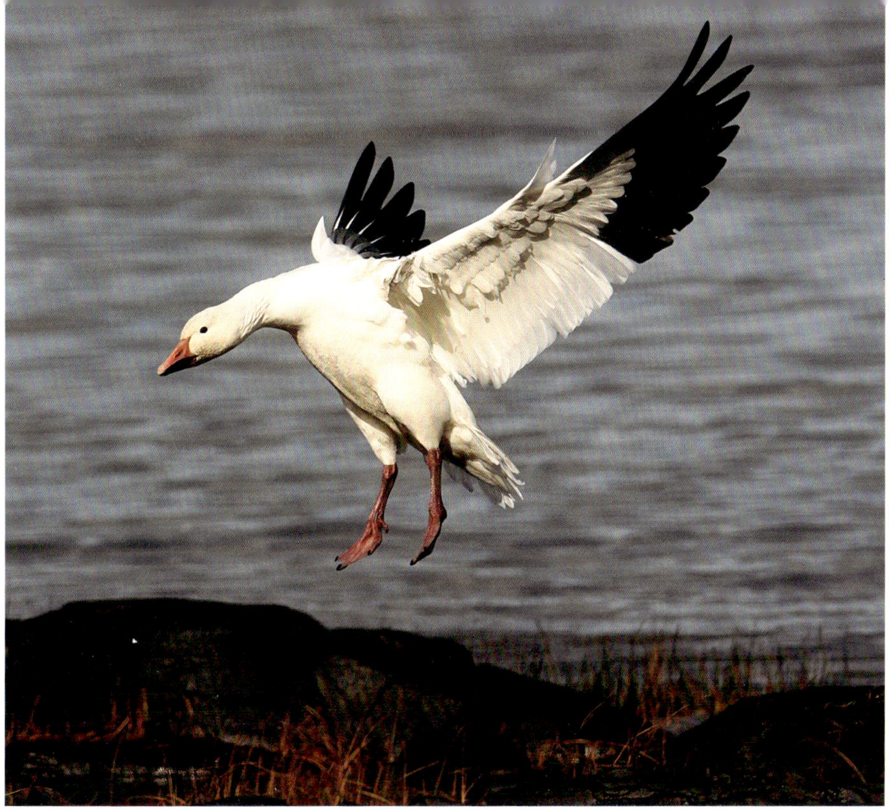

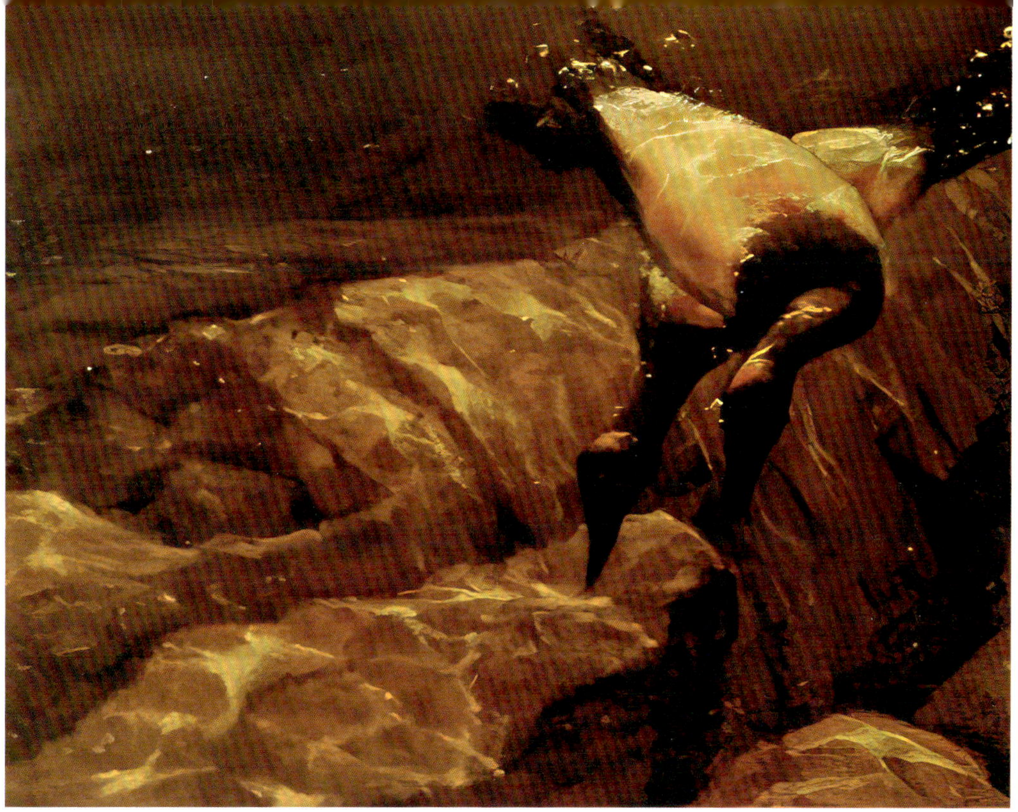

LEFT: Touchdown: snow goose, Montmagny, Quebec. Bonnie Rowell

RIGHT: Canvasback ducks, Quebec. *Water Dance*, oil, Jean-Louis Courteau

Snow goose (*Anser caerulescens*): Nearctic breeders that migrate through the boreal forest and points south in huge—and getting huger—flocks. If you visit a flyway stopover to view the spectacle of thousands of snow geese in one spot, I strongly advise wearing a broad-brimmed hat. When scared or suddenly remembering an appointment, they take off en masse. It is an incredible sight, but they also will "drop ballast" to help gain altitude. It will literally rain goose guano.

Ducks

Ducks punch above their weight in human affairs. Large, meaty, and beautiful, they spur sporting goods companies, art, and gastronomy, and in the 1930s, they even shaped conservation history. Hunters, alarmed by Dust Bowl–era population crashes, were a driving force in the creation of the US National Wildlife Refuge System and Ducks Unlimited. Ecologically, ducks may be no more important than sandpipers, but when it comes to inspiring people to value nature, they play an outsized role.

They fall into two subfamilies:

- *Dabblers* (subfamily Anatinae): Feeding at or near the surface, dabblers "tip up" tail-first to reach vegetative matter and small invertebrates on the bottom of shallows. Their legs are positioned mid-body so that their feet can keep them in position to feed on the bottom.
- *Divers* (subfamily Aythyinae): Foraging on the bottom in deeper water or catching prey in direct pursuit, divers have legs that are positioned astern to provide better thrust underwater.

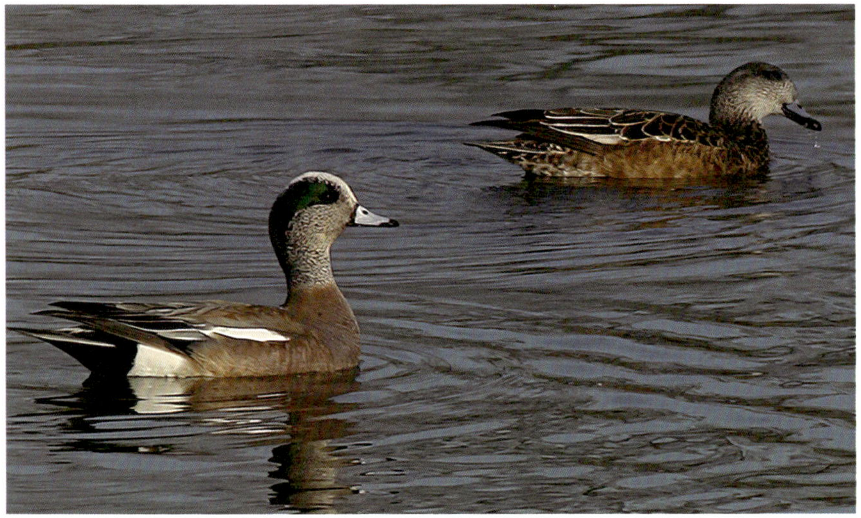

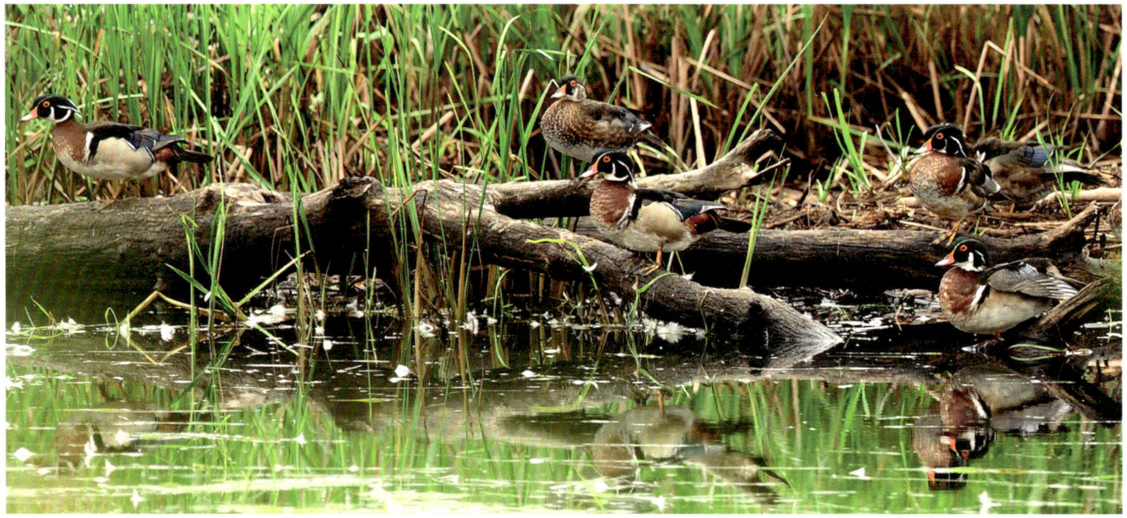

LEFT: Widgeon pair: American wigeon drake and hen.

RIGHT: Wood duck drake stag party, Île Bizard, Quebec. Bonnie Rowell

An easy visual distinction between the two is when they take off. Dabblers, with their legs positioned in the middle of their bodies, can jump straight up into flight like they are late for a meeting. Divers, their legs being farther back like loons, need a run across the water surface to build up speed to take off.

Dabblers

Mallard (*Anas platyrhynchos*): Holarctic in distribution; breed widely across the boreal forest.

American wigeon (*Mareca americana*) and Eurasian wigeon (*Mareca penelope*): "Baldpates"; formerly in genus *Anas*.

Green-winged teal (*Anas carolinensis*): Nearctic in distribution; weigh about 0.75 pounds. Along with the bufflehead (a diver), they are our smallest ducks.

Blue-winged teal (*Spatula discors*): Nearctic in distribution.

Northern pintail (*Anas acuta*): Holarctic in distribution.

Gadwall (*Mareca strepera*): Holarctic in distribution.

Wood duck (*Aix sponsa*): Only 5 percent are boreal nesting, but they are North America's answer to the spectacularly beautiful Mandarin duck.

Divers

As far north as most of our journeys have been and, to miss the height of the bug season, as late in the summer (and early fall), the average duck has the sense to head south well ahead of us. The exceptions are some of the divers. We have often seen goldeneyes and common mergansers, some with young still in attendance and sometimes just the young, since they tend to leave after their parents.

Common goldeneye (*Bucephala clangula*): Holarctic in distribution. With 83 percent of the North American population nesting in the boreal forest (and likely a similar proportion in Eurasia), common goldeneyes are the quintessential boreal forest diving duck, nesting throughout the biome worldwide.

Barrow's goldeneye (*Bucephala islandica*): Nearctic in distribution (western North America including Alaska and western Canada).

Bufflehead (*Bucephala albeola*): Nearctic in distribution, with nearly 90 percent nesting in the boreal forest. Generally tied with the green-winged teal as North America's smallest duck; also known as butterball and the spirit duck. Most ducks are seasonally monogamous, but not the little bufflehead. As with swans, buffleheads mate for life. And they are cavity nesters, strongly favoring abandoned flicker holes, facilitated by their small size.

MERGANSERS ARE distinguished from other divers by their long, narrow, saw-like serrated-edged bills that make them look like flying gars. While the serrations are not teeth, they perform the same function in securing the ducks' general diet of fish.

Common merganser (*Mergus merganser*), aka goosander in Europe: Another cavity nester that uses the same sites as wood ducks. After breeding, the males depart. This is why we have often met only female and young common mergansers on trips. They range around the globe, occupying all of the boreal forest except for northern Alaska and northeastern Siberia. Over 50 percent of North American common mergansers nest in the boreal forest.

Red-breasted merganser (*Mergus serrator*): IUCN Red List status is Least Concern globally but threatened in Europe. Red-breasted mergansers have a wide global range that extends farther north than the common merganser and spans the entire boreal forest. The North American population uses the forest for about 46 percent of its nesting and typically winters in saltwater on the coast.

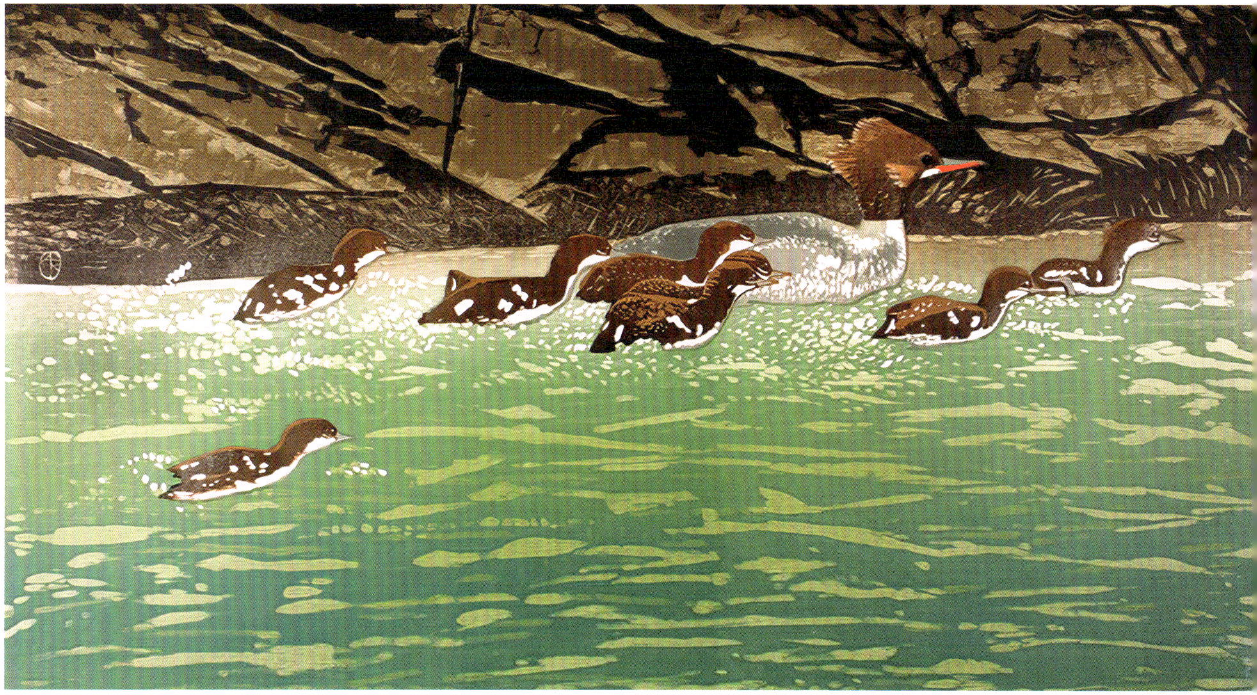

Common merganser hen and her brood, Lake Superior North Shore expedition. *Merganser Brood*, woodcut print, Andrea Rich

Red-breasted mergansers' plumage makes them look like the punk rockers of the waterfowl world. Both sexes have a shaggy, double crest that often looks like they just got up, bright red bills (thinner than the common mergansers'), and bright red eyes, which stand out dramatically from the drakes' dark green heads. The hens are dressed similarly as the common merganser hens, but they are a bit smaller and have shaggier crests. The definitive field mark is the gradual transition on their necks from gray body to reddish-brown head (common merganser hens have a sharp line).

Hooded merganser (*Lophodytes cucullatus*): The smallest and most smartly dressed North American merganser, with over half nesting in the boreal forest from Alaska to Nova Scotia.

TOP: Hoodie drake. *Spring Snow*, graphite, Cole Johnson

BOTTOM: Hoodie drake driving off a rival, Île Bizard, Quebec. *Marsh Missile*, acrylic, Rob Mullen

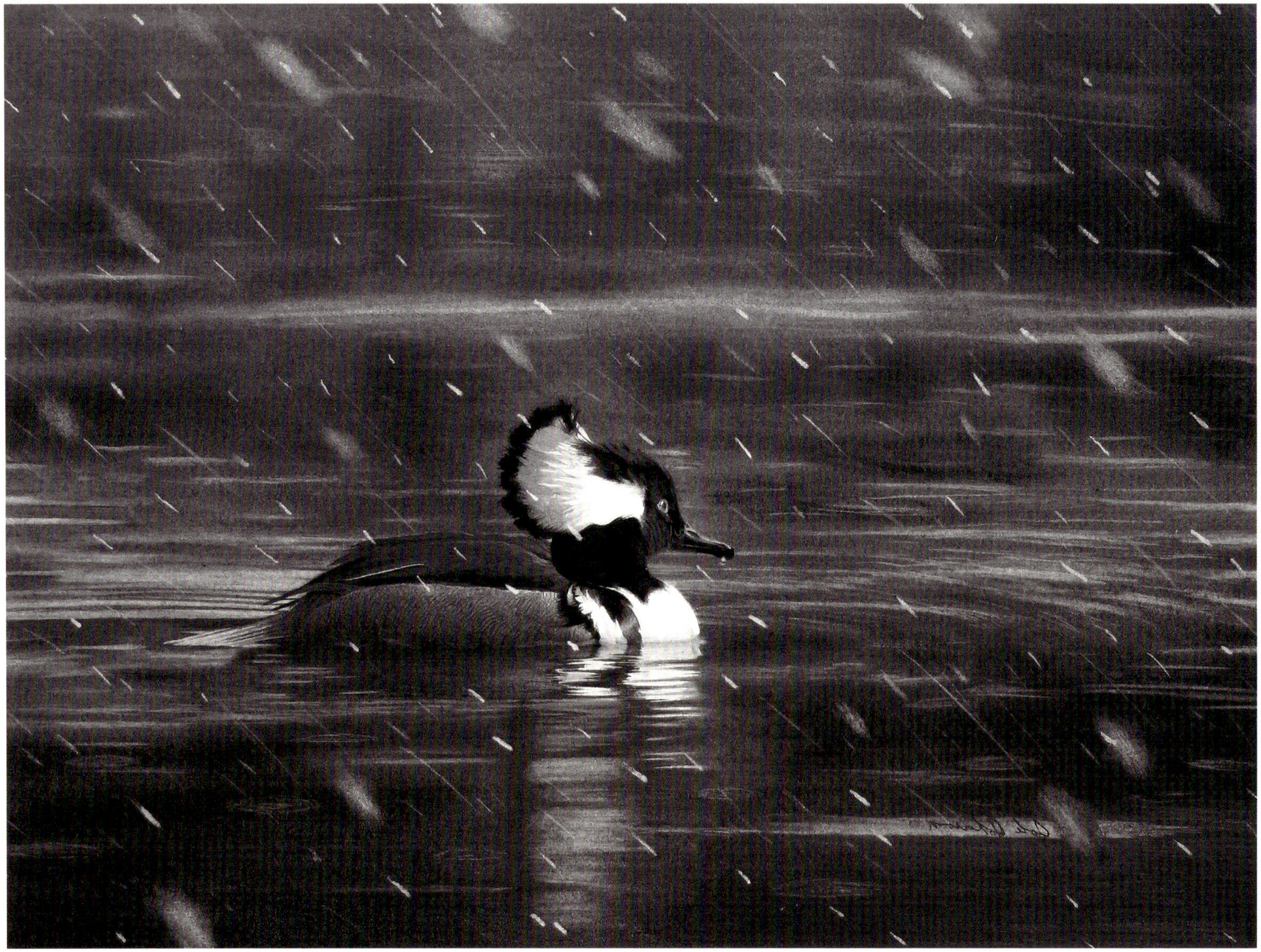

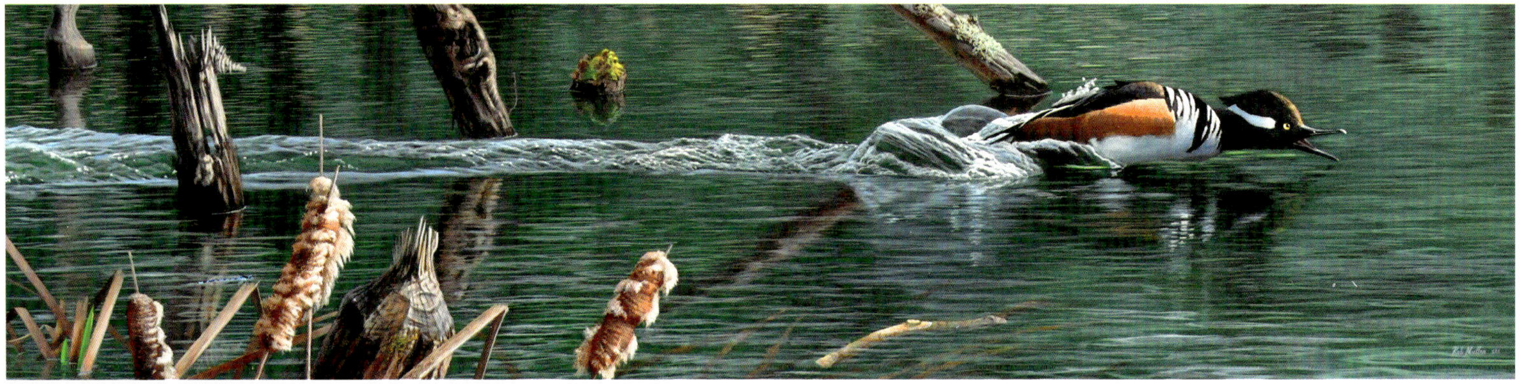

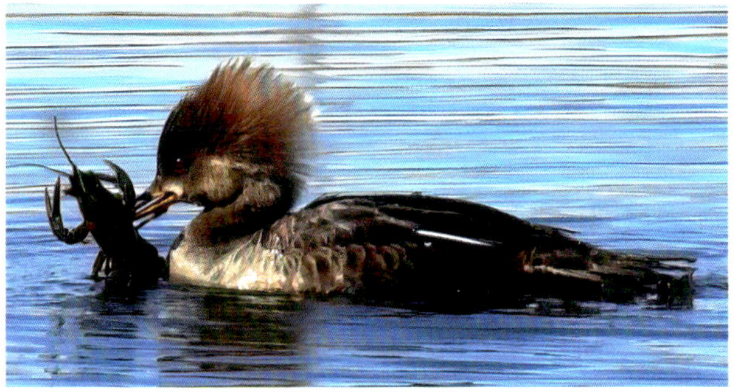

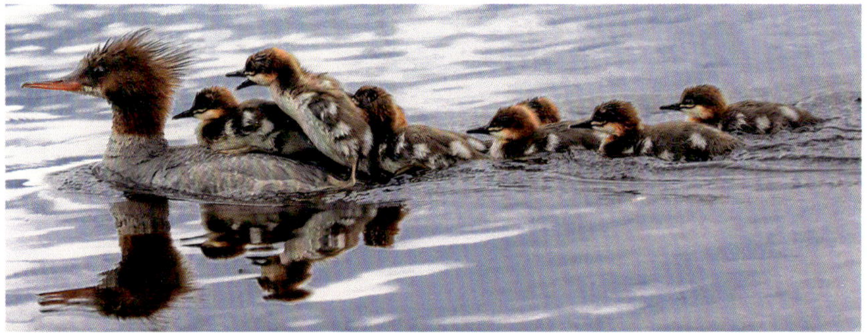

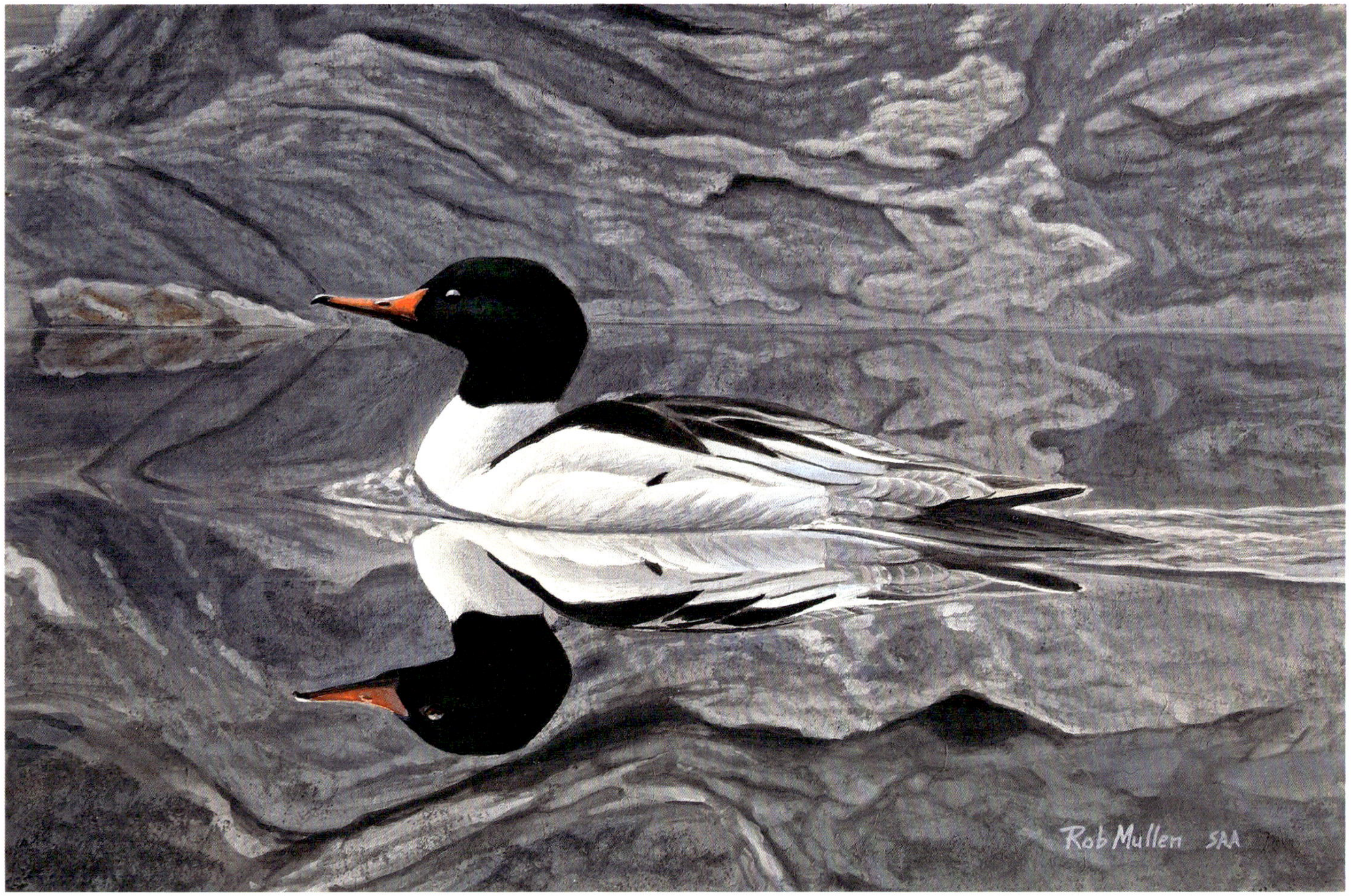

CLOCKWISE FROM TOP LEFT:
Biting off more than you can chew: hooded merganser hen, Île Bizard, Quebec. Bonnie Rowell

All aboard: common merganser hen with ducklings, Algonquin Park, Ontario. Steve Bruno

Common merganser drake, April dawn on Preston Pond. *Cutting a Wake*, acrylic, Rob Mullen

TOP: Bufflehead drakes and hens. Charcoal, Cole Johnson

BOTTOM: Harlequin hen and ducklings possibly washed out to sea by heavy July storms, including the one that battered us off the coast of New Brunswick (*Skidbladnir*'s Journey Home, chapter 3), Chaleur Bay, Chandler, Quebec, July 2023.

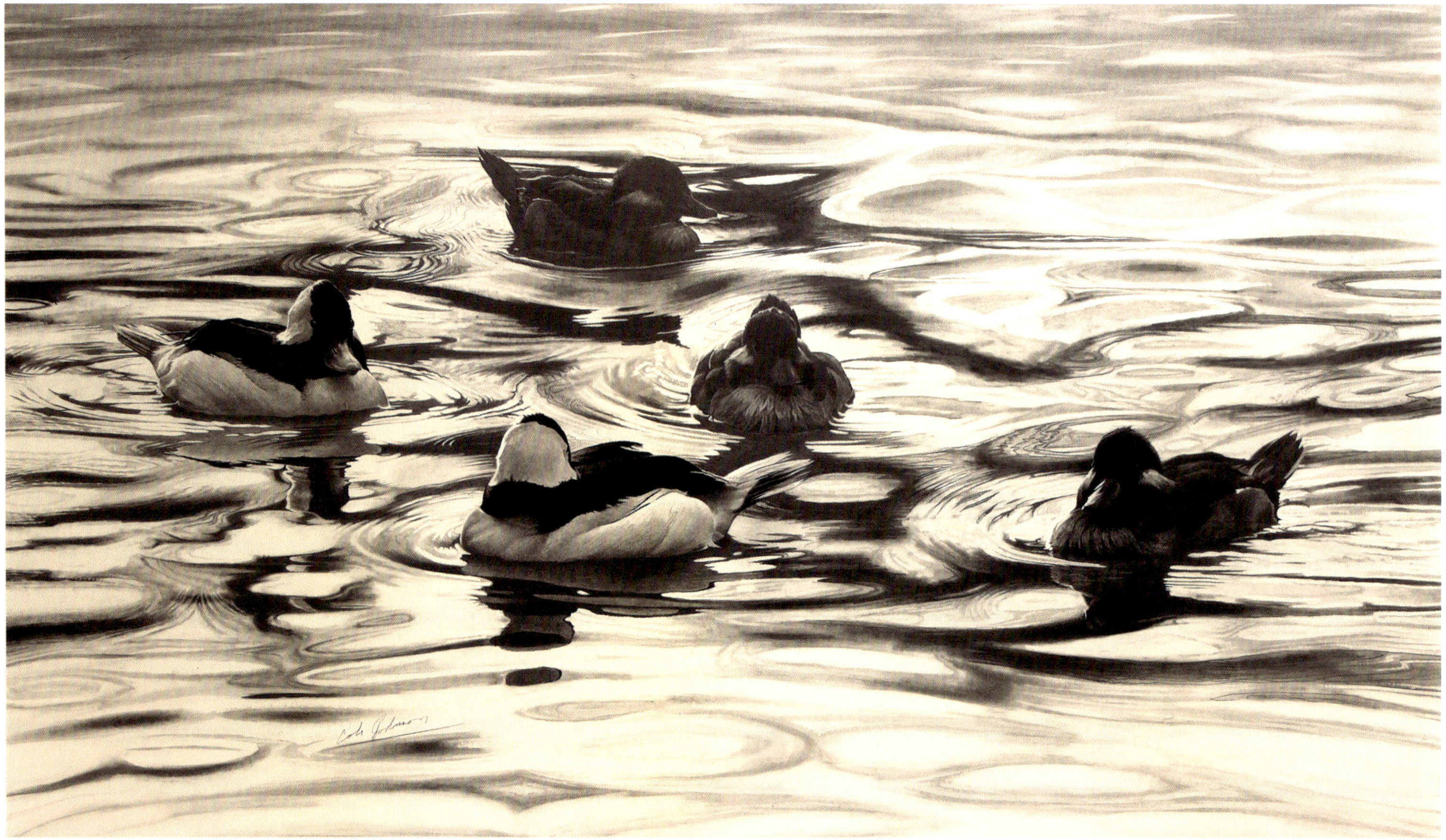

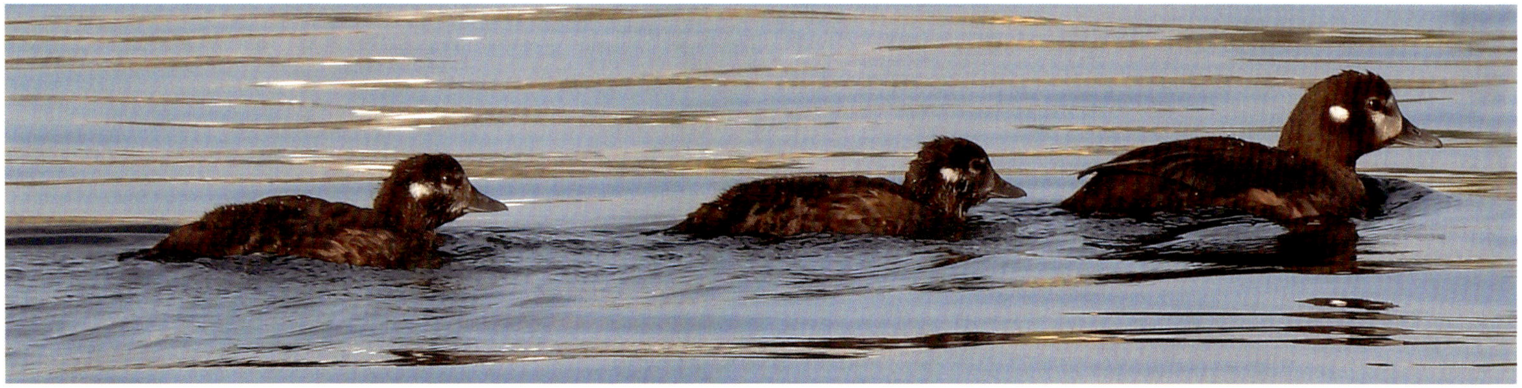

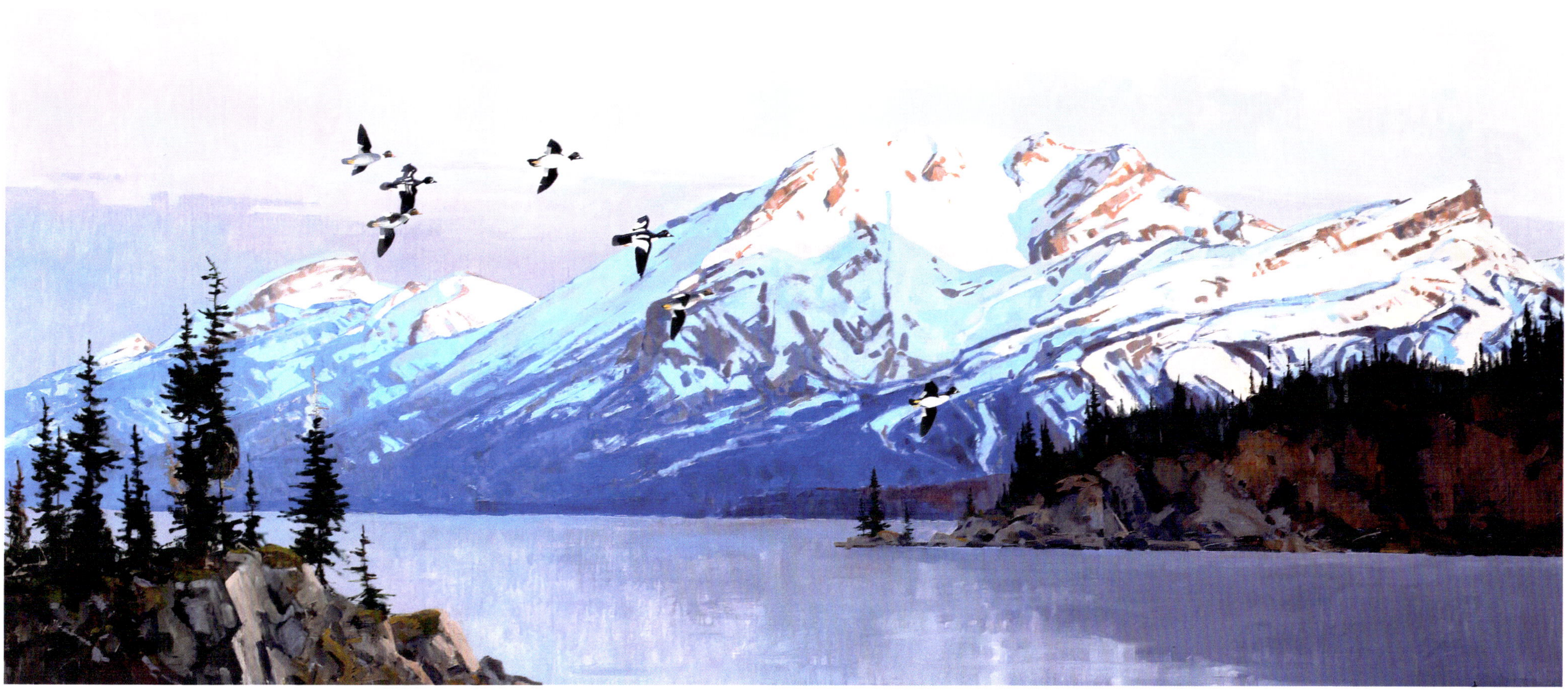

Specialists

Harlequin duck (*Histrionicus histrionicus*): IUCN Red List status is Least Concern globally, though the eastern North American breeding population is declining. They are a small sea duck in eastern Siberia and northwestern North America, with a smaller population in northeastern North America (the declining one).

The drake's dramatic breeding coloration is among the most flamboyant of any duck. And the remarkable plumage isn't the only distinctive trait. Displaying a penchant for the wild side, Harlequins eschew the usual quiet ponds and back bays for the chaotic, noisy violence of white-water rapids as breeding sites and even prefer tempestuous rocky shores on the coast for winter quarters, for which the drakes depart as soon as mating is over (unsurprisingly, they appear to suffer more broken bones than any other bird). Despite its almost garish patterning, the colors and patterns of the drakes remarkably blend in with the turbulence. The hens also bear white spots on their heads, which can be confused for the spray and spume of a rapid.

Common goldeneyes, Brooks Range, Alaska. *Golden Sunset Retreat*, oil, Greg McHuron

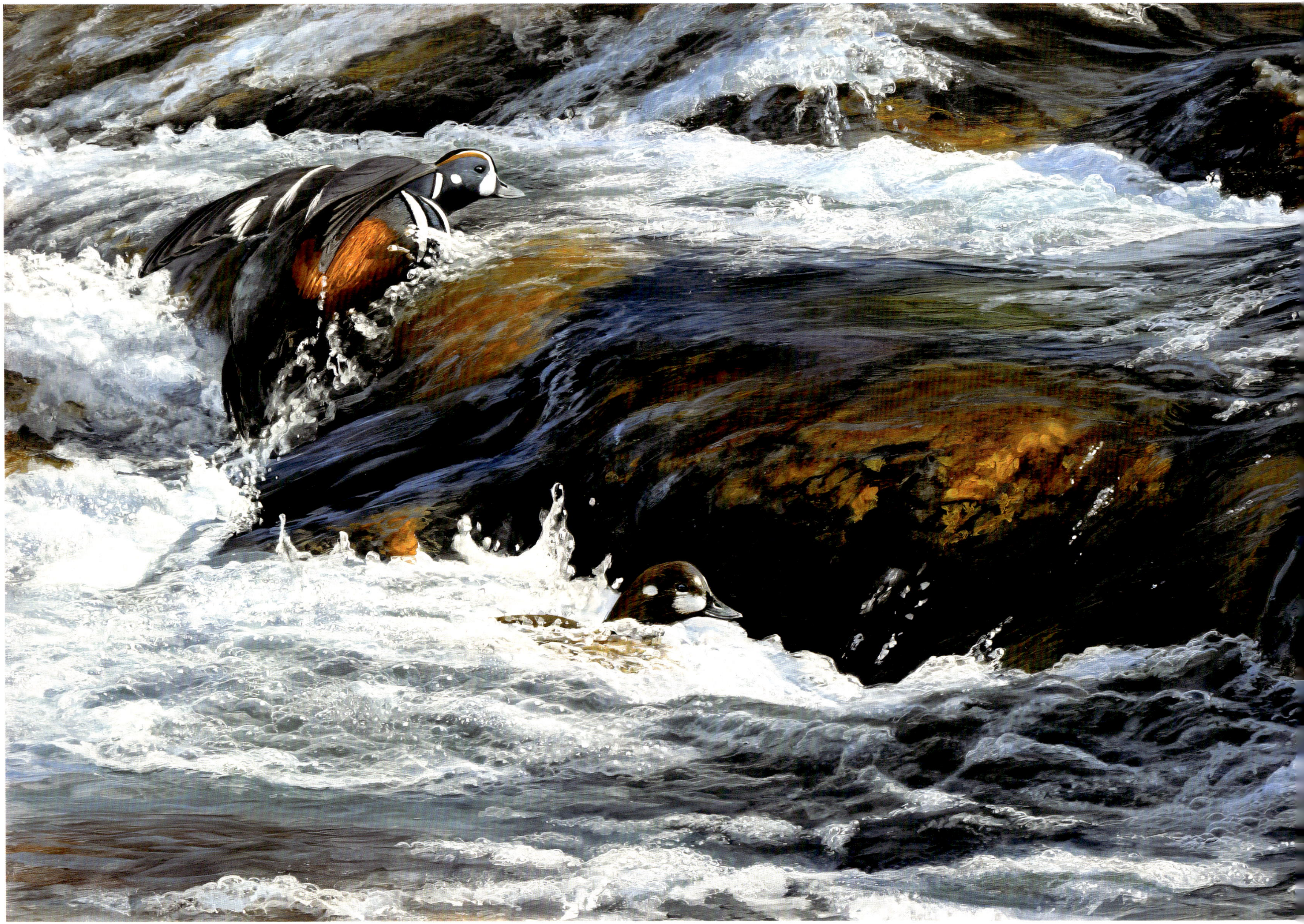

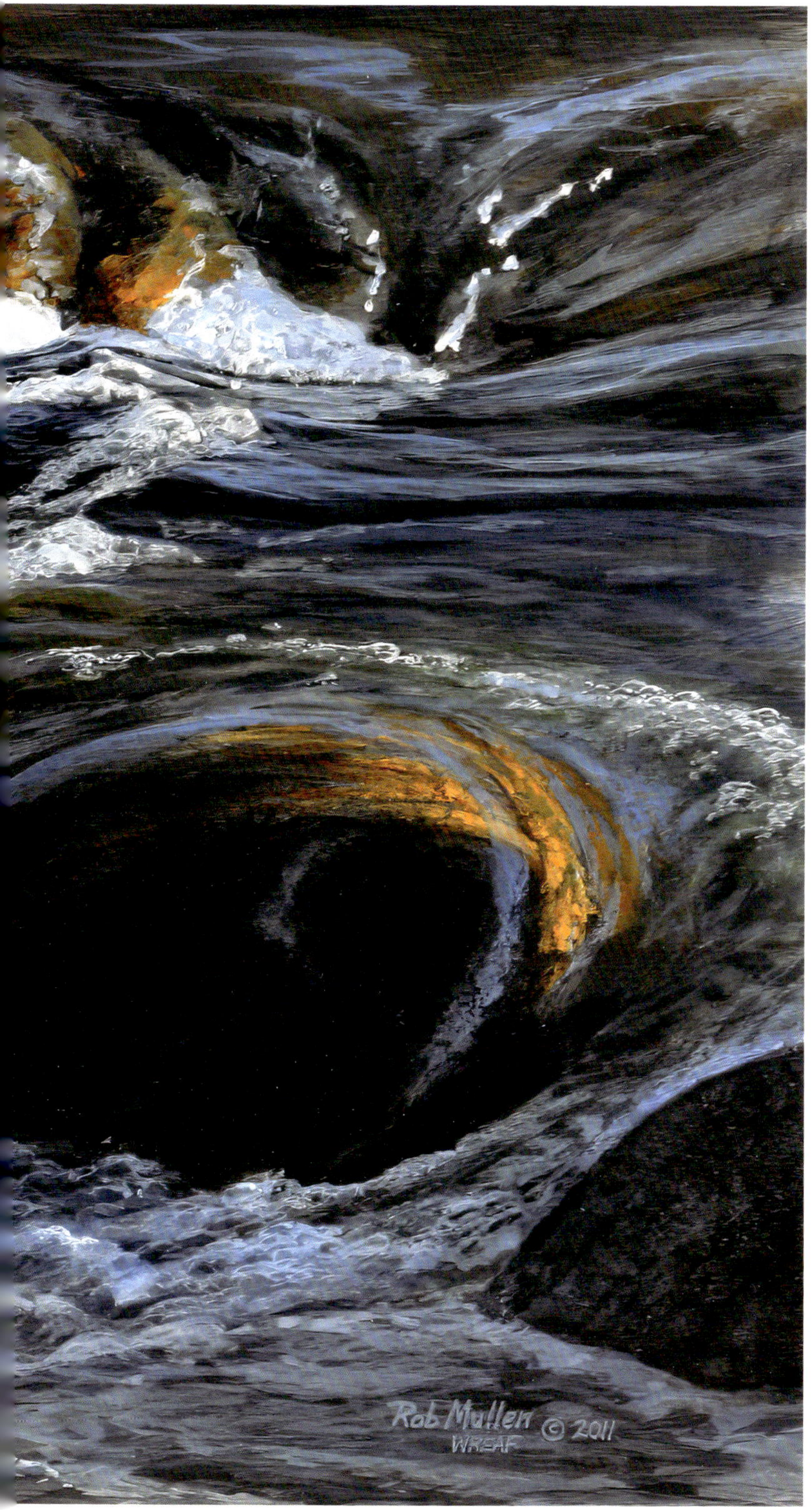

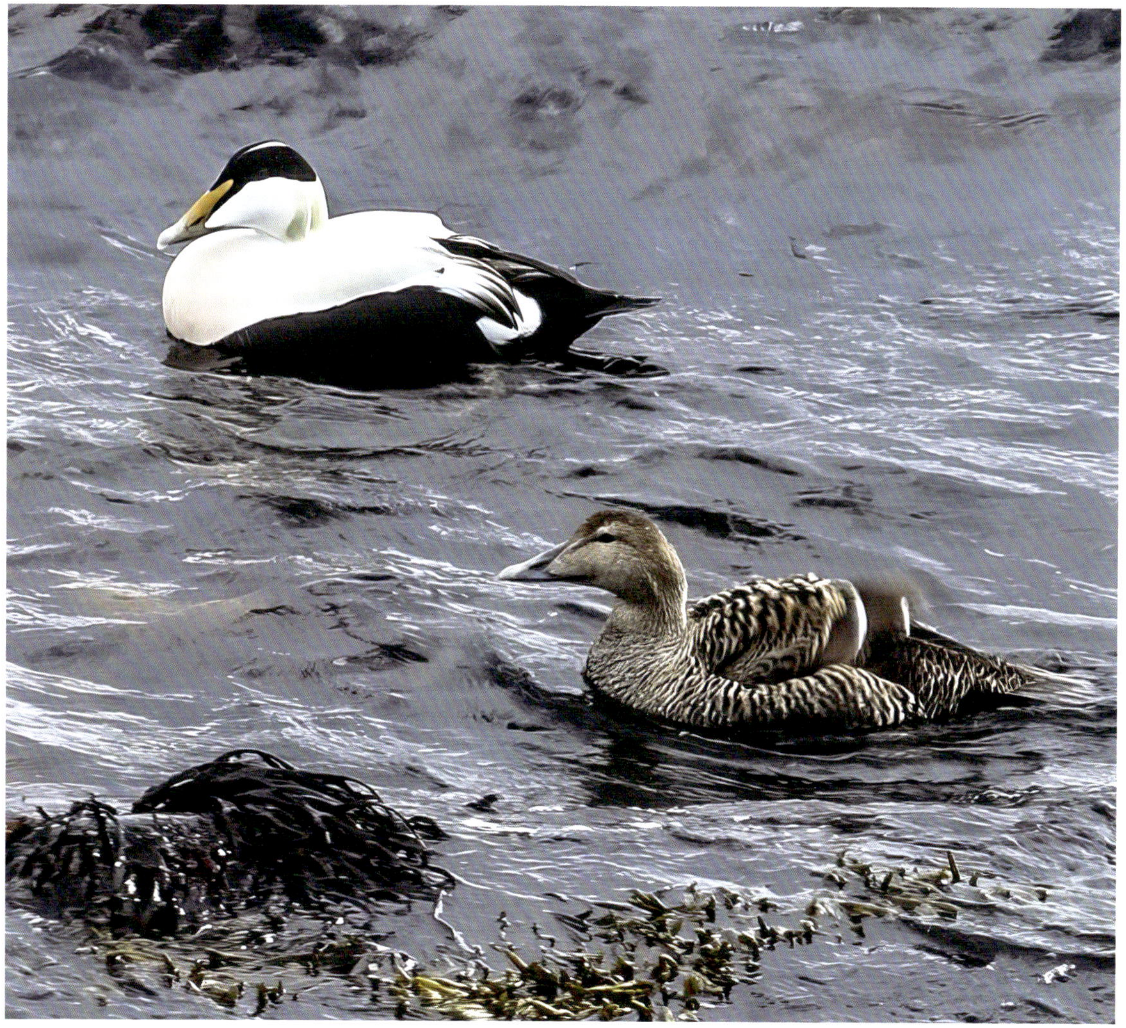

Common eider ducks, Labrador.
Steve Bruno

FACING: Harlequin ducks, Kamestastin River, Labrador. *Turbulent Currents*, acrylic, Rob Mullen

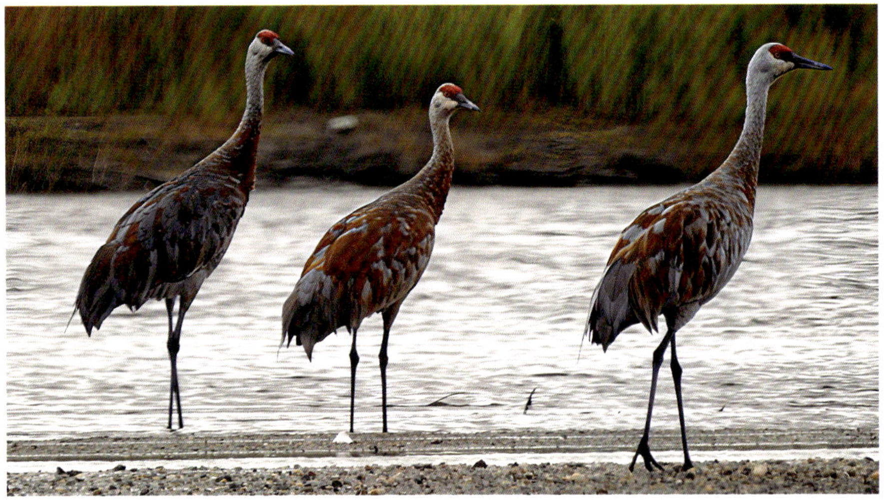

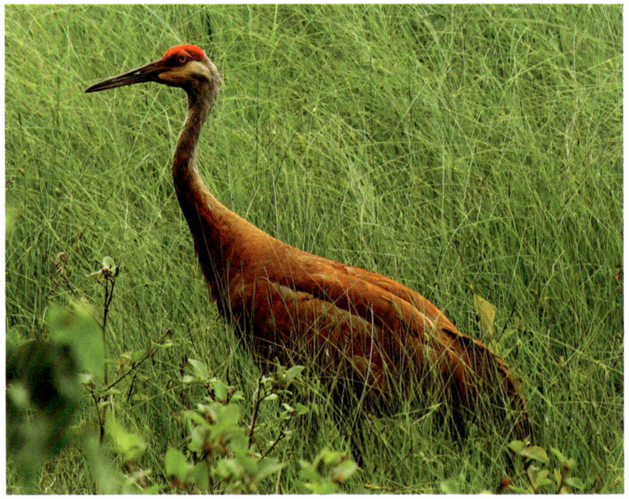

LEFT: Lesser sandhill cranes on the Kobuk River, 2015.

RIGHT: Greater sandhill crane, Upper Peninsula, Michigan. Bonnie Rowell

Common eider (*Somateria mollissima*): IUCN Red List status is Near Threatened. Their wingspan is over three feet; their weight, over five pounds. They are the largest sea duck and second largest duck in the world, behind only the Muscovy duck. Coastal birds, common eiders use the margins of boreal forest habitat for breeding in Alaska, Hudson Bay, Atlantic Canada, Maine, Scandinavia, and Siberia. Chief among their favored foods are mussels, which, like many diving ducks, they eat whole, crushing the shells in their gizzards. Eiders are famous for their down, which they use for their nest.

Marsh Birds: Order Gruiformes, families Gruidae and Rallidae (cranes, rails, and kin)

We already covered the famous whooping crane, so continuing with a "keystone species" for this book (the birds I met on September 11, 2001, on the Missinaibi), I'd like to introduce sandhill cranes.

Canadian sandhill crane (*Antigone canadensis rowani*): They have a wingspan of up to six-and-a-half feet and breed in central Canada. This is the subspecies of the birds that were feeding by my canoe on the Missinaibi River. Whether they are a separate subspecies from the greater sandhill cranes (*Antigone canadensis tabida*) that breed in the United States is under debate.

Lesser sandhill crane (*Antigone canadensis canadensis*): With a standing height up to four-and-a-half feet and a wingspan of up to six feet, lesser sandhill cranes breed from eastern Siberia through Alaska and northern Canada.

Size differences aside, all subspecies are similar in appearance (greater sandhills' bills can be one-and-a-half times longer than those of lesser sandhills). In a seeming contradiction to Bergmann's rule, the lesser sandhills that breed in the far north are, as their name implies, the smallest, likely the effect of migration.

The population of sandhill cranes is robust with estimates ranging from 500,000 to 1 million.

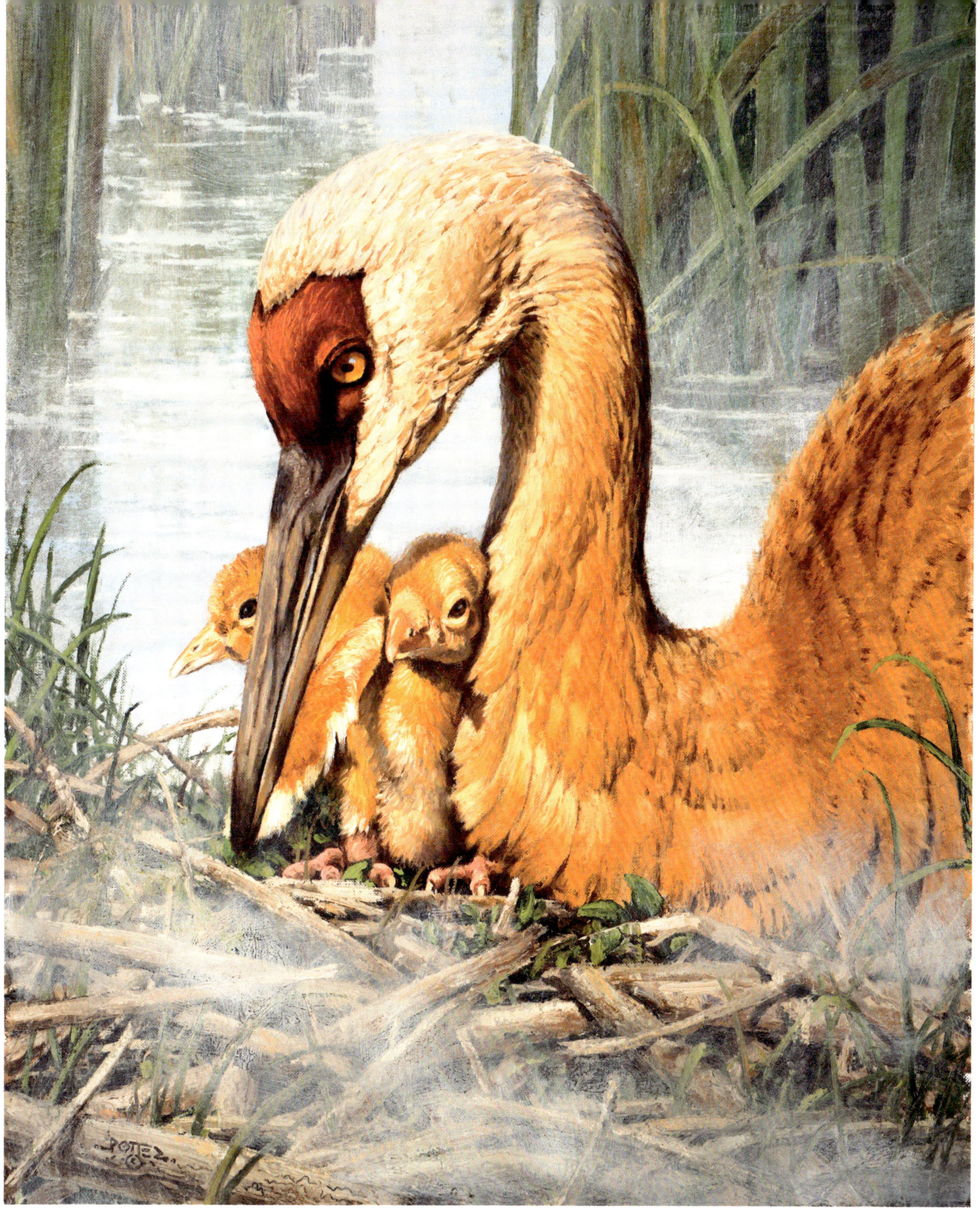

Sandhill crane on a nest.
Nestled In, oil, John Potter

Sandhill cranes on a migration stop in Montana. *Reflections*, oil, Kalon Baughan

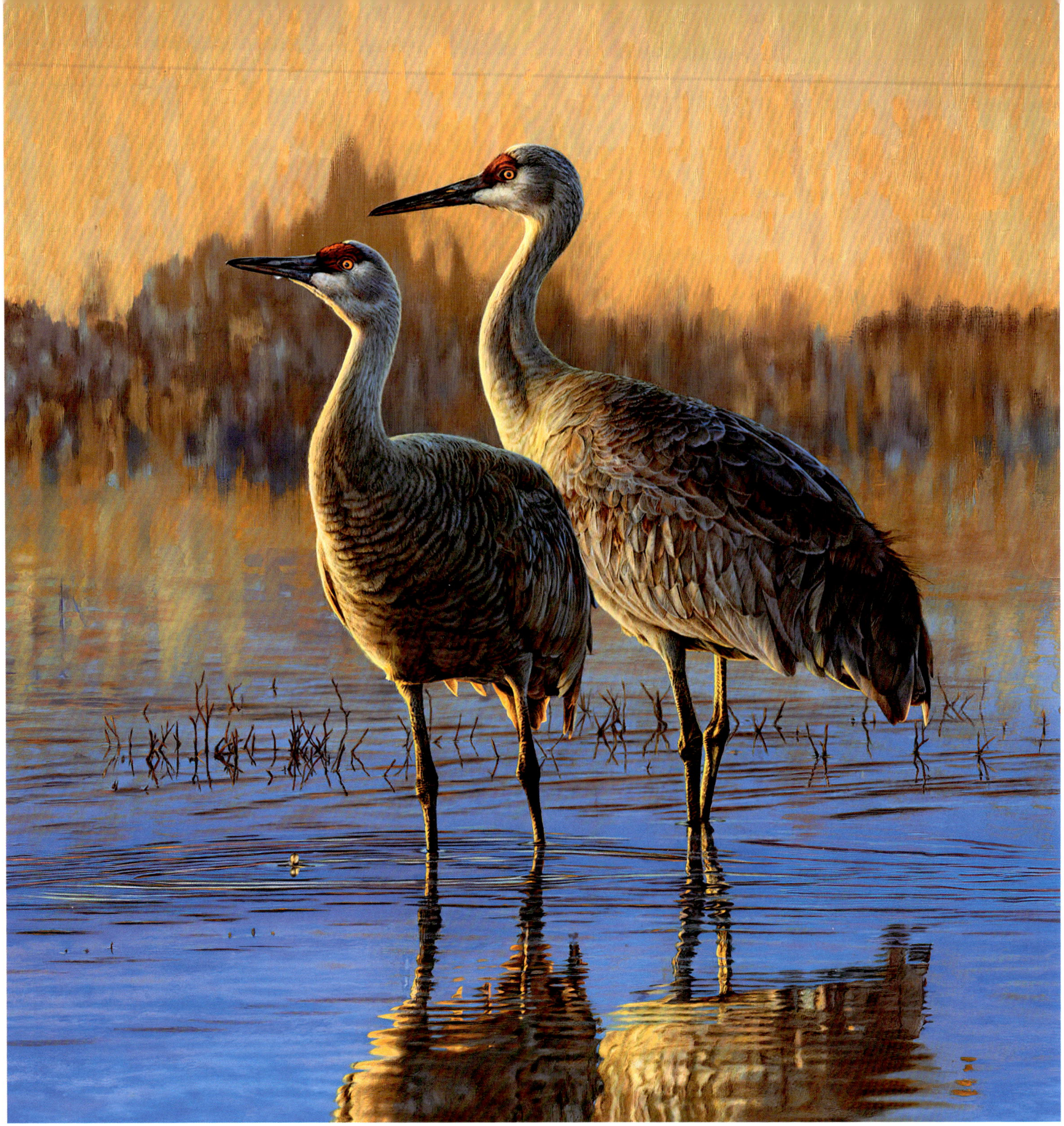

While generally herbivorous if given a choice (exploding snow goose populations compete for waste corn during migration), when necessity strikes, they are readily omnivorous—especially the northern populations (they'll eat all the food that fits). There are, after all, fewer corn fields in the boreal.

Sora (*Porzana carolina*): An estimated 67 percent nest in the boreal forest.

Yellow rail (*Coturnicops noveboracensis*): An estimated 57 percent nest in the boreal forest. While related to cranes, both of these small, short-legged, short-tailed, and secretive marsh birds are hard to see, even if within a few feet. An easy way to identify yellow rails is by the male's territorial/mating call, a staccato monotone of clicks, like two stones tapping together.

Wading birds: Family Ardeidae (herons)

For water-loving birds, most herons take remarkably little interest in the wettest terrestrial biome on Earth. The only two with a significant presence in North America are the American bittern (*Botaurus lentiginosus*) and the great blue heron (*Ardea herodias*) and a complementary pair in Eurasia, the Eurasian bittern (*Botaurus stellaris*) and the gray heron (*Ardea cinerea*).

Most herons, if not showy, are at least rather just "out there" to be seen. Not so, the bittern. Famous for holding its head straight up to disappear into the grass, its territorial/mating calls are, however, unlike the birds themselves, hard to miss. Starting with an odd *tp tp tp tp*, after a second's pause, they continue with a *conk-KA-blonk . . . conk-KA-blonk* as they gulp air and then, like a schoolyard wit, belch it out in a deep, resonant three "note" mating call that can be heard throughout the dense marsh. Naturally, such a singing voice has earned bitterns a variety of wry nicknames, such as "thunder pumper," "marsh boomer," and "bog belcher" (as far as I know, I made that last one up).

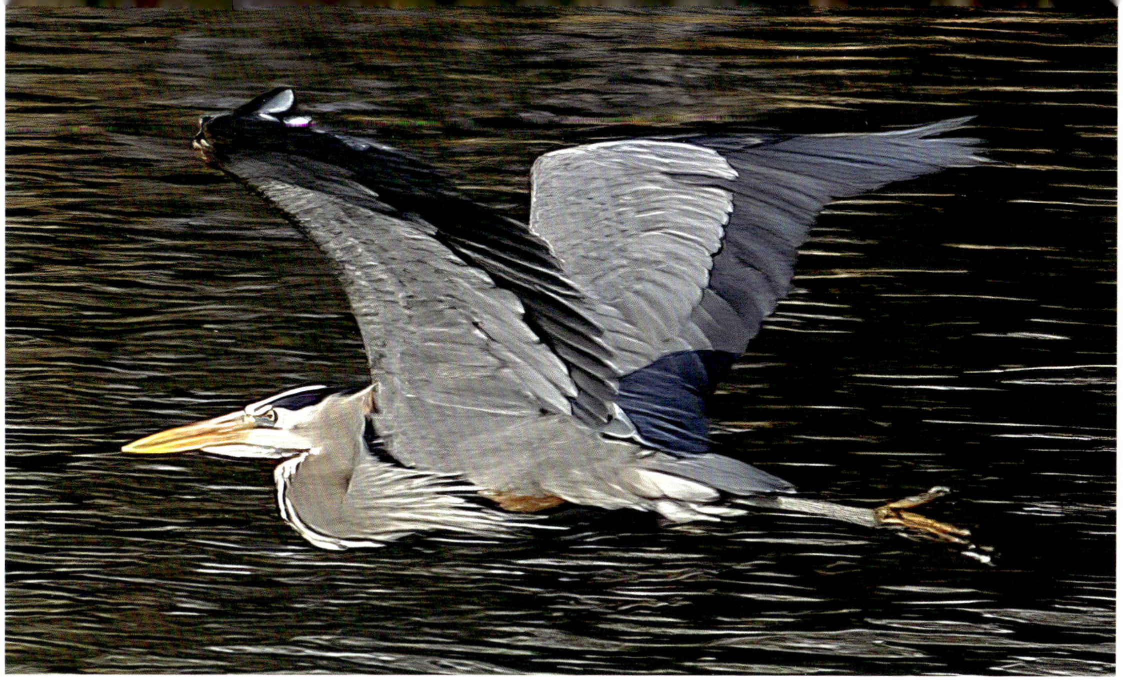

Great blue heron in flight, Lachine Rapids, St. Lawrence River

Shorebirds: Families Scolopacidae (sandpipers) and Charadriidae (plovers)

Whatever reservations herons have about the boreal forest, shorebirds don't share them. At least 60 to 70 species of sandpipers and plovers use the boreal for breeding and/or rest and refueling during migration (many migrate to the tropics). About a third of them worldwide are under various levels of threat according to the IUCN Red List. They have no option but to migrate, and many winter in human-stressed habitats. As of November 2024, though not reassessed by the IUCN since 2018, the slender-billed curlew (*Numenius tenuirostris*), which migrated between western Siberia and Morocco, has been declared likely extinct.

Great blue heron in rushes—Fletcher Lake, Algonquin Park, Ontario. Steve Bruno

FACING, CLOCKWISE FROM TOP LEFT:
Lesser yellowlegs, Lake Superior.

Solitary sandpiper, Île Bizard, Quebec. Bonnie Rowell

Semipalmated sandpiper on the shore of Lac Wiyâshâkimî (Clearwater Lakes).

Marbled godwit on the Upper Missouri River Breaks National Monument, Montana, expedition. Bonnie Rowell

Spotted sandpiper in breeding plumage, Labrador coast. Neil Simmons

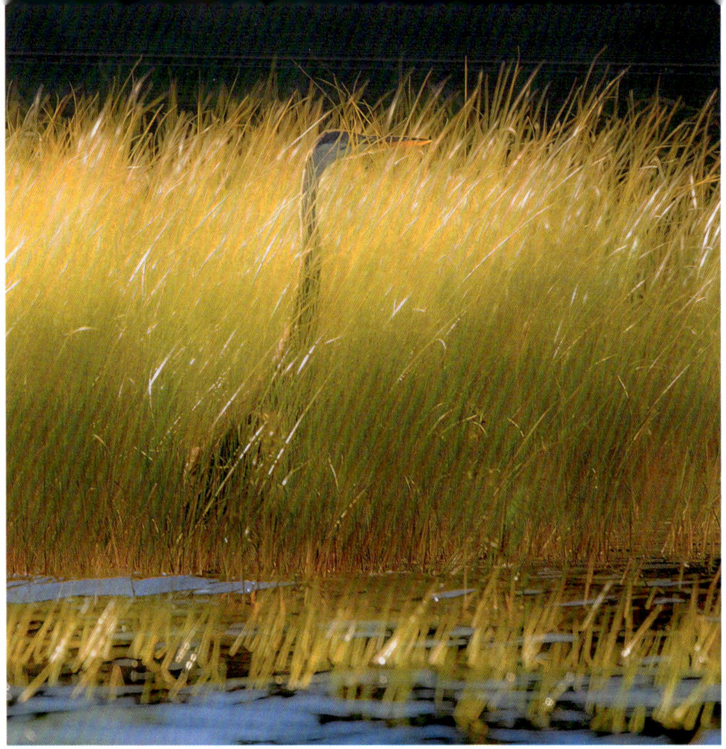

However, with populations in the hundreds of thousands and millions, even with a sizeable proportion of the species under varying degrees of threat, shorebirds still provide significant nutrient and resource transport between ecosystems.

The following are some representative species we've met on expeditions:

Greater yellowlegs (*Tringa melanoleuca*): IUCN Red List status is Near Threatened (Least Concern as of November 1, 2024). About 71 percent of the greater yellowlegs' population nests in the boreal forest.

Lesser yellowlegs (*Tringa flavipes*): IUCN Red List status is Vulnerable (Least Concern as of November 1, 2024). More lightly built and delicate than greater yellowlegs; 94 percent nest in the boreal forest.

Both yellowlegs are Nearctic breeders that nest primarily in the boreal forest and winter from the southern, coastal US to Mexico, Central America, and South America, making them, in large part, neotropical migrants.

Solitary sandpiper (*Tringa solitaria*): Nearctic breeders, 85 percent of solitary sandpipers rely on the North American boreal forest for their nesting habitat. One of my favorite birds simply for the company they provided on several trips. They were not quite as in-your-face-friendly (aka "food-mooching") as Canada jays but notably tolerant when I was alone.

Spotted sandpiper (*Actitis macularius*): Nearctic breeders with about 73 percent relying on the boreal forest for their nesting habitat. They are the most widely distributed sandpipers in North America and are plain little birds until the breeding season, when they adopt their namesake spots. A remarkable and entertaining fact about spotted sandpipers is their cute yet comic bobbing as they walk. I thought the vertical motion might be a means of correcting for diffraction parallax when targeting a tasty morsel underwater (as when owls and other predators move their heads side to side to judge distance), but their heads don't participate in the dance move, and they do it even when feeding on shore. It is fun to watch anyway, and a great means of distinguishing them from similar-looking birds.

Marbled godwit (*Limosa fedoa*): IUCN Red List status is Vulnerable. Nearctic in distribution. Mostly Great Plains nesters, about 13 percent of marbled godwits breed in the boreal forest, mostly around Lake Winnipeg and a disjunct outpost along the west shore of James Bay.

Semipalmated sandpiper (*Calidris pusilla*): IUCN Red List status is Near Threatened. About 30 percent of the population of this American bird breeds in the boreal forest and most rely on it for migration. They winter in the southern US and into South America.

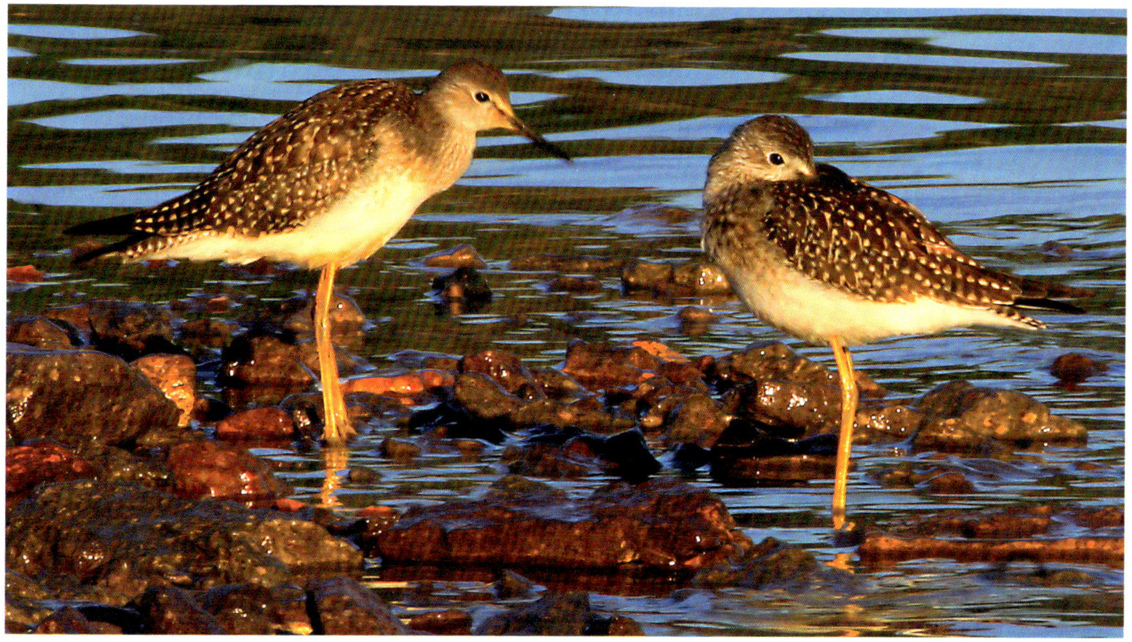

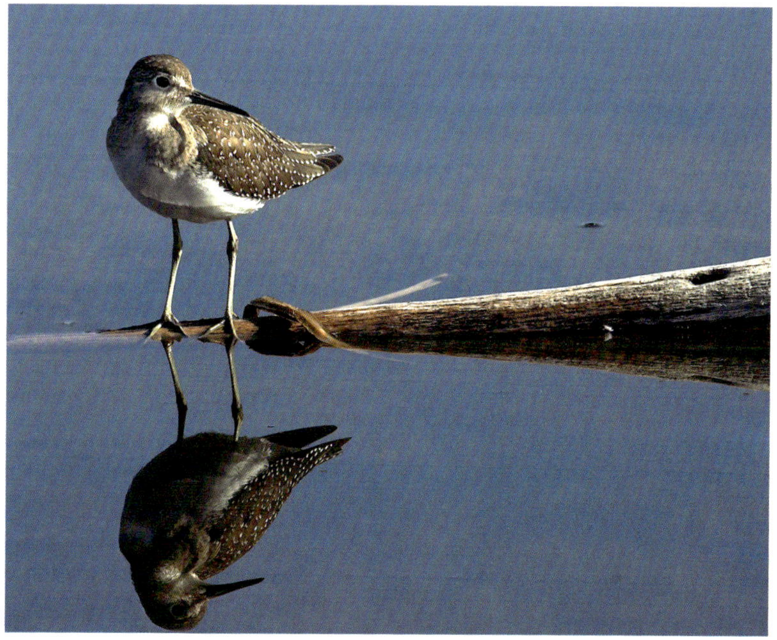

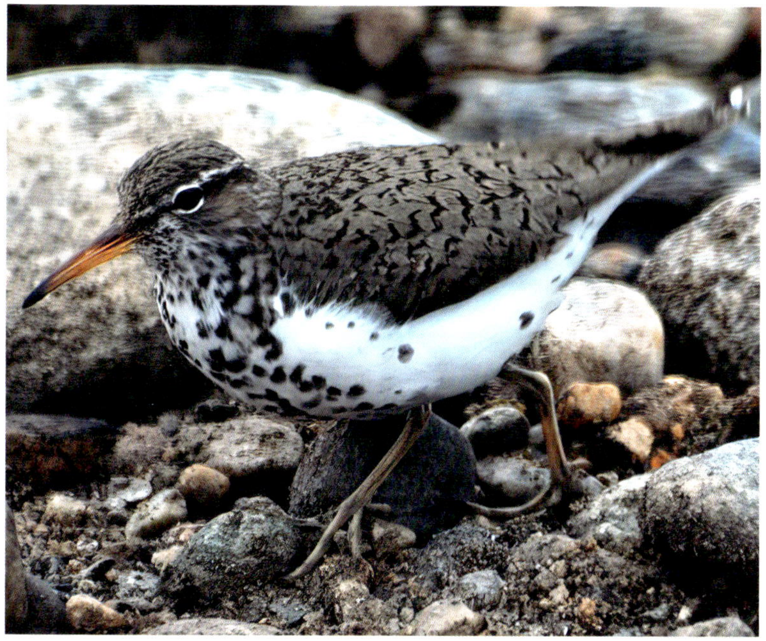

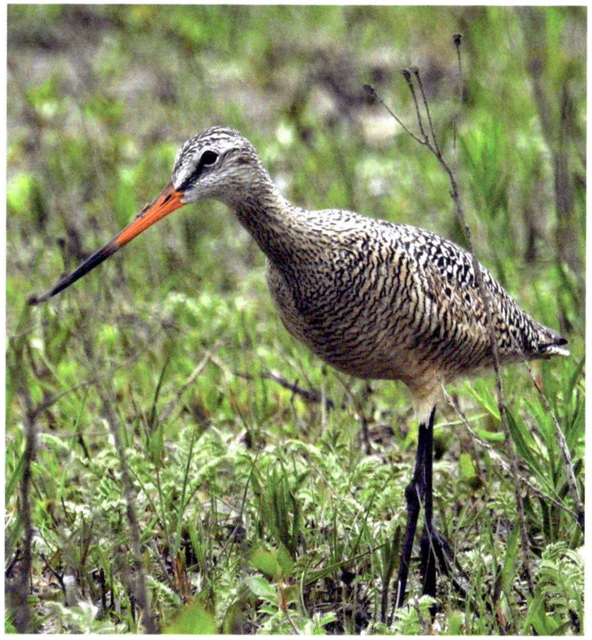

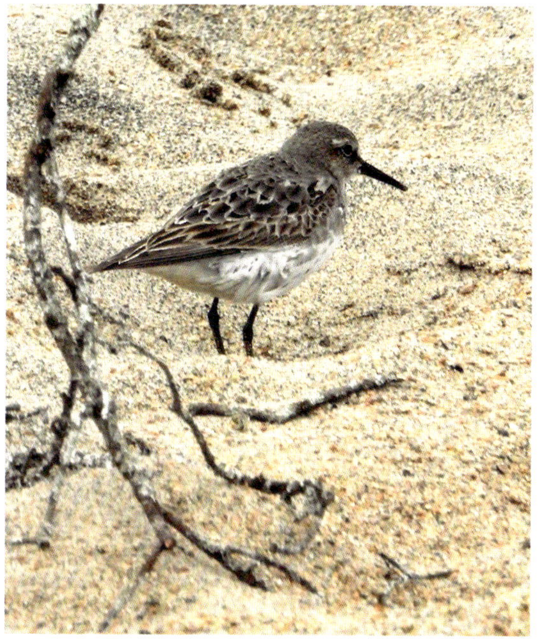

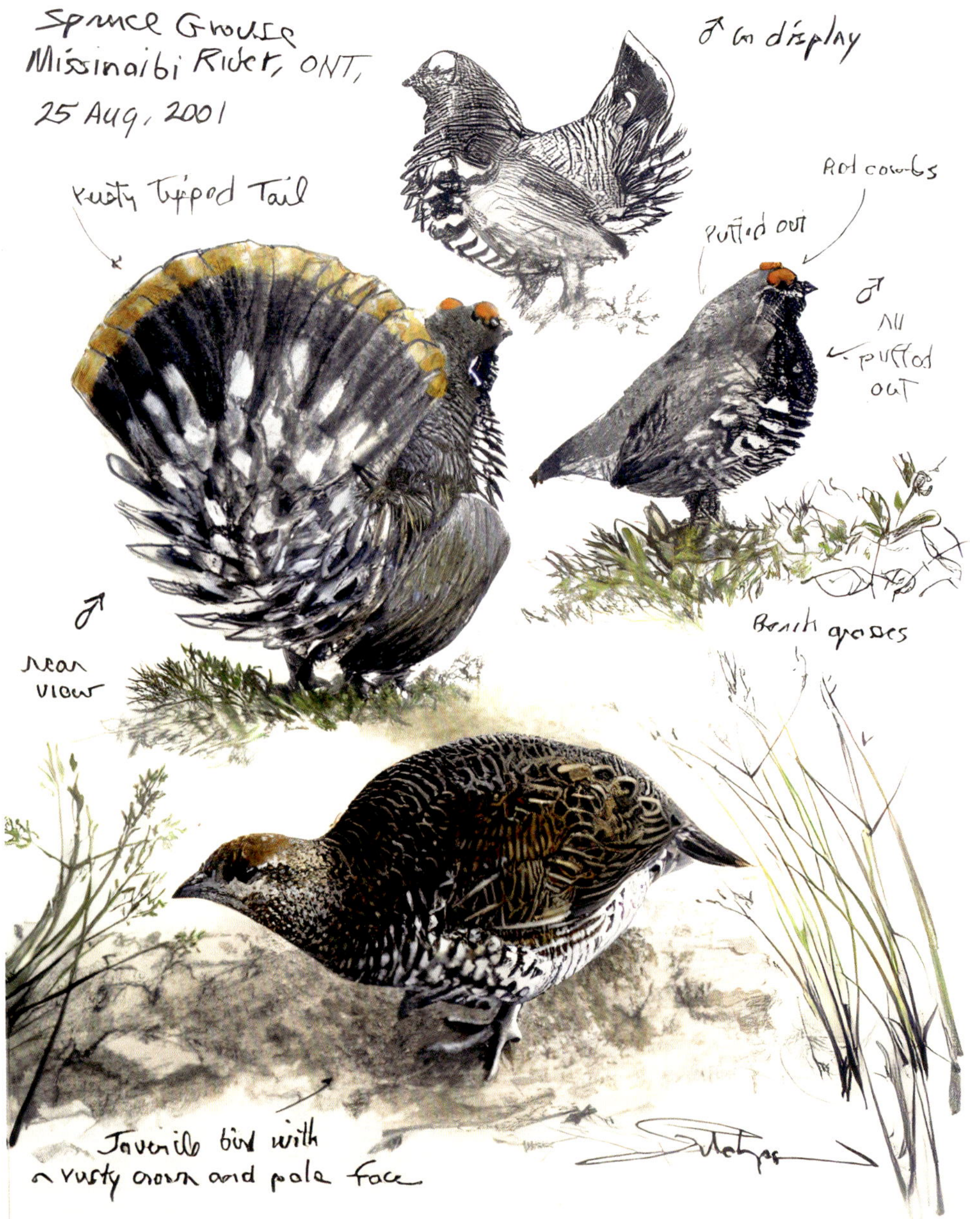

Upland game birds: Family Phasianidae (grouse, ptarmigan, partridge, and quail)

Now, we are getting into some tough birds. They don't flee but face the brutal storms, deep snow, and cold of winter. Then again, their food supplies are not buried under several feet of ice, and they aren't built for long-distance flight. With behavioral adaptations such as burrowing into the snow at night and morphological adaptations such as feathered feet and crops that can hold up to a tenth of their body weight in food for overnight survival, these birds can winter in place. Consequently, they live full time in the largest, most pristine wilderness habitat on Earth and provide a refreshing change from the lists of endangered species elsewhere.

Of the 20 or more species of year-round, boreal forest residents, only two Eurasian species, the Siberian grouse (*Falcipennis falcipennis*) and the Japanese quail (*Coturnix japonica*), are assessed as Near Threatened by the IUCN (9–10 percent overall and 0 percent for North America). Worldwide, of the 166 or so species that live elsewhere, the IUCN lists the conservation status of 59 as either Near Threatened or worse (about 36 percent). That is quite a benefit to living in the boreal forest in the Anthropocene (the age of human geophysical influence on climate and environment).

There are eight common North American species, five of which are endemic. Two of them have kept us company on many journeys. There are about ten endemic Eurasian species. The two Holarctic species (willow and rock ptarmigan) are common to both continents.

Two remarkable Eurasian game birds are the western capercaillie (*Tetrao urogallus*) and the black-billed capercaillie (*Tetrao urogalloides*). The western capercaillie can weigh up to 16 pounds and is the largest grouse in the world if you don't count turkeys. In the boreal forest, they range from Scandinavia through much of Siberia. Its cousin, the black-billed capercaillie, is somewhat smaller and is found through the rest of the eastern Eurasian boreal forest.

An easy-to-spot species in North America are the spruce grouse (*Canachites canadensis*); 87 percent of the population nests in the boreal forest. They're easy to spot not because they are brightly colored—they aren't—but because they don't hide. Known as "fool's hens" for their absurdly tame behavior, they have been regular companions on many of our trips. John Pitcher and I were treated to a pointed display on the first of these art trips on the Missinaibi in 2001.

August 24, 2001, early evening: First night's camp, about ten miles north of Lake Missinaibi, Missinaibi River, Ontario

After an easy start and early camp, John and I set up to do some painting. Intent on our work while the fire burned to cooking coals, we were startled by small, dark fluffy things that came piling out of the brush and ran around us and even under my legs as I sat trying to focus on catching the evening light. Bizarre as that was, moments later, they were joined by a larger one—apparently Mom. "Spruce grouse," exclaimed John.

To complete the cozy domestic scene, the male showed up at the beach's edge but did not join his mate and their chicks running around us like the circus was in town. John, an inveterate field artist, undertook to sketch the little rascals as they scampered about while telling me that they were often called fool's hens for being so easy to catch. I could have reached down and had them all in the pot in a minute. We didn't, of course; they were too cute. And who would have the heart to betray that trust?

BELOW LEFT: Adult male spruce grouse "hiding" in a spruce, Algonquin Park, Ontario. Bonnie Rowell

BELOW RIGHT: Willow ptarmigan in summer plumage, Noatak River, Alaska. Oil, Linda Besse

FACING: Spruce grouse, Missinaibi River, ink and acrylic refined field sketches, John C. Pitcher

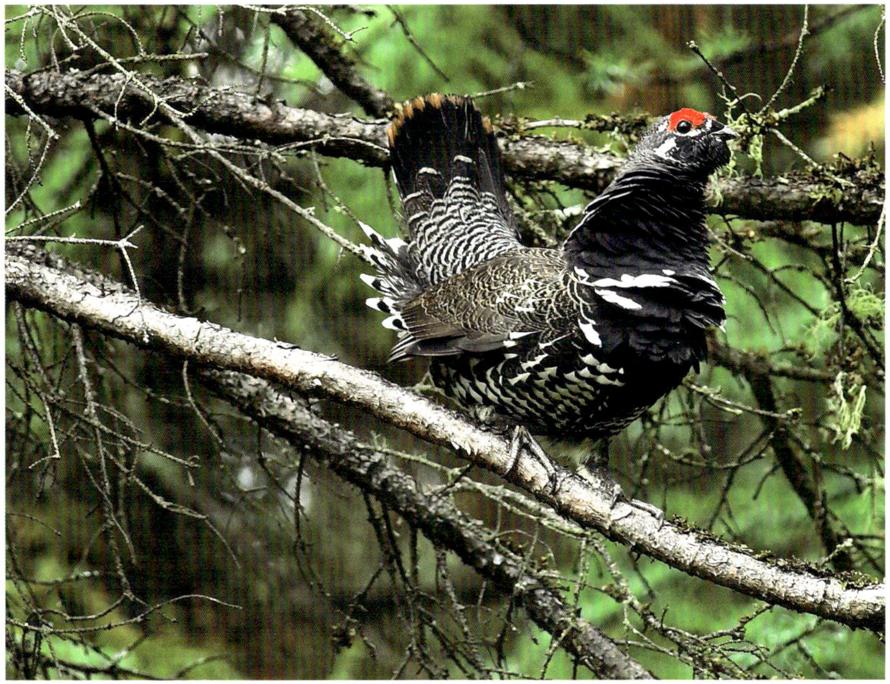

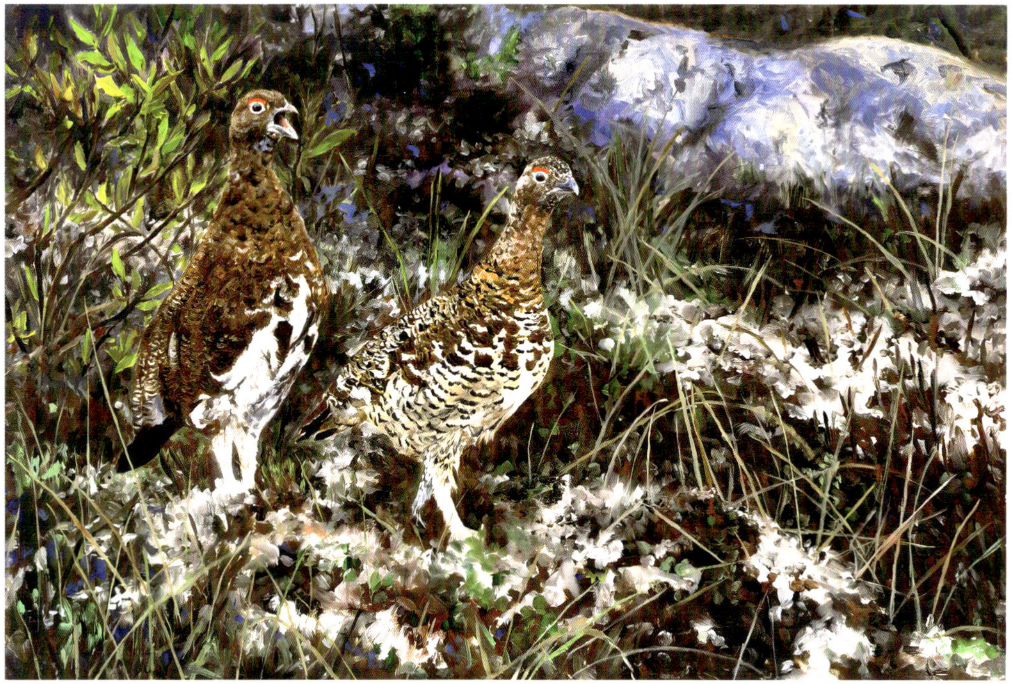

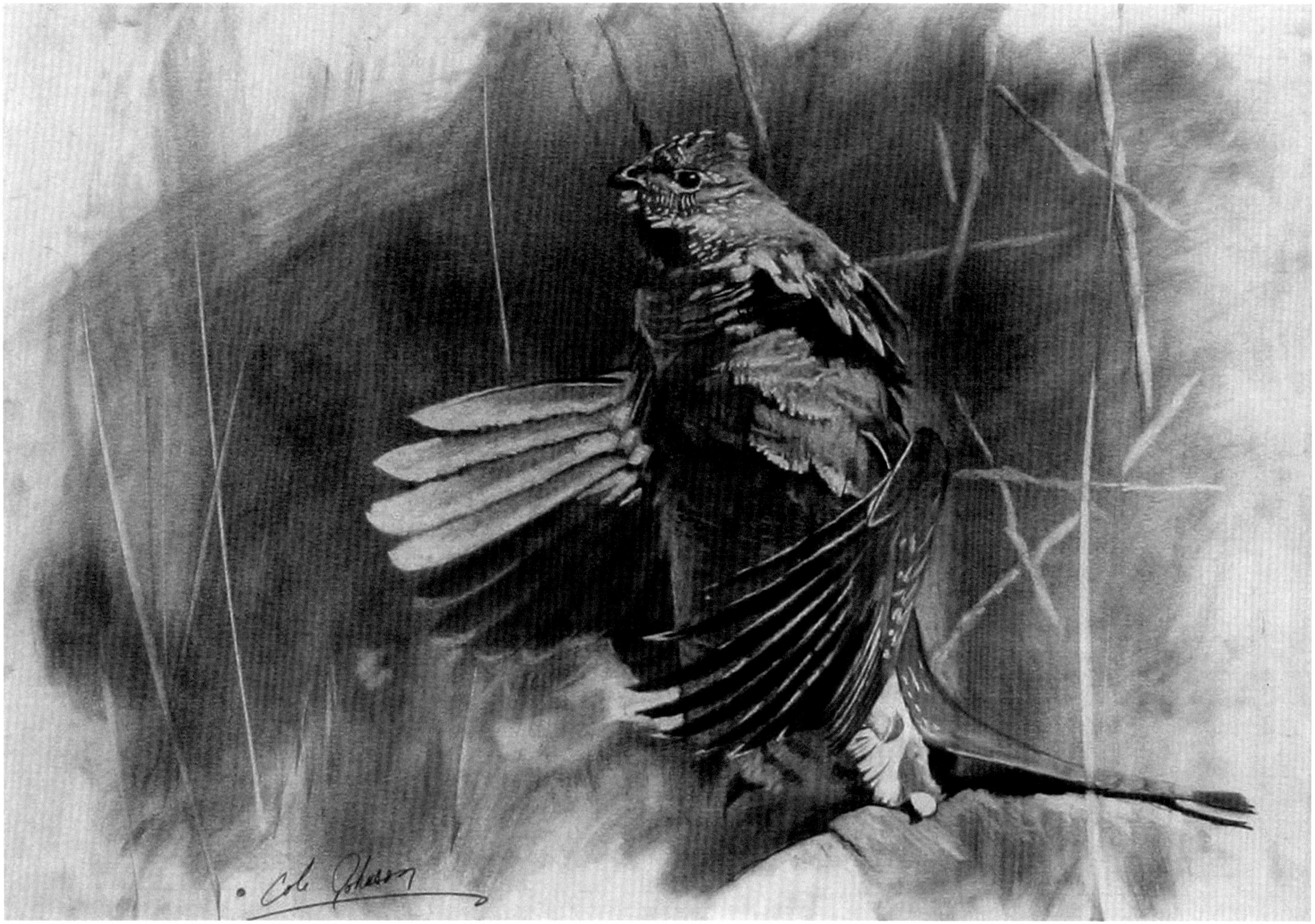

Our other regular companion has been the willow ptarmigan (*Lagopus lagopus*). They are very common with an estimated population of 10 to 30 million. About 24 percent of the North American population nests in the boreal forest. Likely similar in Eurasia.

Ptarmigan, like snowshoe hares, weasels, and Arctic foxes, are famous for not only sticking around for winter but changing to nearly pure white to do so in style (and to blend in). Some black marginal feathers remain on the tail that are most easily visible in flight.

These are the adorable birds that woke Cole and me up each morning in the Labrador Barrens with their gentle clucking calls to each other.

Finally, there is the ruffed grouse (*Bonasa umbellus*). A North American bird, over half the population nests throughout much of the boreal forest. In Vermont, they are called "partridges" by all but birders and biologists. They are famous for their drumming, which sounds like someone trying to start a tractor.

Birds of prey (raptors): Families Accipitridae (hawks and eagles), Falconidae (falcons), and Strigidae (owls)

It's been said humans connect with predators because their forward-facing eyes mirror ours. That may work with mammals, but with birds . . . I'm not so sure.

May 1990: Manhattan

A juvenile female Cooper's hawk (*Astur cooperii*) hit my office window; I rescued her. After a week's rehab on rats, I was invited to release her. Holding a wild hawk by the shanks is something every wildlife artist should try—the fearless, laser-beam ferocity in her eyes left no doubt she thought she could take me. It gives you an idea of the attitude it takes to be a wild hawk.

Having started with Cooper's hawks, we will continue with their clan, the "true hawks."

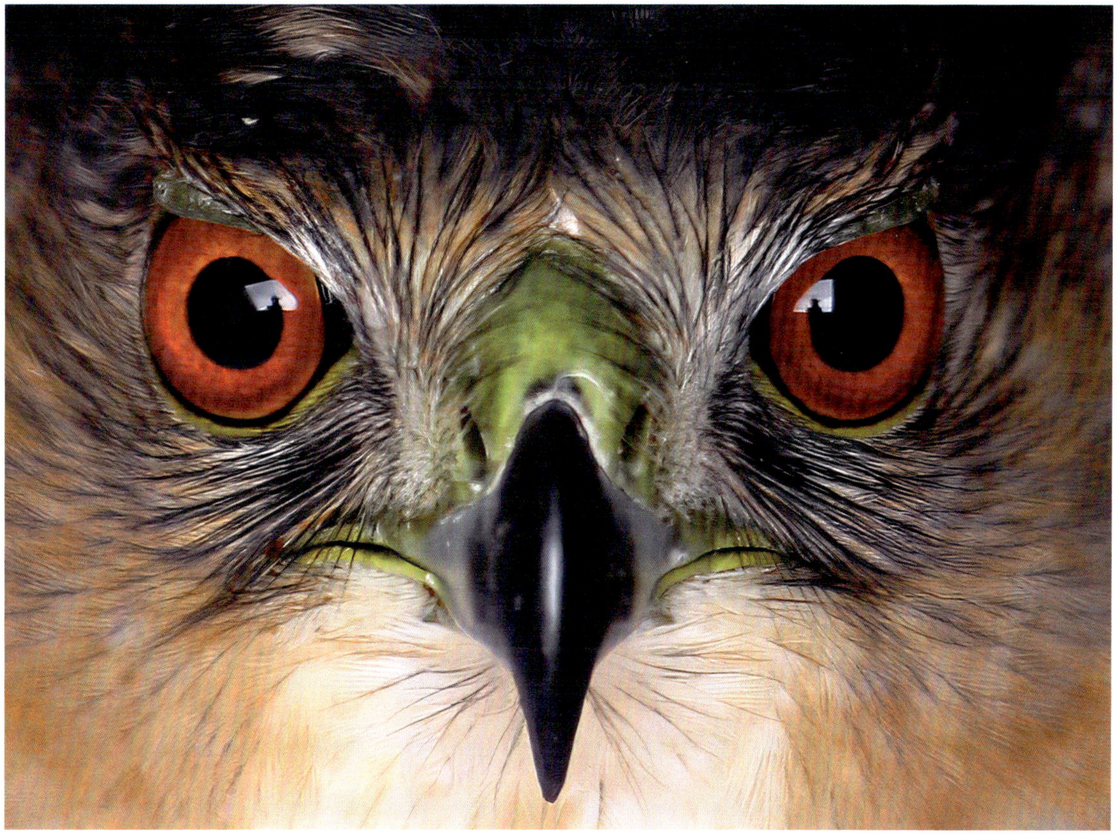

Cooper's hawk being prepped for release, Noyan, Quebec. She was another rescue, 20 years after the Manhattan bird. You can see me reflected in her eyes.

FACING: Male ruffed grouse drumming. *Drummer*, charcoal, Cole Johnson

Subfamily Accipitrinae (true hawks)

Forest specialists with short wings, long tails, and aerial agility, hawks often prey on other birds, frequently taking them on the wing. Females are larger than males, allowing the pair to expand and diversify their food sources by targeting different-sized prey.

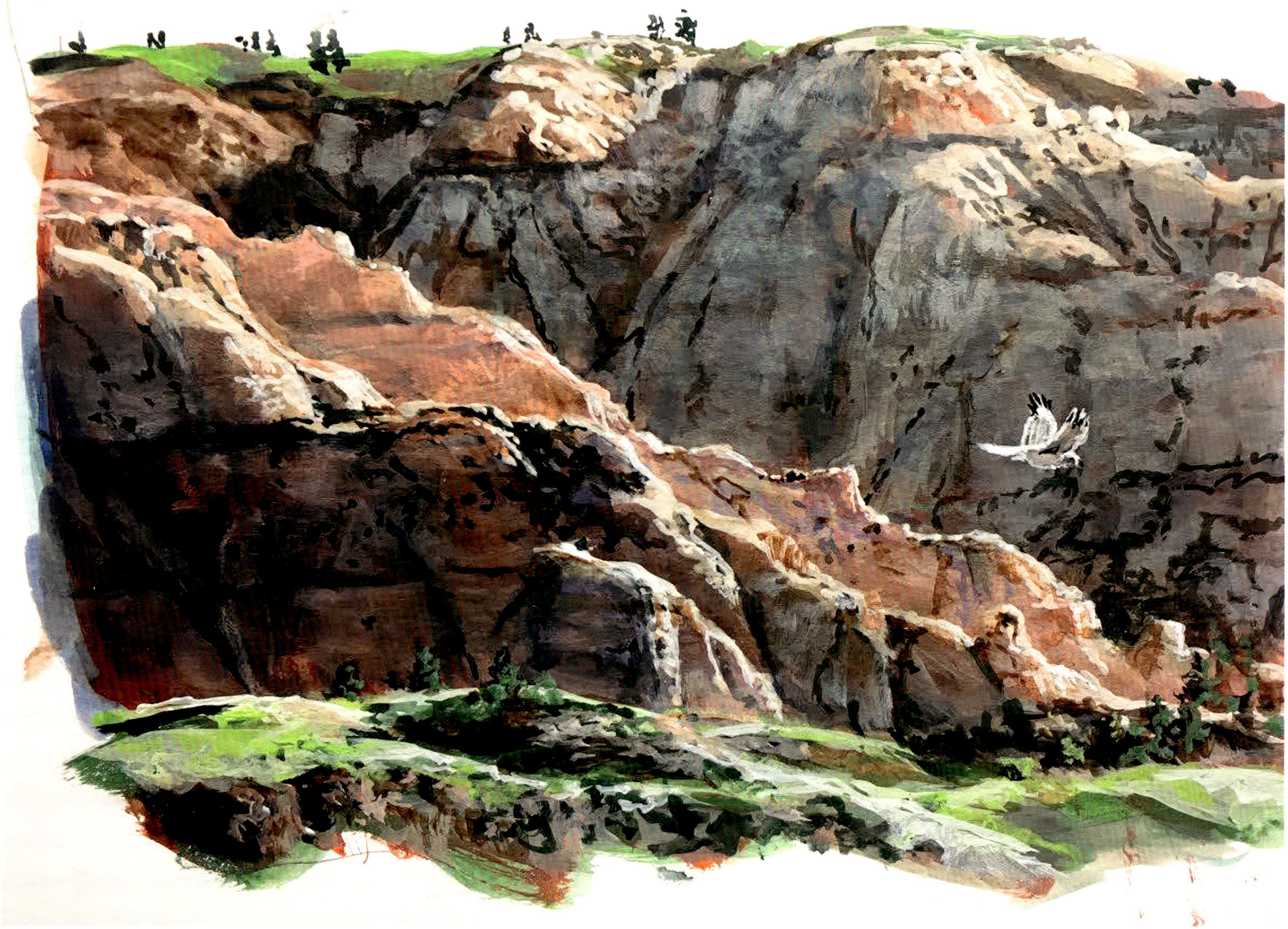

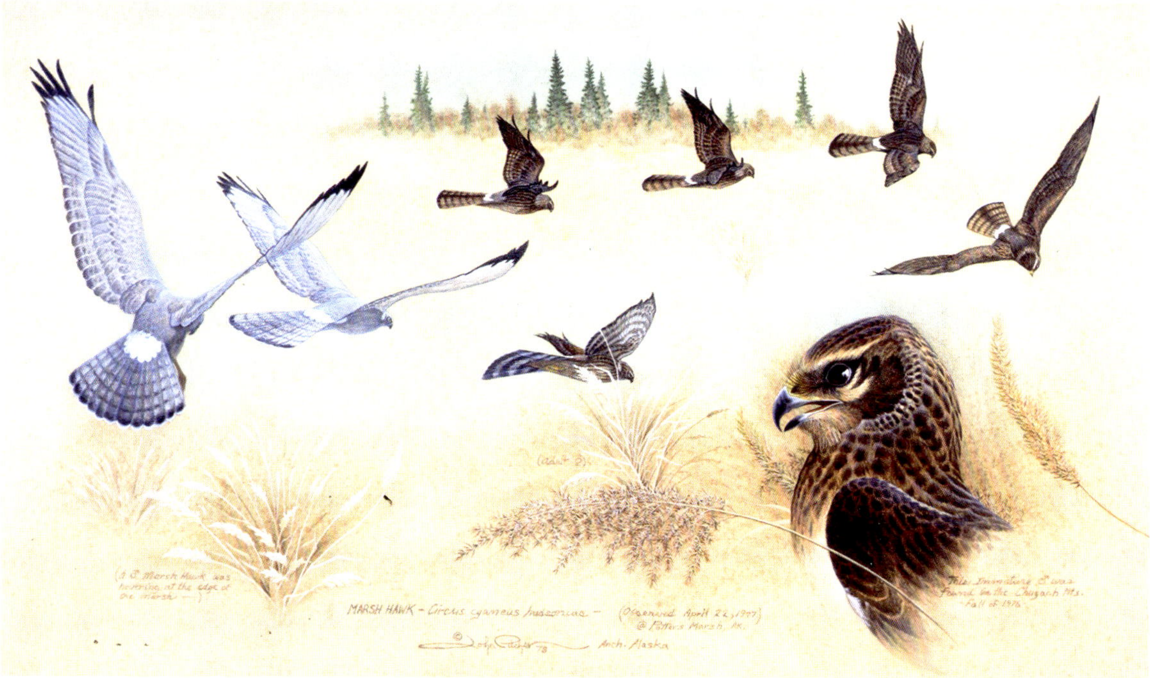

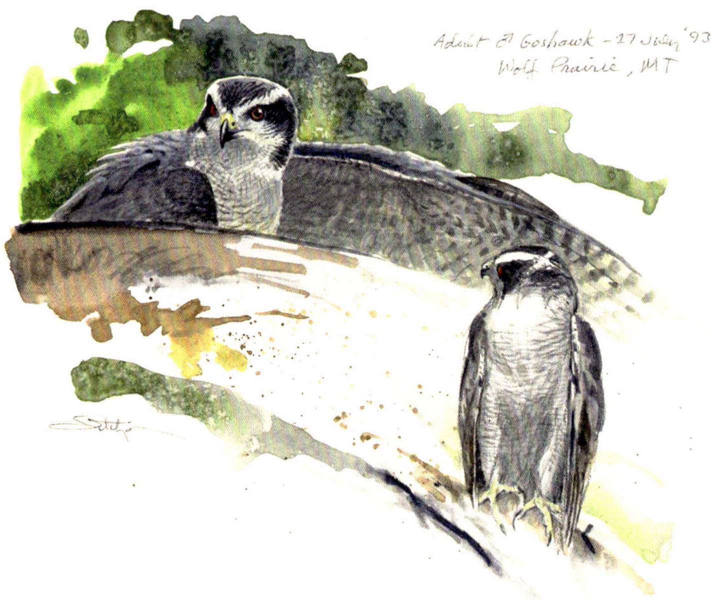

LEFT: Northern harrier. Refined acrylic field study, John C. Pitcher

RIGHT: American goshawk. Watercolor, John C. Pitcher

FACING: Northern harrier on the Missouri. Acrylic, Jan Martin McGuire

Sharp-shinned hawk (*Accipiter striatus*): The smallest North American hawk, sharp-shinned hawks are convergent look-alikes to the Cooper's hawk, but notably smaller and with other, more subtle differences. There is no plumage difference between the sexes of either species, and large female "sharpies" can overlap in size with small male Cooper's, so identification can be confusing.

Sharpies are common and widespread throughout North, South, and Central America, including all of the North American boreal forest.

American goshawk (*Astur atricapillus*): Nearctic in distribution. It is up to twice the weight of a Cooper's hawk, ranging across the boreal forest and much of northern and mountainous regions of North America.

Eurasian goshawk (*Astur gentilis*): Formerly the same species as the American goshawk, the two together were known as the northern goshawk. They range across the Eurasian boreal forest and much of the rest of northern Eurasia and the Himalayas.

Northern harrier (*Circus hudsonius*): Nearctic in range, harriers breed across the boreal forest and much of the northern US and migrate south as far as the edge of South America.

Harriers are notable for their long wings and tails and their almost owl-like facial discs which may aid in hearing prey (mostly small mammals instead of birds) as they slowly circle and crisscross above open areas just above the ground (their genus name comes from this behavior). They love following farmers as they mow hay fields. Along with the usual accipiter

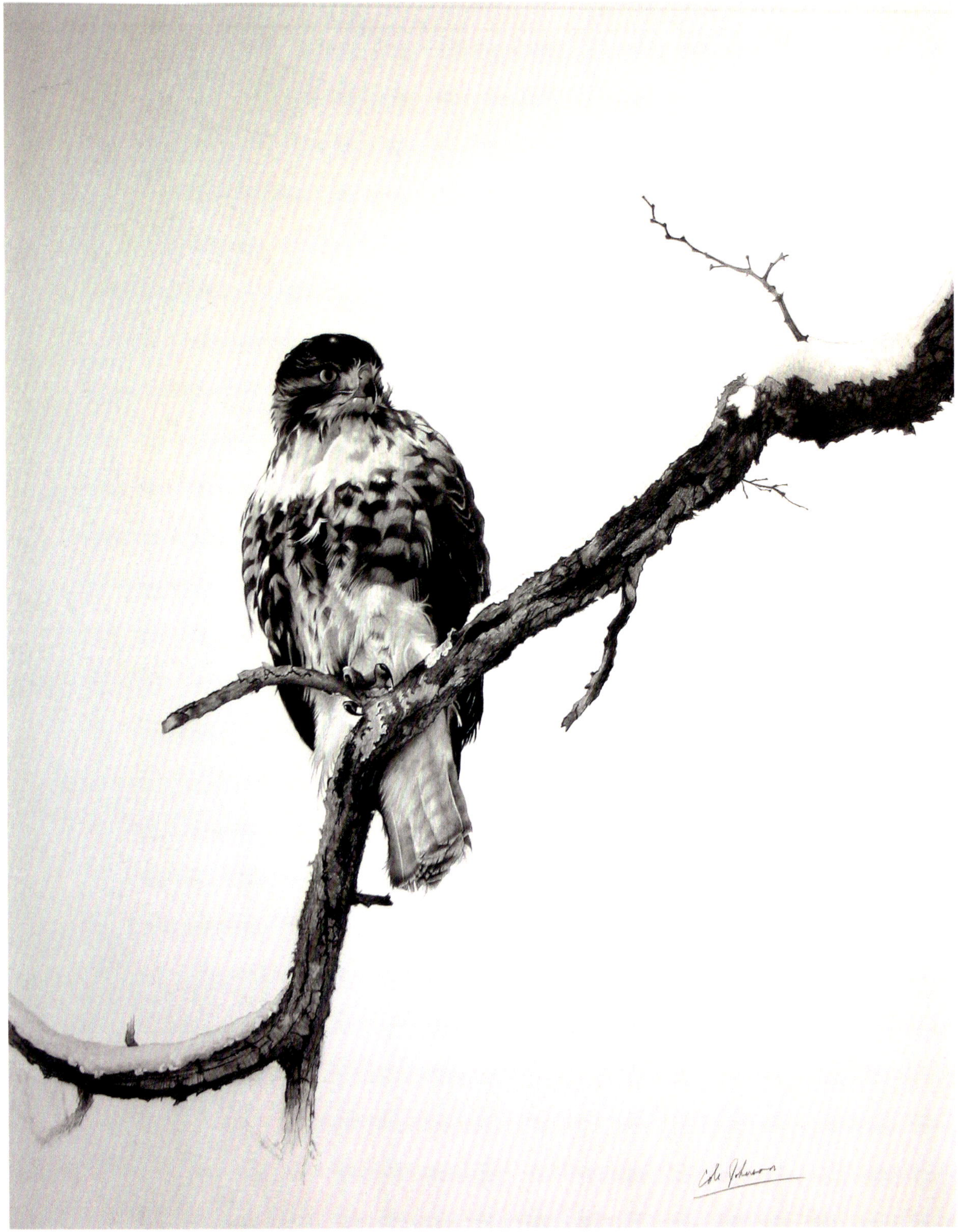

sexual difference is size, immature and female harriers are brown and streaked while mature males have a strikingly light gray plumage with jet-black tips on their primary feathers giving them a "dipped in ink" appearance.

Genus *Buteo* (buteos)

Buteos are large hawks that frequently hunt by soaring or from a high perch (they are frequently seen in trees along highways). They prey on a wide variety of small animals.

Red-tailed hawk (*Buteo jamaicensis*): Nearctic in distribution, red-tailed hawks are one of the most widespread and commonly seen large hawks in North America (they are one of the largest buteos with wingspans that can be four to nearly five feet). They nest throughout all but the northern boreal forest (that is the realm of our next hawk) from southern Alaska to the Maritime provinces of Canada but are not resident there, migrating as far south as Central America. They are not misnamed; despite some plumage variations among several morphs, regional variants, and subspecies across their wide range, all red-tailed hawks have a tail that is strong brick red above and light, reddish off-white underneath. Source of the iconic *CHEEeeuurrr* of movie soundtracks whenever the aura of rugged wilderness is desired—usually accompanied by a soaring eagle.

Rough-legged hawk (*Buteo lagopus*): Holarctic in distribution. They are named for their feathered legs, an adaptation for the cold, and are essentially the same size as red-tailed hawks on average. They range farther north, breeding in the northern boreal forest and the Arctic, and winter in more southern regions of the boreal forest and south of it (they are frequent winter residents in Vermont). While at home soaring and hanging out along highways, rough-legged hawks are the only

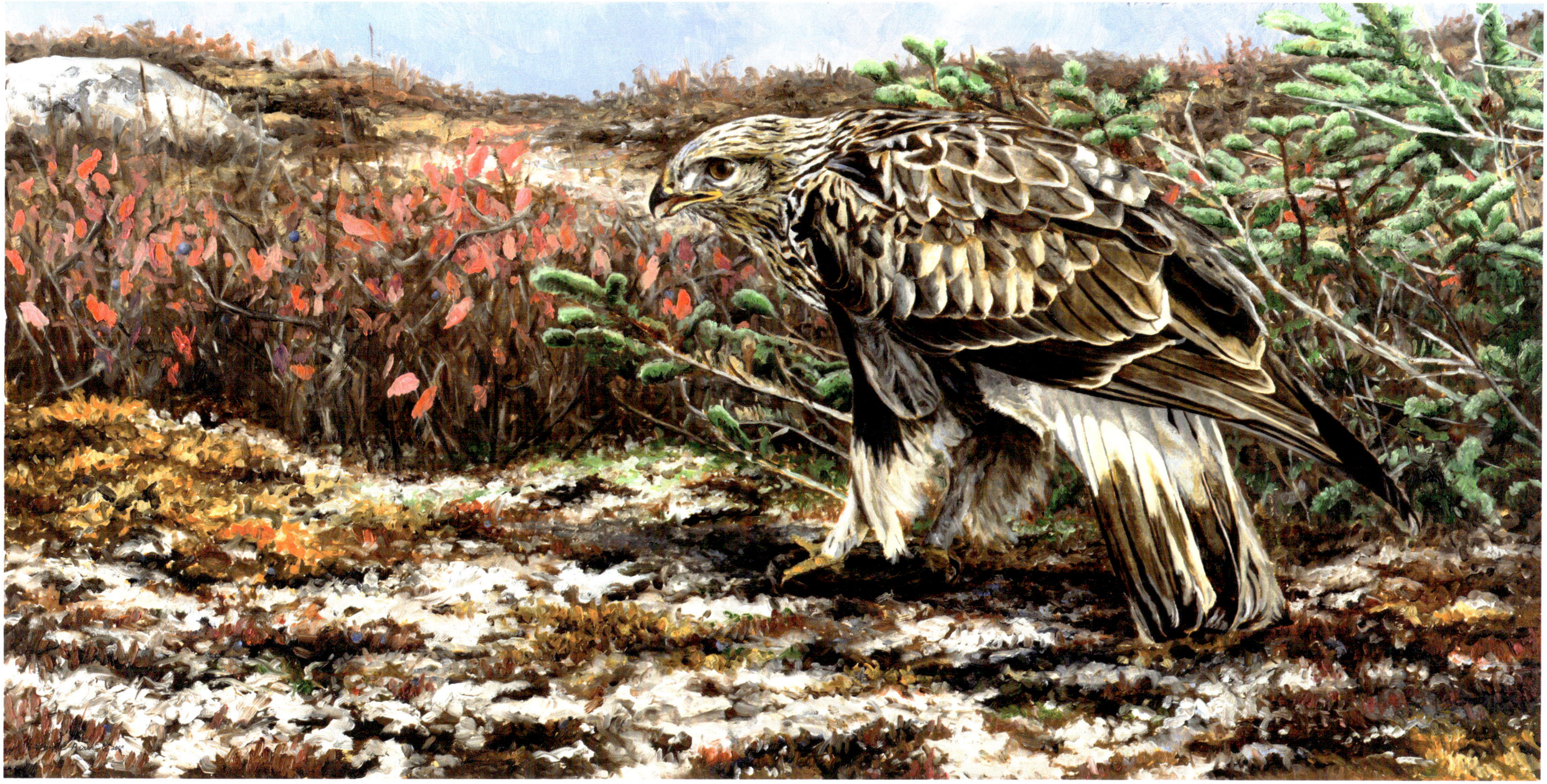

Rough-legged hawk mantling prey, Kamestastin, Labrador. *Guardian*, oil, Linda Besse

FACING: Winter red-tailed hawk. Charcoal, Cole Johnson

CLOCKWISE FROM TOP LEFT:
White-tailed sea eagle, Norway. Steve Bruno

"What's for sup?" "Fish" (inside joke). Ospreys on nest, Otter Creek, Vermont. (The skinny on the joke: It's the full transcript of a typical nightly phone conversation between Bonnie's parents before her father left work.) Bonnie Rowell

Osprey dive from perch, Lake Champlain, Vermont. (Study done while watching the bird from the cockpit of the *Artful Otter*.) *Off to Breakfast*, oil, Rob Mullen

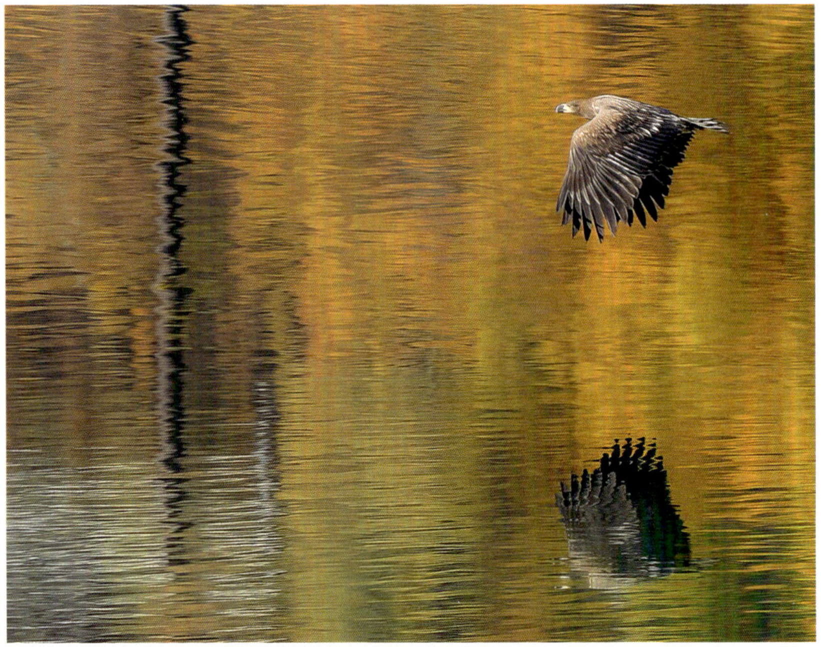

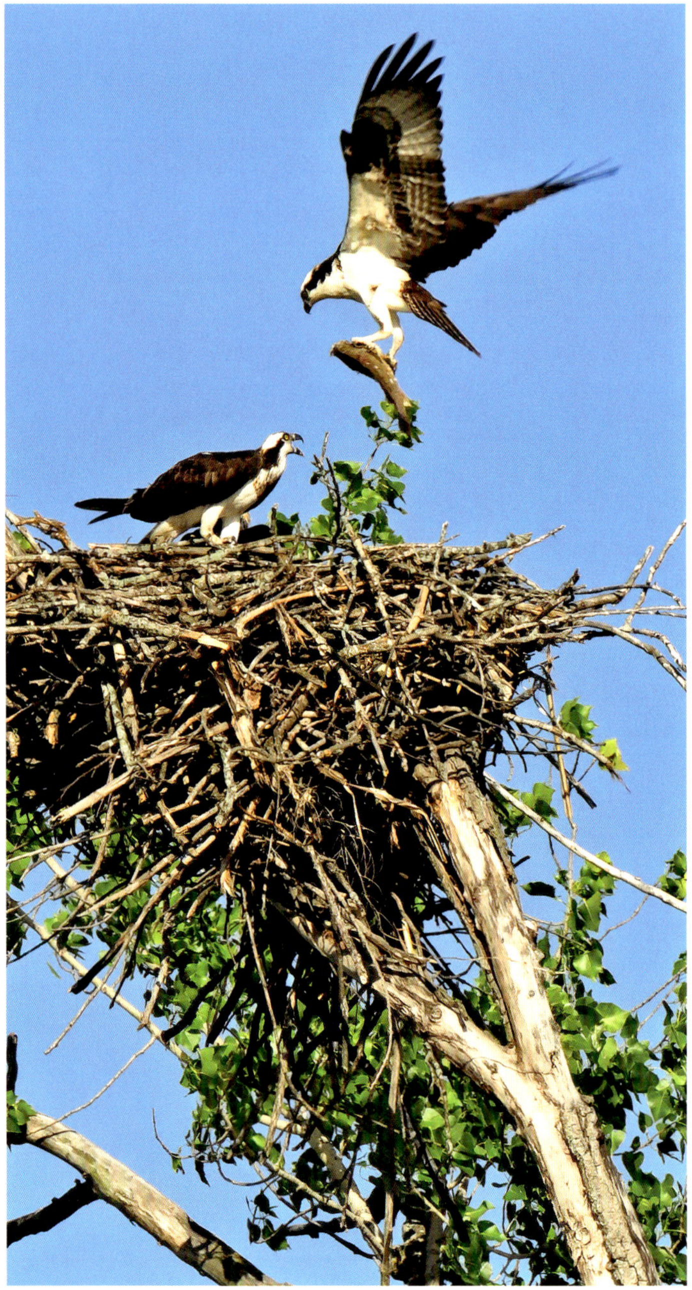

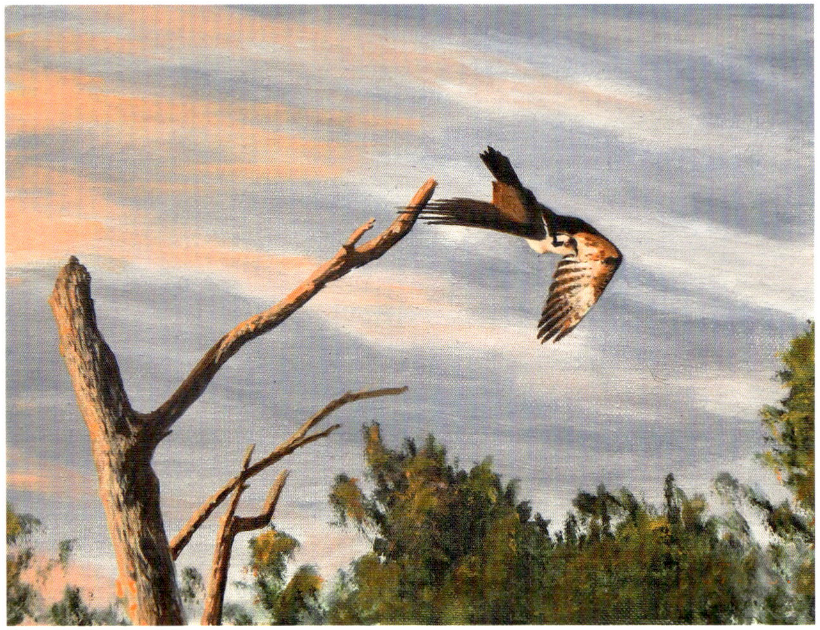

buteo that habitually also hovers to hunt. Humorously known for perching on absurdly small twigs.

Broad-winged hawk (*Buteo platypterus*): The smallest buteo, about the size of a crow, broad-winged hawks are more interior forest hunters. They nest throughout the eastern half of North America, including the boreal forest and winter in Central and South America (neotropical migrants), famously flying south in enormous "kettles" that can contain thousands of hawks.

Recently, one was outflown by a mourning dove and crashed into a maple tree right in front of me while I was tending to dog matters in the yard. The hawk was stunned and sat motionless, wings akimbo at the base of the tree. I had no camera (only the pooper-scooper), so we stared motionlessly at each other for several minutes. Then she recovered. Straightening herself out and gathering what was left of her dignity, she took off. It's tough being a hawk.

Eurasian boreal buteos include the common buzzard (*Buteo buteo*), upland buzzard (*Buteo hemilasius*), and Japanese buzzard (*Buteo japonicus*).

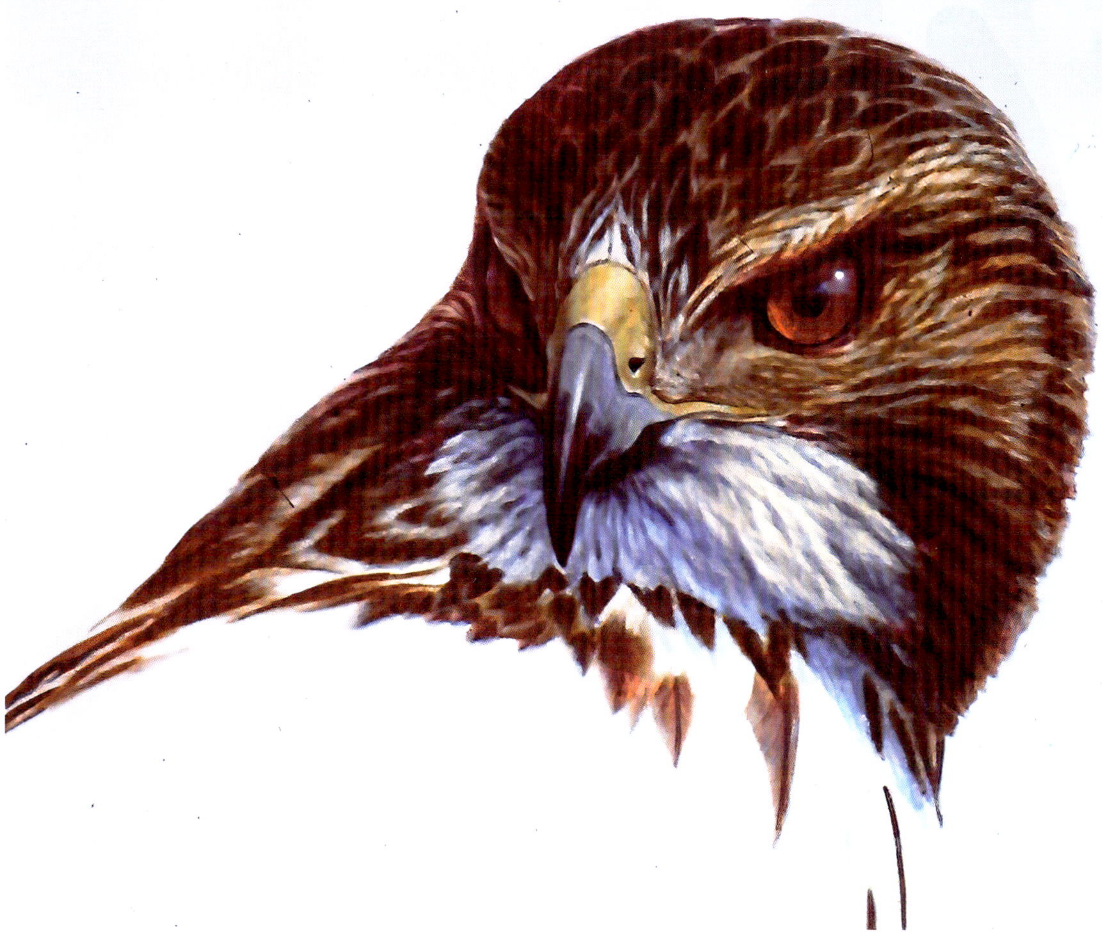

Adult red-tailed hawk head study. Gouache, Rob Mullen

Eagles

Eagles take the hawk design up several notches. As with most raptors, females tend to be larger: Their wingspans can exceed seven feet. Possibly because they are too big to lurk in lonely trees, hoping to go unnoticed by wary prey, they often hunt from high in the air.

There is no better icon of northern wilderness than the golden eagle (*Aquila chrysaetos*), the national bird of Mexico. Resident in much of the boreal forest around the world, this enormous, majestic, and ferocious bird is as predatory as they come. No rotting fish or robbing gulls for dinner (well, maybe if hungry enough).

Closely related to the buteos but enormous, their wingspan can exceed seven feet, and they can weigh 10–15 pounds. Basically, a dark brown bird with a golden nape. The juveniles have a fair amount of white on them and can be confused with juvenile bald eagles, though goldens have a smaller head and bill, and the white on juvenile goldens is usually more patterned. Young bald eagles can look like they forgot to wear their smocks during splatter-painting class.

Osprey (*Pandion haliaetus*): Holarctic, second only to peregrines in range. While only 30 percent of the North American population nests in the boreal forest, it does so throughout

Gyrfalcon, George River, Nunavik. *Mountain Monarch*, acrylic, Robert Bateman

FACING LEFT: Merlin attack on the Noatak.

FACING RIGHT: Stooping peregrine falcon. Charcoal, Cole Johnson

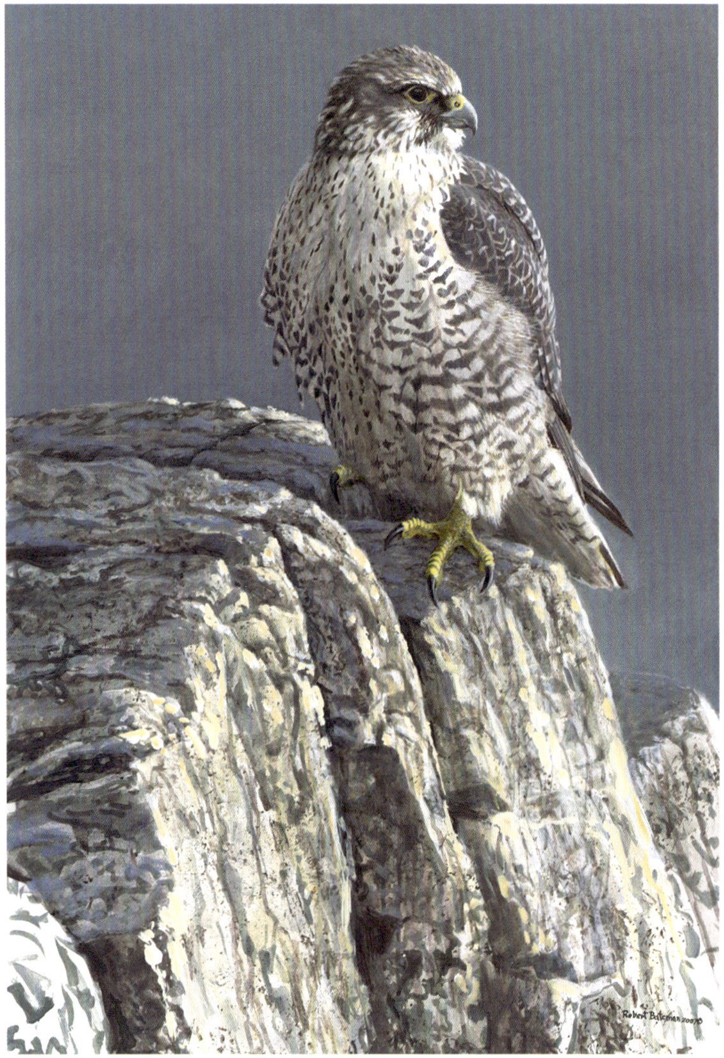

virtually the entire boreal forest. Almost exclusively fish eaters, osprey are perfect companions for canoeists. We have seen loads of them from Labrador to Alaska. Despite their wide range, ospreys were devastated by DDT and, along with many other raptors, faced extirpation across vast swaths of their range if not complete extinction. They rebounded quickly after the poison was banned. I grew up with ospreys (as well as eagles) being very rare, so I was very excited to see my first osprey while canoeing on the Winooski River in Vermont as a teenager, just a few years after DDT was banned.

White-tailed sea eagle (*Haliaeetus albicilla*): The bald eagle's relative in Eurasia, this migrant breeder nests throughout the Eurasian boreal forest. White-tailed sea eagles reside in parts of Siberia, Kamchatka, Norway, Sweden, and southwestern Greenland.

Family Falconidae (falcons)

Long, pointed wings with stiffened primary feathers to resist warping and nostril baffles to break air flow so they don't suffocate are two adaptations of falcons for speed—incredible speed. The most famous, the peregrine falcon (*Falco peregrinus*), is essentially the ultimate smart missile: the fastest animal on Earth, capable of reaching 240 miles per hour in its attack dive. Falcons employ high-altitude surveillance, long-range target acquisition, continuously updated guidance adjustment ("proportional navigation"), and blurring speed to run deadly intercepts on their prey midair, striking a stunning or killing blow with their talons or breastbone. The peregrine is the most widely distributed bird species on Earth, breeding throughout the entire boreal forest. Yet another species nearly wiped out by DDT that has made a remarkable comeback since the poison was banned.

Gyrfalcon (*Falco rusticolus*): Holarctic in distribution, gyrfalcons reside throughout the northern boreal forest and are either resident or short migrators in the Arctic. There are many color variants of gyrfalcons, ranging from nearly pure white to black and every grade of gray and brown in between.

Darker birds are typical in the boreal forest, while white birds are more the norm in the Arctic. With wingspans in excess of four feet, they are the largest falcon in the world. Their adaptations for speed are not as refined as peregrines', and they are not quite as fast, often hunting ptarmigan, shorebirds, and waterfowl in direct, level chases.

Merlin (*Falco columbarius*): Holarctic in distribution. This smallest falcon loves the boreal forest, nesting almost exclusively in the entire biome around the world. Unlike the stolid gyrfalcon, merlins migrate south for the winter.

On the Noatak River (and later the Kobuk River) in Alaska, the local merlin youth had recently fledged and were noisily hectoring their parents for food. They were comically hard to miss. While hiking high above the river, I inadvertently neared a nest or kill site, and two merlins promptly attacked. Hoping to de-escalate the situation, I backed away and tried to get some photos—not easy when being dive-bombed by a brace of irate falcons.

Along with the wide-ranging peregrine, Eurasian boreal forest (Palearctic) species include the common kestrel (*Falco tinnunculus*), Amur falcon (*Falco amurensis*), and red-footed falcon (*Falco vespertinus*; IUCN Red List status is Vulnerable).

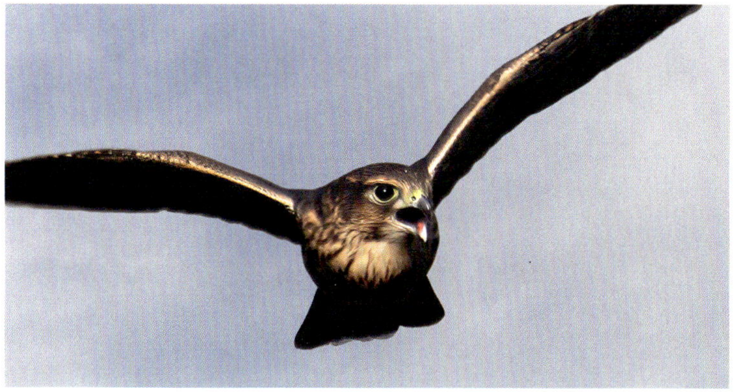

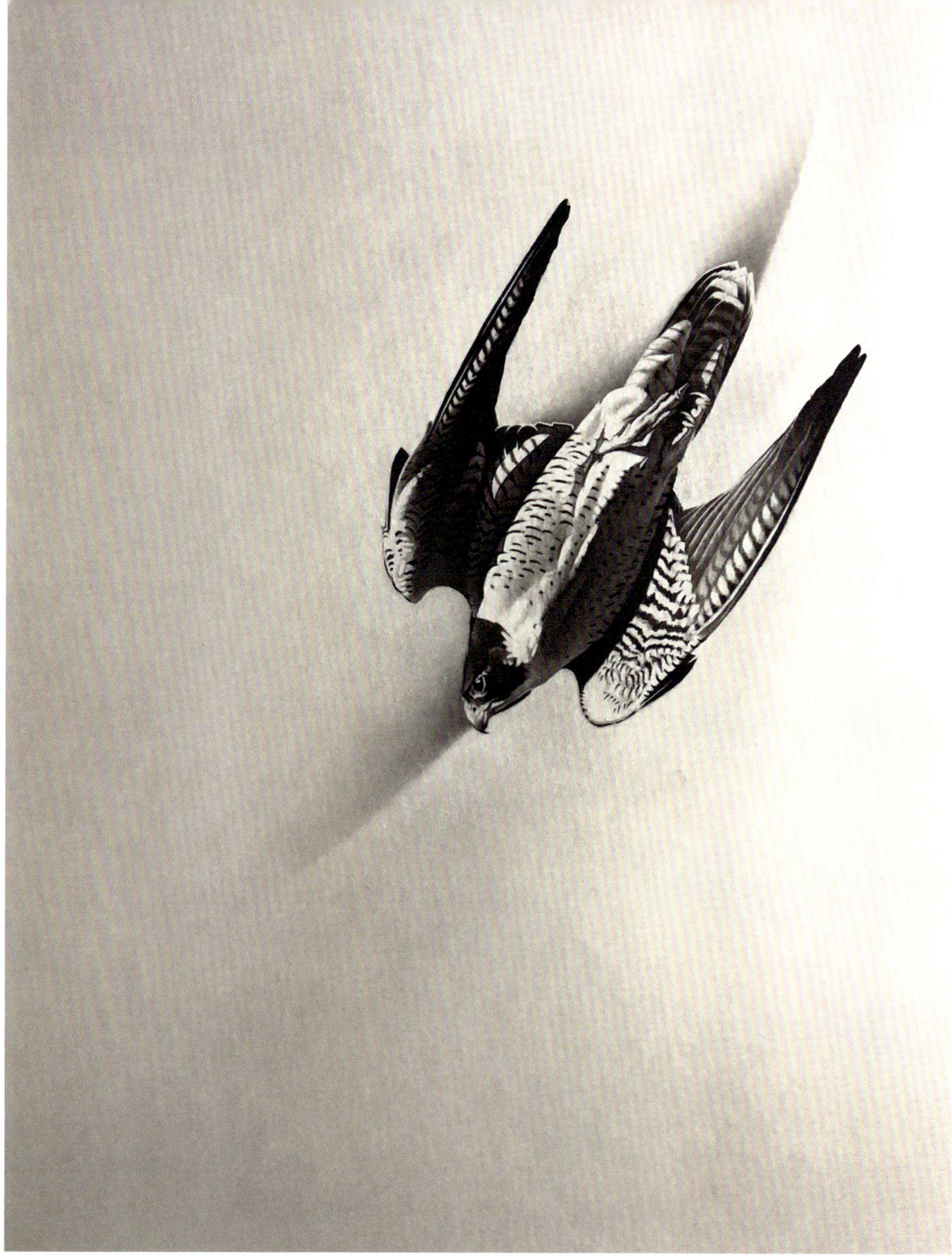

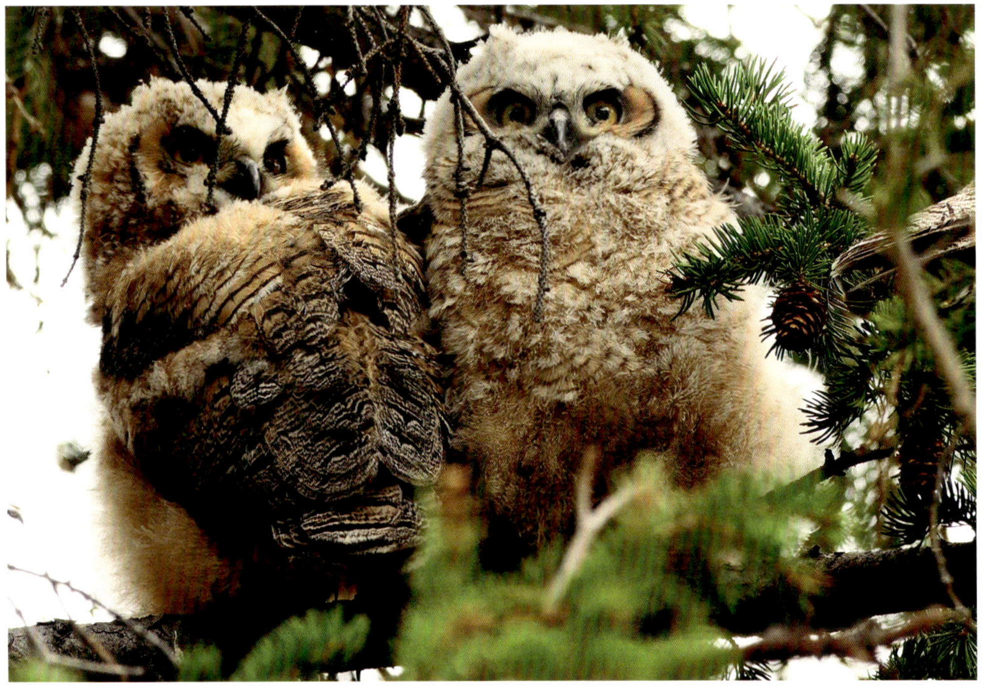

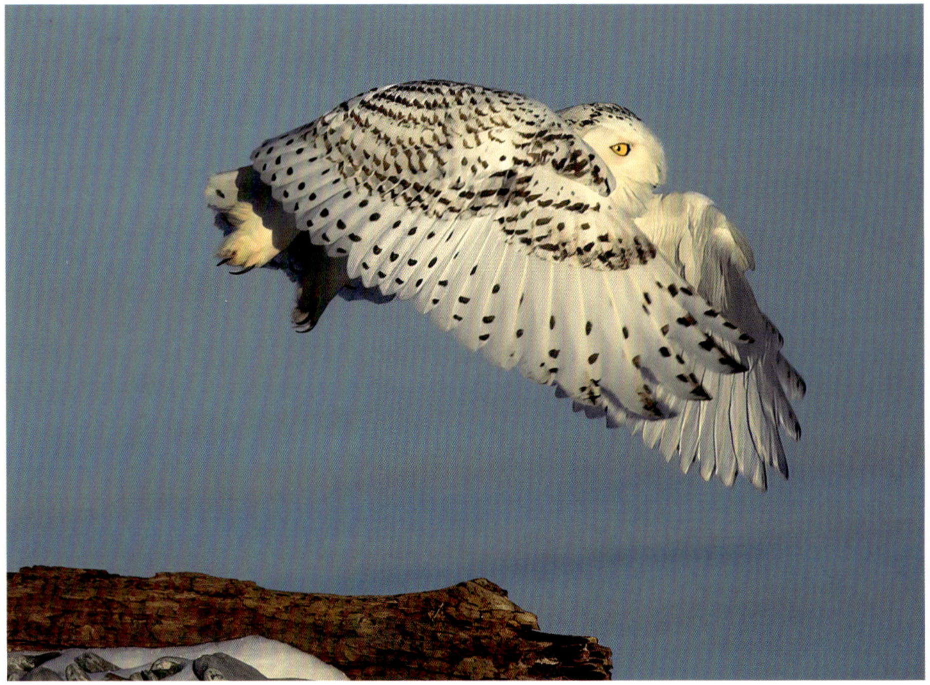

ABOVE LEFT: Great horned owlets, Yellowstone National Park, Wyoming. Bonnie Rowell

ABOVE RIGHT: Yellow eyes: snowy owl, Amherst Island, Ontario. Bonnie Rowell

FACING LEFT: Barred owl. *Snow Ending This Evening, Turning Colder, Chance of Rodents*, acrylic, Rob Mullen

FACING RIGHT: Saw-whet owl, Terra-Cotta Natural Park, Montreal, Quebec. *Eyes in the Night*, acrylic, Rob Mullen

Family Strigidae (owls)

At least 18 owl species range into the boreal forest worldwide, most of them resident or short-distance migrants. Only two are currently listed by the IUCN as Threatened (about 11 percent), compared to 30 percent of non–boreal forest owl species (with four extinct).

Sadly, one threatened boreal forest resident is also one of its most iconic: the snowy owl.

Snowy owl (*Bubo scandiacus*): IUCN Red List status is Vulnerable. Holarctic giants with a wingspan of over four feet and weighing up to four pounds, they are the heaviest owl in North America. Females are white with dark barring; males, almost pure white. They are short-range migrants, shifting to various parts of the boreal forest and Arctic in winter, sometimes farther south in irruption years when lemming numbers crash.

Two other boreal giants rival the snowy owl in North America: The Nearctic great horned owl (*Bubo virginianus*) and the Holarctic great gray owl (*Strix nebulosa*). Great horned owls (with a wingspan of up to five feet) are fierce, non-migratory residents from the boreal forest to South America. Great gray owls, the tallest owl in the world but much lighter than a snowy, wear their bulk in feathers for insulation. About 83 percent of the North American population breeds in the boreal forest and it is likely similar in Eurasia: Great gray owls are northern specialists. I've come close several times but have never seen one in the wild—my "great white whale" of birds.

The Nearctic barred owl (*Strix varia*) is more personal: I grew up with their calls (and still hear them outside regularly)

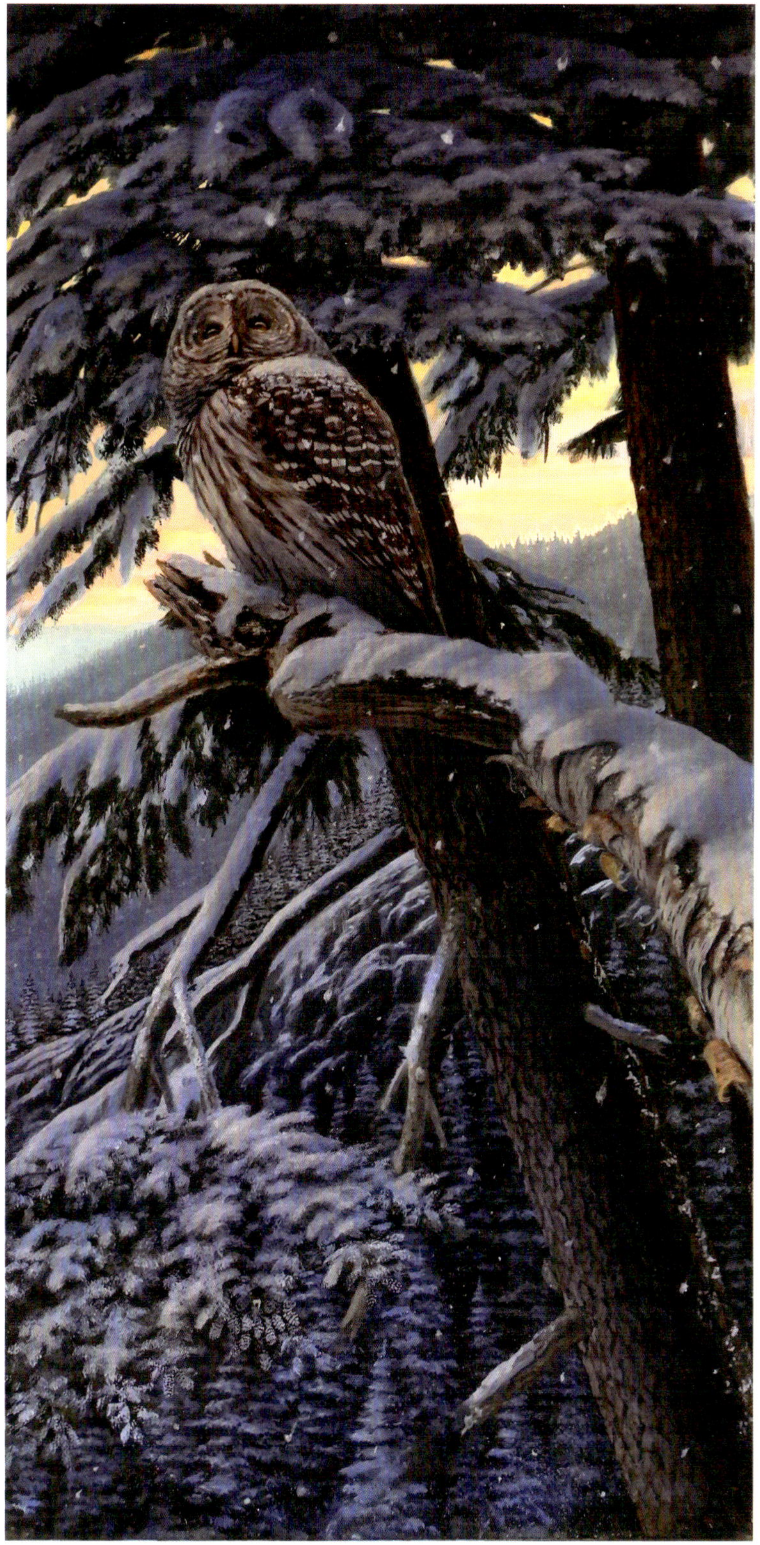

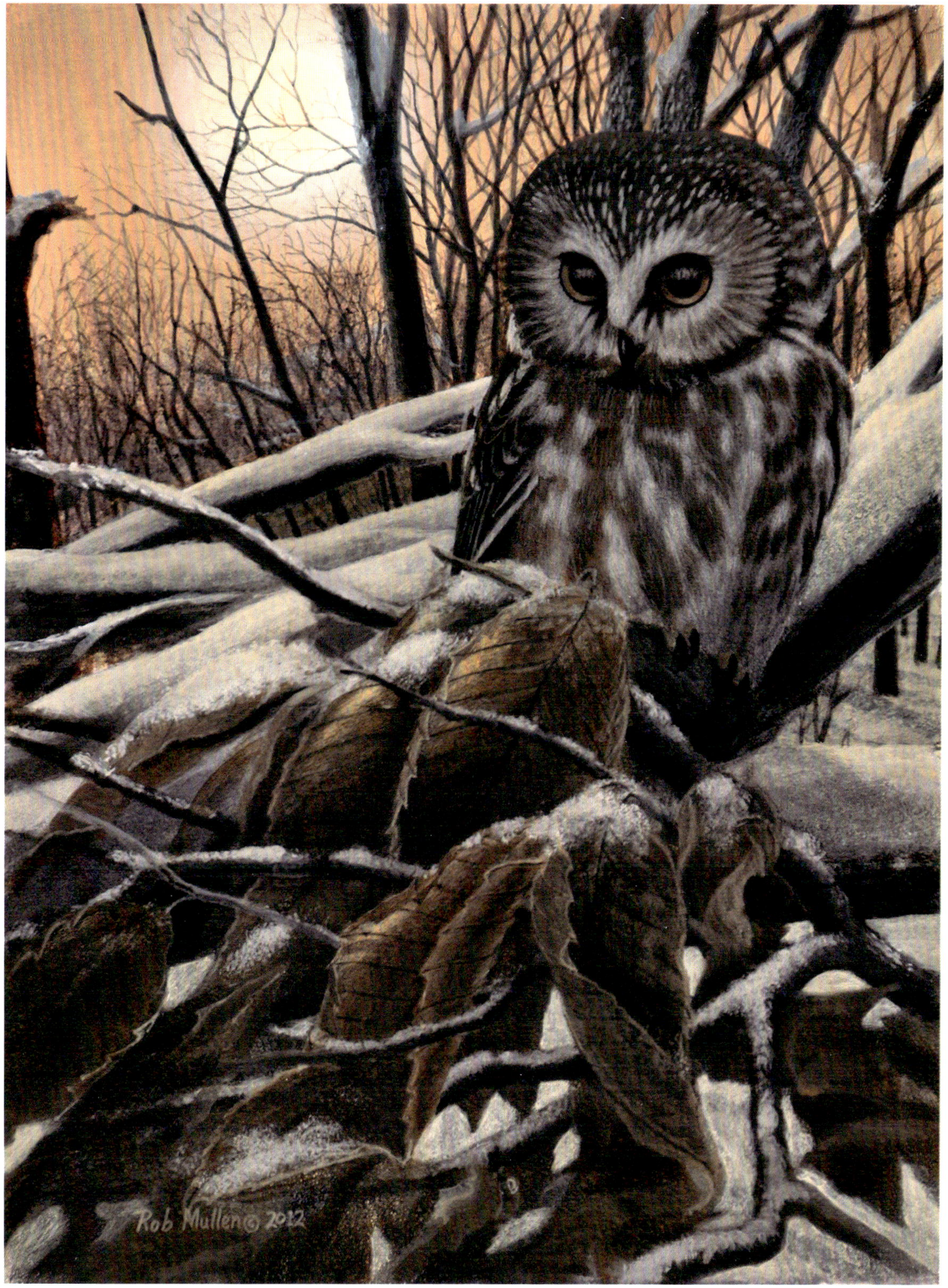
Rob Mullen © 2012

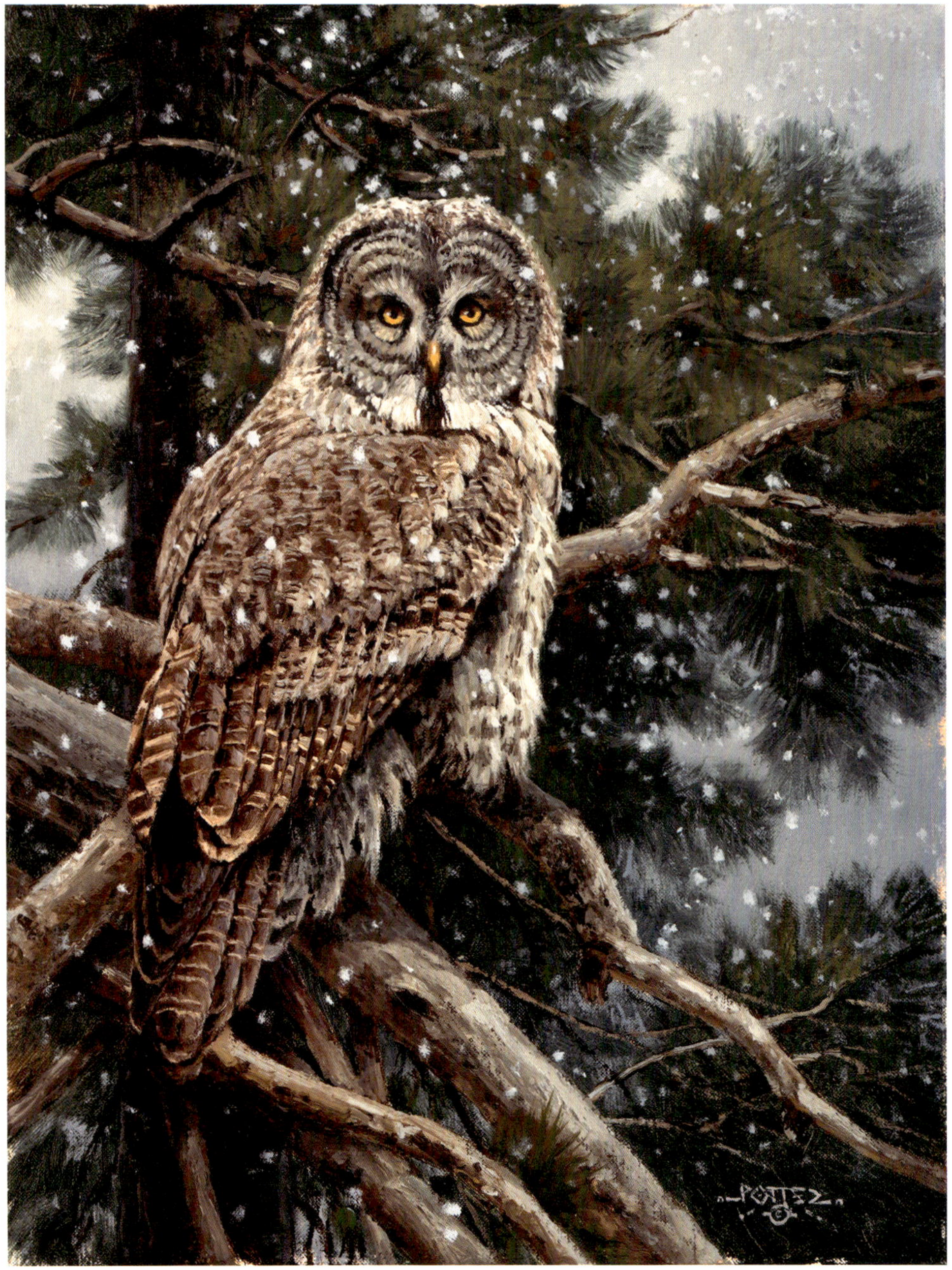

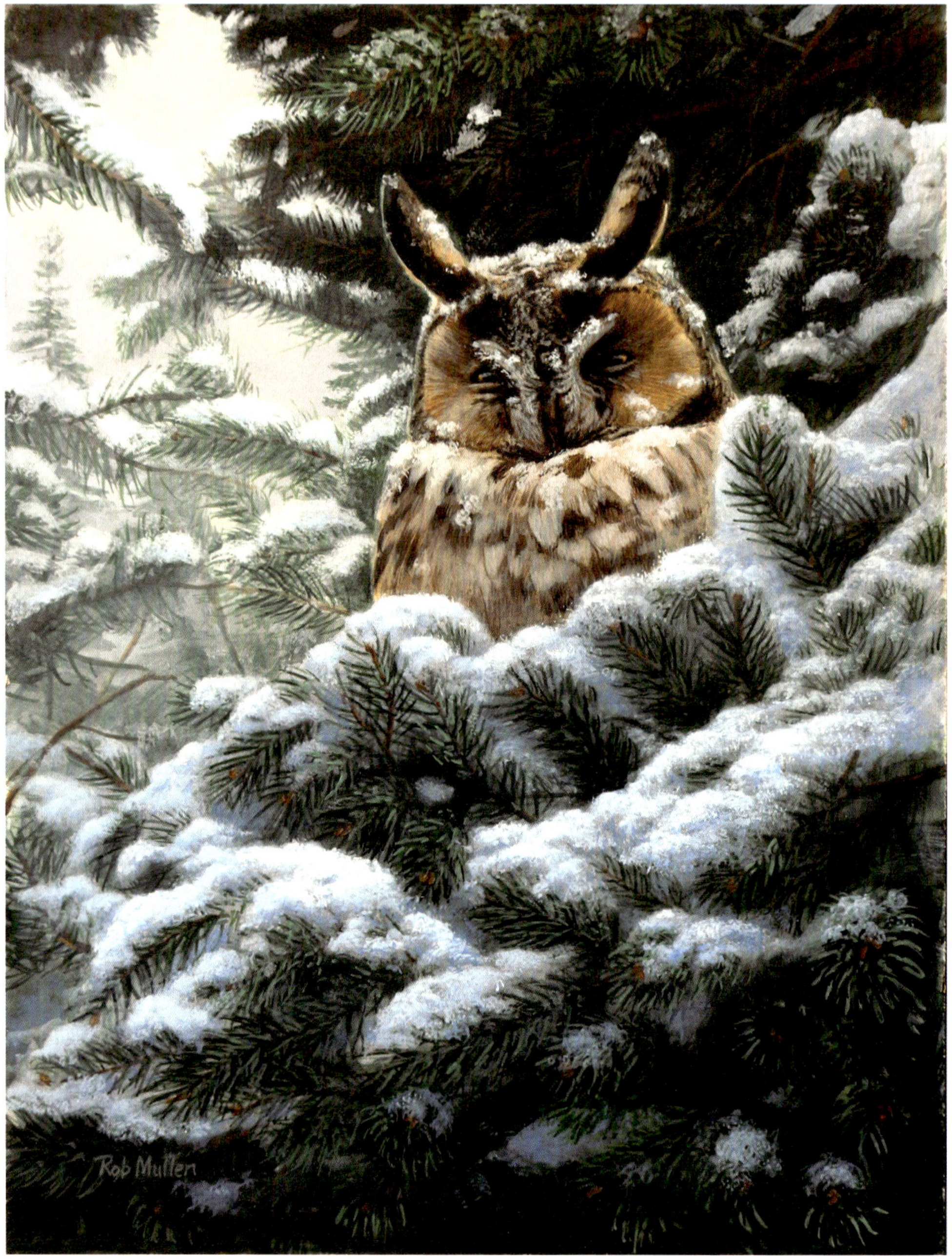
Rob Mullen

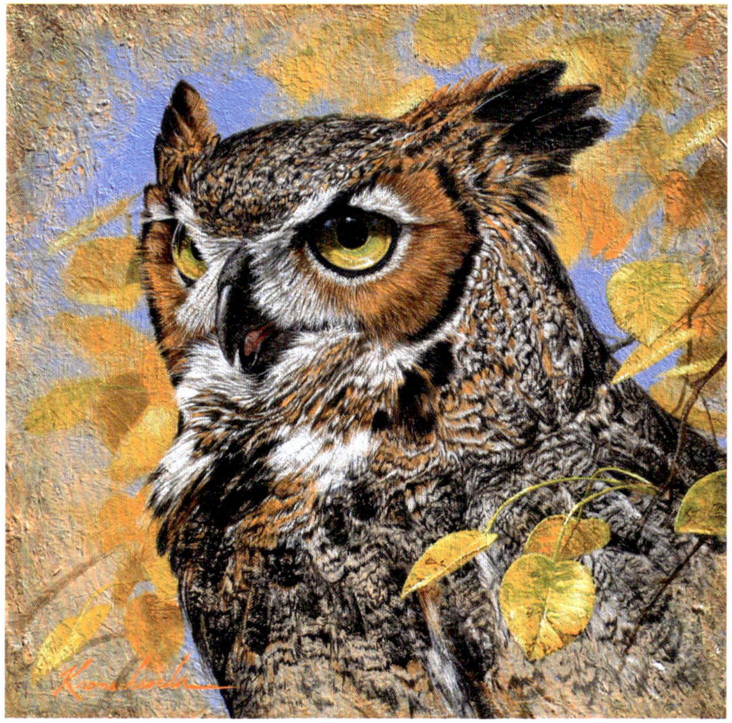

and used to enjoy calling them in, though I now avoid stressing them. They're known for their "Who cooks for you? Who cooks for you-all?" call, but also for sudden, unhinged vocalizations that sound like a banshee in a horror film.

At just six inches tall, the tiny Nearctic northern saw-whet owl (*Aegolius acadicus*) has a call like sharpening a saw with a hand file. Impossibly cute (a favorite image on gifts for your grandmother), it is often oddly tame and can be approached closely. A few years ago, one flew into our garage and stayed for several hours. I talked to her for a long time. She seemed to listen.

The long-eared owl (*Asio otus*) and short-eared owl (*Asio flammeus*) are both Holarctic in distribution and use almost the entire boreal forest worldwide.

The aptly named boreal owl (*Aegolius funereus*) lives almost entirely within the entire boreal forest worldwide, with only a few pockets outside. Tough and mostly resident, it rarely migrates far.

Northern hawk-owls (*Surnia ulula*) break convention: They are diurnal hunters and strikingly patterned. Reminiscent of the rough-legged hawk, they are famous for perching at the apex of the flimsiest spruce tips. My first sighting was on the George River, and the owl was indeed precariously balanced at the tiniest tip of a spruce tree above the recovery pool of a rapid I'd just run. Another Holarctic owl, like the boreal owl, northern hawk-owls live almost exclusively in the boreal forest but do not migrate, so their range map is an almost perfect replica of the boreal forest.

The endemic Eurasian boreal forest owls include the endangered Blakiston's fish owl (*Ketupa blakistoni*), which is restricted to far southeastern Siberia and the Russian island of Sakhalin. Three others are widely distributed throughout the Eurasian forest: the Ural owl (*Strix uralensis*), the Eurasian pygmy owl (*Glaucidium passerinum*), and on the other end of the scale, one of the largest owls on Earth, a close relative of Blakiston's fish owl, the Eurasian eagle-owl (*Bubo bubo*), a resident of most of the forest.

Family Alcedinidae (kingfishers)

With so much water, it is a little surprising that out of over 120 species of kingfisher listed on the IUCN Red List, only two occur in the boreal forest, and only the North American one really commits.

Common kingfisher (*Alcedo atthis*): This small kingfisher is widespread throughout Europe, Asia, and Africa. Some of the population breeds in western Russia and eastern Siberia.

Great horned owl head study. Acrylic, Lee Kromschroeder

FACING LEFT: Great gray owl, Montana. *Soul Searching*, oil, John Potter

FACING RIGHT: Long-eared owl, Ottawa River, Ontario. *Nestled In*, acrylic, Rob Mullen

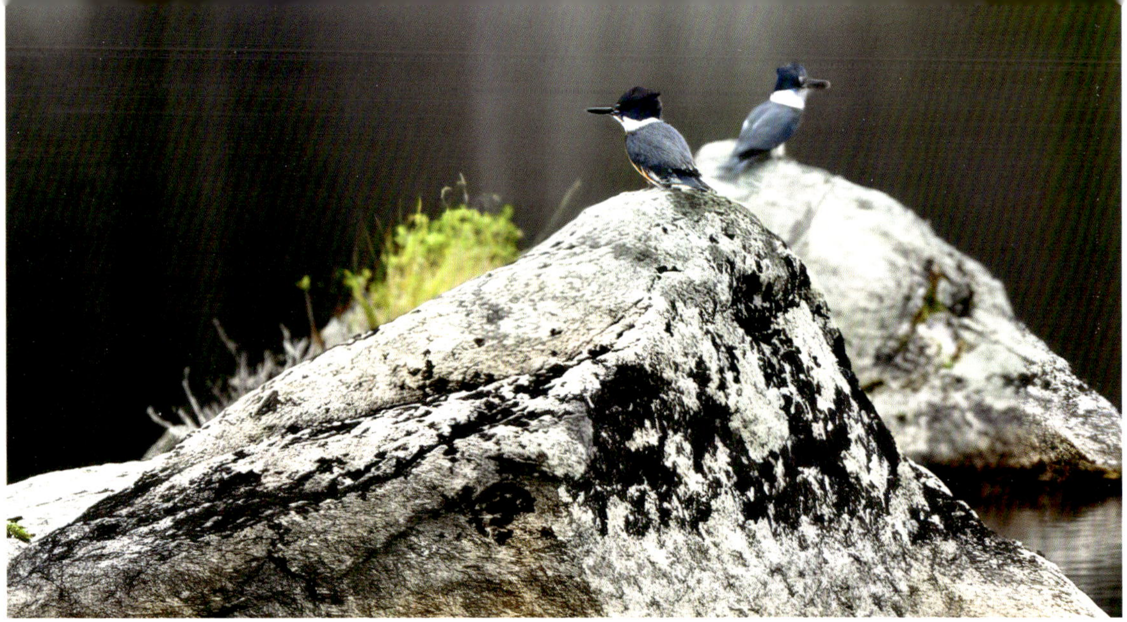

Twins: belted kingfishers, Vermont. Daron Tansley

Belted kingfisher (*Megaceryle alcyon*—genus formerly *Ceryle*): This robust Nearctic bird is more than twice the size of the common kingfisher. It ranges throughout North America but for a slice of northern Alaska and Canada, following the boreal forest tree line.

John Pitcher and I witnessed firsthand what creative aerialists kingfishers are.

August 27, 2001: Upper Missinaibi River, Ontario

John and I watched a life-and-death drama unfold over the river in front of our canoes. Breaking into the serenity of the river, careening out from the bank in wild swerves and arcing dives, a female sharp-shinned hawk was hot on the tail of a kingfisher. The kingfisher was good, but such dogfights are what sharpies are built for. The kingfisher was putting up a valiant fight, but it looked like he'd be outmaneuvered and killed at any moment . . . until suddenly, with a slight, sudden feint to the left, he rolled right and plunged straight into the water!

Witnessing the aerial dogfight had been a wonder. The look on that hawk's face? Priceless.

Family Picidae (woodpeckers)

Of the roughly 240 woodpecker species worldwide, at least 15 live in the boreal forest: seven in North America and eight in Eurasia (several of them familiar backyard visitors). In yet another example of the boreal forest's refuge quality, not one of its woodpecker species is threatened, while nearly 20 of the species' relatives elsewhere are listed as Endangered, Threatened, or Vulnerable by the IUCN.

None are truly Holarctic in distribution, though the three-toed woodpecker was long thought to be until recent genetic work split it into separate Eurasian (*Picoides tridactylus*) and American (*Picoides dorsalis*) species. Most boreal species are year-round residents or only short-distance migrants, except for the Eurasian wryneck (*Jynx torquilla*), which breeds in much of the Eurasian boreal and winters in Africa.

Despite their remarkable—and still not fully understood—adaptations for surviving high-G impacts to their bird brains, woodpeckers are not immune to head trauma. On May 12, 2007, I heard the unmistakable whack of a bird hitting a window. Outside lay a male yellow-bellied sapsucker (*Sphyrapicus varius*; over half the population nests in the North American boreal). He looked dead, but to my surprise, he still had a pulse.

With little hope, I cupped him in my hands to keep him warm and prepared to sit with him to the end. An hour later, after eyeing me warily for a few more minutes, he took off into the woods as if nothing had happened. Since he survived, I decided he wouldn't mind a caricature poking fun at his ordeal.

We've since marked our windows with vertical white dry-erase lines, spaced about four inches apart with a straightedge. They've stopped strikes for years, and we hardly notice them.

Familiar backyard woodpeckers, the downy woodpecker (*Dryobates pubescens*), the smallest North American woodpecker at one ounce, and the hairy woodpecker

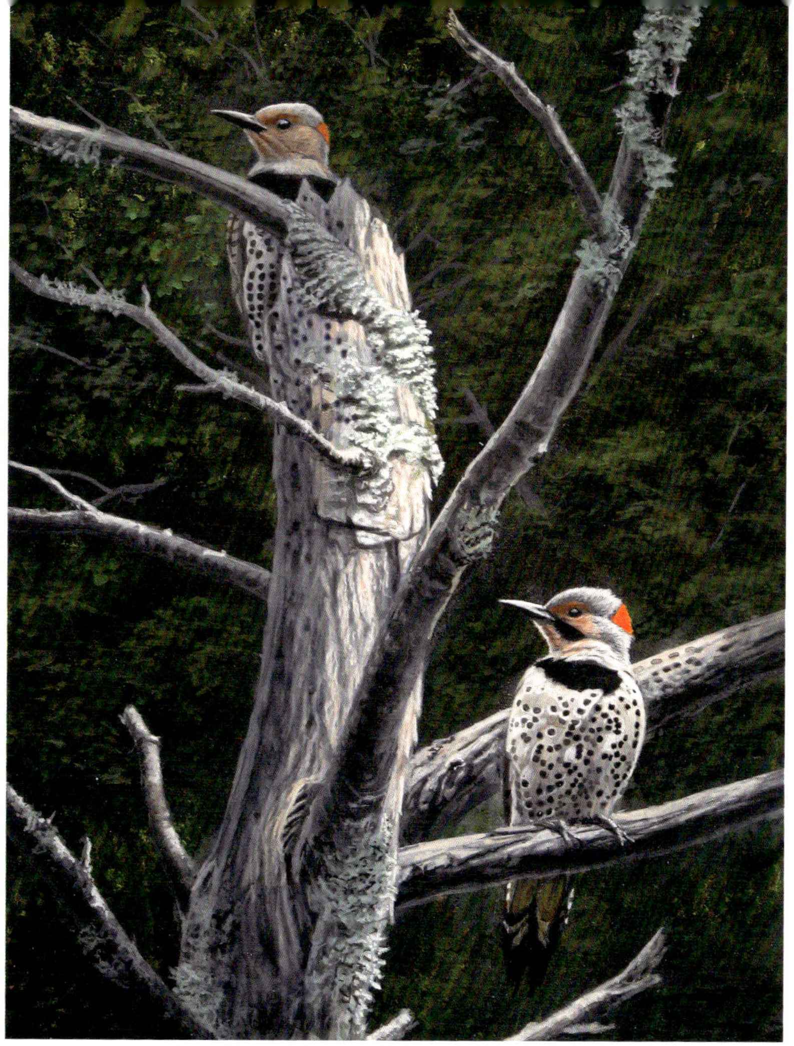

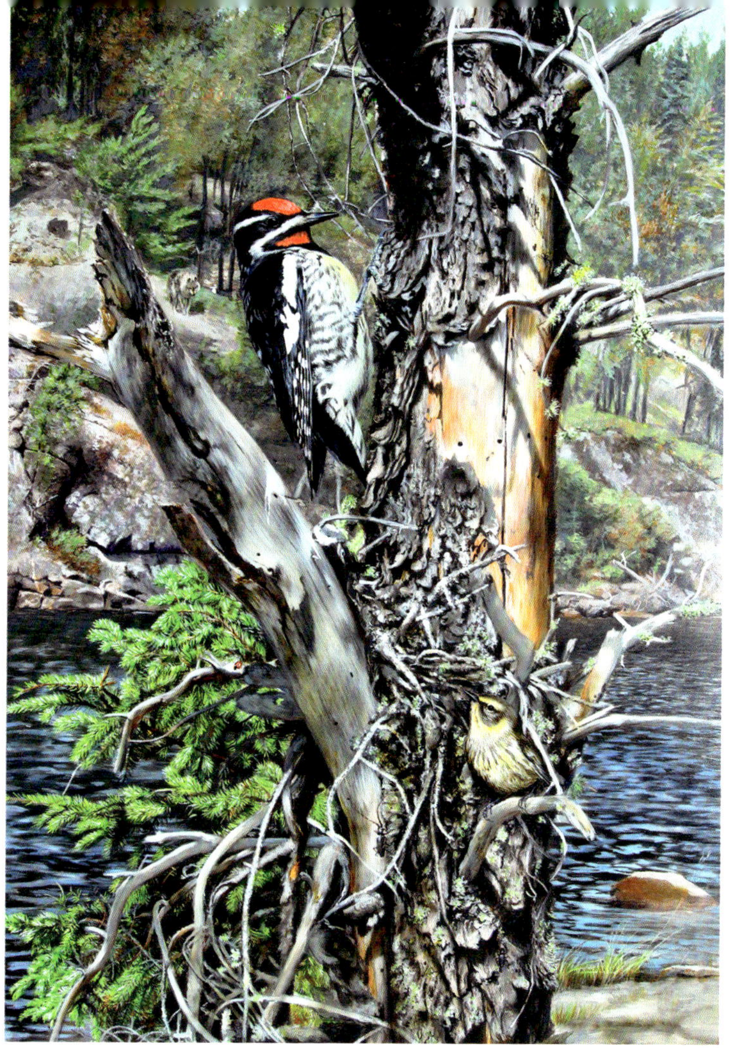

LEFT: Beaver pond complex, Preston Pond Conservation Area. *Forest Flickers*, acrylic, Rob Mullen

RIGHT: Yellow-bellied sapsucker and Cape May warbler with an onlooker, Woodland Caribou Provincial Park. *Voyeur*, acrylic Julia Hargreaves

(*Leuconotopicus villosus*) are common throughout the boreal forest in North America, yet because of their wide distribution, only about 13 percent of downies and 33 percent of hairies nest in the boreal forest. Through convergent evolution, they appear identical except that downies are half the size.

Northern flickers (*Colaptes auratus*) are non-migratory in their US range except for the northernmost tier and are nesting migrants throughout nearly the entire boreal forest. Their return to Vermont in spring is usually first noticed by their staccato, mechanical laughing call.

Pileated woodpecker (*Dryocopus pileatus*)

With a territorial laugh like a flicker's but more maniacal, pileated woodpeckers sound like they belong on an old Jungle Jim or Tarzan movie soundtrack. These crow-sized hammerers with bright red crests are one of the flashiest birds in the Northwoods. They are among the largest woodpeckers in the world, the largest in North America (unless you are holding out for the ivory-billed woodpecker to reemerge someday), and the second largest in the boreal forest.

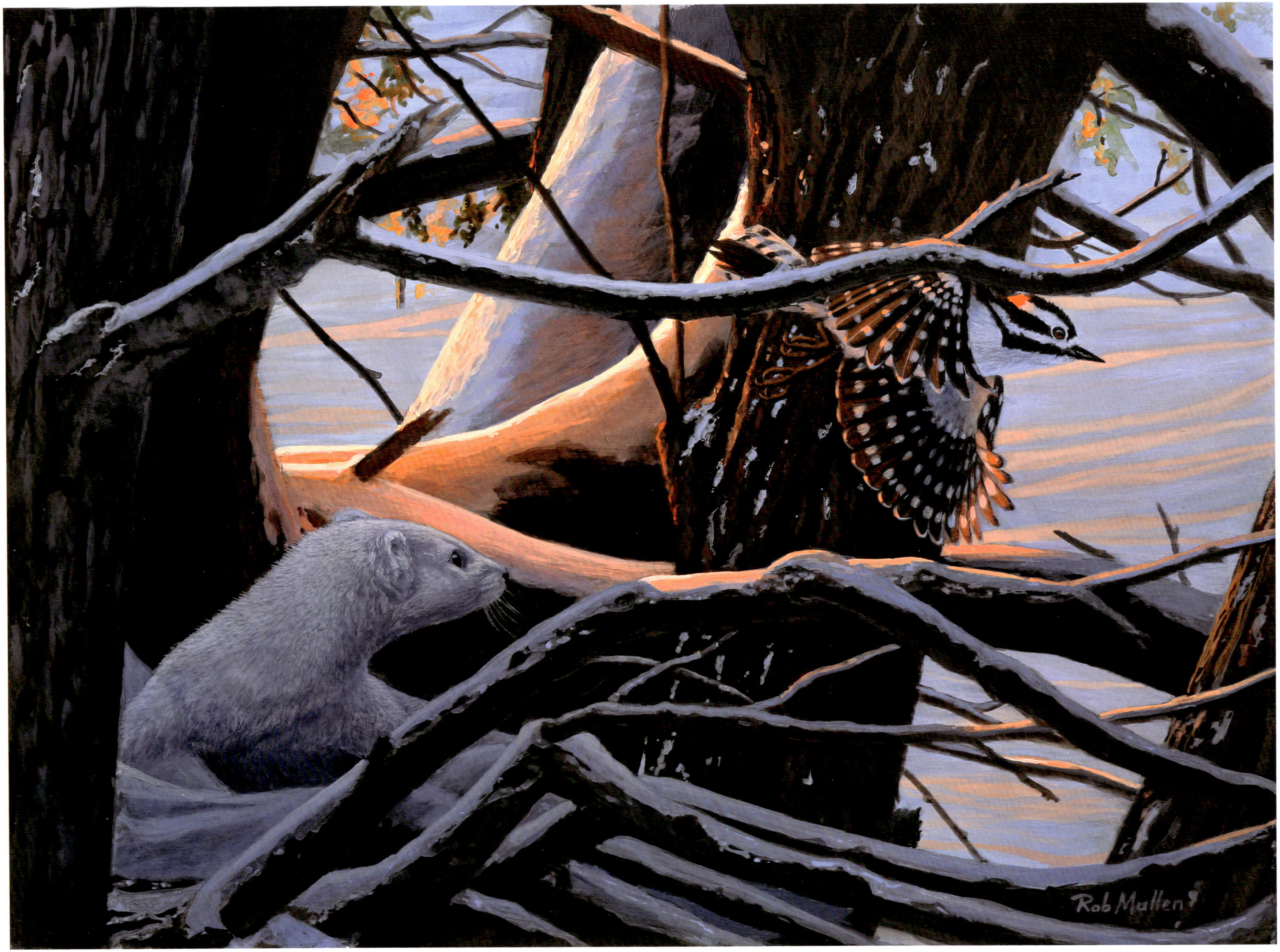
Rob Mullen

Their excavations can be so deep that they become a keystone species, providing shelter for other species in their abandoned nest holes, allowing smaller species access to arthropod prey in the wood that they otherwise couldn't reach, and possibly helping stem insect invasions.

The black-backed woodpecker (*Picoides arcticus*) and the American and Eurasian three-toed woodpeckers (*Picoides dorsalis* and *Picoides tridactylus*, respectively) are boreal forest specialists and, interestingly, are all distinguished by being medium-sized black and white birds with only three toes on each foot. These are hardcore northerners: from 75 to over 90 percent of their respective populations reside across the boreal forest.

The eight Eurasian boreal forest woodpeckers are all residents throughout much of the Eurasian boreal forest, with several occupying the breadth of it. As noted earlier, the exception to full-time residency is the Eurasian wryneck, which migrates to Africa and Asia (and, fun fact, plays dead like an opossum).

The largest woodpecker in the boreal forest, the black woodpecker (*Dryocopus martius*), is a close Eurasian relative of the pileated, and basically a slightly larger pileated that is all black. The male has a full red cap, and the female, like the pileated, has a shorter red cap on the back of her head.

From the loud, hammering woodpeckers, let's move to the more melodic passerines. Where woodpeckers shape the forest physically—excavating nest sites, unlocking insect prey, and even creating shelter for other species—passerines shape it sonically and biotically. They fill the trees with a layered chorus of song to claim territories, attract mates, and soothe blackfly-beset canoeists. Using their strong, perching adapted feet, they also work overlapping layers of the forest canopy, cleansing destructive insects from the trees and understory.

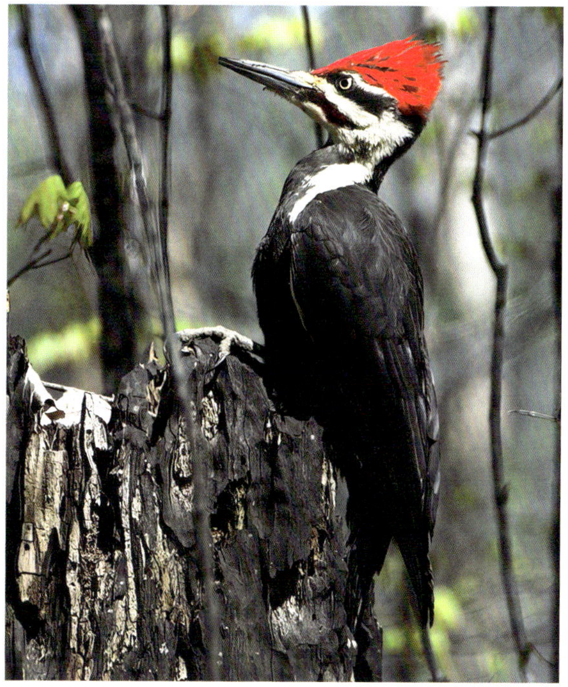

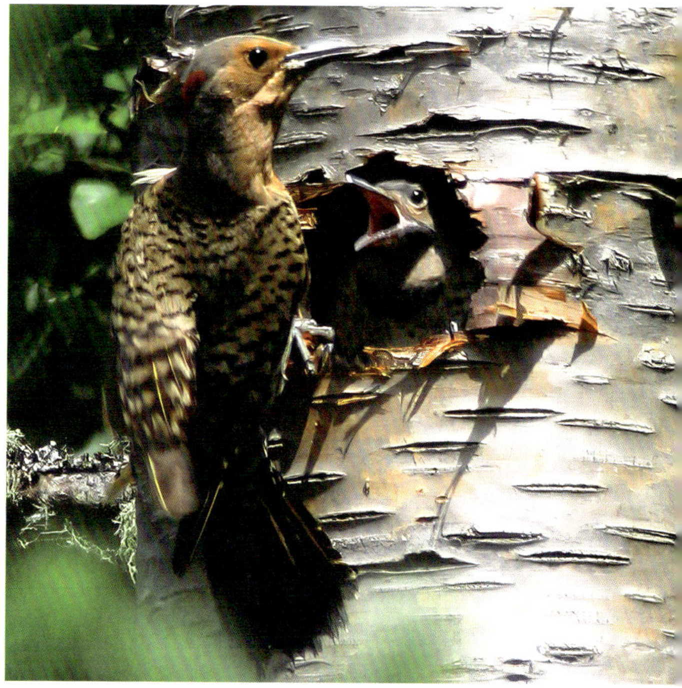

LEFT: Male pileated woodpecker, Quebec. Bonnie Rowell

RIGHT: Flicker family moment: female at the nest cavity (note the formerly namesake yellow-shafted feathers) on an island in Lake Superior where we camped during a thunderstorm.

FACING: A likely American ermine (formerly short-tailed weasel) and downy woodpecker in a cedar snag on Libby's Lookout Hill, West Bolton, Vermont. *Fire and Ice*, acrylic, Rob Mullen

Order Passeriformes (perching pirds)

Suborder Passeri (songbirds)

A key feature that makes passerines perching birds is the structure of their feet. The toes are arranged three forward and one back (an arrangement known as anisodactyl) like many birds but with a tendon that runs from the back of the lower leg to the underside of the toes. This tendon is pulled tight when the leg bends, forcing the toes to flex and grip without muscular effort, allowing them to sleep while perched.

What makes passerines "songbirds" is their ability to exquisitely control their vocal organ, the syrinx, located at the junction of the trachea and the two bronchial branches leading to the lungs. By modulating the vibrating air from each lung

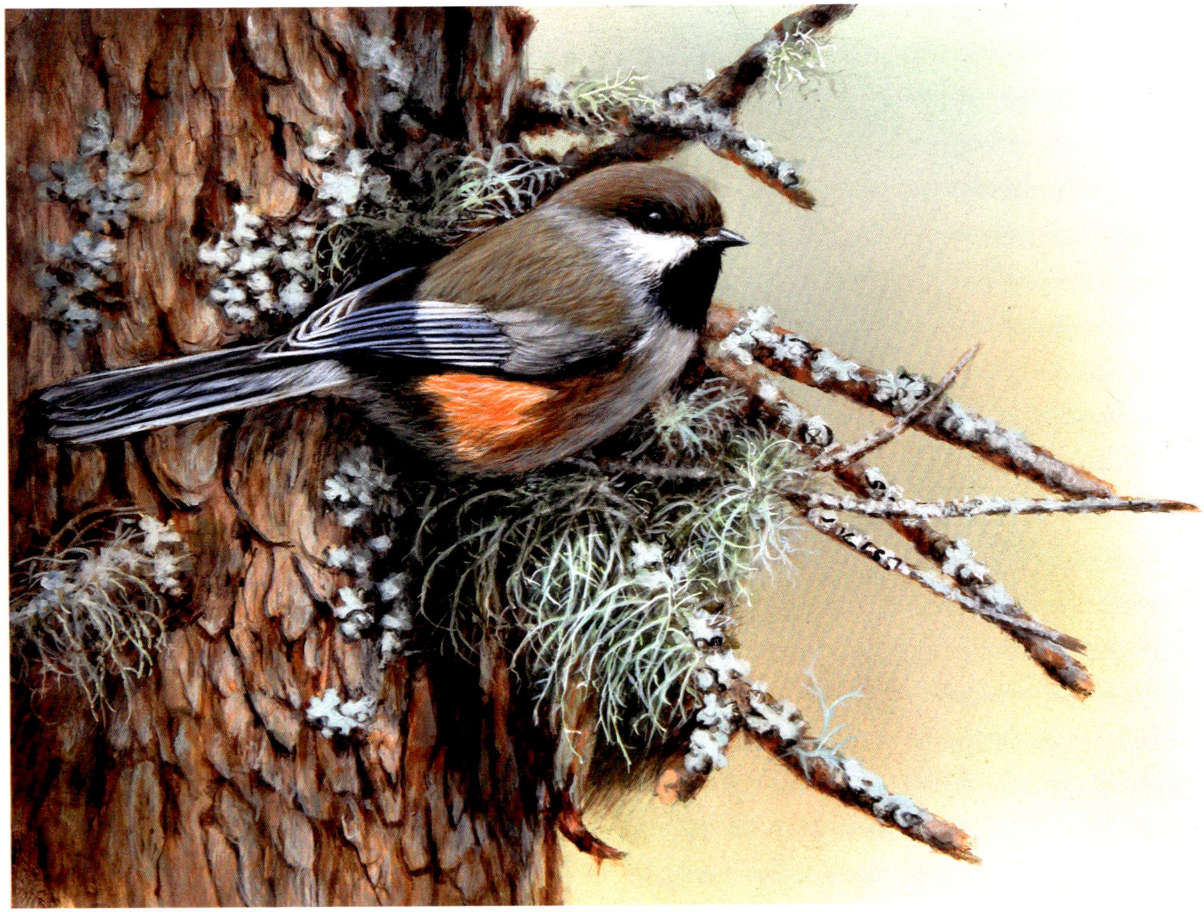

Boreal chickadee study, acrylic, Julia Hargreaves

FACING LEFT: Black-capped chickadee hunting in apple blossoms, West Bolton, Vermont. *Leaf Gleaner*, acrylic, Rob Mullen

FACING RIGHT: Boreal chickadee in Labrador tea. *Tea Time*, acrylic, Rod Lawrence

independently, songbirds create multifrequency songs, most notably the ethereal harmonics of some thrushes.

Passerines are the largest order of birds in the world. More than half the bird species on Earth are passerines, and that proportion holds true in the boreal forest. Globally, there are over 140 families and around 6,500 species, with about 25 families with representative species in the boreal forest, totaling at least 280 boreal forest passerine species worldwide. We will not be able to cover all the families, so we will settle for the main ones and a few representative or particularly interesting species in each. Where to start? With so many options, why not wade in with the most cheerful and outgoing birds in the northern forest?

Family Paridae (chickadees and tits)

Curious and personable, chickadees are often the first to investigate a disturbance and offer advice—especially if you've stopped for a break and toppled over in your overstuffed portage pack like an overturned turtle.

Between three species the entire boreal forest is covered, worldwide: gray-headed chickadees (*Poecile cinctus*, aka Siberian tit) in northern Eurasia, Alaska, and the Yukon; willow tits (*Poecile montanus*) across most of Eurasia; and boreal chickadees (*Poecile hudsonicus*) across the entire North American boreal. However, black-capped chickadees (*Poecile atricapillus*), while not boreal forest specialists, were still 75 percent of my chickadee encounters in the 61 weeks I spent afield on the expeditions in this book.

Two of our local black-caps, Ozzie and Harriet, have trained us to hand out sunflower seeds on demand. We've learned a lot from them over the years. (They first showed up in April of 2021 and are still here today, knock on wood.) They mate for life, stash seeds for later, remember the spots, and don't even mind if our youngest dog sits under them scarfing reject seeds (they can be very discriminating) while they eat out of my hand—a little ad hoc commensalism.

Family Tyrannidae (tyrant flycatchers)

Exclusively New World, most of this family are tropical residents. Eleven species migrate north to breed in the boreal forest, all returning to Central or South America, making them neotropical migrants.

Flycatchers are generally drab but agile insect hunters; members of the genus *Empidonax* are notoriously difficult to

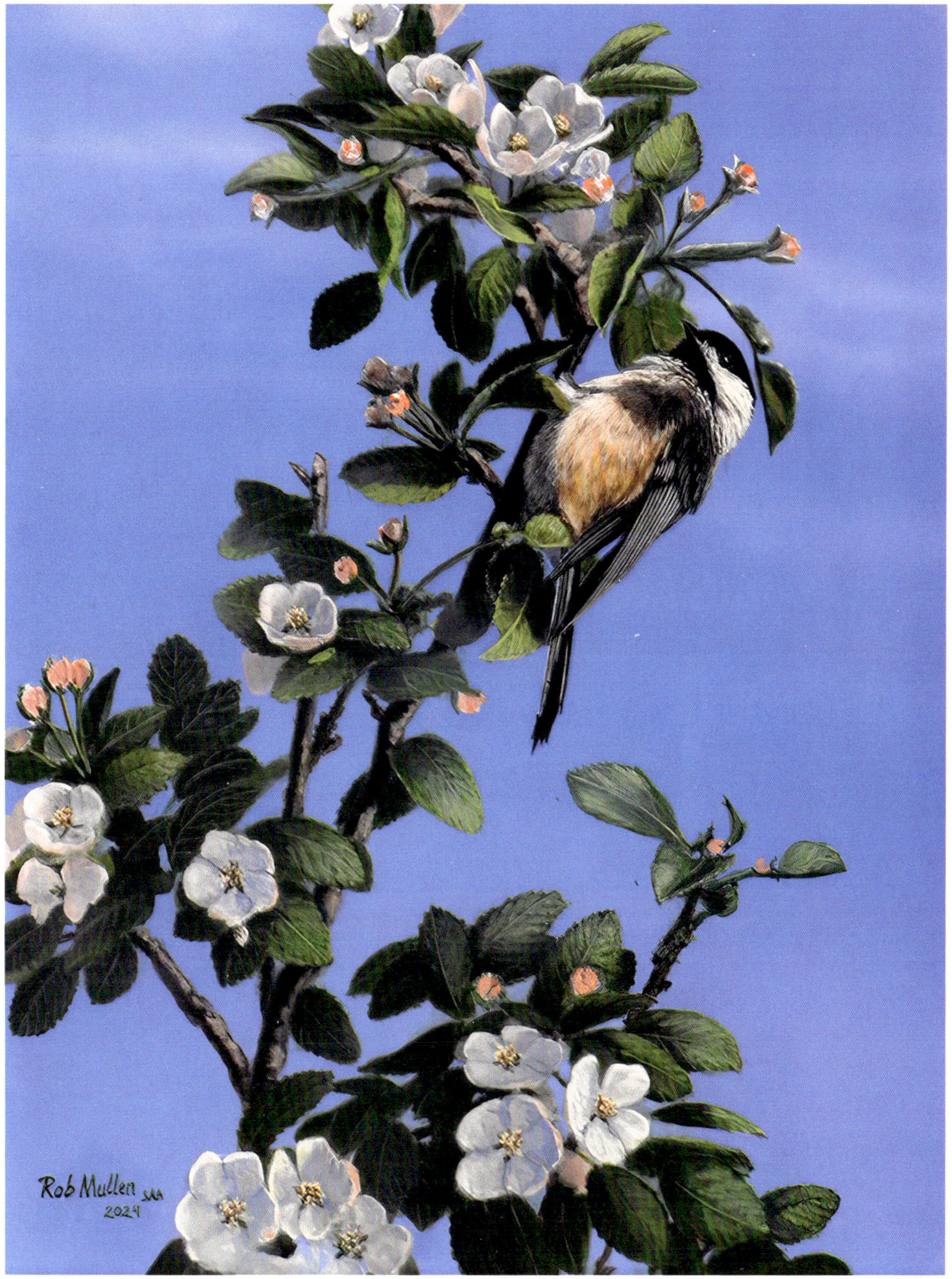
Rob Mullen SAA
2024
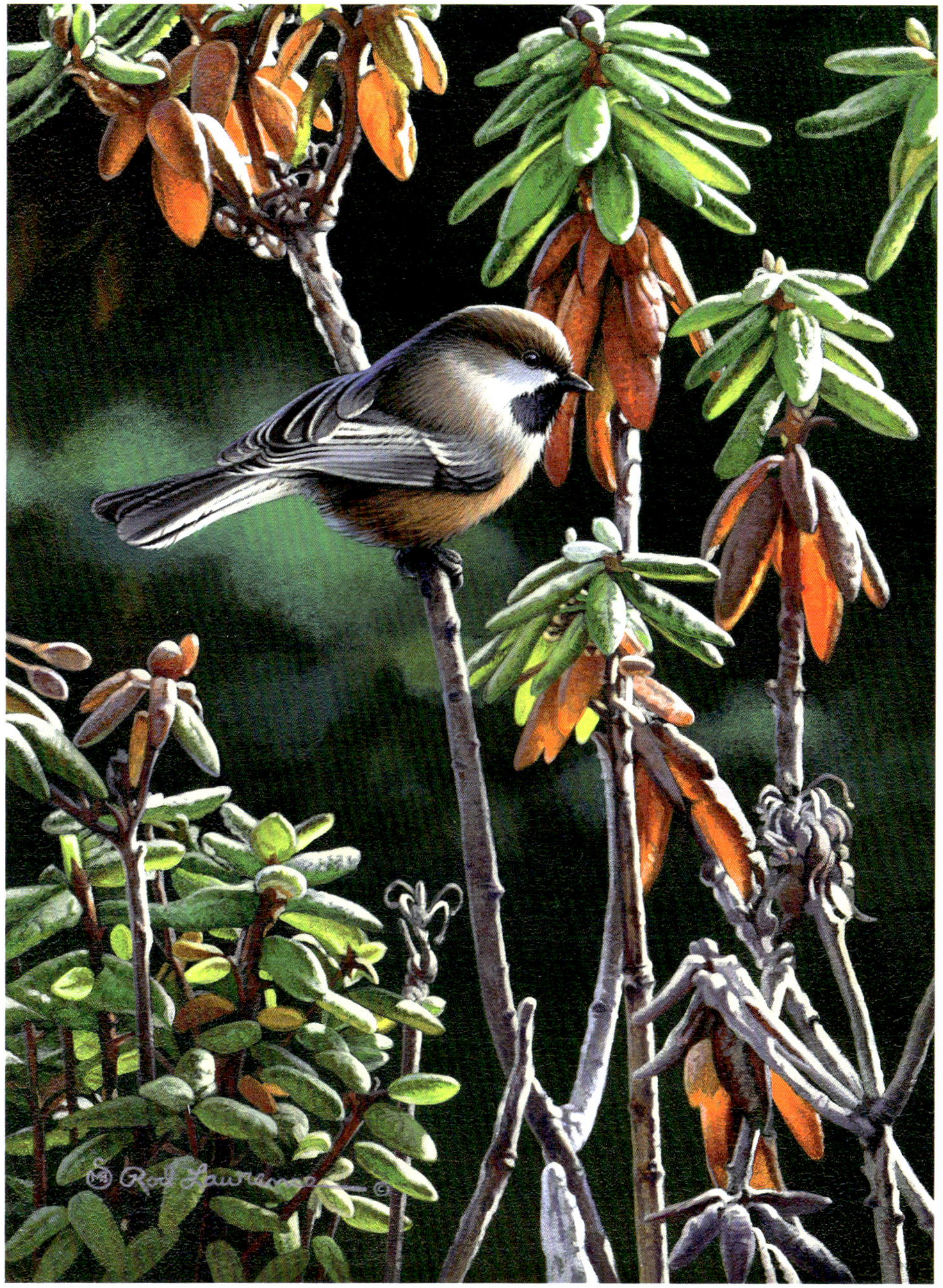

CLOCKWISE FROM TOP LEFT:
Busy phoebe parent with food at our bird bath, spring 2016.

Philadelphia vireo. Peter Riley

Red-eyed vireo in boreal habitat, Seney National Wildlife Refuge, May 2010.

Least flycatcher, Baxter State Park, Maine. Daron Tansley

Blue-headed vireo. Peter Riley

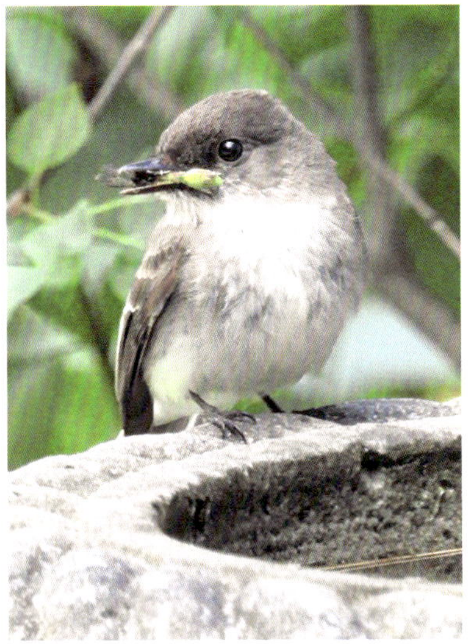

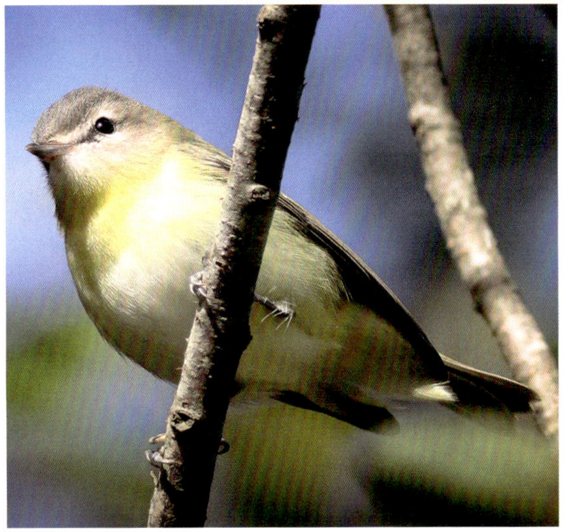

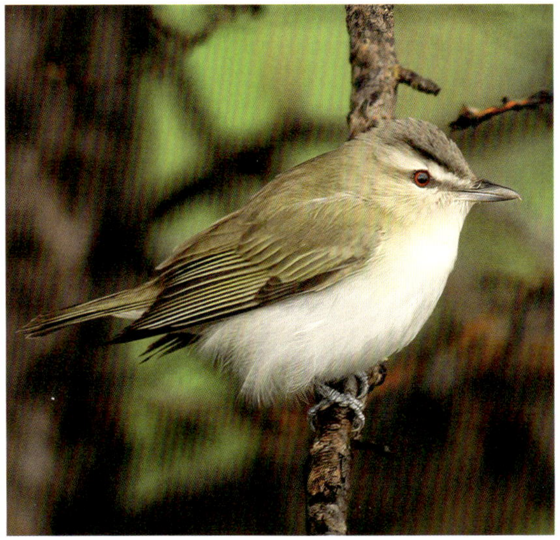

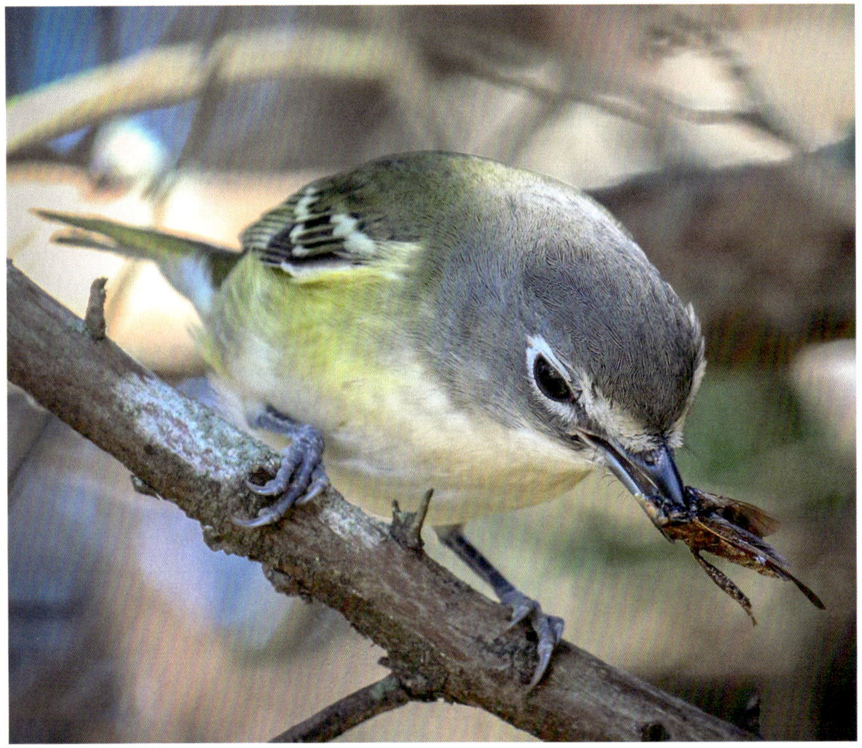

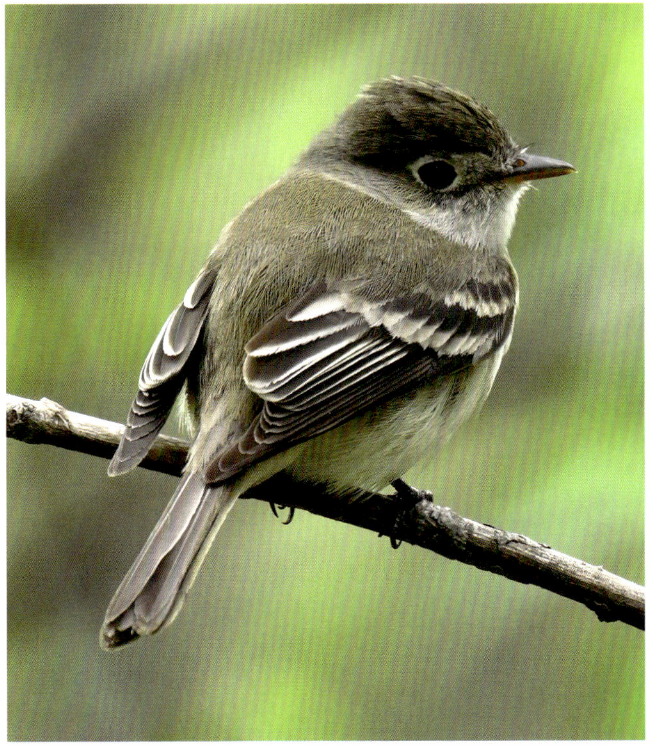

identify. The eastern phoebe (*Sayornis phoebe*) is a welcome exception, flicking its tail as a helpful field mark. One pair nested above our front door, tolerating my daily photos of eggs, hatchlings, and fledglings . . . until all four simultaneously launched over my head like fighter jets at an airshow—a successful fledging!

Boreal regulars include olive-sided flycatchers (*Contopus cooperi*), yellow-bellied flycatchers (*Empidonax flaviventris*), least flycatchers (*Empidonax minimus*), and alder flycatchers (*Empidonax alnorum*).

Family Muscicapidae (old world flycatchers)

Unrelated to tyrant flycatchers, these Eurasian songbirds number 340 species, with 34 in the boreal forest. Notables include the Siberian rubythroat (*Calliope calliope*), thrush nightingale (*Luscinia luscinia*), and European robin (*Erithacus rubecula*), plus two species that also occur in North America: the northern wheatear (*Oenanthe oenanthe*) and the bluethroat (*Luscinia svecica*). The IUCN assesses all the boreal forest species as Least Concern. Of the 306 non-boreal species, 18 percent are assessed at some level of threat or extinction (two).

Families Laniidae (shrikes) and Vireonidae (vireos)

Shrikes, with their hooked bills, prey on small vertebrates, including other songbirds, sometimes impaling them on thorns as a form of impromptu larder. Of the nine boreal species, the northern shrike (*Lanius borealis*) is the most dependent on the boreal forest. Vireos, close relatives, are insect specialists and neotropical migrants; boreal breeders include blue-headed vireos (*Vireo solitarius*), Philadelphia vireos (*Vireo philadelphicus*), and red-eyed vireos (*Vireo olivaceus*). Red-eyed vireos are relentless singers with possibly the most boring birdsong in the woods—and they start early, right outside our bedroom window.

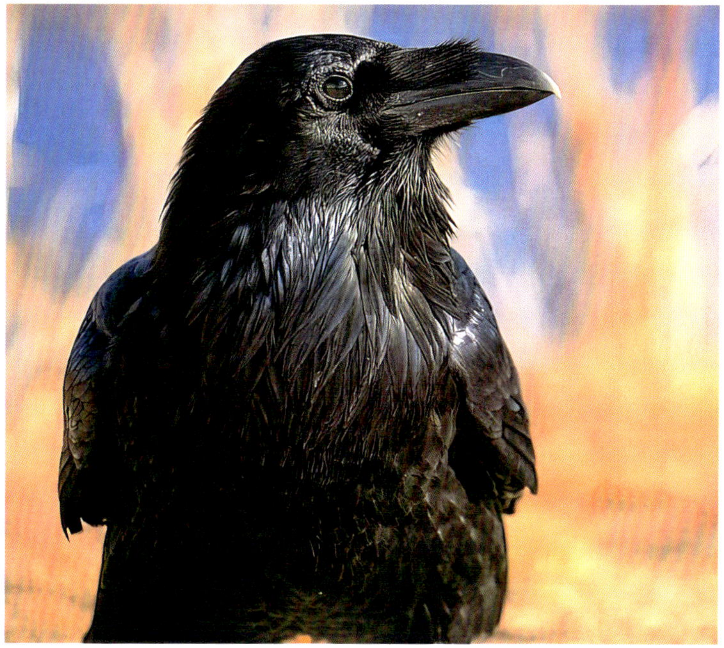

Northern raven, Algonquin Provincial Park. Steve Bruno

Family Corvidae (ravens, crows, jays, and magpies)

Among the world's smartest birds, corvids can count (even understanding the concept of "zero"), use tools, hold grudges, have a sense of humor, and feature prominently in myths worldwide (the chief Norse god Odin's two most trusted counselors were ravens). The northern raven (*Corvus corax*), weighing up to three-and-a-half pounds, is, believe it or not, a songbird; the world's largest and an icon of boreal wilderness worldwide. Local raven dialects seem to vary (to me anyway), from *gronk*s in the east to more creative metallic clanks in Alaska, but basically, if you hear any really strange sound in the woods, and have no idea of who made it, consider a northern raven as a suspect. Note: *Corvus corax* is usually called the common raven. There is nothing common about them, and as northern raven is also used, we will refer to them as northern ravens.

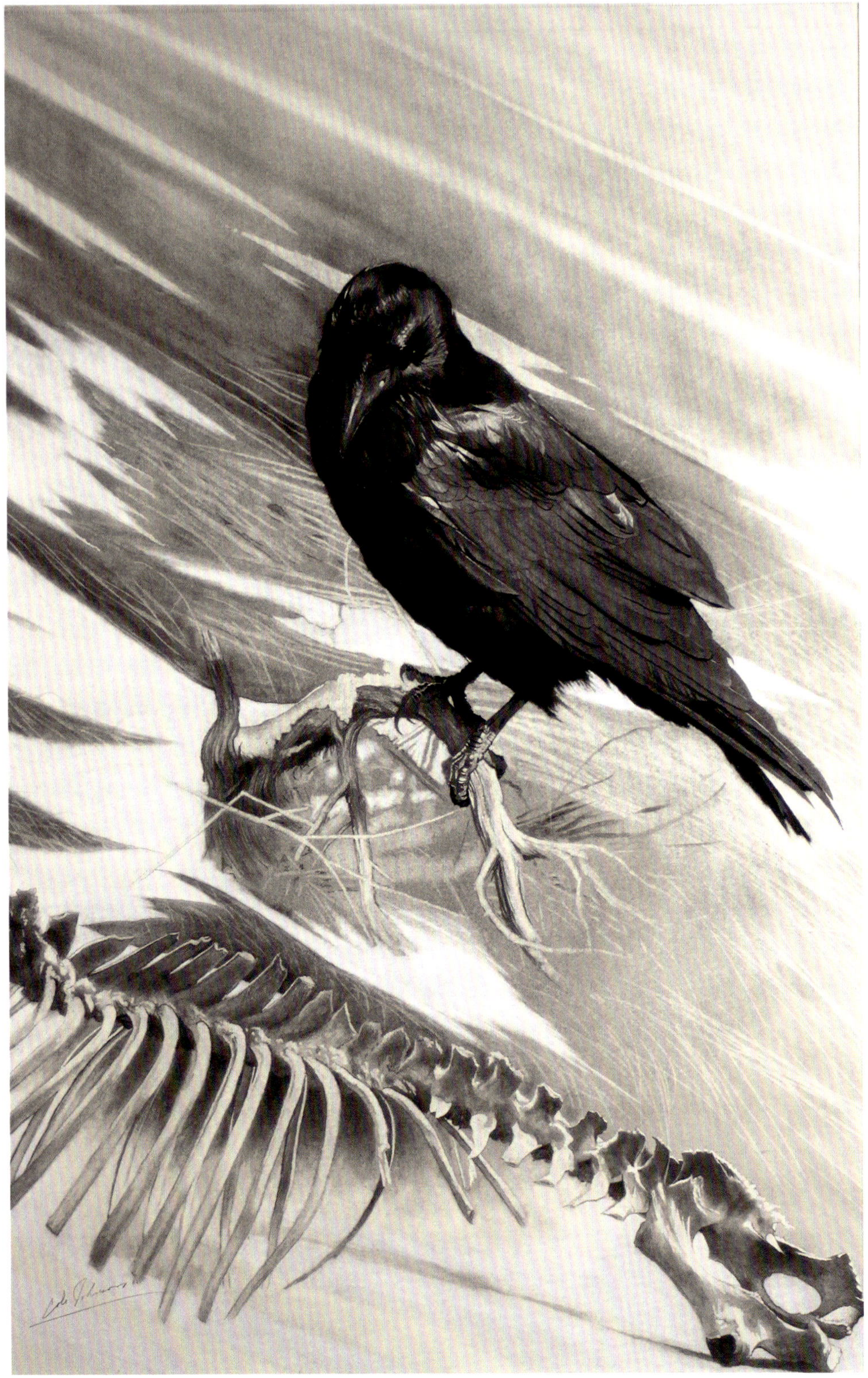

Canada jays (*Perisoreus canadensis*) are frequent companions in camp and along the portage trail, following travelers and helping themselves to proffered (or unsecured) food. Black-billed magpies (*Pica hudsonia*) are more common in western North America. If we had been more superstitious in the Upper Missouri River Breaks, they could have occupied much of our time since, according to ancient and reliable lore, if you see one, it is bad luck not to spot a second. Eurasian birds include the Siberian jay (*Perisoreus infaustus*) and northern nutcracker (*Nucifraga caryocatactes*), both of which occupy almost the entire Eurasian boreal forest.

Families Sittidae (nuthatches) and Certhiidae (treecreepers)

Whereas many songbirds such as chickadees and warblers are leaf-gleaners who forage in the foliage, nuthatches and creepers work the trunks of trees: the creepers head up and the nuthatches head down.

American (*Certhia americana*) and Eurasian (*Certhia familiaris*) treecreepers occur throughout much of the North American and Eurasian boreal forest, respectively. They are well camouflaged and quiet year-round residents, so they can be hard to see despite being common. For a refreshing change, none of the 16 species of treecreepers worldwide are assessed as Threatened on the IUCN Red List.

Nuthatches, on the other hand, are not quiet, frequently erupting into strident conversations of short, nasal notes. In North America, the red-breasted nuthatch (*Sitta canadensis*) is the most widespread in the boreal forest, with the white-breasted nuthatches (*Sitta carolinensis*) occurring in the more southerly reaches of the biome. In Eurasia, the Siberian (*Sitta arctica*) and Eurasian (*Sitta europaea*) nuthatches occupy much of the forest between them. Worldwide, there are about 32 species, of which seven (22 percent) are at some level of IUCN-recognized conservation risk. None of the six boreal forest species are at risk (0 percent).

Northern raven. *Darkness There and Nothing More*, charcoal, Cole Johnson

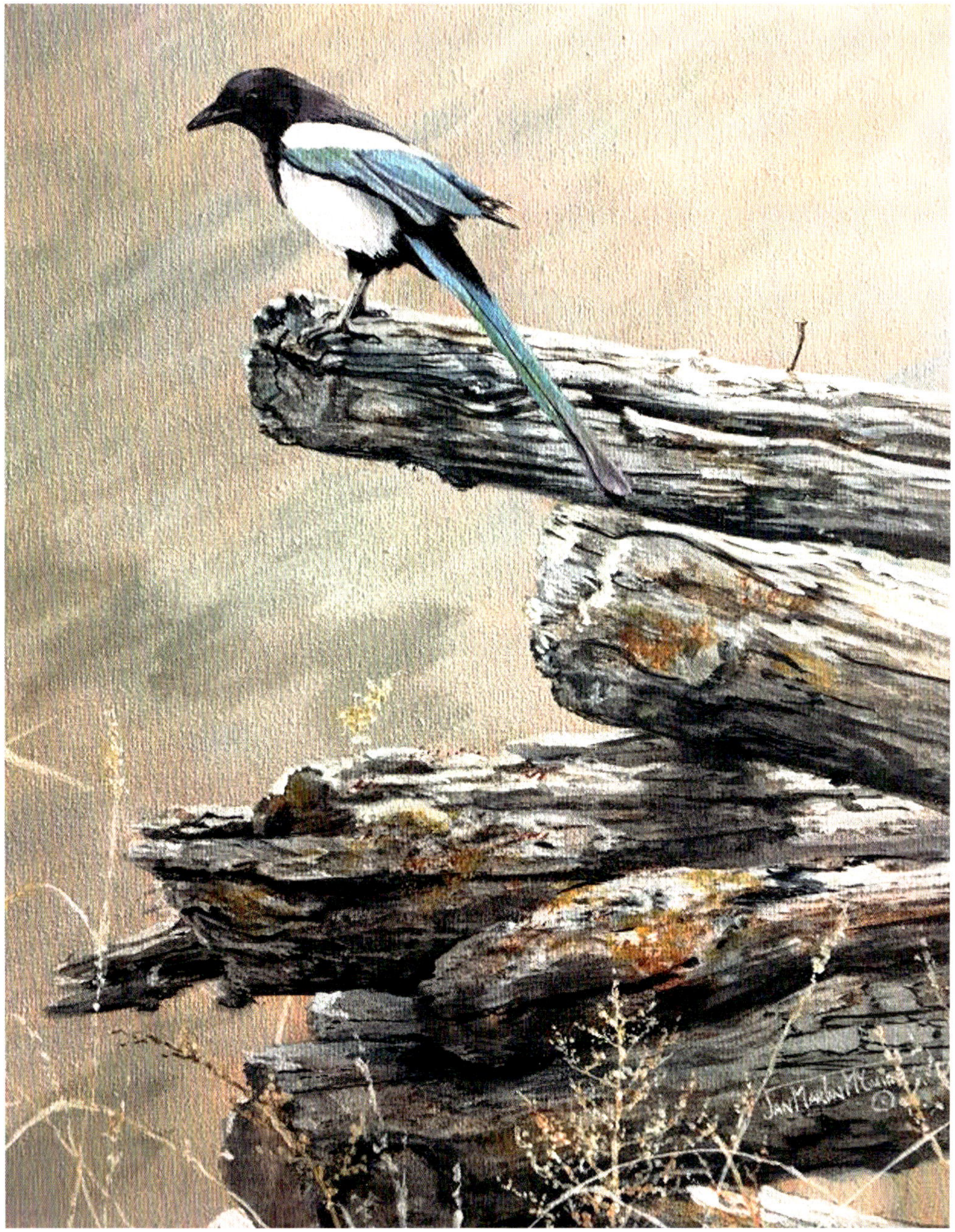

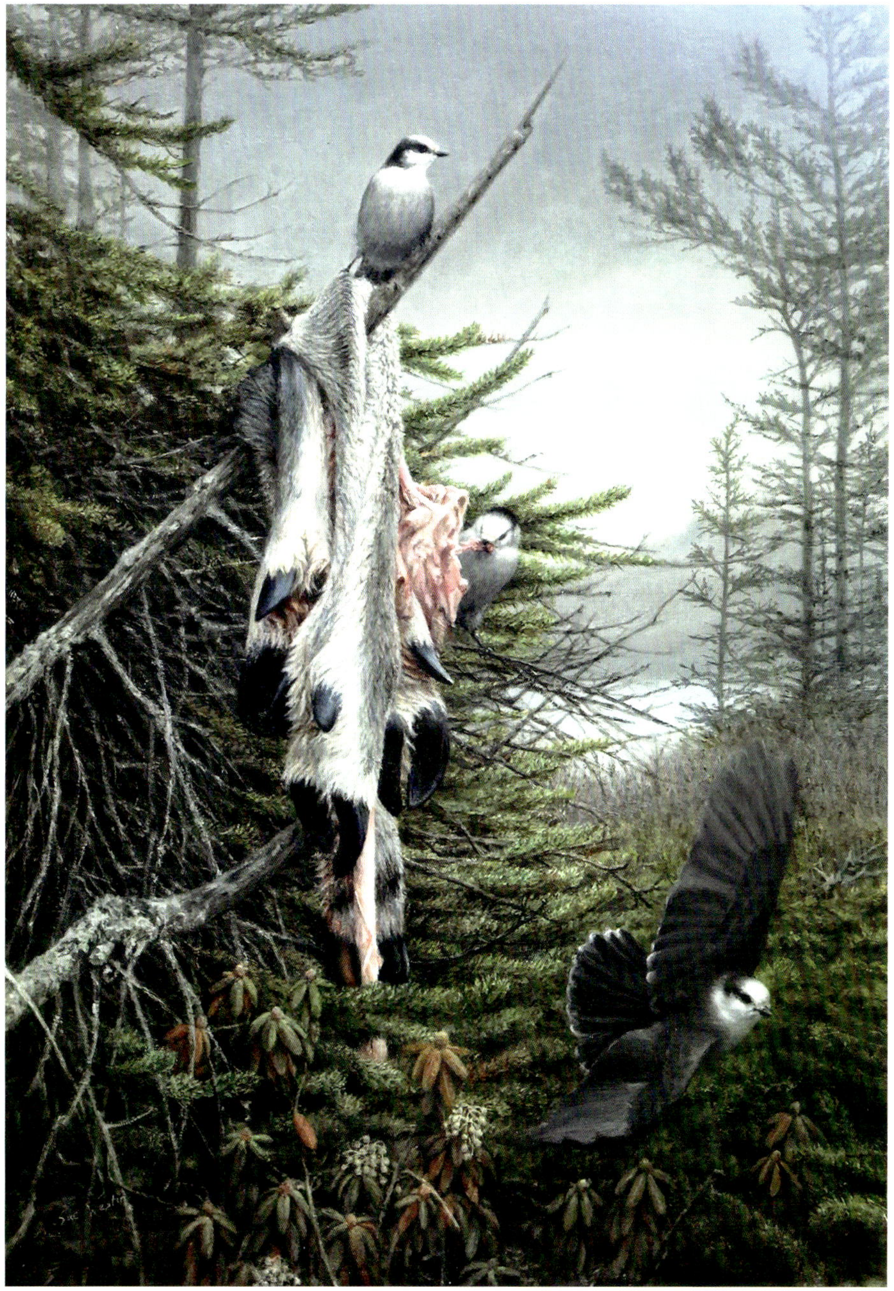

LEFT: Magpie, Missouri River Breaks. *On the Fence*, acrylic, Jan Martin McGuire

RIGHT: Canada jay and Innu offering. *The Gift Tree*, acrylic, Sue Westin

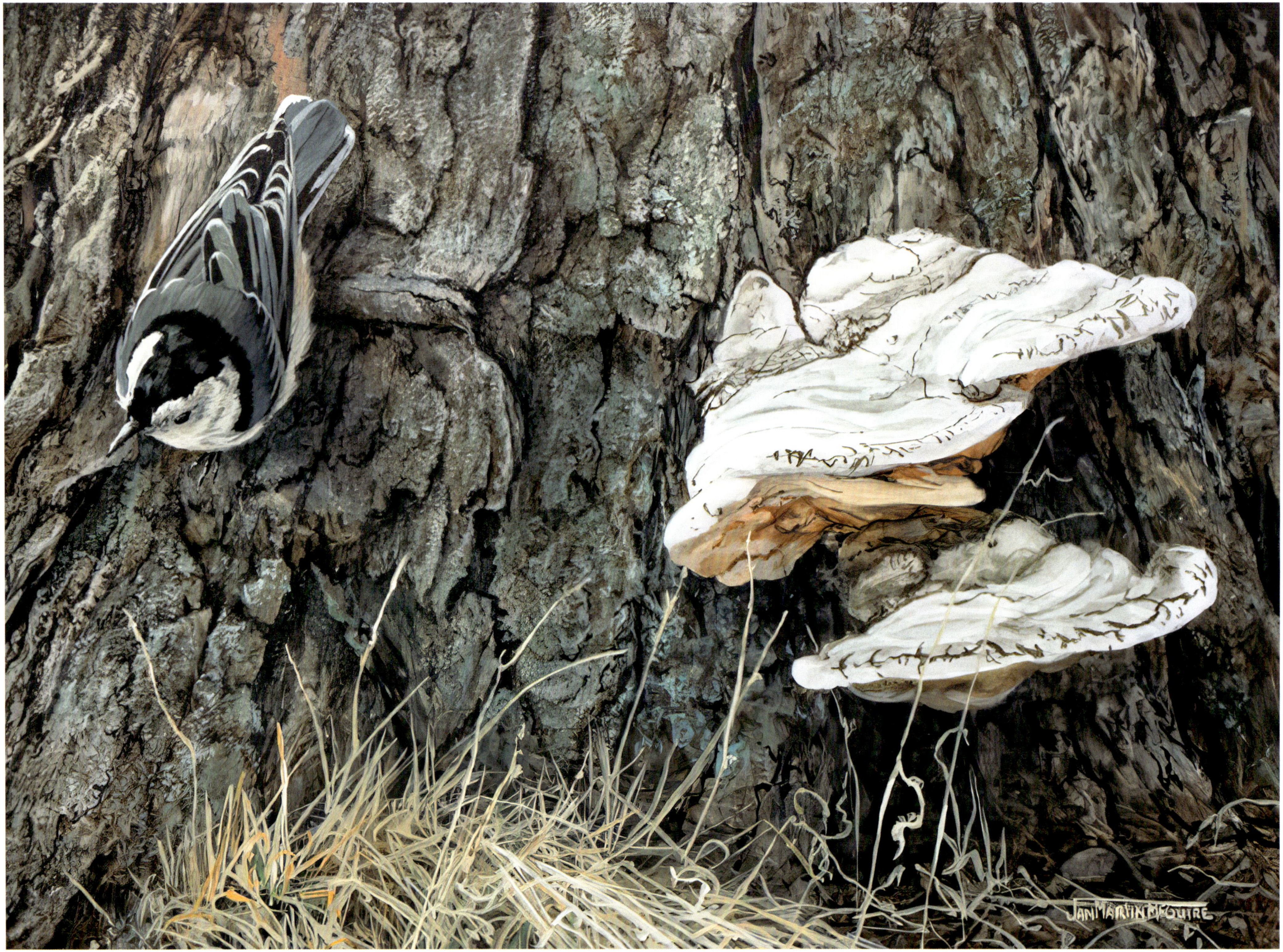
JAN MARTIN MCGUIRE

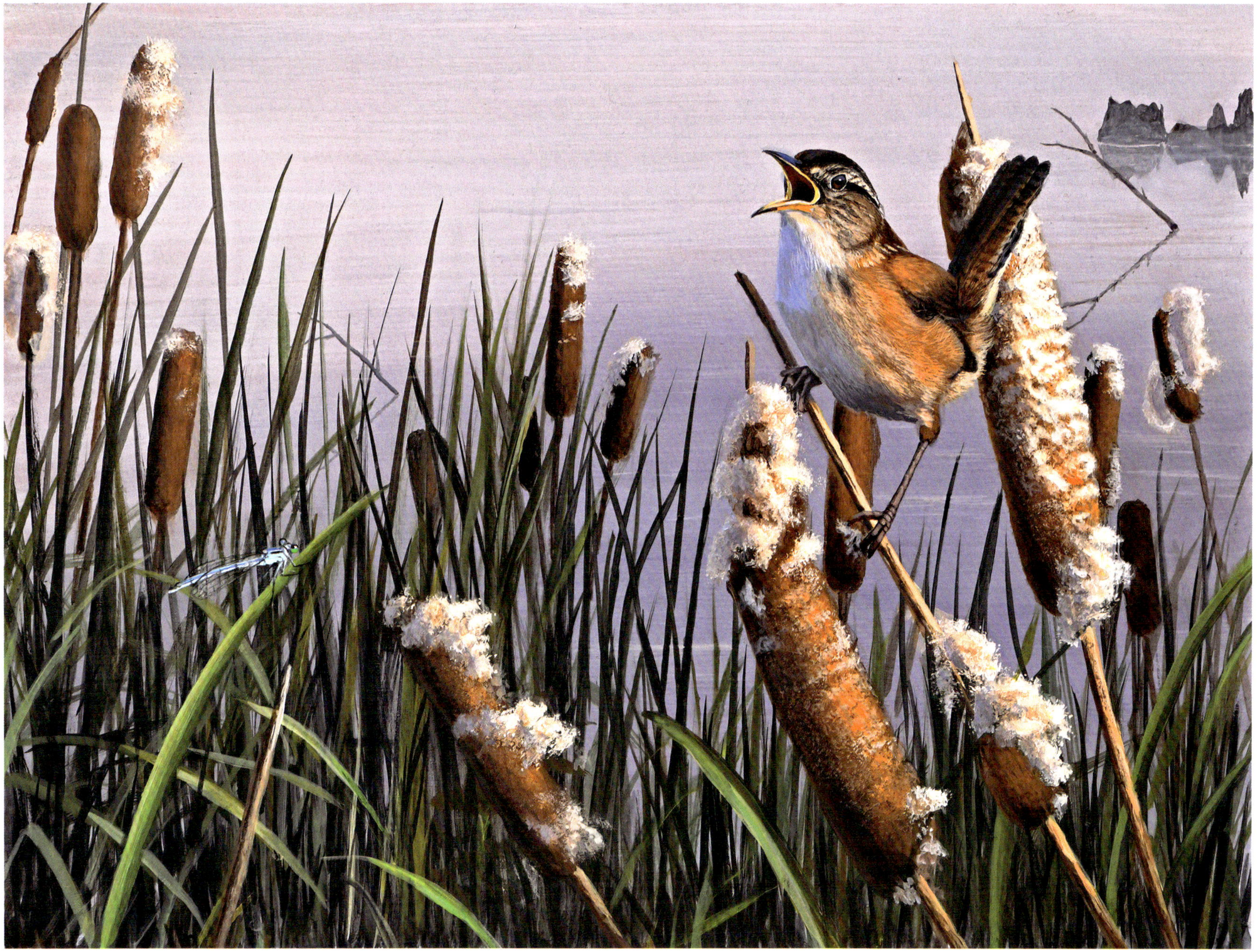

Marsh wren. *Marsh Morning*, acrylic, Rob Mullen

FACING: White-breasted nuthatch study. Acrylic, Jan Martin McGuire

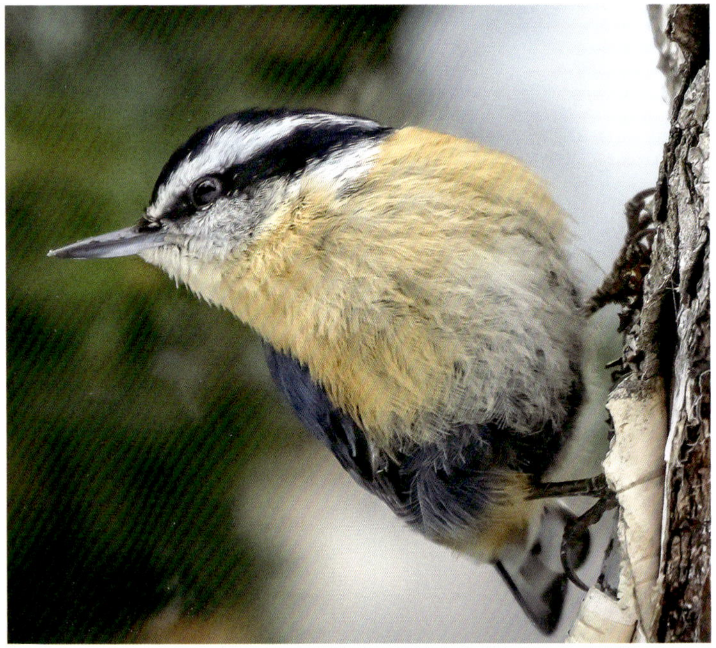

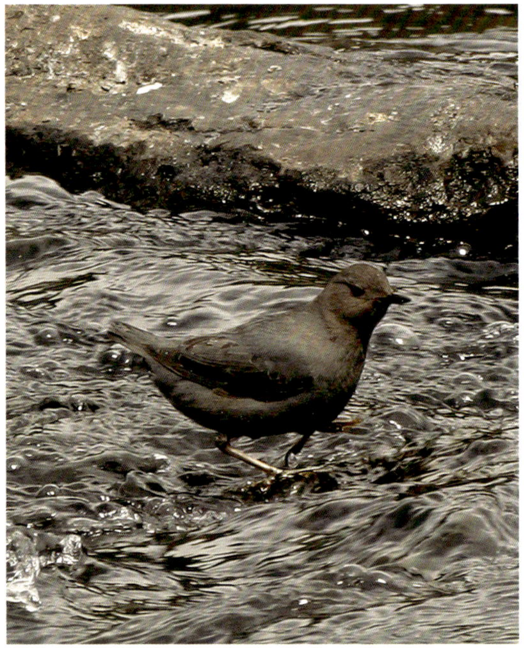

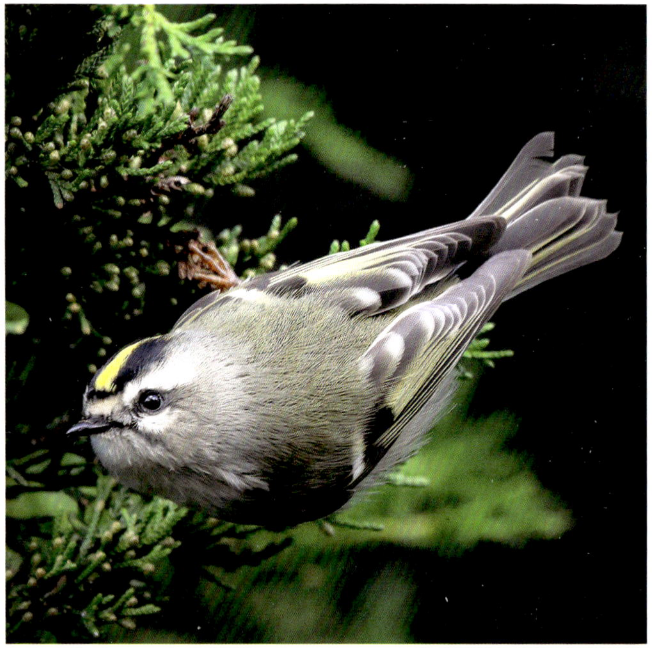

LEFT: Red-breasted nuthatch. Peter Riley

CENTER: American dipper, Yellowstone National Park, May 2010.

RIGHT: Golden-crowned kinglet. Peter Riley

Wrens (family Troglodytidae) generally skip the trees and hunt on the ground hidden in the understory. They are usually small, roundish birds that carry their tails upright. Fewer than 10 of the 126 species worldwide use the boreal forest. The northern or Eurasian wren (*Troglodytes troglodytes*) and the winter wren (*Troglodytes hiemalis*; considered *Troglodytes troglodytes* until 2010) are the main boreal forest breeders with others occupying smaller portions of the biome. All migrate, and no boreal forest species are threatened. Of the 126 species worldwide, 19 percent are threatened, including two that are extinct.

Dippers (family Cinclidae) are the only songbirds that swim and walk underwater, inhabiting cold streams and feeding on small invertebrates. The American dipper (*Cinclus mexicanus*), also known as a water ouzel, ranges into boreal forest habitat in Alberta and Alaska. In Eurasia, the white-throated dipper (*Cinclus cinclus*) is found in the boreal forest of Scandinavia and western Russia, while the brown dipper (*Cinclus pallasii*) is found in far eastern Siberia, Sakhalin, and the Kamchatka Peninsula.

Kinglets (family Regulidae) are tiny canopy foragers that are among the smallest passerines in the world. Like the dippers, there are few species in the boreal forest, but unlike dippers, they are widely distributed and migrate out of the far north for winter. Nearly 70 percent of the ruby-crowned kinglet (*Corthylio calendula*) population breeds throughout the entire North American boreal forest and migrates to southern US, Mexico, and Central America for the winter. The golden-crowned kinglet (*Regulus satrapa*) doesn't range as far north to breed but doesn't fly as far south in the winter. It is even a year-round resident in Alaska and New England.

Family Turdidae (stop snickering) (thrushes)

The most hauntingly beautiful songs in the north are sung by thrushes. I grew up listening to hermit thrushes (*Catharus guttatus*), Vermont's state bird, 75 percent of whom nest in the boreal forest; wood thrushes *(Hylocichla mustelina)*; Swainson's thrushes (*Catharus ustulatus*), also nearly 75 percent boreal forest nesters; and veeries (*Catharus fuscescens*). When I started traveling to art shows and for canoe trips, I was stopped in my tracks by the multi-resonant song of varied thrushes (*Ixoreus naevius*), a resident of the boreal forest in Alaska and far western Canada.

Naturally, the thrush North Americans see most often is the ubiquitous American robin (*Turdus migratorius*). Robins seem to have missed a few of the required singing lessons for thrushes. They don't sing badly; they just don't do it at the same level. Be all that as it may, on the north slope of the Brooks Range in Alaska, I gained a special appreciation for robins in an example of how birds, with their wide-ranging habits, can connect us to wild places.

July 25, 2013, 12:30 p.m.: Headwaters of the Noatak River

As Dirk Nickisch and his bush plane were lost to sight amidst the mountains, I was left alone, 80 miles north of the Arctic Circle. It was an awesome feeling in every sense of the word. As I got busy organizing gear and setting camp to keep my mind from contemplating my situation, I almost laughed out loud when I spotted my first animal of the trip. It was not a curious grizzly (like the year before) but a robin. The familiarity was a tonic. Suddenly, instead of being a lonely speck in a vast wilderness, wondering what I had done, I felt like I was sorting gear in my backyard.

Interestingly, of the 192 species of thrushes on the IUCN Red List, 24 percent are assessed at some level of risk or extinct (four species). Of the 27 species with a significant presence in the boreal forest worldwide, one species is Near Threatened and another is Vulnerable (together, 7 percent of the boreal species).

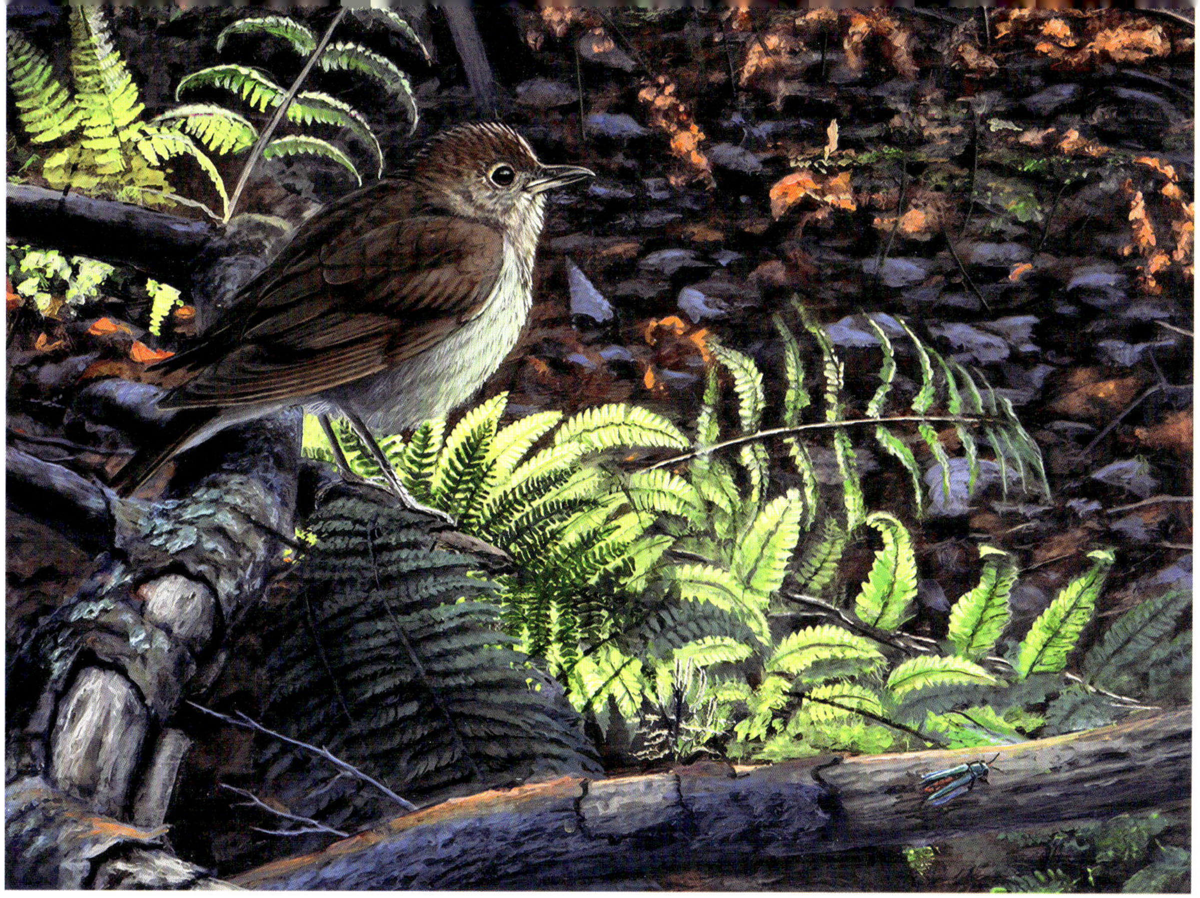

Veery, Île Bizard, Quebec.
Veery Light, acrylic, Rob Mullen

Family Motacillidae (wagtails and pipits)

Slender ground feeders, about 20 species breed in the boreal forest. American pipits (*Anthus rubescens*) are widespread and often the only companions for lonely canoeists hiking over remote mountain passes.

Of the 67 species IUCN lists worldwide, four are Near Threatened, five are Vulnerable, and three are Endangered (together, 18 percent). Of the nearly 20 species that breed in the boreal forest, one (less than 5 percent) is Vulnerable: Sprague's pipit (*Anthus spragueii*). It is mostly a Great Plains resident.

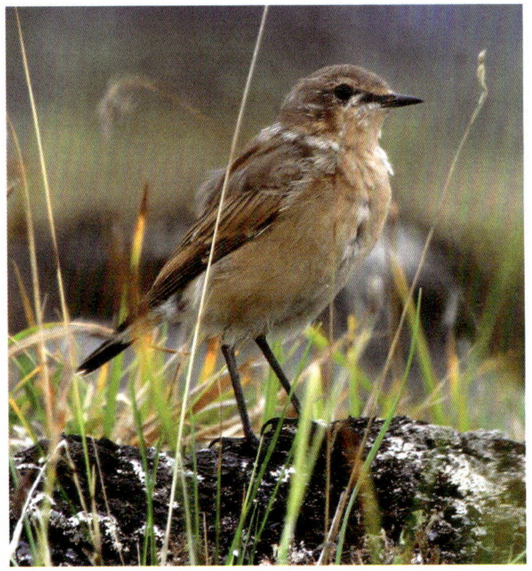

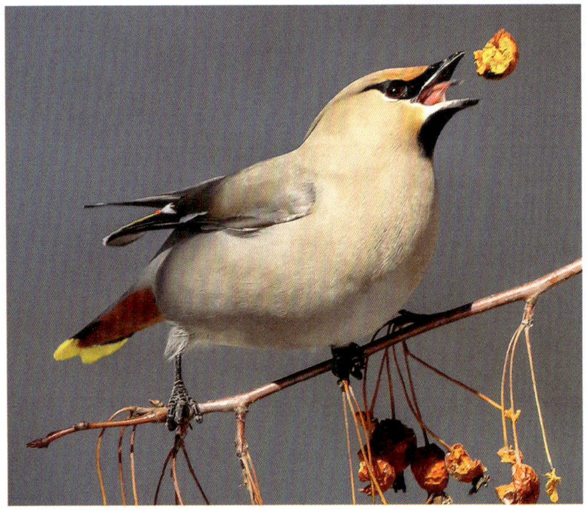

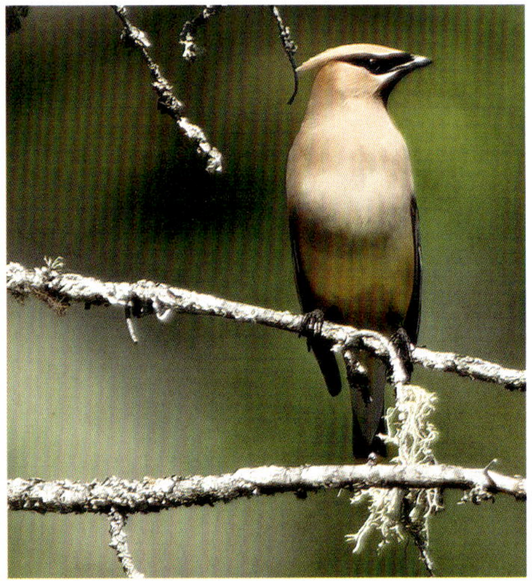

LEFT: American pipit in Angiaak Pass, Alaska.

CENTER: Bohemian waxwing eating berries. Peter Riley

RIGHT: Cedar waxwing. Bonnie Rowell

FACING, CLOCKWISE FROM TOP LEFT:
Magnolia warbler. Peter Riley

Cape May warbler. Peter Riley

Palm warbler. Peter Riley

Bay-breasted warbler. Peter Riley

Yellow-rumped warbler ("butter-butt") male on cattails. An example of how, unlike most warblers who "leaf glean," butter-butts often feed like flycatchers, sitting on a perch and waiting for a passing insect to buzz by too closely.

Blackpoll warbler male singing. USFWS

Blackburnian warbler, Magee Marsh, Ohio. Bonnie Rowell

Yellow warbler in the alders with its lunch, Alaska.

Family Bombycillidae (waxwings)

These three sleek, berry-loving species are all boreal breeders: not as flashy as some, but neat and well put together. Bohemian waxwings (*Bombycilla garrulus*) are Holarctic in distribution and often mix with cedar waxwings (*Bombycilla cedrorum*, Nearctic) and Japanese waxwings (*Bombycilla japonica*, Palearctic; IUCN Red List: Near Threatened).

Family Parulidae (wood warblers)

Among North American birders, wood warblers—which exist only in North America (Nearctic)—are the celebrities of the songbird world: small, energetic, dazzlingly patterned, and often challenging to identify. Spotting them can be addictive; I've seen seven species in a single tree. Many of their admirers suffer for them: "warbler neck" is real. Among others, I'm not that manic—unless I happen to be driving by Magee Marsh Wildlife Area on Lake Erie during fall migration when it can be filled with exhausted warblers "falling out" after crossing the lake for a refueling stopover en route to the tropics. I've seen many warblers among over 80 species identified there in a day.

A yellow warbler came to my rescue while stranded in the tangle of Alder Hell (Mirkwood for J.R.R. Tolkien fans) in Angiaak Pass in Alaska's Brooks Range. Not a physical rescue—that was Dirk Nickisch and Danielle Tirrell airdropping two weeks of dinners from their plane, but the little bird did a good job of lifting my spirits with his friendly tolerance.

While a comprehensive list would be cumbersome to include, here is a probably non-exhaustive list of wood warbler species relying on the boreal forest for at least 70 percent of their nesting habitat:

- Palm warbler (*Dendroica palmarum*): 98 percent of the population nesting in the boreal
- Tennessee warbler (*Leiothlypis peregrina*): 97 percent
- Connecticut warbler (*Oporornis agilis*): 93 percent
- Cape May warbler (*Setophaga tigrina*): 83 percent
- Blackpoll warbler (*Dendroica striata*): 82 percent
- Bay-breasted warbler (*Dendroica castanea*): 79 percent
- Mourning warbler (*Geothlypis philadelphia*): 75 percent
- Magnolia warbler (*Setophaga magnolia*): 74 percent

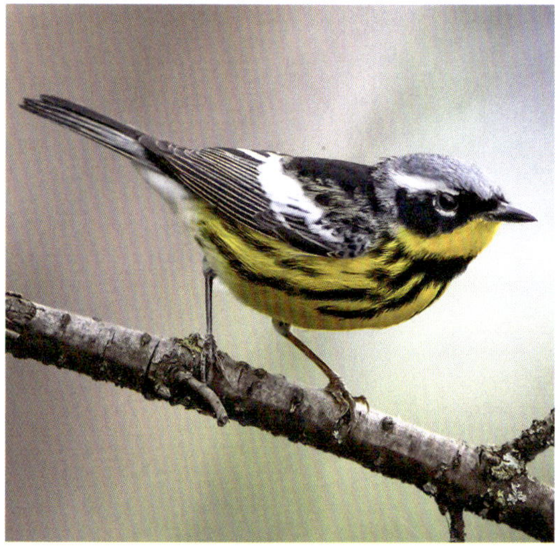
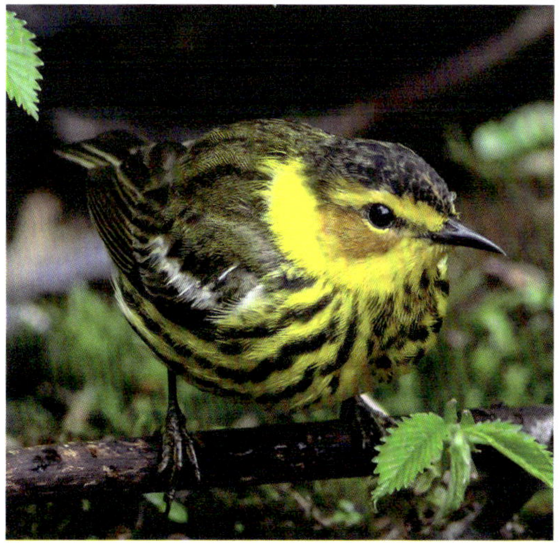
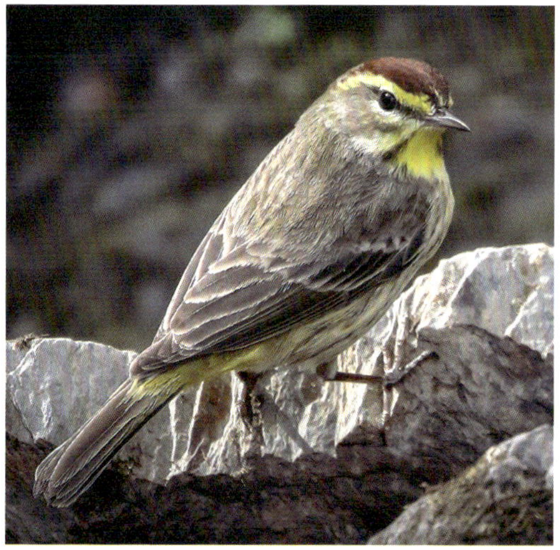
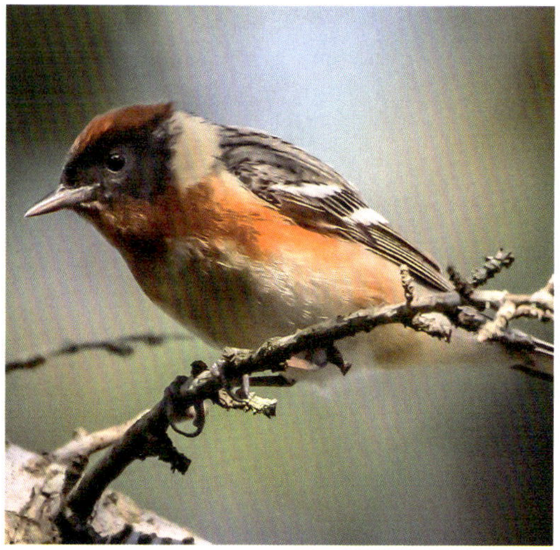
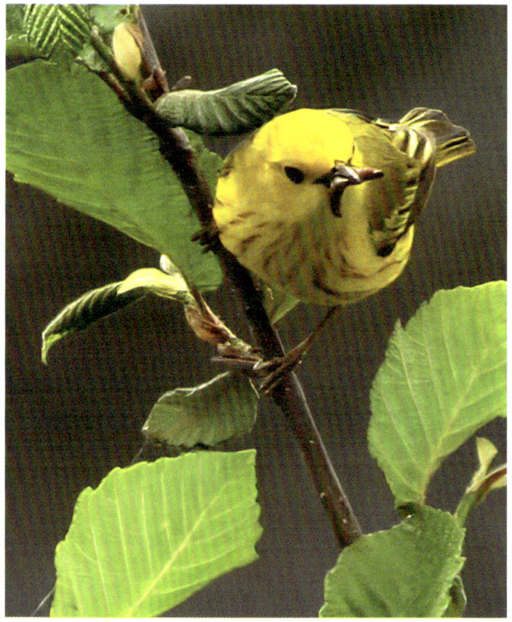
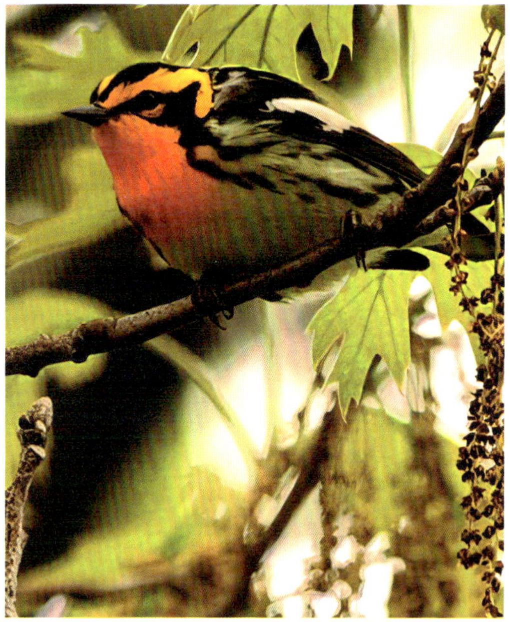
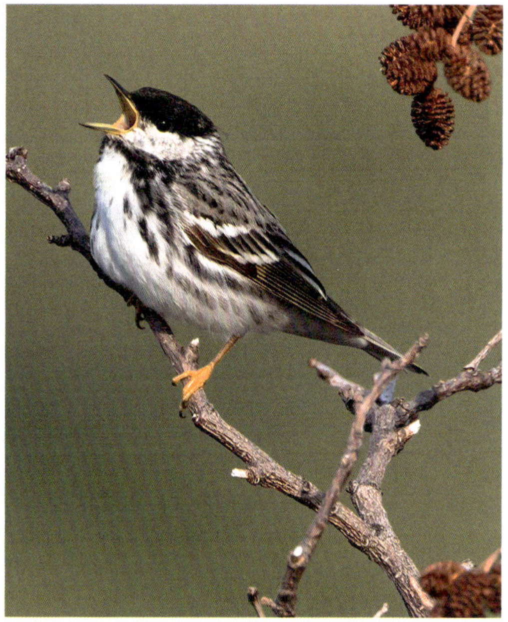
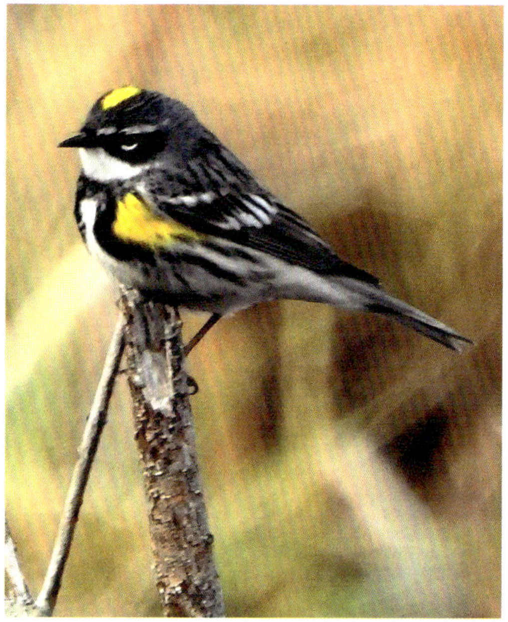

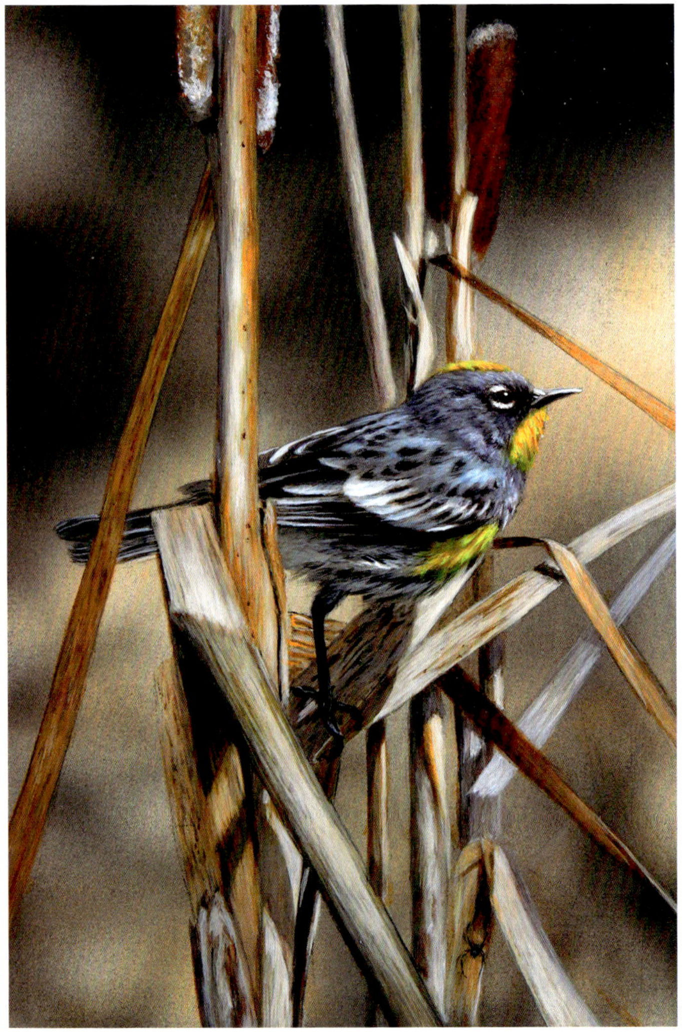

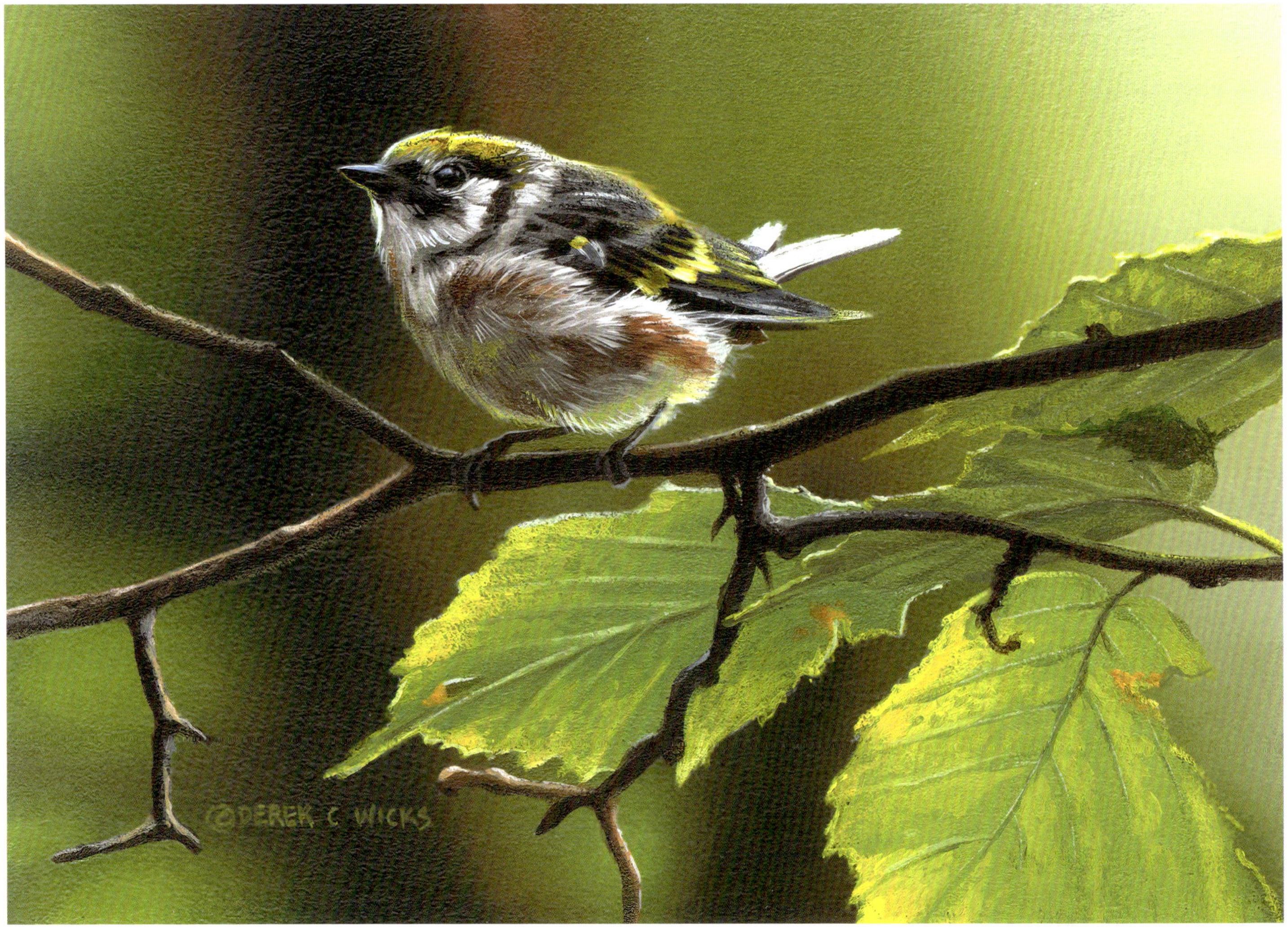

LEFT: Male yellow-rumped warbler.
Acrylic, Julia Hargreaves

RIGHT: Chestnut-sided warbler.
Acrylic, Derek Wicks

Most warblers are canopy foragers, deftly gleaning insects and other small arthropods from leaves. Various species have favored foraging zones, such as the brilliantly colored blackburnian warbler (*Setophaga fusca*), who prefers insects high in the forest canopy; the common yellowthroat (*Geothlypis trichas*), who prefers wetland margins; the yellow-rumped warbler (*Setophaga coronata*), who prefers open areas; and even a few like ovenbirds (*Seiurus aurocapilla*) and the northern waterthrush (*Parkesia noveboracensis*), who forage in the understory (despite their names, like the common yellowthroat, they are all wood warblers). Working together, the overlapping zones ensure that the entire forest canopy and ground cover are patrolled for harmful insects, promoting forest health and productivity. This is an example of an ecological service, provided for free by birds simply eating. More on ecological services in chapter 8.

Warblers are classic neotropical migrants. Of the roughly 125 species of New World warblers, most are resident in the tropics, and virtually all the others spend winter there. Confirming the pattern we've noted in other families of birds and even some mammals, their conservation status varies depending on their use of the intact ecosystem of the boreal forest versus ecosystems that have been degraded by human development.

- Boreal forest nesting neotropical migrants: 1 of 26 species (4 percent) at risk.
- Non-boreal forest nesting neotropical migrants: 5 of 22 species (23 percent) at risk.
- Neotropical residents (non-migratory): 15 of 71 species (21 percent) at risk.

Species relying on the boreal forest face lower risks. Habitat loss in temperate and tropical zones pressures both resident species and shorter-distance migrants year-round. Long-haul migrants who forge north to the boreal forest take on the hazards of a longer journey but get a break in the summer. For millennia, the migration trade-off worked for all migrants. Today, only the long-distance migrants benefit from a safe, productive breeding ground. The boreal forest, as an enormous, intact ecosystem, is increasingly vital for sustaining these A-list songbirds.

The only boreal forest nesting wood warbler that is in trouble is the blackpoll warbler (*Setophaga striata*; IUCN Red List status is Near Threatened). The Alaska populations fly over 6,000 miles each way to wintering grounds in South

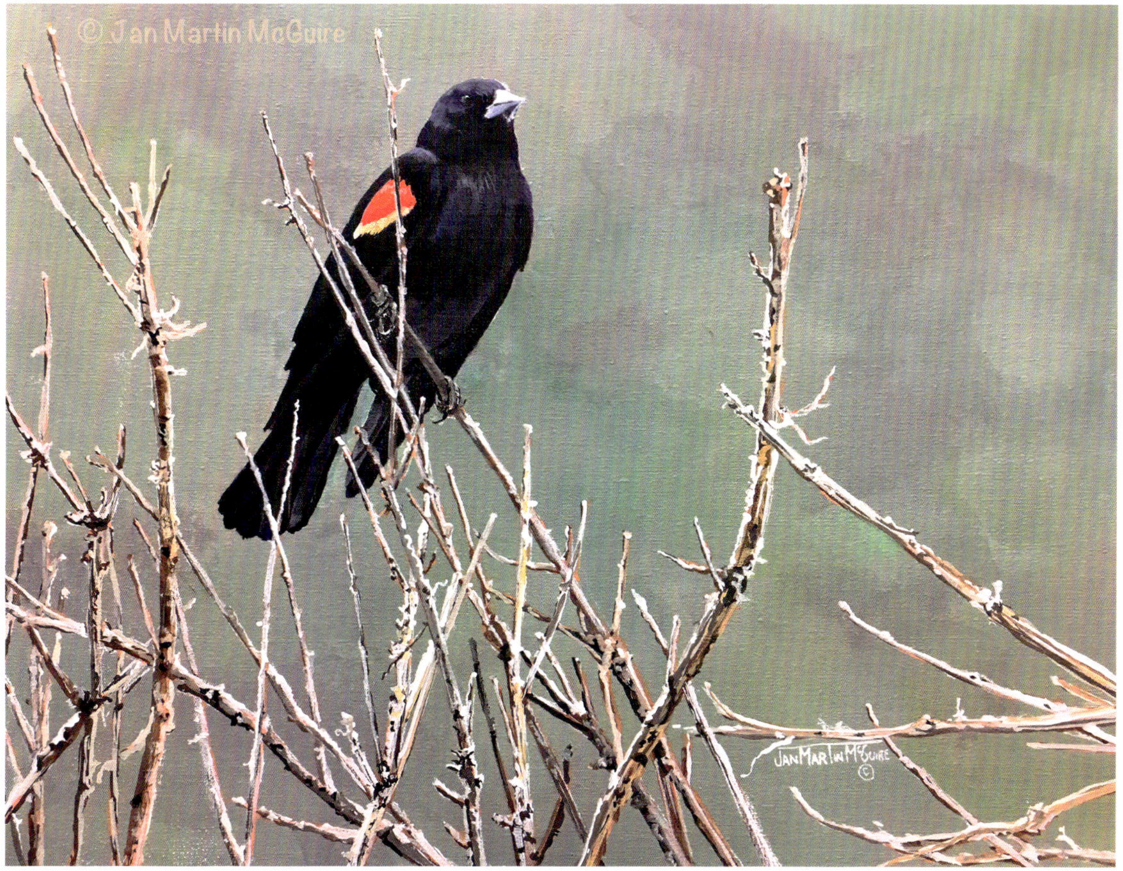

Red-winged blackbird, Missouri River Breaks. Acrylic, Jan Martin McGuire

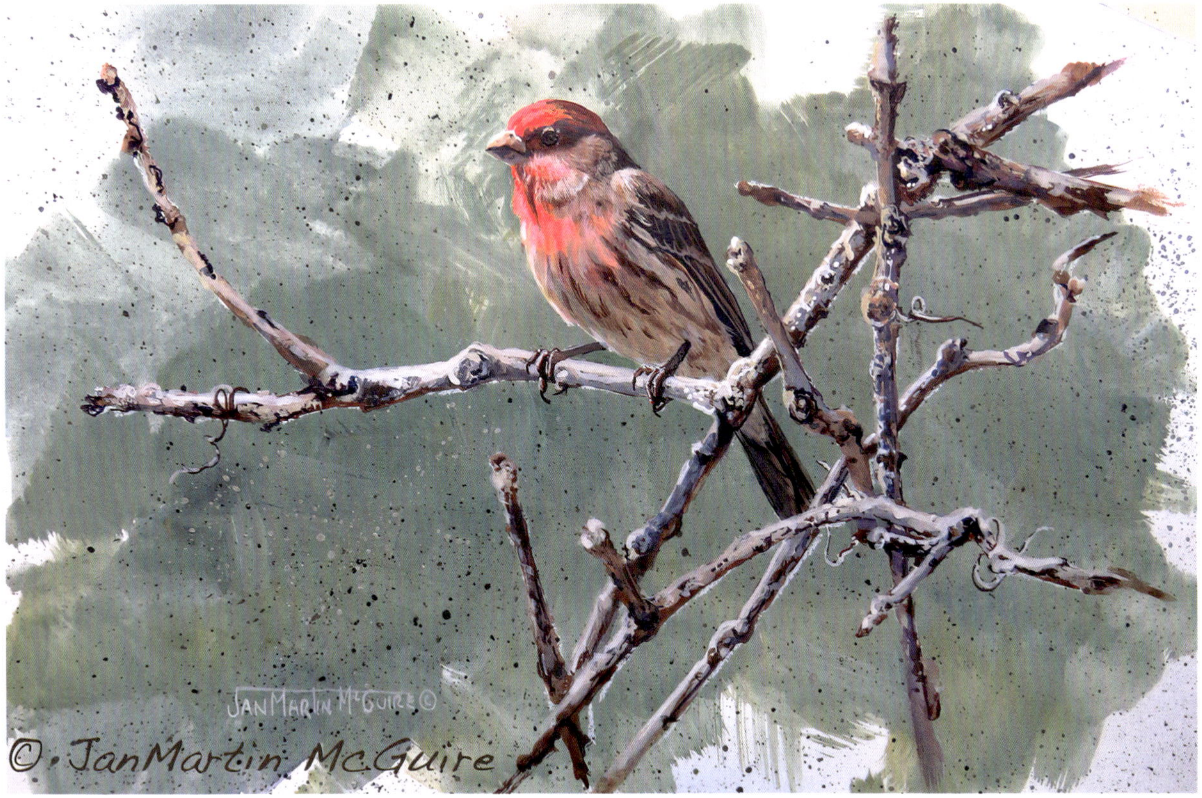

Purple finch. Acrylic, Jan Martin McGuire

America, while eastern populations, not to be outdone, during the fall migration, undertake the longest non-stop, over-water migration of any songbird in the world, flying up to 1,700 miles over the Atlantic Ocean before making landfall on their way also to South America. (They take a more leisurely, overland route back north in the spring.)

Family Passerellidae (New World sparrows, juncos, towhees)

Often dismissed—even by birders—as "little brown jobs," sparrows are abundant ground foragers and key seed dispersers. White-throated and white-crowned sparrows (*Zonotrichia albicollis* and *Zonotrichia leucophrys*, respectively), swamp sparrows (*Melospiza georgiana*), song sparrows (*Melospiza melodia*), and dark-eyed juncos (*Junco hyemalis*) are all major boreal breeders. Their populations, numbering into the millions (and in some cases, tens and hundreds of millions), make these small birds major ecological service providers.

There are 148 species, about 18 of which range into the boreal forest. Doing the same extinction risk calculation as for other families yields a similar percentage overall. Twenty-five species, or 16 percent of non-boreal forest species, are at risk (one extinction: the Bermuda towhee). However, only one of the 18 boreal forest species is at risk, rated by the IUCN as Near Threatened (about 6 percent).

The white-throated sparrow can seem misnamed at first, given the striking white crown on its head. However, after seeing your first white-crowned sparrow, you'll understand.

Family Icteridae (blackbirds)

Few range far north. Two boreal species are at risk: rusty blackbird (*Euphagus carolinus*), rated as Vulnerable, and common grackle (*Quiscalus quiscula*), rated as Near Threatened. The red-winged blackbird (*Agelaius phoeniceus*) is a familiar spring arrival.

Family Fringillidae (finches)

I've saved the biggest for last: The IUCN Red List includes 229 species worldwide for this cosmopolitan family. Boreal forest finches are tough birds; none are neotropical migrants. Of the 30 species that inhabit the boreal forest, most are year-round residents or migrate only short distances: 4 are Holarctic in distribution, 6 are endemic to North America, and 20 are endemic to Eurasia.

The American goldfinch (*Spinus tristis*) and the purple finch (*Carpodacus purpureus*) are possibly the most familiar at bird feeders across North America, and both nest in the boreal forest.

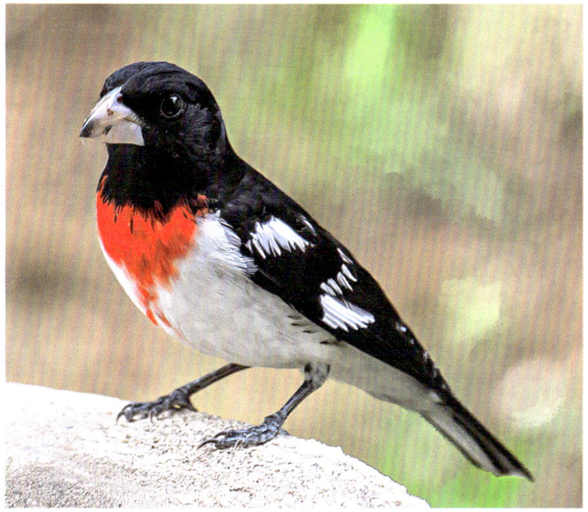
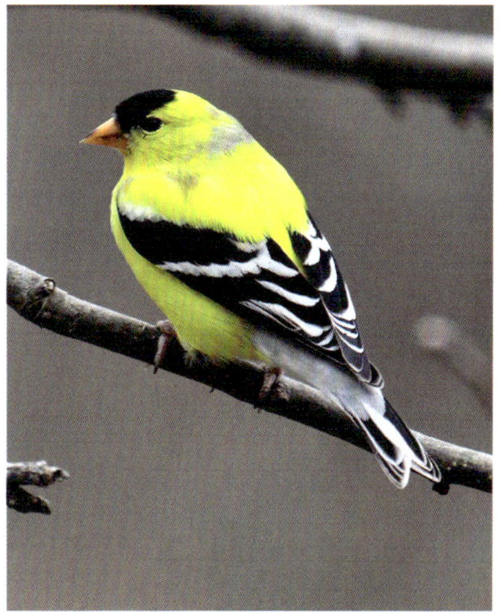
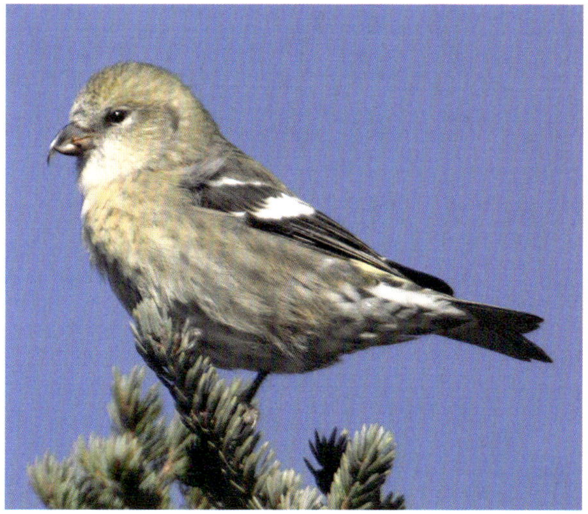
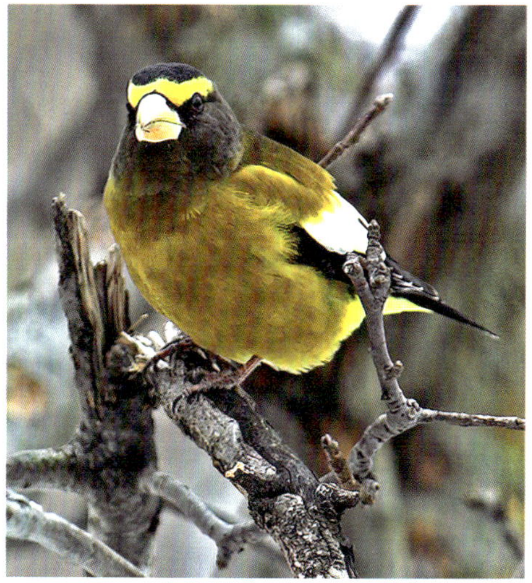
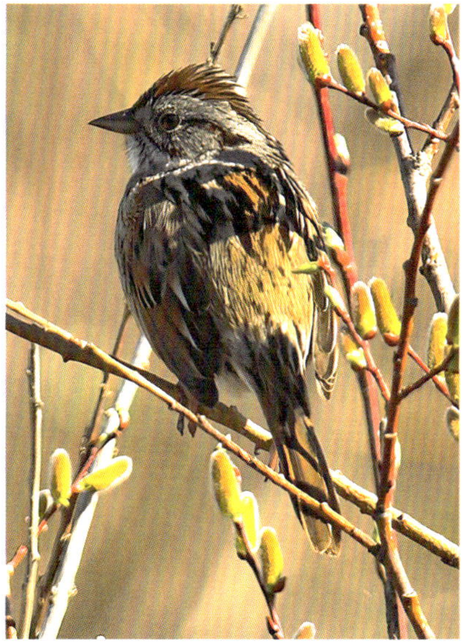
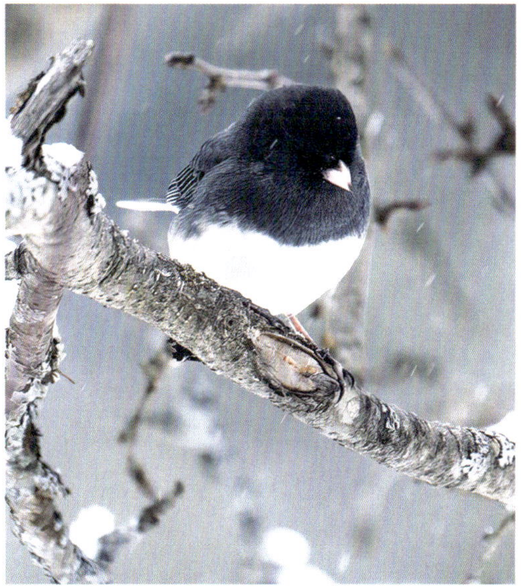
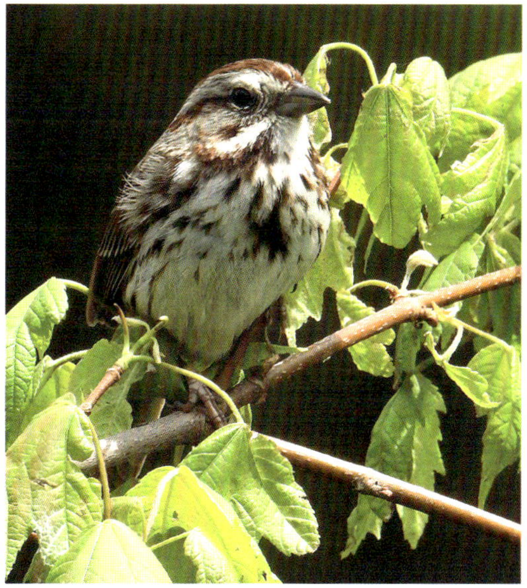
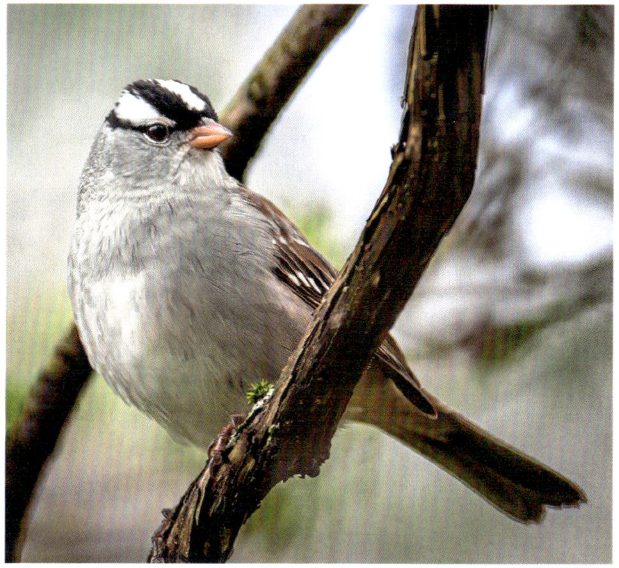

CLOCKWISE FROM TOP LEFT:
Rose-breasted grosbeak. Steve Bruno

Goldfinch. Bonnie Rowell

Crossbill in Kamestastin, Labrador. Jay J. Johnson

Evening grosbeak, West Bolton, Vermont. Bonnie Rowell

White-crowned sparrow, Vermont. Peter Riley

Song sparrow, Preston Pond, Vermont.

Male junco. Steve Bruno

Swamp sparrow. Bonnie Rowell

LEFT: Pine grosbeak, Ontario. Steve Bruno

RIGHT: Redpoll. Bonnie Rowell

FACING: White-throated sparrow. *Brook Bath*, acrylic, Rob Mullen

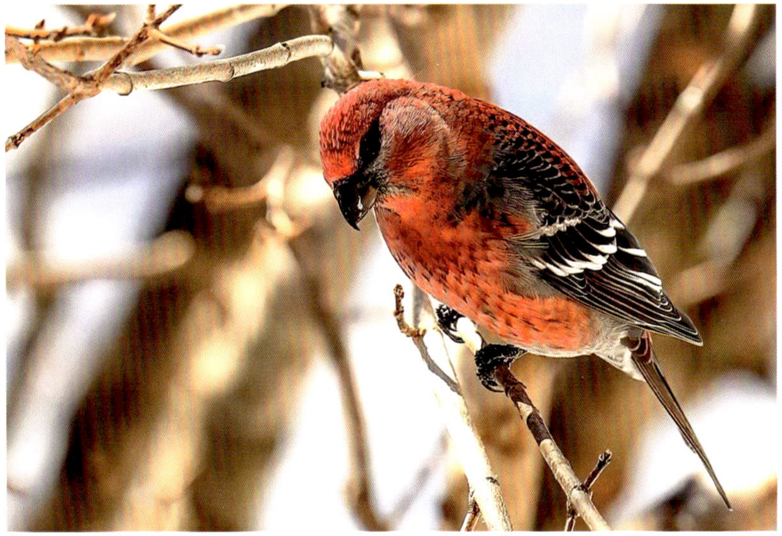

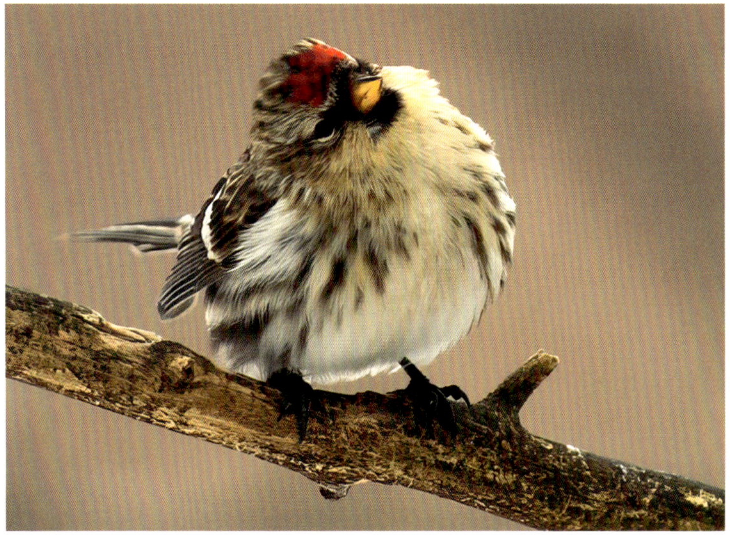

Pine grosbeak (*Pinicola enucleator*) is a boreal forest bird if ever there was one—resident across the biome worldwide. Some wander south in winter, including into Vermont (not exactly a tropical escape).

A note on common and Latin names: The rose-breasted grosbeak (*Pheucticus ludovicianus*), a melodious singer, is not a finch like other grosbeaks, but is actually in the cardinal family—the only member of that family in the boreal forest.

Redpolls (*Acanthis flammea*) may be even more cold-tolerant than pine grosbeaks, occupying the northern swath of the boreal forest worldwide and much of the Arctic.

White-winged or two-barred crossbills (*Loxia leucoptera*) range across the entire North American boreal forest, northern New England (year-round in some areas), and much of the Eurasian boreal forest except Norway. Their namesake crossed bill tips are a specialized adaptation for prying open conifer cones to reach the seeds.

One species of special interest to me is the evening grosbeak (*Hesperiphona vespertina*). The IUCN Red List reassessed this formerly abundant species as Vulnerable. Once a gluttonous, daily winter fixture at our bird feeders, they vanished for years before returning in January 2025—about 50 birds at once—and have stayed (except for one mysterious day's absence) into late April. They are the only boreal forest finch species at risk.

Of the 199 non-boreal forest species in the family, 36 are at risk and 22 are extinct (together, 29 percent of the total; most of the extinctions were on the Hawaiian Islands). In the New World boreal forest, only the evening grosbeak is at risk (10 percent of boreal forest finch species), while none of the 20 Eurasian boreal species are threatened (0 percent).

Taking all these facts into consideration, we see a pattern across bird families (and some mammals) that underscores the boreal forest's role as a vast refuge. This "biodiversity coldspot" deserves the attention of not only conservationists and governments, but all of us.

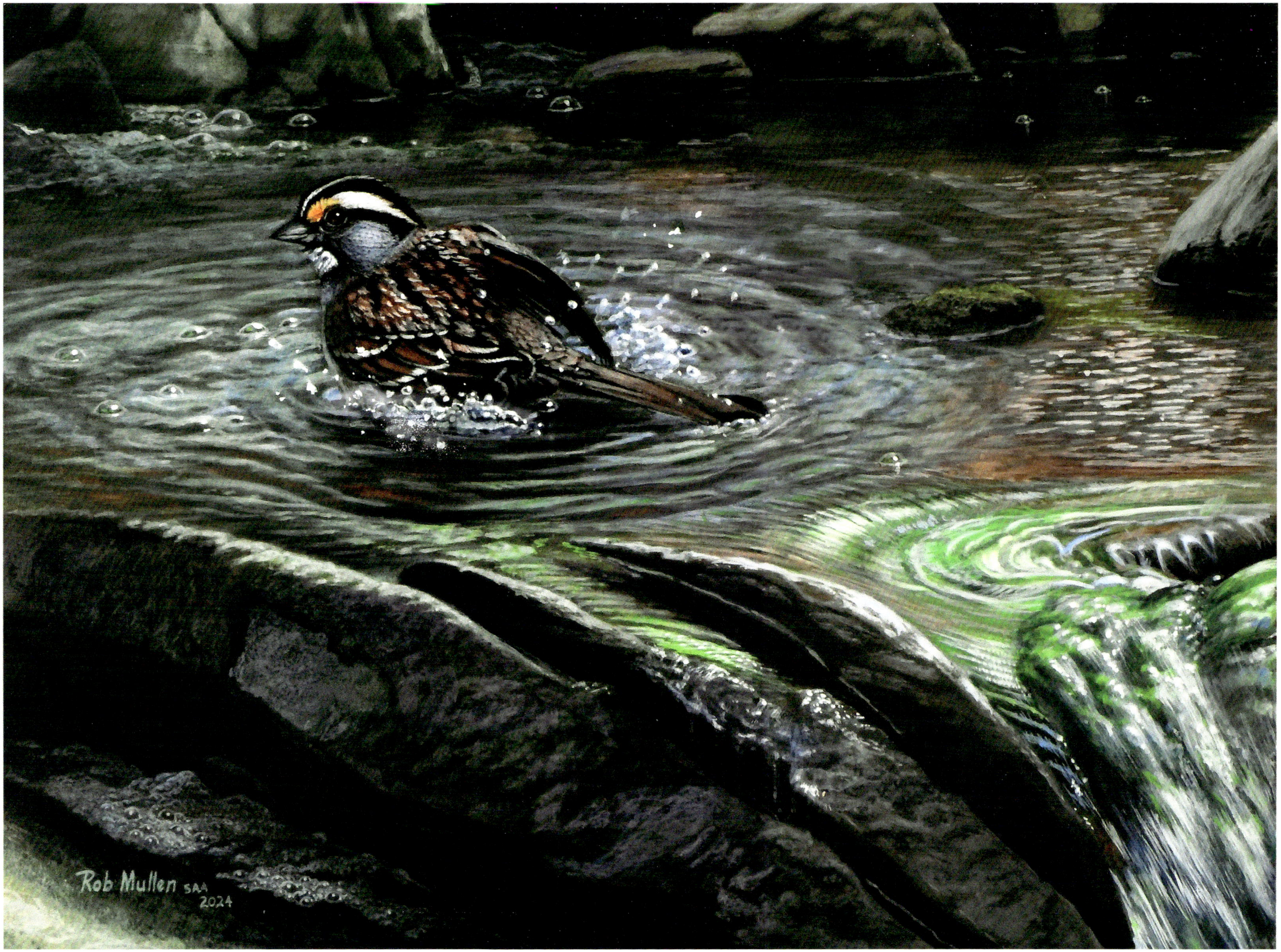
Rob Mullen SAA
2024

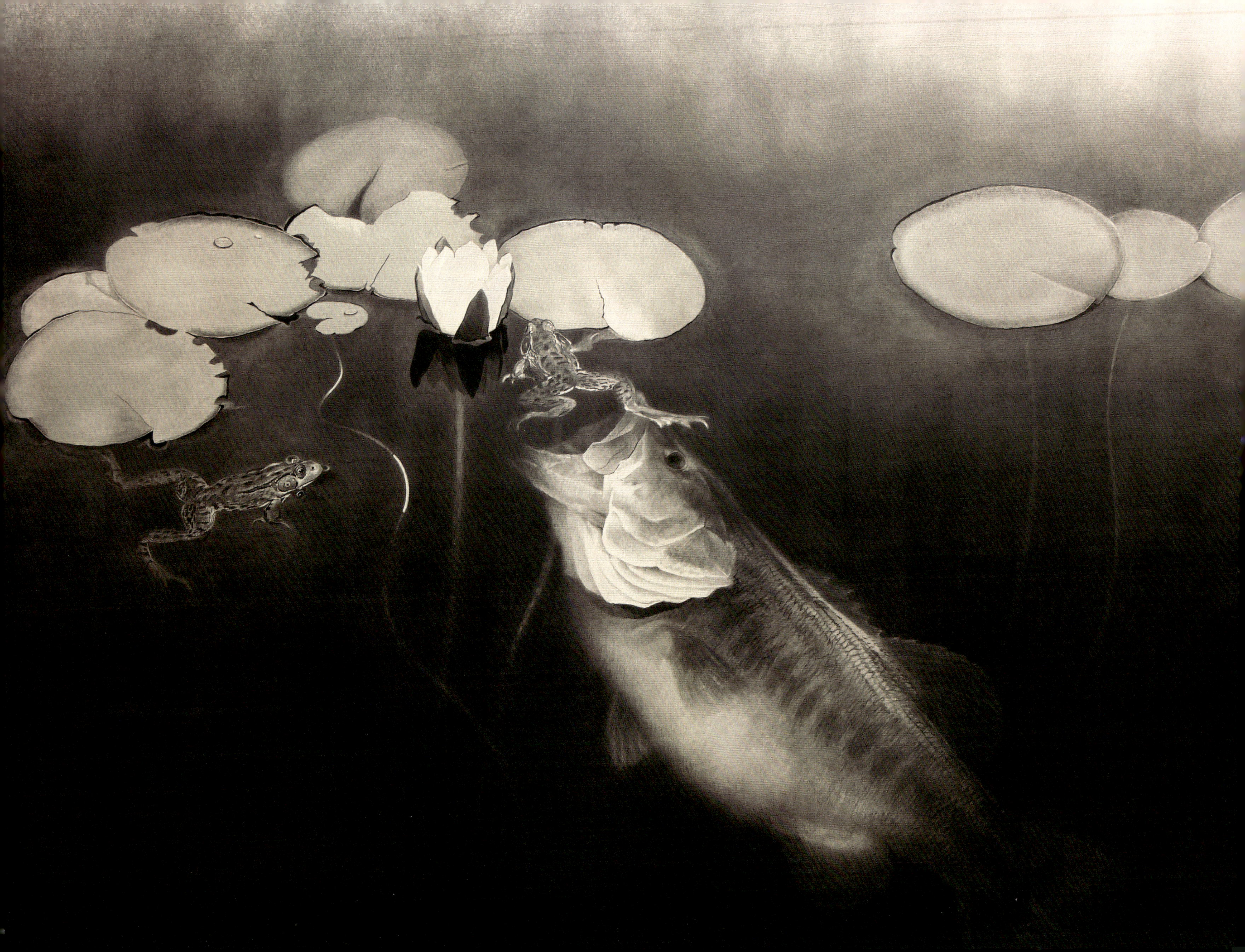

FACING: Largemouth bass and frogs.
And Then There Was One, charcoal,
Cole Johnson

6

Lower Vertebrates: Fish, Reptiles, and Amphibians

WINTER IN THE NORTH is not kind to cold-blooded animals. Consequently, there are relatively few reptiles and amphibians in the boreal forest, let alone the tundra farther north. However, there are fish. Teeming multitudes of them. They are critical components of the food chain, driving nutrient cycling and transport on a massive scale. The reason is water, which, if deep enough, stays liquid under a protective layer of ice.

Fish

Probably the only reason fish aren't better represented among the iconic species of the biome such as caribou, wolves, and bears is that they are underwater (though salmon leap that barrier).

As noted in chapter 5, fish are the most diverse vertebrate group worldwide but only the second-most diverse vertebrates in the boreal forest.

Fewer than two hundred species of fish inhabit the boreal forest. However, some inhabit it in great numbers, contributing critically to the ecosystem's functionality and productivity, human sustenance, cultures, and economies.

Loon Rise, shown on page 146, is emblematic of the publicity problem that fish face: some showboat bird hogging the sunlight and making them easy to miss.

Winter 1986: 2 West 46th Street, Manhattan (my work studio)
As I planned my first big Canadian canoe trip, my NYC-raised friends and colleagues were concerned about the danger of wolves and bears. With a sage surety born of little knowledge and less experience, I assured them that blackflies and mosquitoes were the most dangerous animals in the Canadian wilderness.

None of us figured on fish.

September 7, 2001: Missinaibi River nearing Bell's (or Bull Moose) Bay, Northern Ontario
With the river's dramatic plunge off the Canadian Shield through the maelstrom of Thunderhouse Gorge, Stone Rapids, what some call Hell's Gate, and Long Rapids behind me

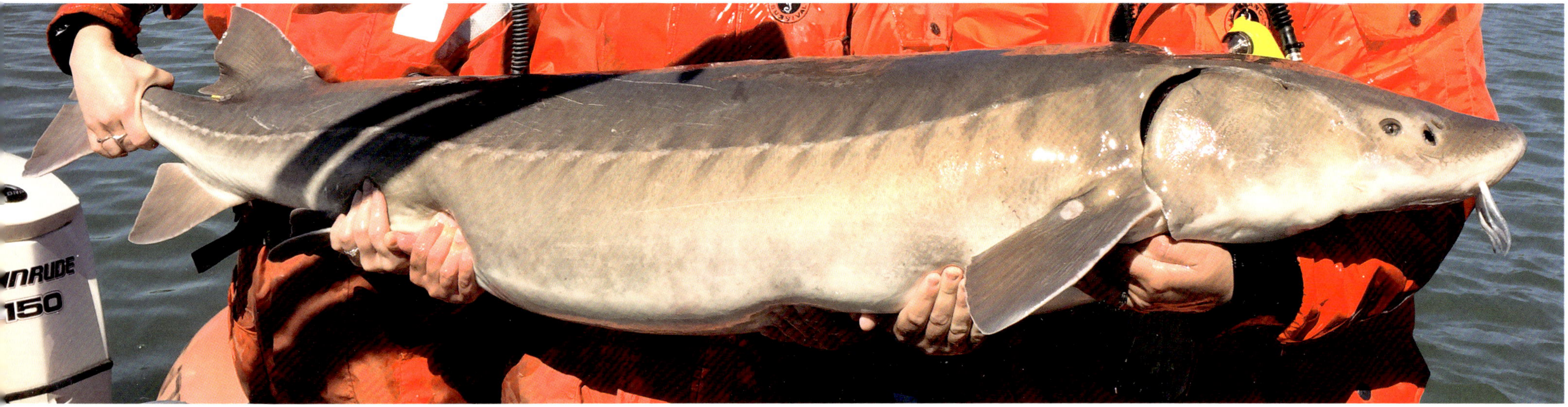

Lake sturgeon and hard-working USFWS fisheries biologists. Justin Chiotti/USFWS

and safely onto the quieter Hudson Bay Lowlands, I felt on top of things. There was a light, warm rain falling, but I was comfortable, the river was beautiful, and there were probably no more portages. I was not thinking about fish. I certainly wasn't thinking about what the violent rapids and waterfalls just behind me kept out of the upper river, where I no longer was. So . . .

As I serenely coasted past an eddy pool to my right, the water under my bilge suddenly heaved and boiled. "What the . . . ?!" With that half thought, the rapidly bulging mass erupted into a geyser as a huge dark shape rocketed upward like a Poseidon missile and crashed back into the water with a thunderous crack and exploding spray. Concurrent with my heart-stopping shock, my poor brain was frantically trying to figure out what was happening. As the water settled down, it dawned on me. A sturgeon!

Others surfaced a few more times, sometimes near the boat and sometimes leaping downstream. It was awesome! I'd never seen a sturgeon in the wild. It did call fish to mind though, and I put a troll line out and had walleye for dinner. Four months later, the danger I had been unthinkingly oblivious to came crashing home on the airwaves.

Late January 2002: Interstate 4 between Orlando and Lakeland, Florida

I was driving down for the Florida Wildlife and Western Art Exposition and scanning for a decent radio station when I heard a news teaser. I stabbed the scan button. Three local fishermen on the Suwannee River had been seriously injured and hospitalized by . . . leaping sturgeons landing in their boats! The damage that two or three hundred pounds of flying fish crashing into you could do had never hit me. Good thing my sturgeons hadn't either.

Those Florida fish were gulf sturgeons (*Acipenser oxyrhynchus desotoi*), a subspecies of the Atlantic sturgeon, and close relatives of my buddies on the Missinaibi, the lake sturgeon. Both species are famous leapers that grow to dangerous sizes for anyone unfortunate enough to be in their flight path.

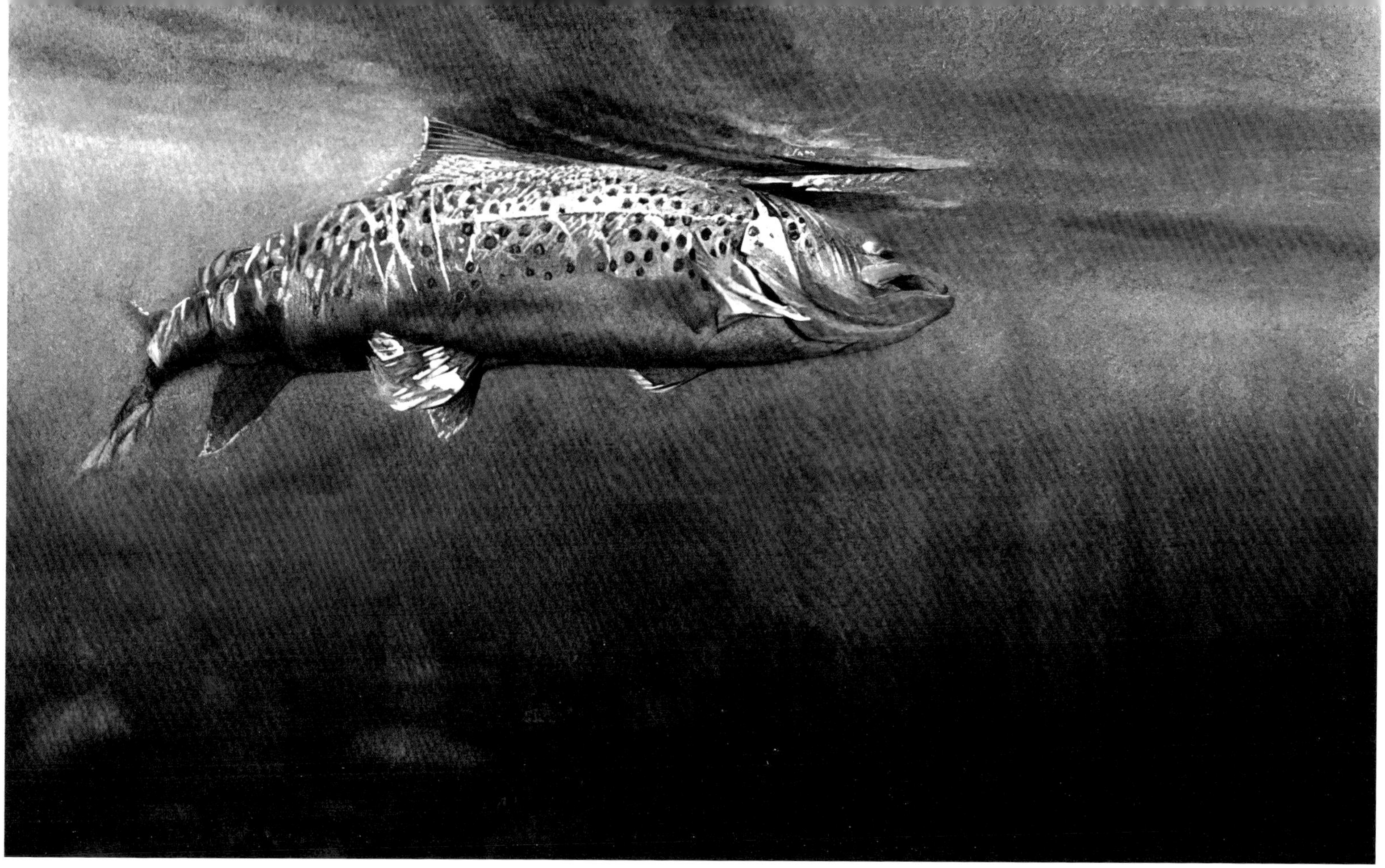

Brown trout. Graphite, Cole Johnson

In 2015, jumping sturgeons seriously injured eight people and killed one in Florida.

Sturgeons are "living fossils." Specimens, virtually unchanged from those of today, were nosing about for tens of millions of years before T. rex first stomped onto the stage. Big, armored with bony scutes, tolerant of a wide range of temperatures and salinities, and with indiscriminate tastes in cuisine, they prospered through unimaginable eons. They were still incredibly abundant just a couple hundred years ago.

Then us. In less than 0.001 percent of their time on Earth, we have slaughtered and dammed them into the world's most endangered family of fish. Not just an endangered species. An entire taxonomic family is on the brink.

In the Eurasian boreal forest, the Siberian sturgeon (*Acipenser baerii*), Sakhalin sturgeon (*Acipenser mikadoi*), and Amur sturgeon (*Acipenser schrenckii*) are all critically endangered. Others, including the stellate or starry sturgeon (*Acipenser stellatus*), are critically endangered and extinct in the boreal forest. Amazingly, the stellate sturgeon and the critically endangered beluga are still two of the main sources of caviar, an essential dietary staple of not billions but billionaires.

The situation in North America is moderately better.

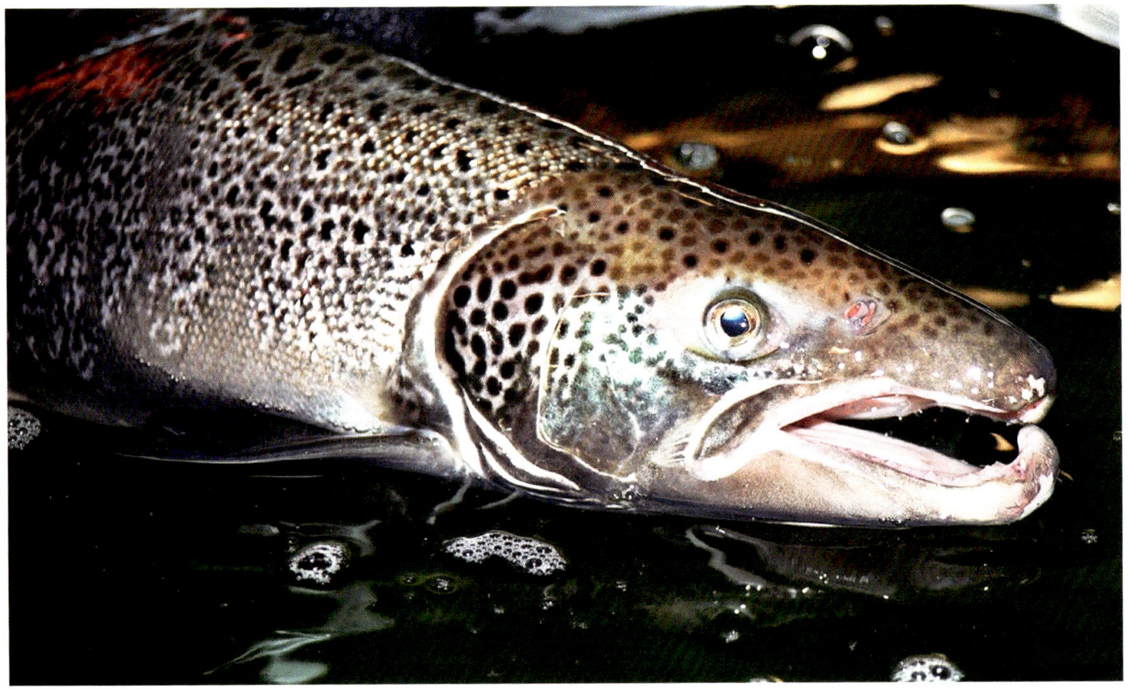

Male adult Atlantic salmon.
Peter Steenstra/USFWS

Atlantic sturgeon (*Acipenser oxyrinchus*): IUCN Red List status: Vulnerable. An anadromous species that ranges from the Gulf of Mexico to the St. Lawrence River and the Labrador coast. Probable spawning rivers in Labrador and along the St. Lawrence would qualify it as the largest fish found in freshwater in the North American boreal forest. They are typically six to eight feet long and over 200 pounds; the largest individuals can be 15 feet and over 800 pounds.

Lake sturgeon (*Huso fulvescens*): IUCN Red List status: Endangered. The second-largest fish in the North American boreal forest. A mostly freshwater species, breeding adults still mimic the anadromous migration pattern of their relatives, living in large rivers and lakes and spawning in small tributaries. The Missinaibi population may venture into the brackish waters of the Moose River estuary as it joins with James Bay.

Despite the endangered status of lake sturgeon, Ontario allows a no-limit fishing season for them on the Missinaibi that excludes only their spawning season. Some Cree I met in Moose River Crossing warned that hooking a sturgeon from a canoe would mean being towed around the river for the rest of the day.

Sturgeons are the largest fish inhabiting the boreal forest. But arguably the most important—for humans anyway, economically and culturally—is a group that rises to the level of icons of the North: salmonids.

Family Salmonidae (salmon, char, grayling, taimen, and whitefish)

Salmon

Pacific salmon, such as Chinook and chum or dog salmon, are famous for their selfless devotion to their spawn, expending all in their heroic ascents to the spawning beds where they were born, only to die after mating. At the other end of the spectrum of reproductive strategies are most other vertebrates, including humans, and the Atlantic salmon, which can reproduce multiple times in a lifetime.

Salmonids are modern, fully bony fishes, but retain the sturgeon's primitive separation of the pectoral (shoulder) and pelvic (hip) fins, like the limbs on an amphibian. They also sport a fleshy adipose fin instead of a second, rayed dorsal fin. Depending on size and circumstances (such as not having to contend with orcas), they are typically mesopredators or apex predators, consuming a wide variety of prey. (As evidenced by the fervid fly-fishing community, many eat insects.)

Found in northern waters around the globe, again, like sturgeons, many salmonids tolerate a wide range of salinities and migrate between large bodies of water (major rivers and lakes or the ocean), where they spend much of their lives, and smaller, gravel-bottomed streams or rivers, where they spawn.

Migratory fish are part of the exchange of nutrients and resources between ecosystems. Ecosystem services such as these will be covered in chapter 8.

Atlantic salmon (*Salmo salar*): IUCN Red List status: Near Threatened. Atlantic salmon range from New Jersey to Baffin Island and Greenland, Iceland, the UK, Scandinavia, and northwestern Russia. They are the third largest salmonid behind the Siberian taimen (*Hucho taimen*) and Chinook salmon (*Oncorhynchus tshawytscha*). Half of North America's remaining Atlantic salmon spawn in the boreal rivers of Quebec, Newfoundland, and Labrador. They carry ocean nutrients inland and return forest nutrients to the sea.

There are landlocked populations in New England (including Lake Champlain) and eastern Canada, commonly called *ouananiche* in Quebec.

Our Atlantic salmon photo (facing page) is credited on the USFWS site as being in Vermont. The only Boquet River I know of is across the lake from us in New York. I suspect that the caption is in error and meant New York state.

Brown trout (*Salmo trutta*): "Brownies" are a popular game fish native to Eurasia that have been widely introduced and are now common throughout the US. In the boreal forest, they range throughout Scandinavia and western Russia.

The term "trout" is not a phylogenetic (related by uniquely common traits) designation, but an informal name for salmon species that spend much of their lives in freshwater.

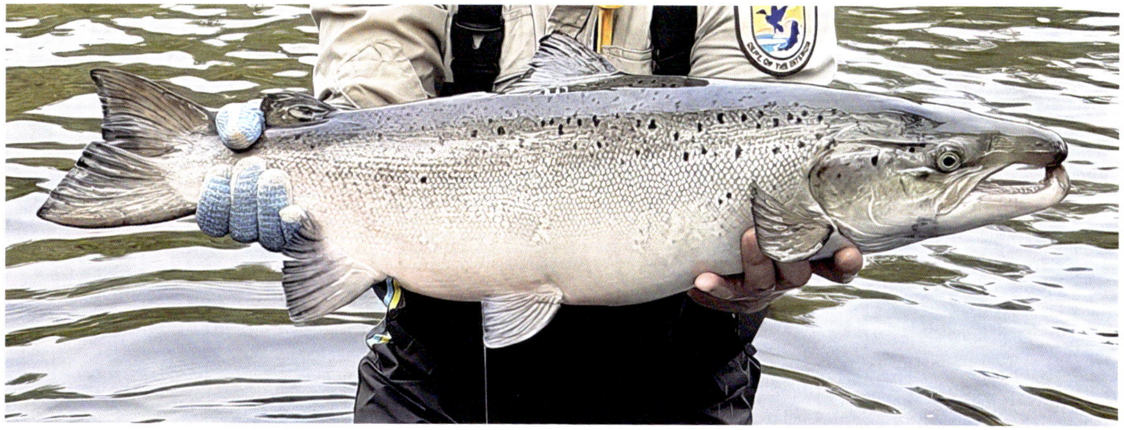

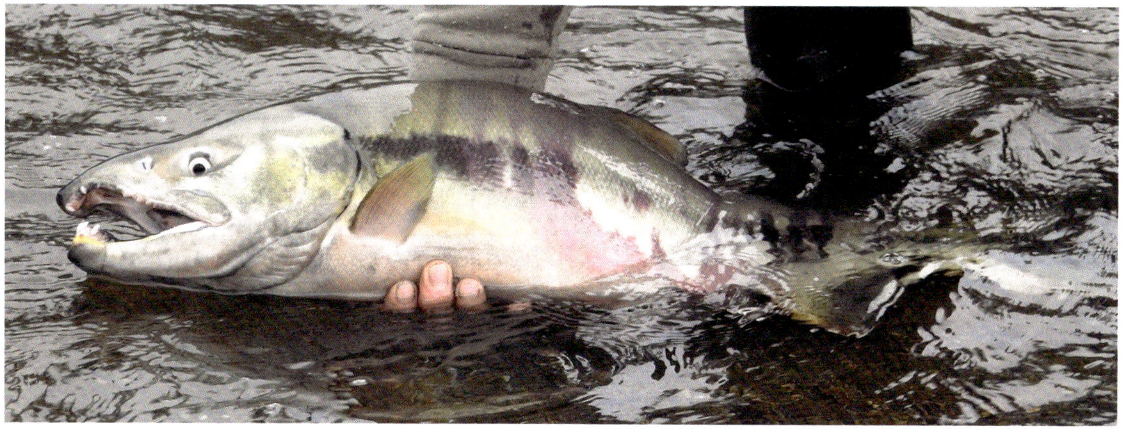

TOP: Landlocked salmon, Boquet River to Lake Champlain. USFWS

BOTTOM: The chum (dog) salmon Dan Berns caught on Beaver Brook, a tributary of the Kobuk River, Arctic Alaska. (The fish was released.)

Chinook salmon (*Oncorhynchus tshawytscha*): Chinooks commonly reach two to three feet long and weigh 10–50 pounds. Exceptional specimens can be four to almost five feet long and weigh over 100 pounds. They are the largest sea-run salmonid and the second-largest salmonid in the world. Native to the North Pacific basin, they ranged from Ventura, California, to Kotzebue, Alaska, in North America and from northern Japan to eastern Siberia in Eurasia. Chinook from the Yukon River in Alaska undertake the longest migration of any salmon, with spawning grounds near Whitehorse, Yukon.

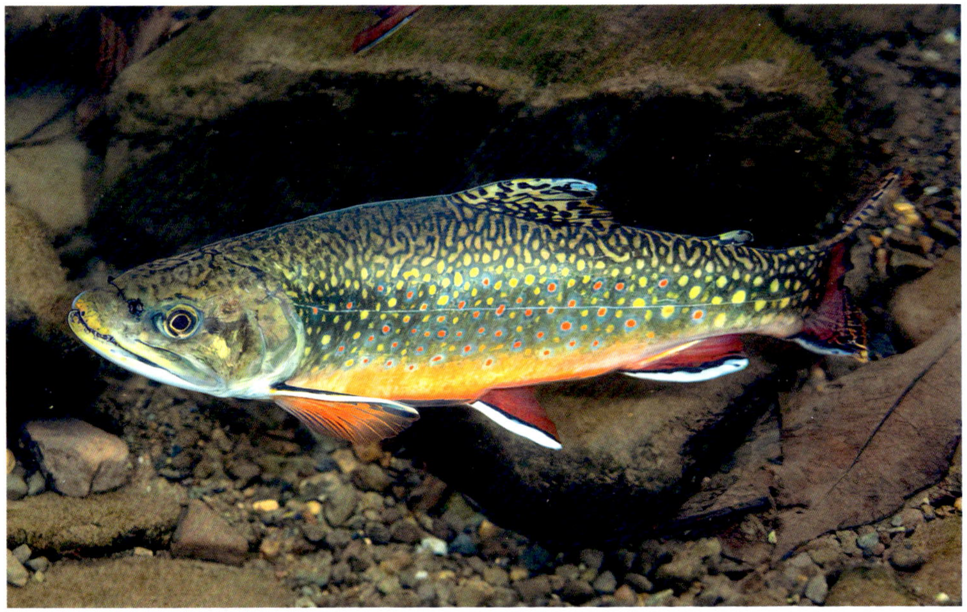

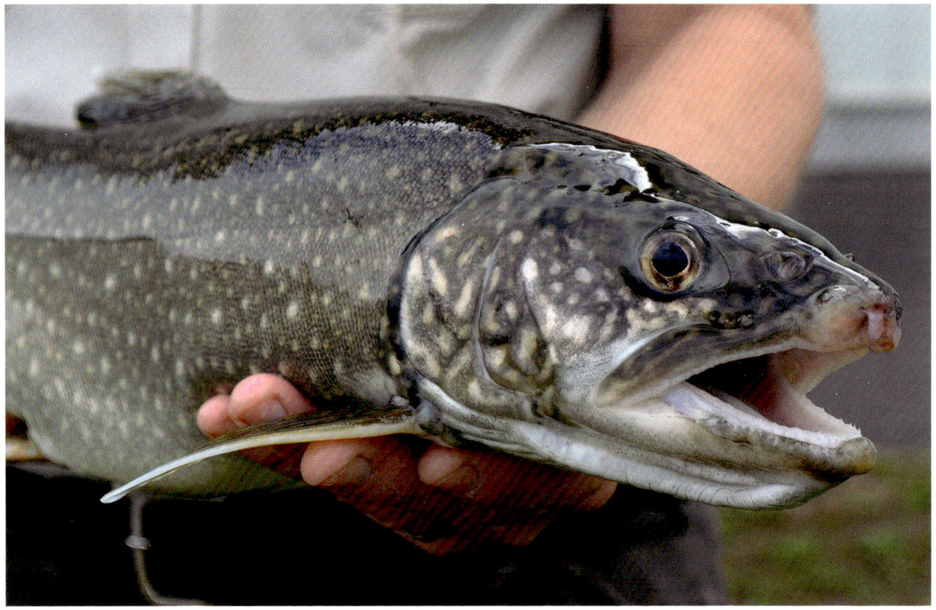

LEFT: Brook trout in Seneca Creek, West Virginia. Ryan Hagerty/USFWS

RIGHT: Lake trout. USFWS

They have suffered significantly from habitat degradation and reduced access to spawning grounds in much of their native range. They continue to fare better in Alaska and the Kamchatka Peninsula in Russia (boreal forest) and have been introduced elsewhere.

Chum or dog salmon (*Oncorhynchus keta*): They have the largest natural range of any Pacific salmon, including much of the boreal forest in Alaska and Yukon. They are not preferred by sport anglers in North America (hence "chum" and "dog"), though the Inupiat are fond enough of them. When we visited Iñupiaq Elder Ricky Ashby on the Noatak River in 2012, we helped him set his gill nets, and he treated us to a dog salmon dinner. They were delicious, but then, it is difficult to judge on a canoe trip: all food tastes better, especially after over a month in the field.

Genus *Salvelinus* (the chars)

While the genus is circumpolar and its 54 species (subject to debate) are northern, cold-water-loving fish, the ranges of many of the species are very limited. The two most widespread species, the Arctic char and the lake trout, cover most of the North American boreal forest.

Arctic char (*Salvelinus alpinus*): Arctic char have the distinction of being the northernmost freshwater fish in the world, with populations on Ellesmere Island in the Canadian High Arctic.

They can be anadromous (migrating between fresh and salt water), landlocked, or a bit of both. As is typical with salmon, the sea-run populations tend to be larger and can be over 20 pounds.

We saw plenty of them on both the Noatak and Kobuk rivers, though they don't range far south in the boreal forest.

Lake trout (*Salvelinus namaycush*): The largest of the char, lake trout (Lake Superior char, according to the IUCN) can reach over four feet and weigh up to 100 pounds. A denizen of deep, cold, well-oxygenated lakes, they are a classic boreal forest fish, ranging through most of the North American forest, and an important economic force. The recreational fishery on the Great Lakes alone is worth about $7 billion a year and supports over 75,000 jobs.

Brook trout (*Salvelinus fontinalis*): The smallest of the char are native to eastern North America. They may be anadromous or purely freshwater, with the freshwater populations being the larger (the reverse of the usual pattern in salmonids). Freshwater fish may reach two feet in length and weigh over 10 pounds. In the boreal forest, they range through much of Ontario and all of Quebec and Labrador up to the southern half of Hudson Bay, James Bay, and Ungava Bay.

Genus *Thymallus* (graylings)

Famous for their large, sail-like dorsal fins, several species of these northern fish range into the boreal forest. Most have restricted ranges in eastern Siberia such as in Lake Baikal. Two boreal species are widely distributed.

Arctic grayling (*Thymallus arcticus*): Strikingly beautiful fish when alive (their iridescence fades quickly if killed), with a flamboyant dorsal fin, Arctic graylings range from the western

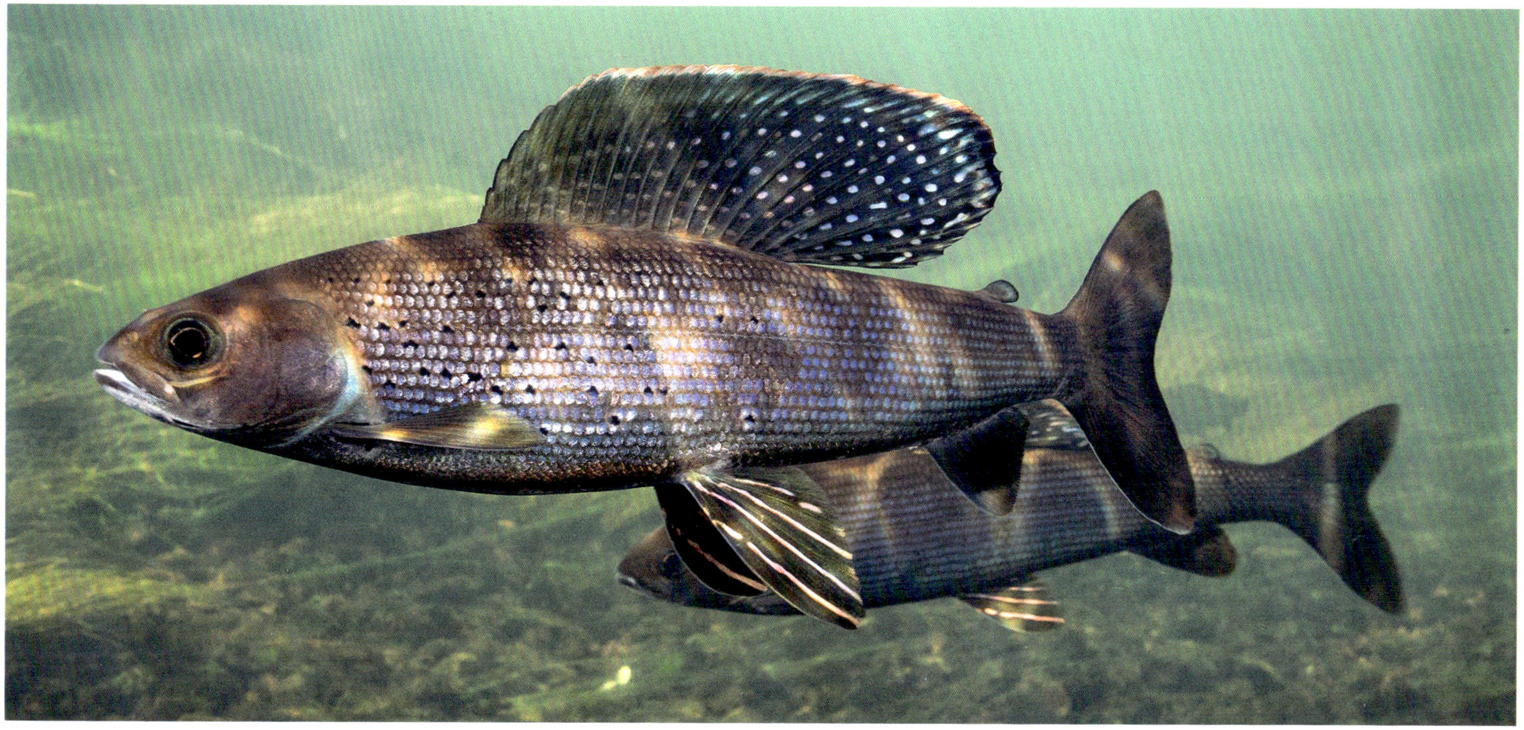

Arctic grayling showing its natural color underwater. USFWS

Dan Berns, our fisherman on the 2015 Kobuk expedition, with a sheefish.

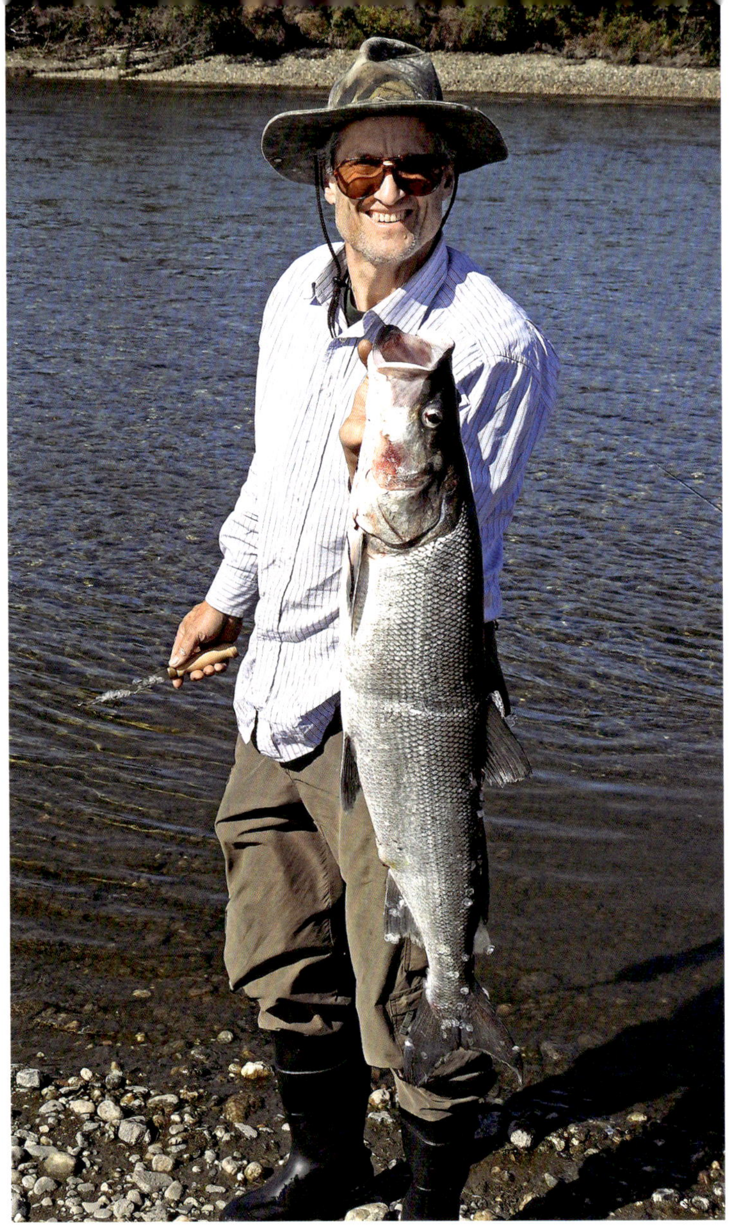

shore of Hudson Bay through all of Alaska and eastern Siberia. Large ones can be up to 30 inches long and weigh eight pounds, though we never saw one close to that size. I could often see them swimming beneath my canoe on the Reed River, high in the Brooks Range in Alaska.

European grayling (*Thymallus thymallus*): Similar in general appearance to the Arctic grayling, this fish is found in northern Europe from the Ural mountains in Russia, westward through Scandinavia and the UK.

THE WHITEFISH genus *Coregonus*, with nearly 70 species, is the most diverse in the salmonid family. Whitefish are critically important to commercial fisheries and for Indigenous people. On our 2015 trip on the Kobuk River in Alaska, the fishing camps upstream of the Iñupiaq village of Kobuk were busy. Thousands of whitefish were being dried and smoked on racks in the village. Farther downstream, in Ambler, the drying racks held many thousands more.

At the end of our Kobuk River journey, we entered the 2,176,000-acre Selawik National Wildlife Refuge. Among the several whitefish species, there is a giant. It is the refuge's namesake: *Selawik* derives from the Iñupiaq name for "place of the sheefish."

Sheefish (*Stenodus nelma*): The largest whitefish, sheefish grow to five feet and weigh 60 pounds. They are mostly anadromous, with some populations migrating more than a thousand miles up the Yukon River. They have been a staple of Indigenous people on both sides of the Bering Sea for thousands of years. Another name for the species is *inconnu*, bestowed on them by French voyageurs accompanying explorer Alexander Mackenzie, referring to "unknown fish." We saw large schools of them on the Kobuk; one was enough for the entire crew's dinner.

If sheefish are the fifth-largest salmonid behind the lake trout and the Atlantic and chinook salmon, that just leaves the largest: the Siberian taimen.

Siberian taimen (*Hucho taimen*): IUCN Red List status: Vulnerable, Palearctic. The Siberian taimen, also known as

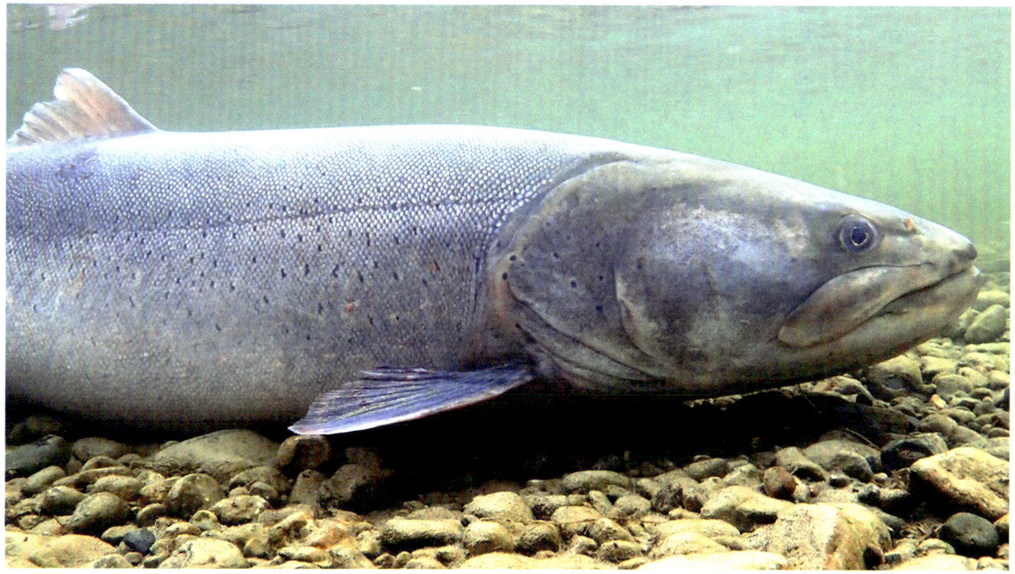

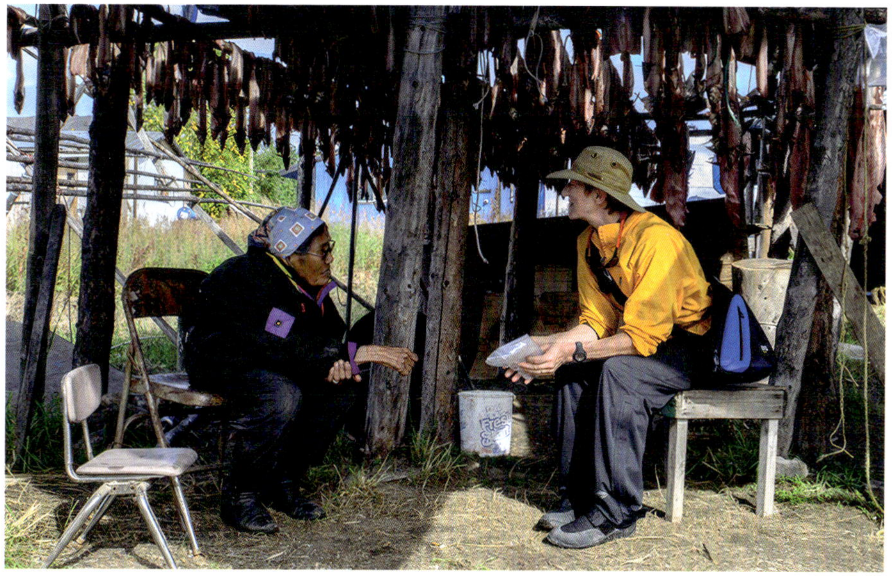

LEFT: Taimen underwater, Siberia. Dr. Matt Sloat/Wild Salmon Center

RIGHT: 2015 Kobuk expedition artist Julia Hargreaves talks with an Iñupiaq woman in Kobuk under racks of drying whitefish. Stephen Gorman

"river tigers," are the largest salmonid on Earth. They grow to six feet, and researchers cite reports of them being close to and over 200 pounds. Aggressive apex predators with non-discriminatory tastes, they dine on small mammals and waterfowl along with their main fish menu (including adult salmon). However, their enthusiastic appetites can lead to a fatal eating disorder. A large taimen was found choked to death with a slightly smaller taimen stuck in its throat; remarkably, the result of a head-on collision. One could infer that the smaller one thought it had a chance.

They are migratory in freshwater, take five to seven years to reach sexual maturity (up to nine in colder water), and can live to nearly 30.

Their dependency on large prey (notably salmon, which have been heavily poached for roe), slow maturation, and migratory travels make them vulnerable to overexploitation and habitat degradation (such as from dams, pollution, and siltation from mining and logging). These factors have reduced four of the five species of taimen (two are boreal forest species) to Threatened or Critically Endangered on the IUCN Red List. On the bright side, the catch-and-release angling world has recently discovered this magnificent animal, arguably the finest fighter a fly-fisher can find (and, until recently, never heard of). In cooperation with local governments and fishing outfitters in Mongolia and Russia and with organizations like the Wild Salmon Center in Portland, Oregon, and the IUCN, conservation efforts and research on taimen are on the rise (some are on hiatus due to Russia's invasion of Ukraine).

Sakhalin taimen (*Parahucho perryi*): IUCN Red List status: Critically Endangered, Palearctic. Restricted to Sakhalin, an island on the Russian coast of the Sea of Japan, and Hokkaido, an island in Japan, the Sakhalin taimen is the only anadromous species of taimen.

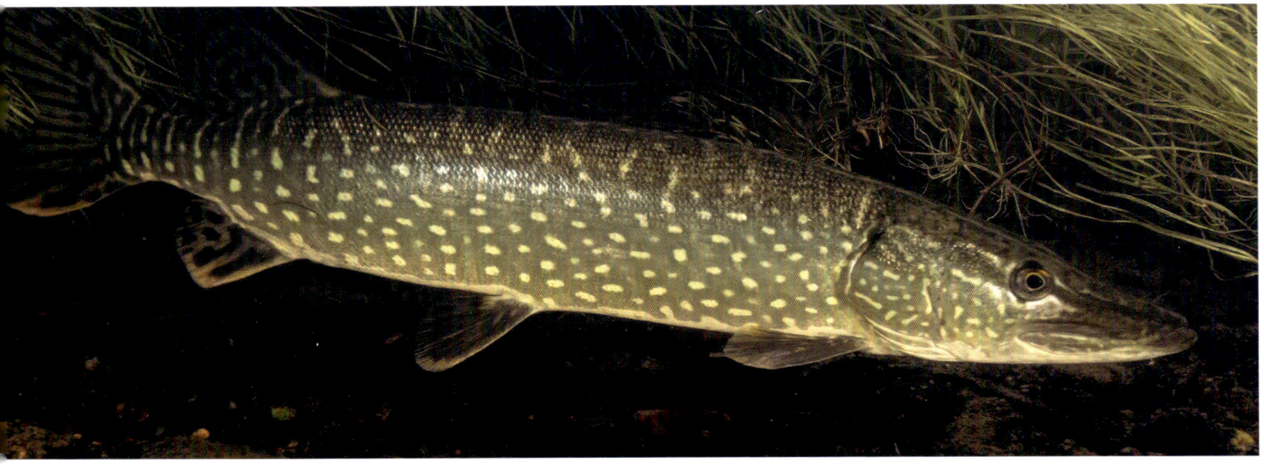

Young northern pike.
Ryan Hagerty/USFWS

FACING: Walleyes on the hunt.
Charcoal, Cole Johnson

Family Esocidae (pike)

While chain pickerel (*Esox niger*) and muskellunge (*Esox masquinongy*) may sneak into the very southernmost fringe of the North American boreal forest, only the northern pike is a significant boreal forest fish. It makes up for its cousins, however, by occupying virtually the entire ecosystem worldwide.

Northern pike (*Esox lucius*): This lurking ambush predator is widely distributed throughout the northern hemisphere in North America and Eurasia and common in many waters. We have frequently encountered pike.

The pectoral and pelvic fins are in the same primitive arrangement as the salmon. However, the dorsal fin is set far astern, paired with the anal fin and a large caudal fin (tail) to provide explosive acceleration, which is how they hunt, lurking in cover until some unfortunate fish wanders too close.

Pike can attain enormous sizes (though not a large as muskellunge), with the current record measuring 57 inches long and weighing 55 pounds. Such beasts are apex predators.

Family Percidae (perch)

In this large and important family, the pelvic fins are placed forward just under the pectoral fins (a more evolutionarily advanced design for greater positional stability and directional control), and the dorsal fin is divided into two sections, separate or nearly so. The forward dorsal has spines, and the aft dorsal fin may have one or two spines but otherwise soft fin rays. The anal fin has one or two spines. Most species are North American with a few Eurasian species.

Walleye (*Sander vitreus*): Native to much of the North American boreal forest except for far northeastern Quebec, Labrador, and Alaska, the walleye is an aggressive mesopredator. I know at least one person who considers it the best-eating freshwater fish in the world (I remind you of the caution of judging food on canoe trips). Walleyes are very common and easily caught by trolling just off the weed edge of the shaded bank late in the day with a Daredevle or Rapala minnow. Dinner is almost guaranteed by the time you make camp. I've caught more walleyes than any other fish on canoe trips, but the last catch made an impression. I have nothing against fishing and haven't sworn it off, but I find more reasons not to bother now.

August 9, 2016: Missinaibi River

As the sponsor of our journey, Destination Ontario asked us to get photos of "fishing fun." I don't fish for fun, but I can't deny that it can be, so I had no problem with the assignment. Late that afternoon, I put a troll line out with a Rapala on it (I like the Rapala lures because they float, so when you slow down, they don't snag on the bottom). Sure enough, not five minutes later, I had a walleye on. Our photographer, Steve Gorman, got into position. Steve is a pro, so he quickly had the shots, and it was time to "dispatch" (kill) our dinner.

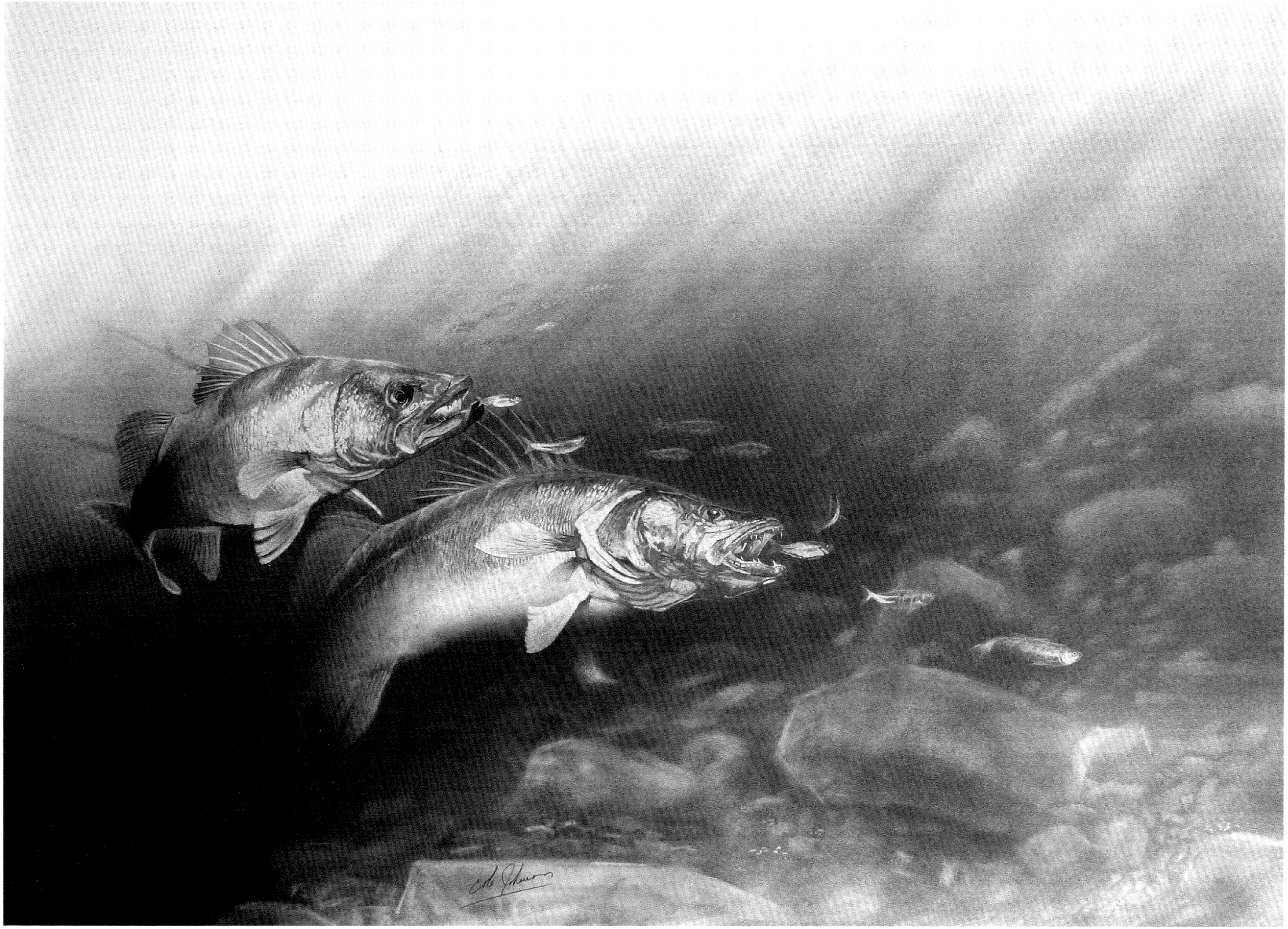

I slipped my heavy Uncle Henry folding knife out to stun it. I enjoy a fresh fish dinner as much as the next person, but this was a relatively short, easy trip with ample food supplies. I looked at him. I put him back in the water and watched him breathe. The hook was not far in and would be easy to remove. I slipped the hook out and he swam off. We had mac and cheese, or chili, or some such for dinner. It was delicious. Canoe trips.

Yellow perch (*Perca flavescens*): A miniature, somewhat flattened version of the walleye (record size is four pounds), the yellow perch is native to the central North American boreal forest. Otters love them. I am ambivalent. When we were kids, every February our gr andparents would visit from Proctor, Vermont, and we'd all head out onto the ice of Keeler Bay on Lake Champlain to put in our year's supply of fish dinners. They were almost invariably yellow perch— hundreds of them. The fishing action could be intense, if a bit gruesome, but then my brother Frank and I had to clean all the fish. And not just clean them, but our grandfather insisted that we fillet them too. It was hours of work in the cold garage with wet, slimy fish and sharp knives. To this day, I'm a bit burned out on yellow perch—and filleting.

Otters eating yellow perch and lounging on the rotting spring ice, Preston Pond, Vermont.

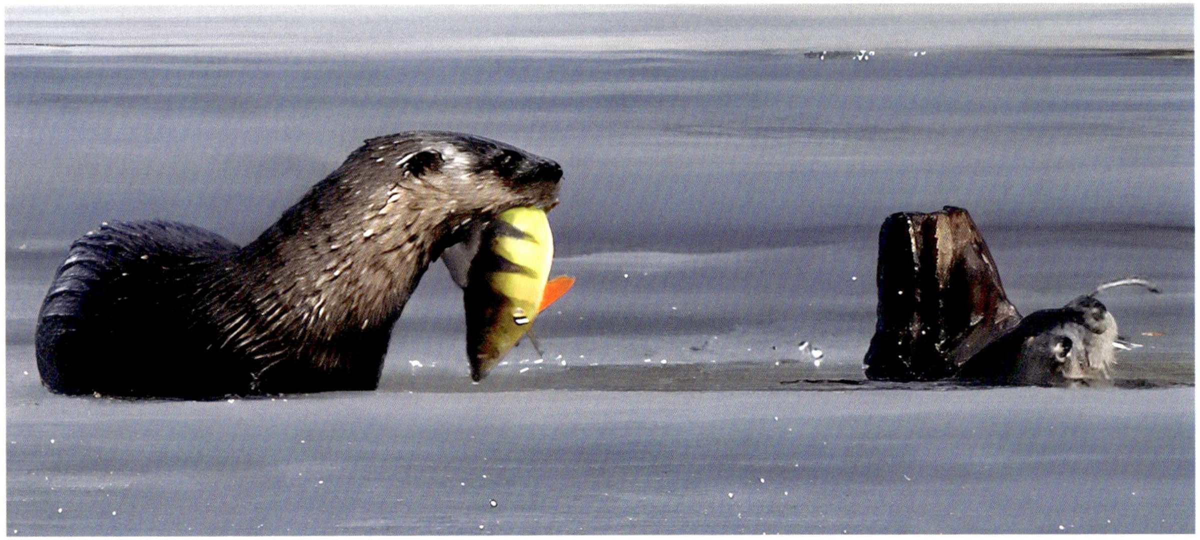

Eurasian perch (*Perca fluviatilis*): Ranging across the entire Eurasian boreal forest, Eurasian or European perch are so like yellow perch that whether they are separate species is in question.

Family Centrarchidae (sunfish and bass)

Bass tend to be sensitive to cold water. While a few pumpkinseed sunfish (*Lepomis gibbosus*) may try to make a liar out of me, only the smallmouth bass (*Micropterus dolomieu*) and, to a lesser extent, the largemouth bass (*Micropterus salmoides*) are present significantly in the southern boreal forest.

Family Cyprinidae (minnows)

So far, almost every fish in this chapter has been a reasonably large mesopredator or apex predator. They spend a lot of time eating each other, even their own species: the whole big fish, small fish thing. However, in a food chain, somebody must form the links that connect the flow of energy and nutrients from primary producers (plants) to the mesopredators and apex predators. That is the job of primary and secondary consumers (plant eaters and small predators). In the myriad aquatic environments of the boreal forest, both roles (with some role overlap and omnivores included to diversify things) include invertebrates and small fish. The fish are often cyprinids, or as they are more familiarly known, minnows. There are many species of this diverse group in the boreal forest.

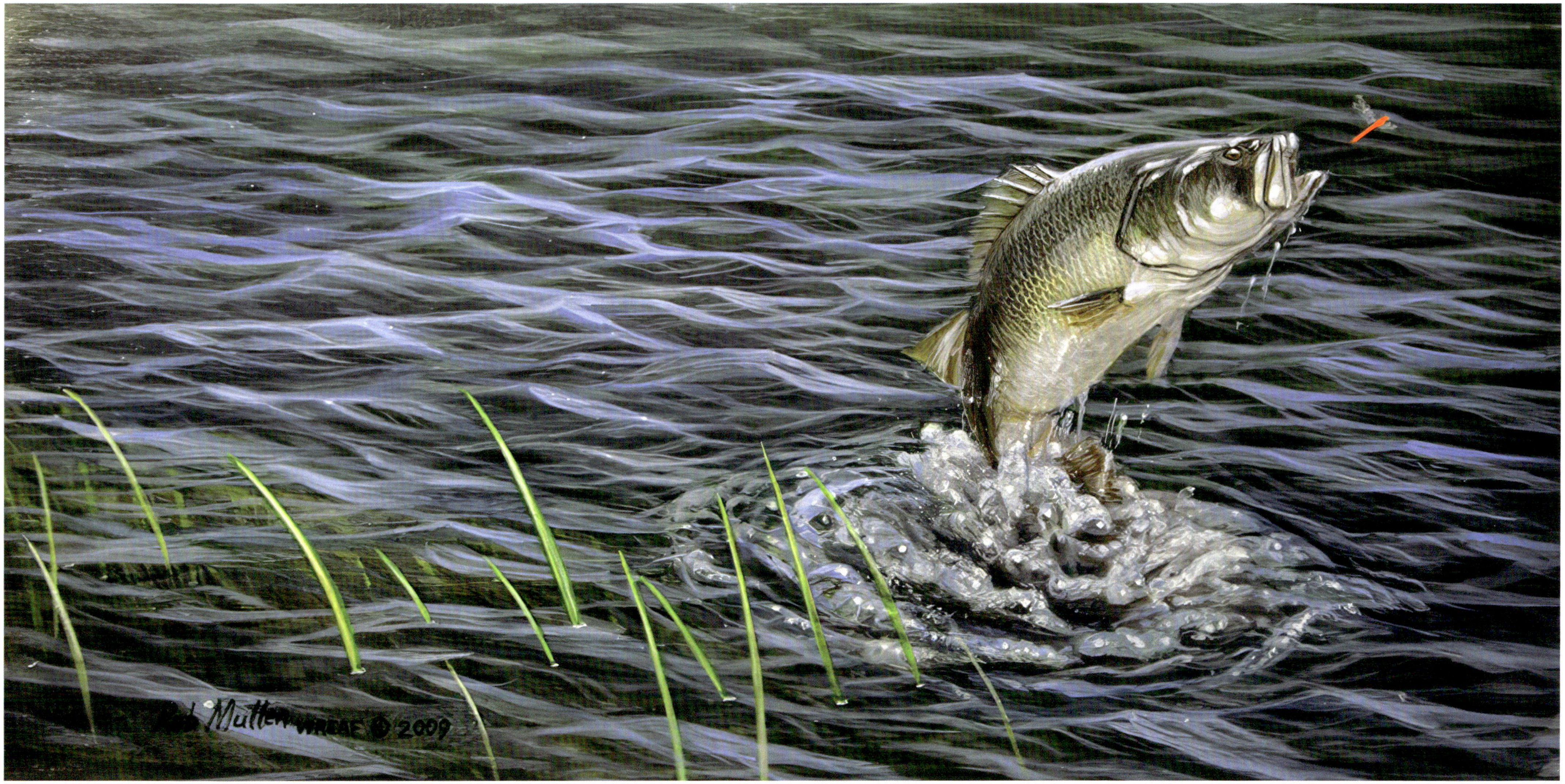

Smallmouth bass and damselfly, Missinaibi River. Note: The pectoral fins are missing. I forgot them. *Damsel in Distress*, acrylic, Rob Mullen

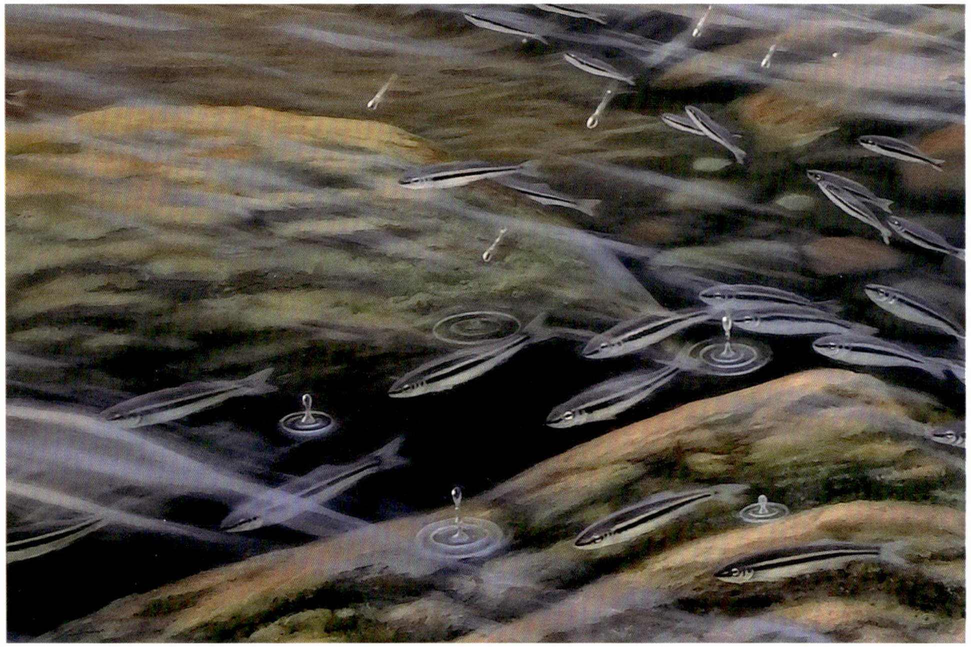

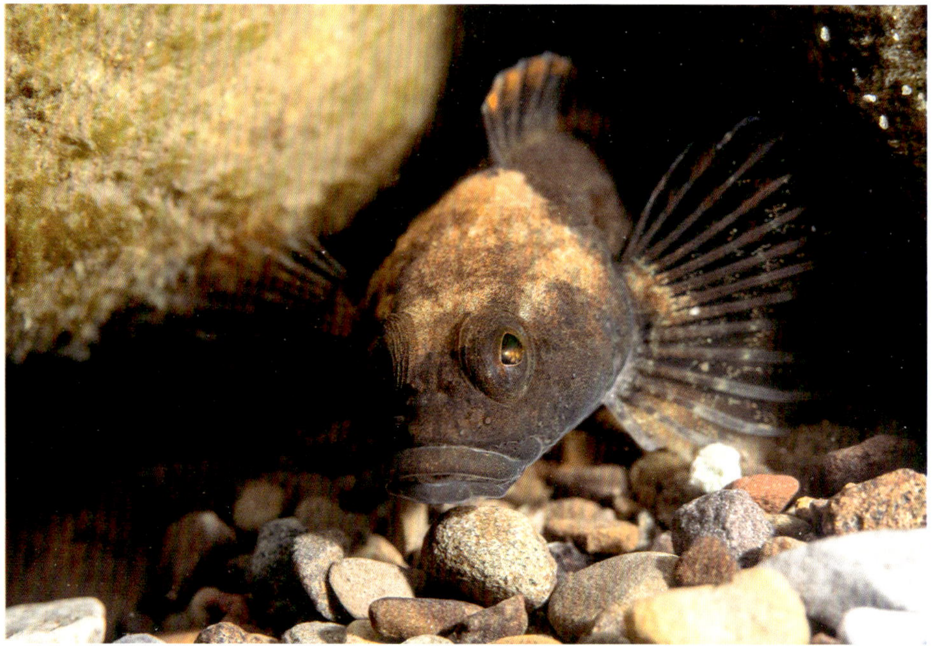

LEFT: Eastern blacknose dace, detail from *Loon Rise*. This color pattern is found in many species of dace, making identification tricky (and I could be wrong about them being blacknose dace). Acrylic, Rob Mullen

RIGHT: Mottled sculpin. Ryan Hagerty/USFWS

While the cyprinid family includes some enormous species of carp, most cyprinids in the boreal forest are less than four inches long and feed on plants, or small animals, usually invertebrates that larger predators overlook or simply can't be bothered with. The very young (fry) of large predatory fish are of similar size and ecological function but grow out of it. Minnows (mostly shiners and dace) fill this niche all the time.

As minnows are prey and need far fewer resources and space per individual than a 10-pound walleye, their populations can be enormous. That is fortunate because they have a lot of work to do (eating) and sacrifices for the greater good to make (being eaten).

Among the lesser-known services various minnows provide is a hedonistic hygiene treatment. I had read of expensive spas offering "all-natural" exfoliation of your feet using schools of minnows. The same luxurious, if rather ticklish, pleasure can be had in almost any sun-warmed shallow pool on even the wildest, most remote wilderness river . . . for free. Just take your stinky canoe boots off, kick the bottom up some, and they will come "running." There's no accounting for taste.

Family Cottidae

Slimy sculpin (*Cottus cognatus*) and mottled sculpin (*Cottus bairdii*)

In the same order as perch (Perciformes), these bottom-dwelling ambush predators max out at six inches, feeding mostly on invertebrates, small fish, and occasional plant material. The slimy sculpin (*Cottus cognatus*) occurs across the boreal forest in North America and eastern Siberia, while the mottled sculpin (*Cottus bairdii*) ranges from Hudson

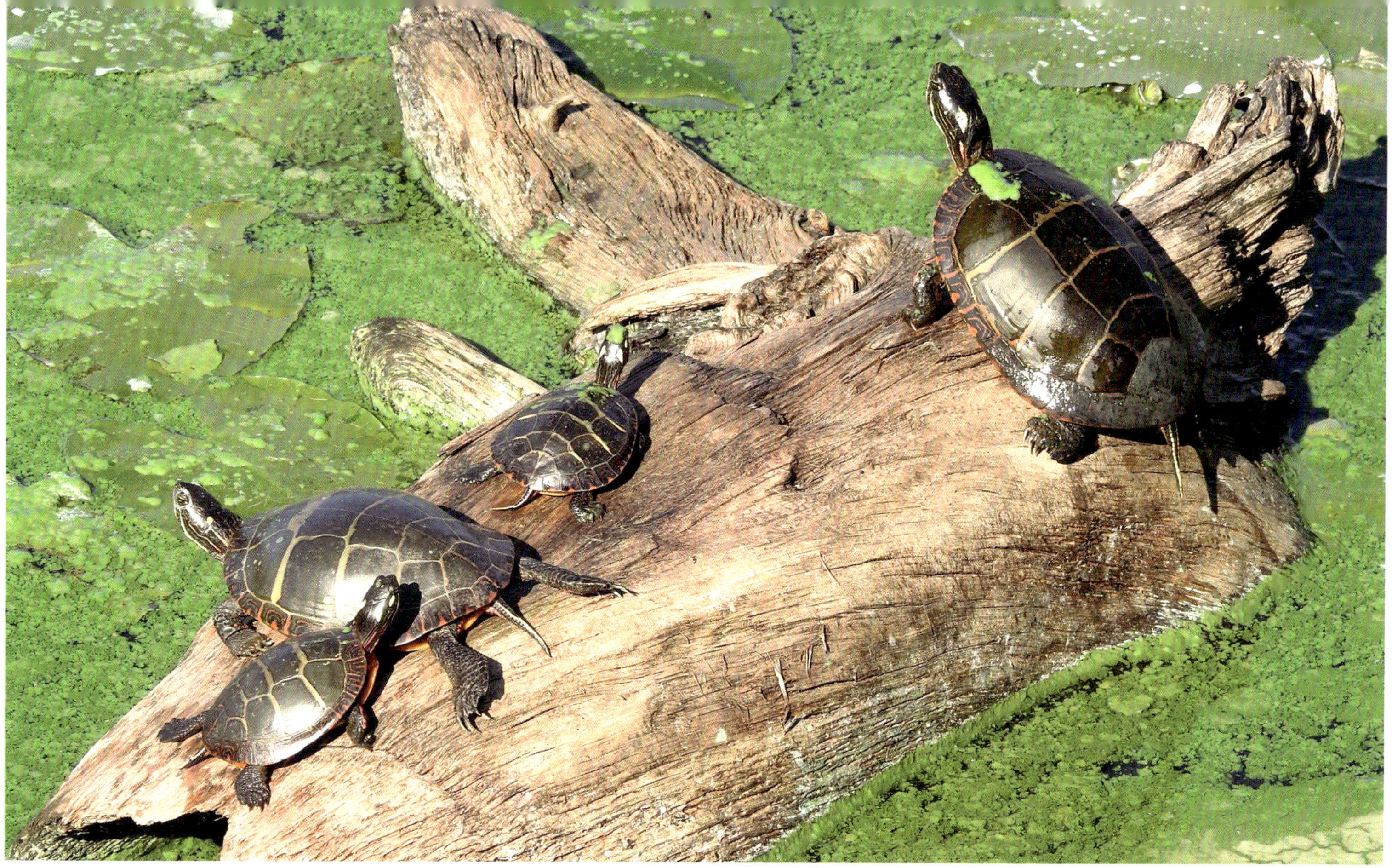

Painted turtles, Île Bizard, Quebec. Bonnie Rowell

Bay and Ungava Bay to Georgia and west in the boreal forest to Saskatchewan.

The slimy sculpin was probably the species I was excited and amazed to find as a kid in the brook next to our house. That such an exotic-looking animal lived right near our house amazed me. One took up residence in my aquarium with the dace. Forty years later, as we set camp below Helen Falls on the George River, south of Ungava Bay, I found 21 baby sculpins stranded in a pool by a drop in the river's water level. The aurora borealis was waxing as the sun sank, so I grabbed my canteen cup and rescued all the sculpins, releasing them in a quiet eddy of the river below the falls. Then I joined the rest of the crew and watched the lights.

Reptiles

Turtles apparently have the greatest difficulty with the cold. This may be because they brumate (lay dormant) beneath the ice, absorbing oxygen through their capillary-rich cloaca (the single vent for intestinal, sexual, and urinary systems in reptiles, amphibians, and birds). Even though they greatly reduce their already low metabolisms, they are large animals, and in the boreal forest, the ice locks them in for four months or more. Oxygen at the bottom can run low. In North America, only the common snapping turtle (*Chelydra serpentina*) and the painted turtle (*Chrysemys picta*) range into the boreal forest and not far at that, restricting

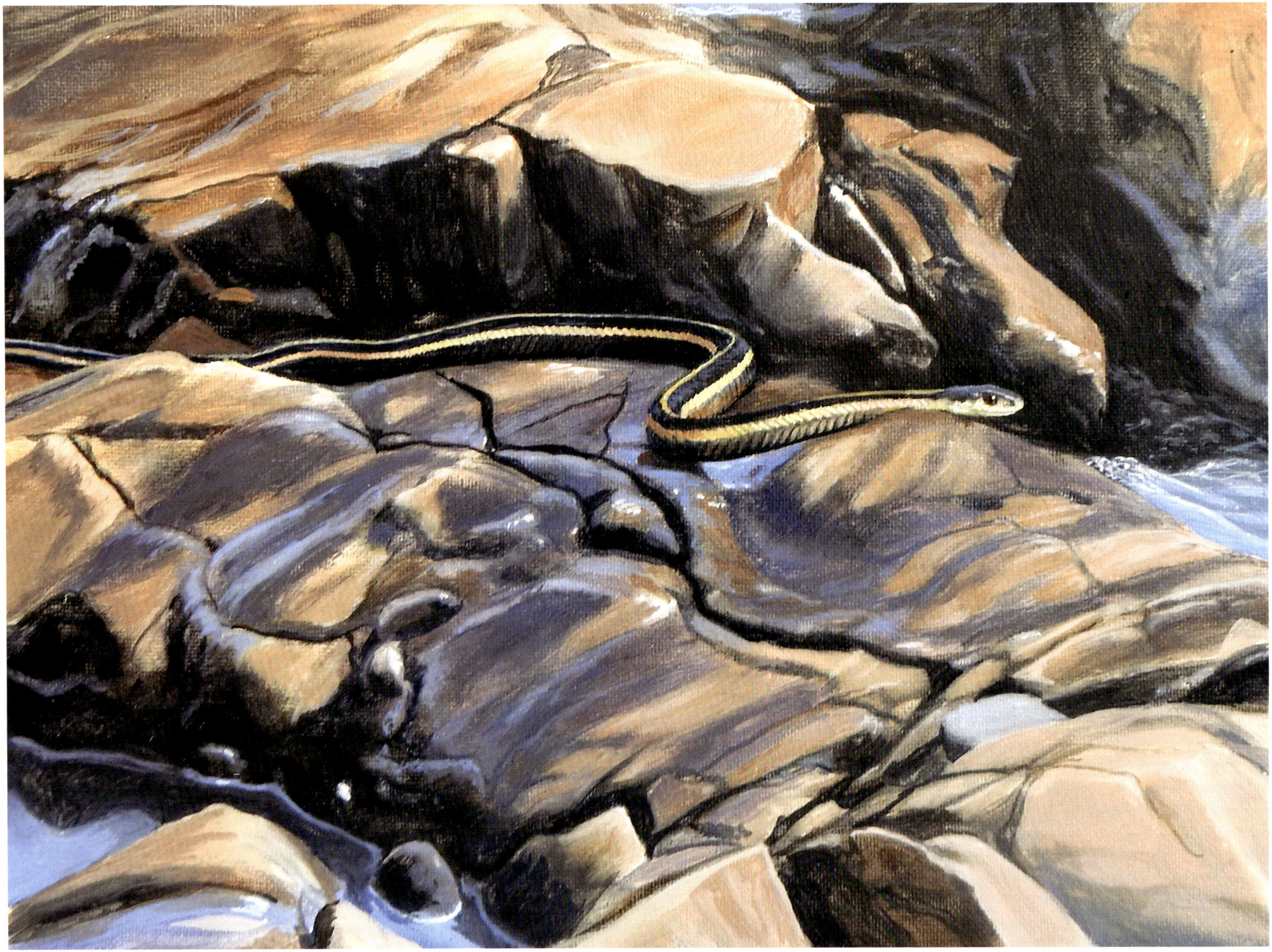

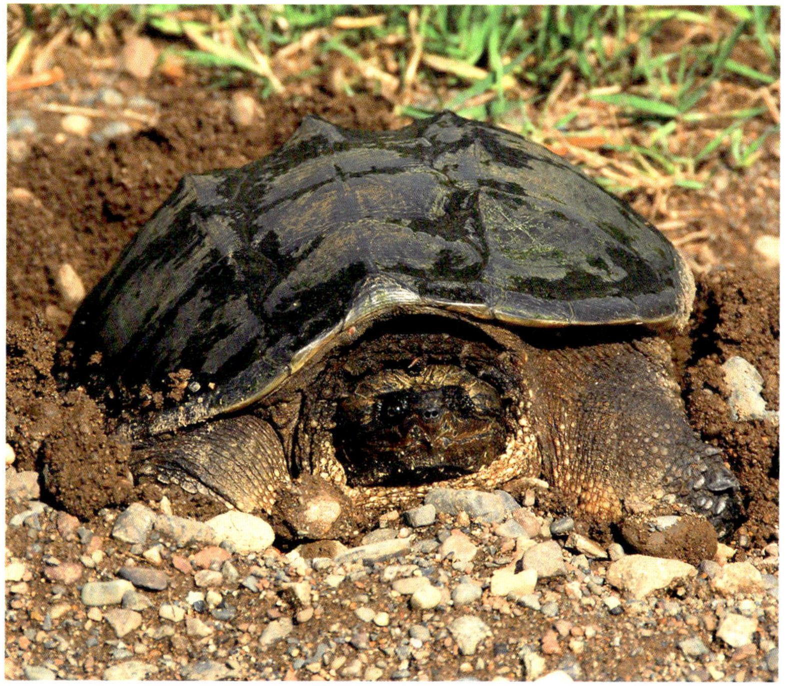

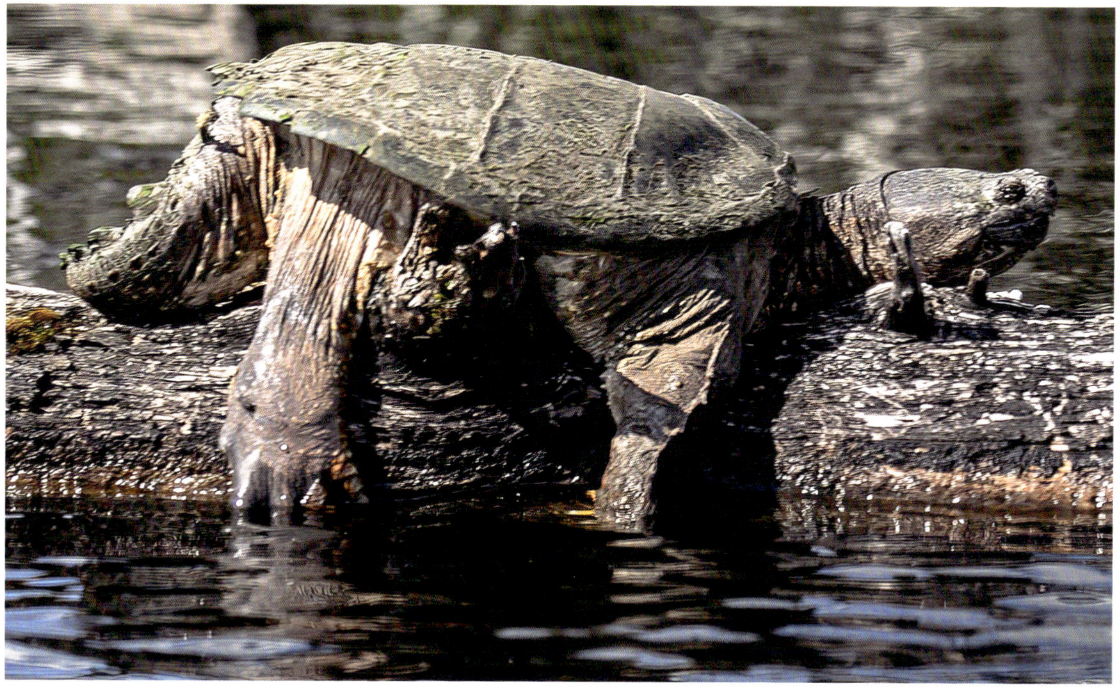

themselves to the southern edge. One species of turtle, the European pond turtle (*Emys orbicularis*) inhabits the eastern Eurasian boreal forest, and it is rated as Near Threatened by the IUCN.

Snakes are better represented. The common garter snake (*Thamnophis sirtalis*) and the smooth green snake (*Opheodrys vernalis*) occupy the North American boreal forest.

In Eurasia, five or six species range at least marginally into the boreal forest. However, only the common European adder or viper (*Vipera berus*) is widespread throughout Scandinavia and along the southern edge of Siberia.

Lizards are least accommodating to the cold. There are none in the North American boreal forest, and only one, the viviparous lizard (*Zootoca vivipara*), in Eurasia. However, it is widespread except in northeastern Siberia. A tough lizard.

Amphibians

Salamanders get around. Most of our local Vermont salamanders range into the southern boreal forest, including the spotted and blue-spotted salamanders, the red-backed salamander, and the eastern newt. We are left to believe the range maps since we've never seen a salamander on our trips.

Frogs are tougher than one might think, but then they have been around for hundreds of millions of years so maybe it shouldn't be surprising.

Wood frogs (*Lithobates sylvaticus*) are a famous example of a northern frog capable of freezing and reviving due to antifreeze in their body tissues. They are found throughout the entire North American boreal forest and all the way to the arctic coast of Alaska.

LEFT: Mother snapping turtle laying eggs, Seney National Wildlife Refuge, Michigan.

RIGHT: Basking snapping turtle, Algonquin Park, Ontario. Steve Bruno

FACING: Garter snake, Voyage of the Artful Otter expedition. *Shoreline Hunter*, oil, Rob Mullen

The odd couple: Red-backed and spotted salamander. Caught out by a sudden freeze after a weird warm spell, they were our overwintering house guests. They had a large terrarium, but interestingly, they were most often together.

FACING, CLOCKWISE FROM TOP LEFT:
Trail buddy: Hikers' favorite beasts—red eft (terrestrial juvenile stage of the eastern newt), Wolfrun Natural Area, West Bolton. Susan C. Morse

American toad, Bell's Bay on the Missinaibi River, Northern Ontario. The toad's pebbly skin texture is excellent camouflage for this largely terrestrial frog.

Young female and adult male mink frogs (maybe), Lac des Loups Marins, Ungava Peninsula. Young female on the top, adult male on the bottom. Generally, males have larger tympanic membranes (ears).

Lake Superior North Shore expedition. *Boreal Chorus Frog*, oil, Patricia Pépin

Female wood frog. Susan C. Morse

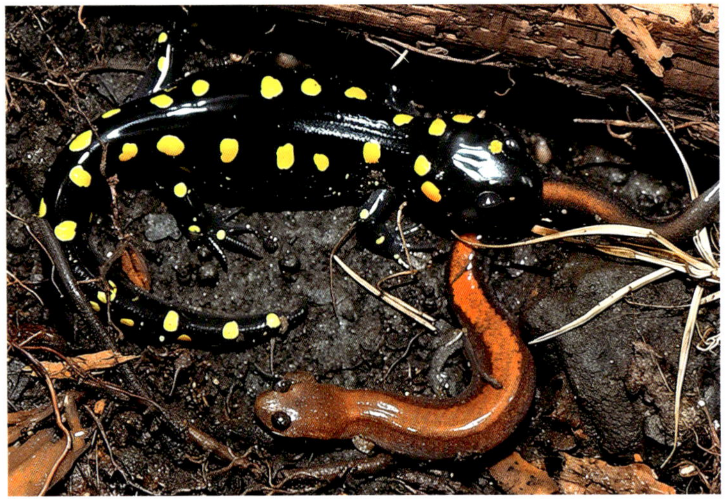

The boreal chorus frog (*Pseudacris maculata*) ranges from Quebec (and northwestern Vermont—though there is worry that it is has recently become extinct in the state) and Ontario to Great Slave Lake in the Northwest Territories and south to Arizona and New Mexico. Talk about adaptability!

Another example of a tough and adaptable amphibian is the humble toad. The American toad (*Anaxyrus americanus*) occupies virtually the entire eastern half of North America, from the Gulf of Mexico to the Arctic Ocean.

We found what I thought were somewhat oddly marked green frogs (*Aquarana clamitans*) while paddling Lacs des Loups Marins between Hudson Bay and Ungava Bay. They were so far north of their purported range that I wrote to the Vermont Reptile and Amphibian Atlas Project. The director thought they were probably mink frogs (*Aquarana septentrionalis*), though lightly marked and still 300 miles north of their published range. Trust range maps only so far—or perhaps we discovered a new species?

Note: An issue that affects all the terrestrial reptiles and amphibians in the boreal forest is global warming. They often rely on the insulating effect of the formerly reliable deep snowpack to keep their sometimes-shallow hibernation shelters at survivable temperatures (even wood frogs can't tolerate deep subzero cold). However, formerly unheard of, increasingly common winter rain reduces and degrades the snow's insulating power. Since deep subzero temperatures typically return (a destructive pattern we see here in Vermont too), the survival of hibernating reptiles and amphibians can be jeopardized.

Scientific names are important. The format of genus followed by species name provides a universal language that identifies the organism, regardless of varying common names, and an indication of its relationships (or lack thereof) to other organisms.

Along with their universal nature, for over 200 years, scientific names were stodgily reliable and didn't easily or frequently change. They still don't easily change (taxonomists are famous for vigorous debate), but genetic research, especially into genomes (an organism's genetic blueprint and therefore its relatives) has revolutionized the frequency over the last 20 years.

A considerable challenge in this book has been to stay current. The humble mink frog is an excellent example. The American Museum of Natural History tracks the changing scientific names of species. The mink frog's scientific name changed twice between 1854 and 1987 and neither event made the news. The first draft of the human genome was published in 2003 and armed with new toys (tools), biologists were off and running. Between 2005 and 2021, the little frog's name changed seven times! This general pattern holds for many species in this book. I'm doing my best.

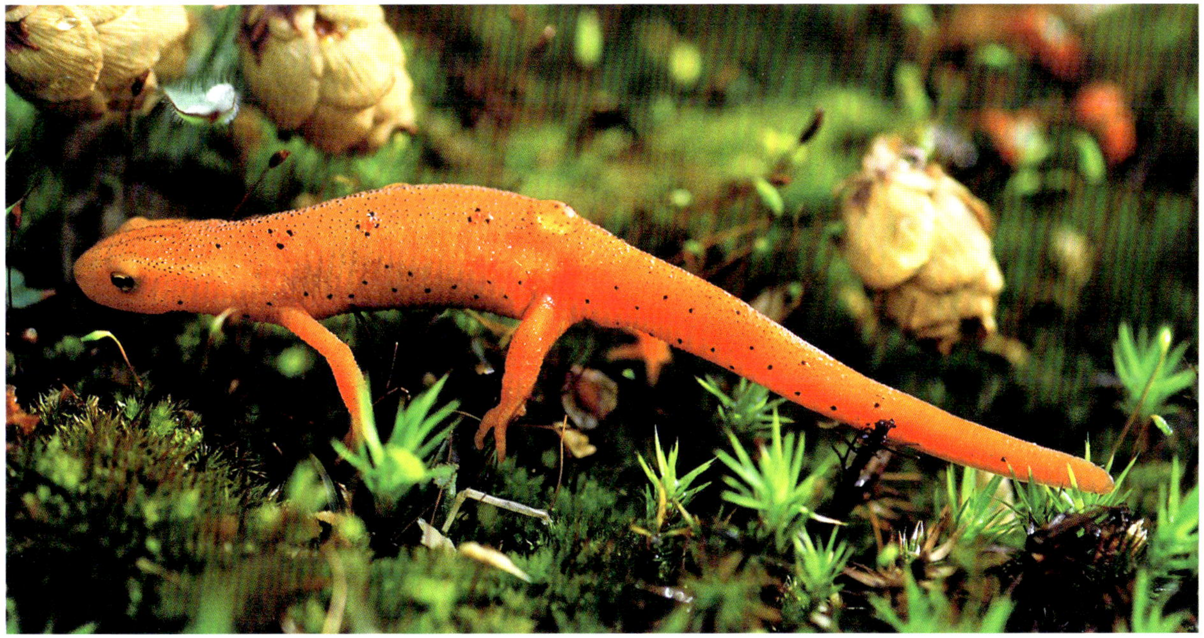
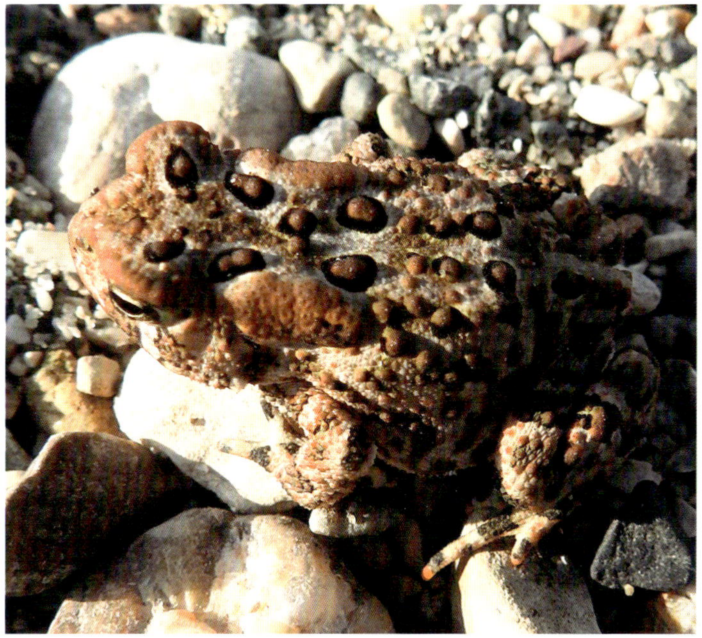
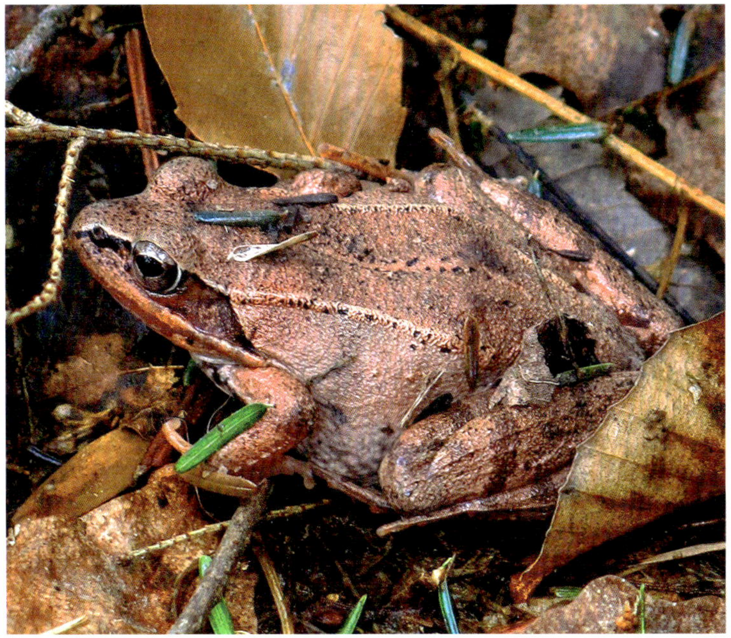
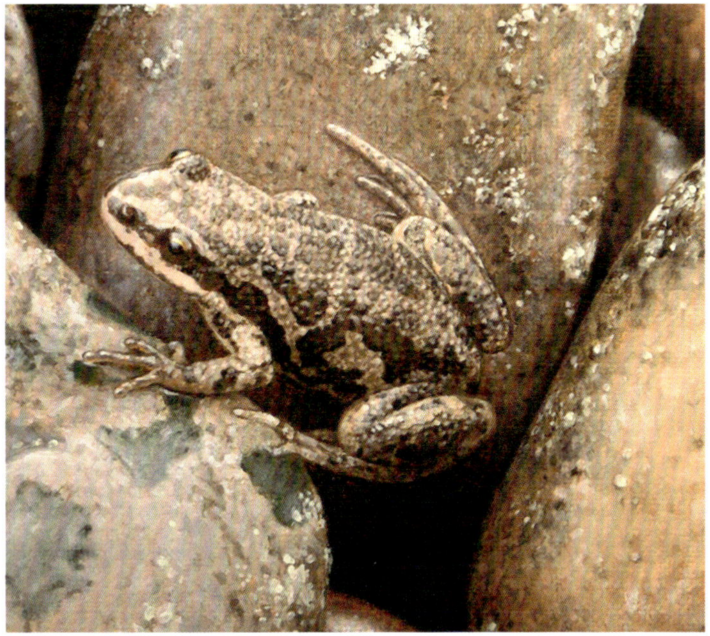
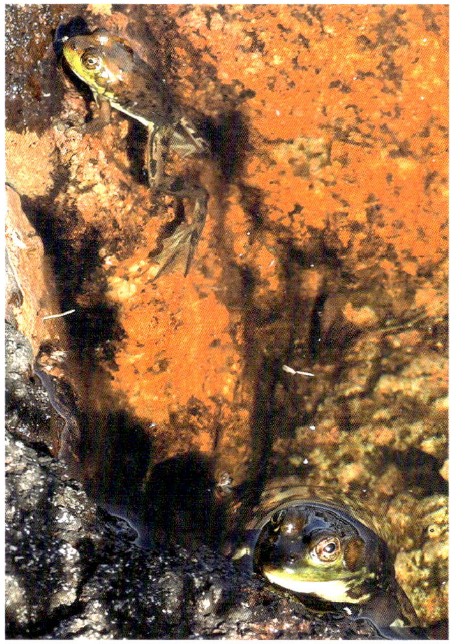

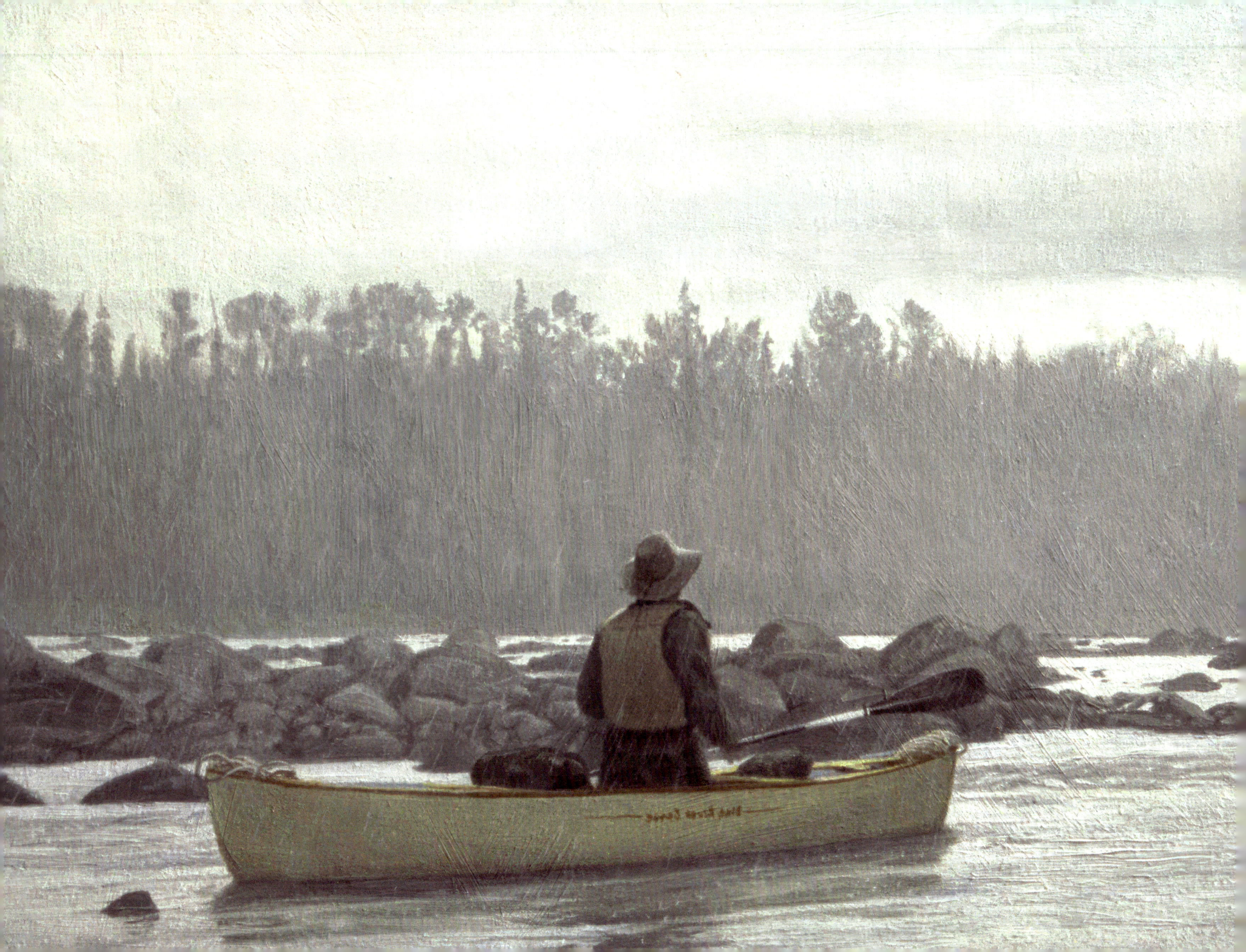

On the Missinaibi River, 2001.
The Path Less Taken, acrylic,
John C. Pitcher

PART THREE

HUMANS IN WILDERLAND

Jordanna Benuen in the guide cabin at the Innu camp at Kamestastin Lake, October 2008. Jordanna and her grandfather Tony Jenkinson accompanied us during our two-week stay.

FACING: Missinaibi River, August 2001. *Land of the Canoe*, acrylic, John C. Pitcher

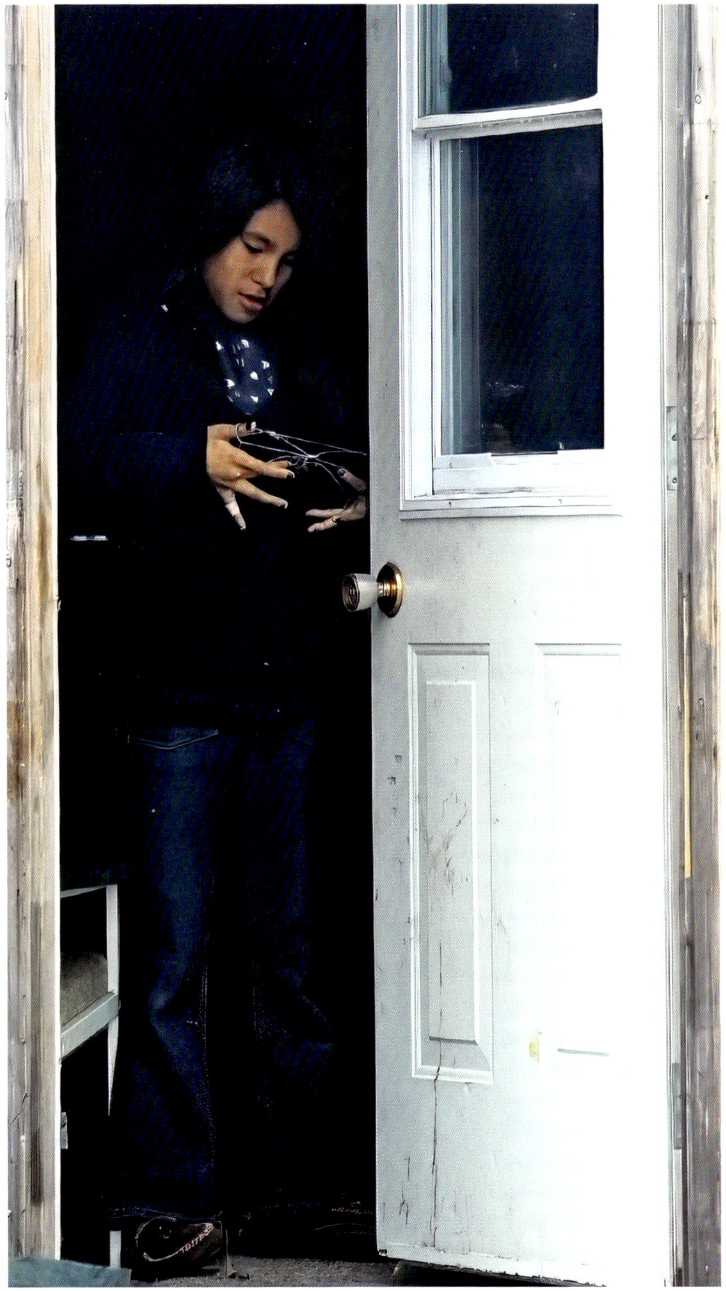

JORDANNA BENUEN would join our forays into the field when her grandfather Tony Jenkinson led us, and her preparations were a poignant contrast to ours. With food and photo gear, binoculars, Gore-Tex boots, and warm clothing, we were ready within 30 minutes. Jordanna would wait until we were leaving, throw on her windbreaker and rubber boots, grab her axe, and be good to go.

The contrast between Jordanna's and our preparations speak to the fallacy that opened this book. To us it was an exciting foray into the unpeopled wilderness. To Jordanna, it was a walk with her grandfather in a land she was learning, her relatives knew, and her ancestors had known for untold generations. But beyond such individualized or cultural perspectives, there is a deeper issue. Whether we call them wildernesses or not, whether they are peopled or not, whether they are boreal forest or not, we need wild lands: intact, productive biomes that preserve the functionality of our biosphere. The forms they take aren't as important as maintaining them, and an insidious threat to maintaining any ecosystem or biome is rapid change. Human civilization could have evolved without a boreal forest. However, it didn't, and now the boreal forest is the largest and therefore among the most critical remaining global wild biomes. We must exercise our vaunted intelligence to maintain it while managing and moderating rapid changes.

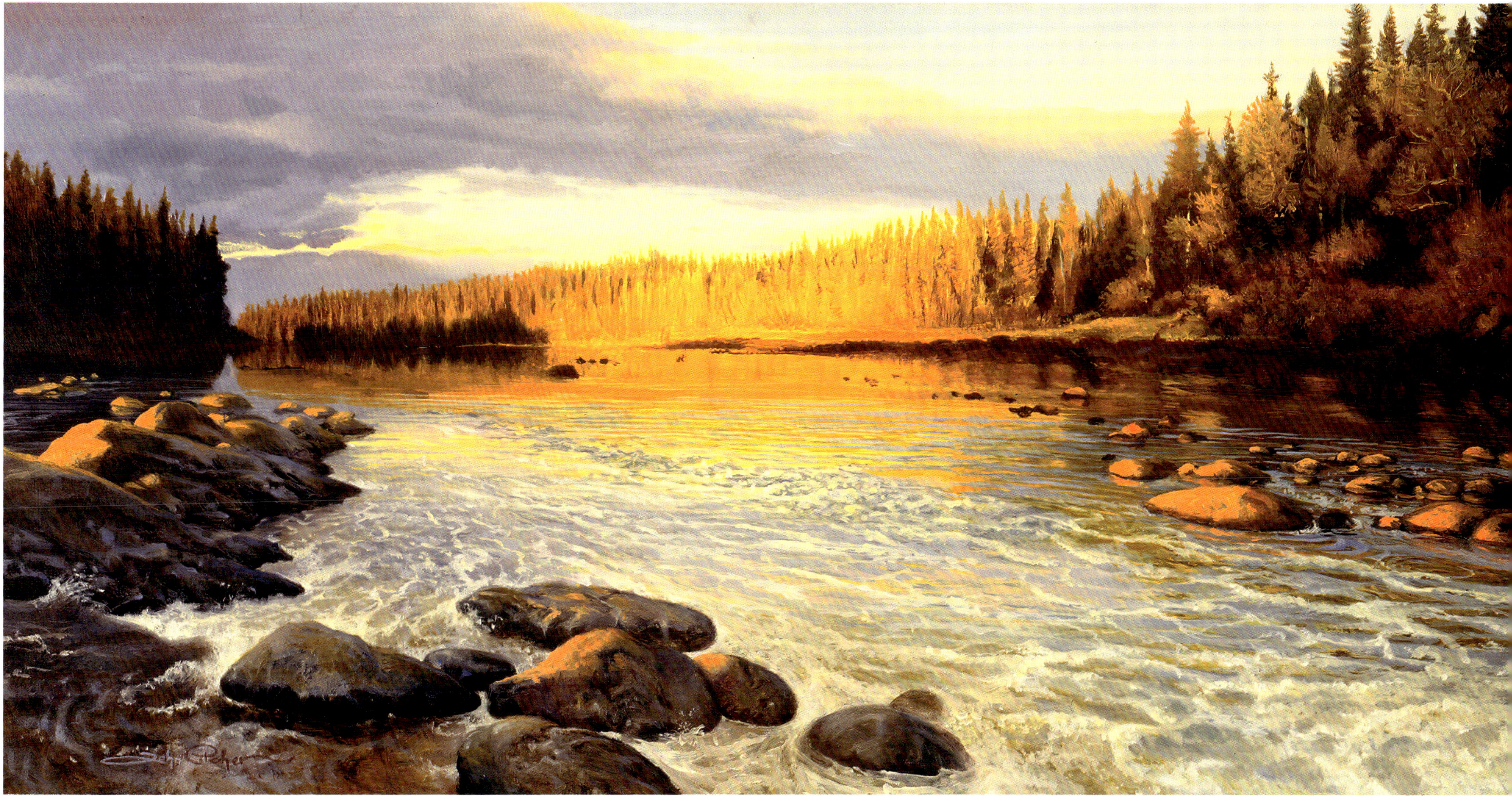

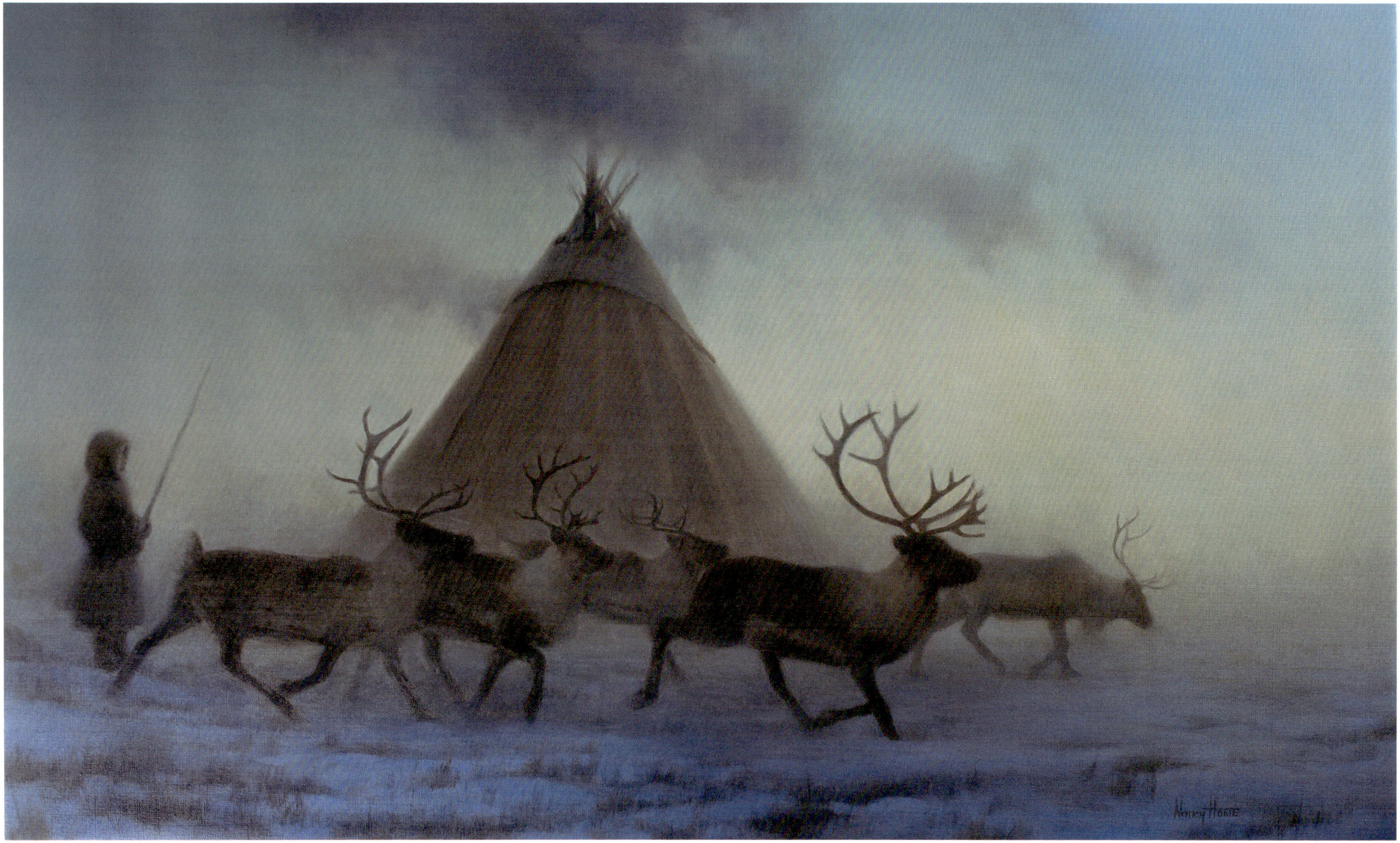

FACING: Saami reindeer herder.
True North, oil, Nancy Howe

7

Cultures of the Boreal Forest

OVER THE YEARS, we've encountered various Indigenous Peoples on our trips.

September 11, 2005: George River Recon, Nunavik (far Northern Quebec)

A day before we finished at Kangiqsualujjuaq, we met some Inuit families gillnetting at the start of the estuary approaching Ungava Bay. One of the artists was asking an older man if certain animals were found locally. His reply to each question was "Oh yes, mmm, good," until he was asked about polar bears—that just got an "Oh yes."

Later, as we beached at Kangiqsualujjuaq, we were met by a small crowd of Inuit who came down on their ATVs to greet us. After a home-cooked dinner, several residents came to visit. We heard about the country to the north: about how dangerous and aggressive the black bears were, of travel along the coast in umiaks, and of efforts to reestablish the abandoned settlement of Killiniq on the island of the same name at the very northern tip of the Labrador peninsula on the east side of Ungava Bay (the abandoned village is in Nunavut as the island is divided between Nunavut and Labrador).

October 2009: Cole Johnson and I had to wait for four days in Nain, Labrador, for the next coastal freighter. While there, we had a couple of impromptu art shows as local Inuit carvers heard about us and brought in their work.

August 2012: We were generously hosted for two nights on the Noatak River in Alaska by Iñupiaq Elder Ricky Ashby, who lived alone in a cabin next to the river. We treated him to a freeze-dried dinner the first night, after which he told us stories of the land and the Inupiat for hours. He invited us for a sauna and fresh salmon and blueberries the second night, then more stories that surreally morphed into a very earnest religious sermon.

And as was related in the prologue and in chapter 3, there were the Cree we encountered at Moose River Crossing on September 12, 2001, and again in October 2003, and again at Moose Factory and the interpretive center in 2016, plus Fred Neegan at the beginning of the trip.

Our Inuit guides and staff in Tursujuq National Park and the Umiujaq community in 2018 were wonderful. And Anishinaabe Elders in Red Lake in 2011 were very friendly

History of James Bay Aboriginal People, Cree Cultural Interpretive Centre on James Bay, Moose Factory, Ontario.

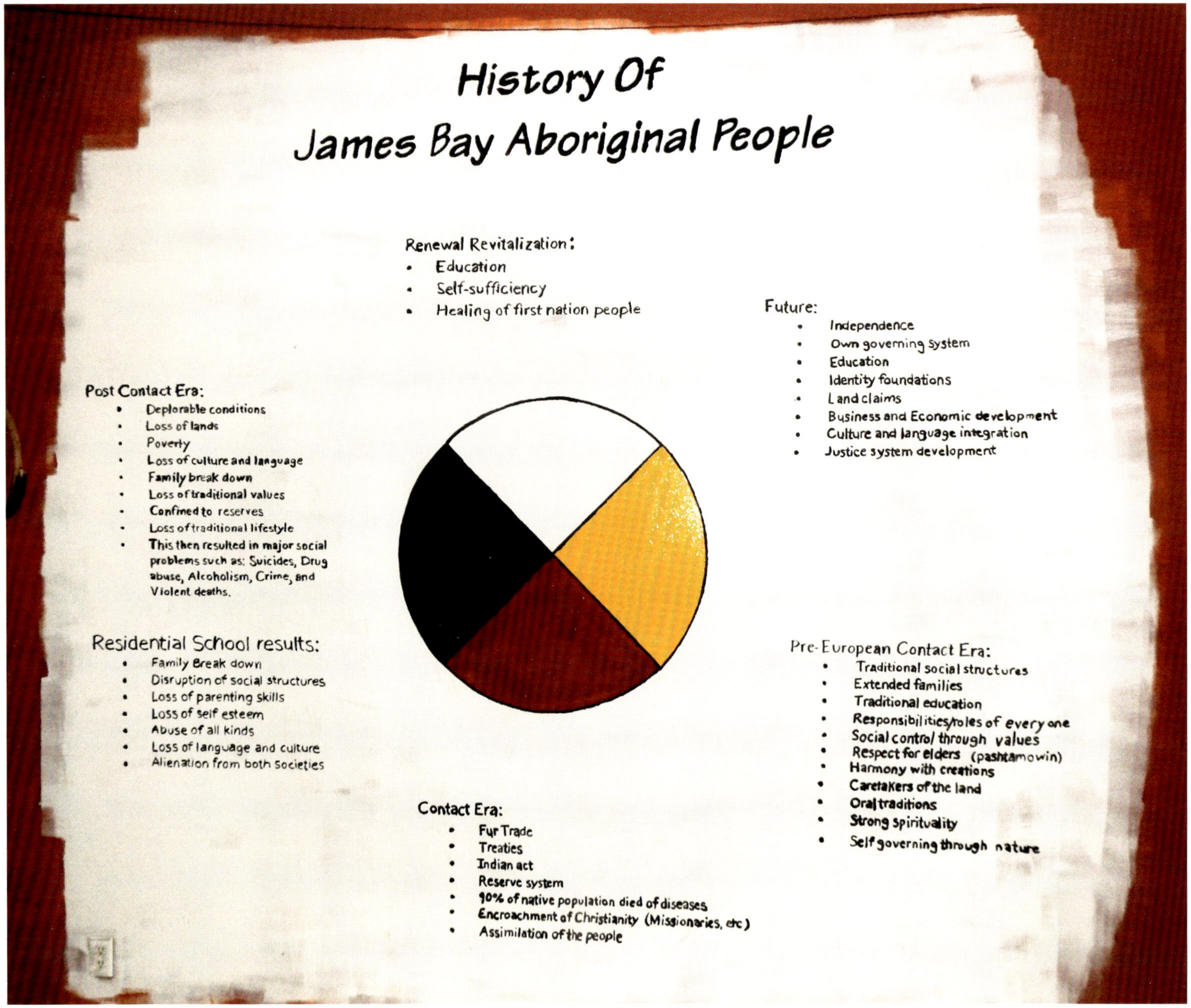

and helpful in discussing possible future expeditions after the forest fire barred us from Manitoba.

However, plans to involve First Nations and Inuit members directly in our expeditions or to work closely with them have most often been thwarted.

Time and funding were never sufficient to return to the Labrador coast with the Inuit artists we met in Nain. The 2011 forest fire quashed a feast with the Anishinaabe of Bloodvein River. We had planned to link up with Innu families coming to Kamestastin to hunt in 2008, but their schedule was dictated by the herds, and the caribou didn't cooperate. I was particularly keen to include Innu youth on the crews in Quebec and Labrador since I was told that the young men would consider it a point of pride to carry the heavy gear over portages. More to the point, in discussing supplies, I was told not to worry: "They will just bring a gun and a rod." I am sure we would have all learned a lot from each other, but that didn't work out either.

These setbacks on the art expeditions were disappointing because the boreal forest and tundra aren't all about wildlife and ecology. Meeting Indigenous people, tramping traditional portage trails, camping at ancient campsites, and traveling with an anthropologist like Dr. Stephen Loring, have all expanded my vision to include the ongoing story of us humans in the forest.

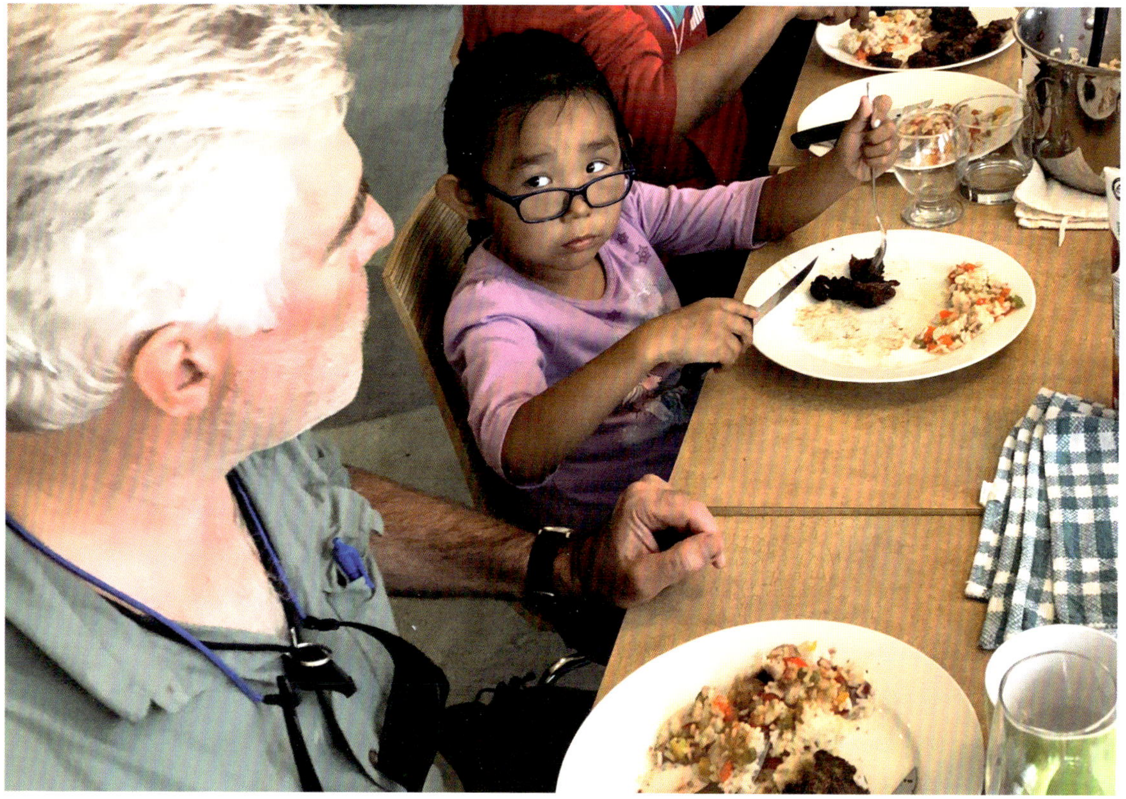

Making some kind of impression on the youth of Umiujaq at dinner. Robin Tapley

First Nations: Here first and longest

When you read about the latest environmental campaign to save this or stop the destruction of that, it is easy to assume that wilderness regions like the boreal forest are the primordial state of things and have just always been there until despoiled (or tamed, depending on your outlook) when humanity came along. However, as Dr. Steven Young mentioned in chapter 3, humans have been around far longer than our current boreal forest and have always been part of it.

The last glacial period of our current ice age was at its height until about 18,000 years ago, when glaciers covered much of what is now the North American and western and northern Eurasian boreal forest. When the ice age ended, the ice sheets didn't disappear like office workers at 5:00 p.m. They receded. Slowly. Lifetimes and millennia passed.

Note: From the beginning of the "end" of the ice age to when the ice retreated from West Bolton, Vermont, took 5,000 years—and even then, we still had a glacier in the backyard.

Our current Boreal "Wilderland" has existed at most for about 10,000 years. Anatomically modern humans (*Homo sapiens*) arose at least 200,000 years ago. Humans successfully expanded to Eurasia at least 50,000 years ago and arrived in the Americas at least 12,000 to 15,000 years ago. As the ice receded, plants, wildlife, and humans followed it northward. Humans were there as the boreal forest moved north from its refugia and took root in Canada, Alaska, and Eurasia. (In unglaciated regions of Siberia, humans were likely there long before the forest reestablished itself.) They still are, and despite the trials and tragedies of colonial expansion, both in Canada and Eurasia, modern Indigenous boreal forest cultures remain and maintain the hunter-gathering heritage we all share.

As Dr. Steven Young explained in chapter 3, most of what is now the eastern boreal forest in Eurasia was not glaciated during the latest "ice age," but was the mammoth steppe. Cold, but ice free, it could have provided tens of thousands more years for human habitation across Siberia than in western Europe or most of North America. The cultural diversity of Siberia bears that out. There are about five major Indigenous language groups across the Eurasian boreal forest and about 40 recognized Indigenous cultures in Siberia (an unknown number succumbed to the Russian conquest), including the Nenet, Yakut, Buryat, Dolgan, Even, Evenk, Selkups, Nganasan, Khanty, Mansi, Koryaks, Itelmens, Ainu, Kets, Yupik, and Chukchi (the last two shared with Alaska). In western Europe, there are the Saami in Norway, Sweden, Finland, and Russia. Just the Saami, but then the ice has been gone only for about 10,000 years.

There are many horrific atrocities within "trials and tragedies." While Canada forced Indigenous people into settlements and residential schools in a policy of assimilation, which the Truth and Reconciliation Commission of Canada has deemed a cultural genocide, the Russian conquest of Siberia (comparable to the "Indian Wars" the US waged) went straight to physical genocide. In 1742, Empress Elizabeth, generally considered one of Russia's more enlightened rulers, ordered her Cossacks to attack the Indigenous Siberian Chukchi and to erase their cultures and extirpate them through war. Reportedly, the Chukchi kept the Cossack major's head as a trophy (Major Pavlutskiy was famous for atrocities against Indigenous Siberians).

Across the boreal forest region in North America, the situation is similar: ice free for only about 10,000 years and with only two major Indigenous language groups between Alaska and Labrador: Athabascan and Algonquian (Inuit-Yupik, shared with Eurasia, alone spans the Arctic and abuts the northern fringe of the forest).

Ancient Indigenous inhabitants of the North American boreal forest have provided some of the most memorable and stirring encounters on our journeys. Smaller than grand vistas and more peaceful than charging bears, but in their own quiet way powerfully impressive, the pictographs left by fellow artist/canoeists hundreds or even thousands of years ago go beyond goosebumps. There is something about being in the presence of the actual thing that links you to the deep past (such as the *Mona Lisa*, a Minutemen musket, or a 40,000-year-old stone axe)—almost like they have a special power. There may yet be a bit of animism in us, at least among those of us who enjoy museums. It is difficult to know what inspired the pictograph artists: messages about hunting or war parties, offerings to the powers-that-be, or Stone Age Banksys, but they are captivating connections, and they are art.

Art is central to this book (along with nature and canoes). It may predate our species and is possibly as old as the

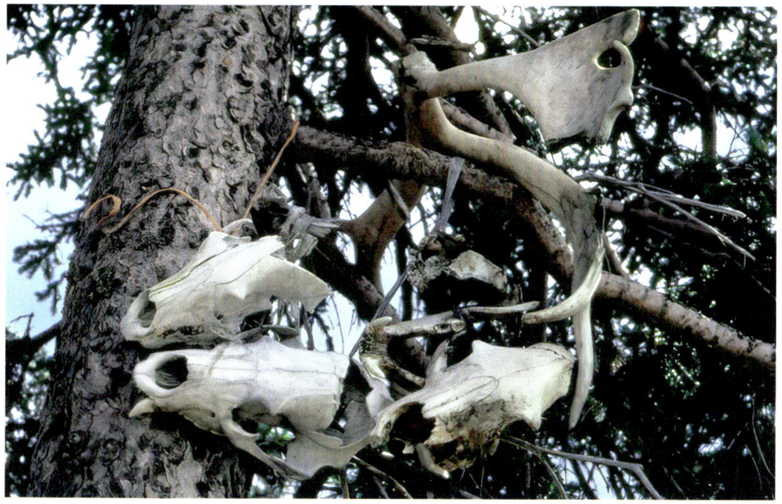

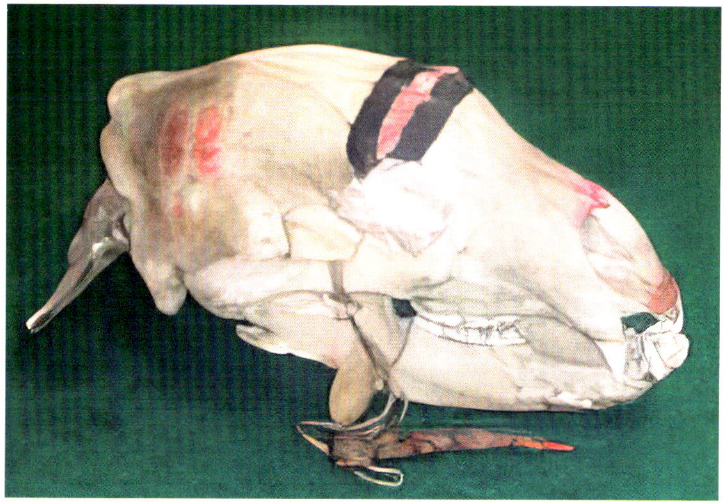

Art as sacred encounter: In many mediums, throughout the ages, people have sought to express their relationship with the divine. For the people of the forest, this relationship is clearly expressed in their treatment of animal remains. *Left:* Black bear skulls and caribou antlers placed in a tree at Joachim Nui's camp in Sango Bay, Labrador, to signal respect for the Animal Masters, 1982. Stephen Loring. *Right:* Painted black bear skull with plug-tobacco offering, collected by Rev. John Cooper, 1934, at the Cree community of Moose Factory, Ontario.

opposable thumb and a sense of self. It is universal among human cultures, from the Stone Age to the space age. Cultures of the boreal forest are no exception.

Art *in* the boreal forest

By Dr. Stephen Loring

Eighteen thousand years ago, the boreal forest of North America existed as a relatively thin band across mid-continent with narrow tendrils extending south along the higher elevations of the Appalachians in the east and the Rocky Mountains in the west. Northward, stretching to the continent's rim and the polar sea, was an impenetrable wall of ice thousands of feet thick. Slowly at first, but gradually faster and faster, an ameliorating climate set in, and as the ice withdrew, the forest advanced, superseding the shrub and tundra vegetation that appeared in the immediate wake of the glacier's retreat.

By 12,000 years ago, if not a little before, the icy mantle to North America was breached by intrepid bands of highly mobile hunters and gatherers whose cultural antecedents and skills had perfected an adaptation to life in the frozen forests of the North. From those first human footsteps to the present day, the forested reaches of northern Canada and adjacent Alaska have been a homeland for Algonquian and Athabaskan (Dene) peoples—Innu, Cree (Eeyou), Mi'kmaq, Ojibwe (Anishinaabe), Chipewyan, Tłı̨chǫ, Gwich'in, and Koyukon (Denaakk'e) among others—who have been and are an integral feature of the forest. Every aspect of their lives, language, demeanor, and disposition reflects the intimacy they share with their forested homeland. And, central to their identity is a profound relationship predicated on respect: respect for the land, respect for the creatures that inhabit it, and respect for the spirits and the inanimate agency that pervade everything in their (our) world.

An enduring tenet in the relationship between Indigenous hunters and their prey is that in exchange for their lives, and their gifts of meat, hide, bone, and antler, you honor their spirit (Art as sacred encounter, above). It is not so much that one is a clever or skilled hunter so much as a devout one. Put

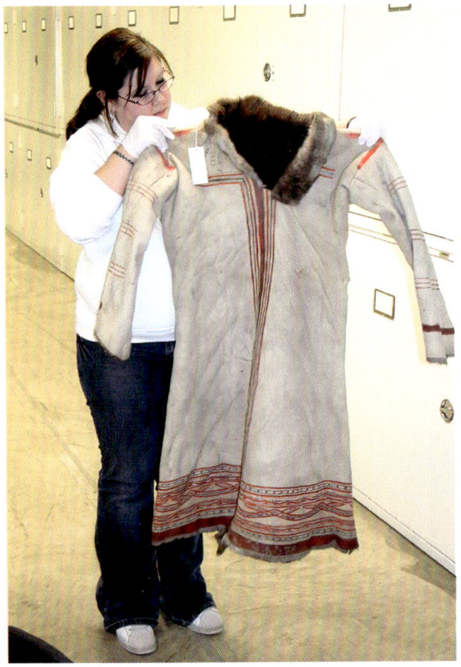
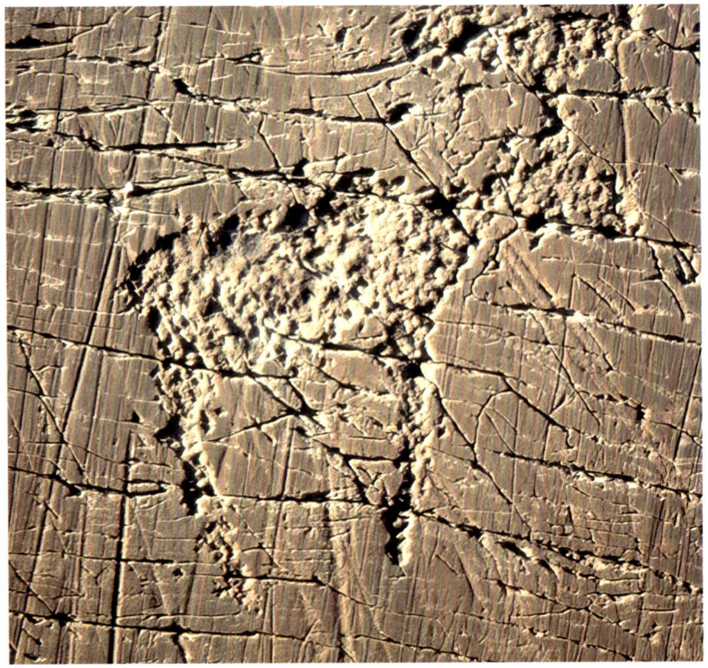
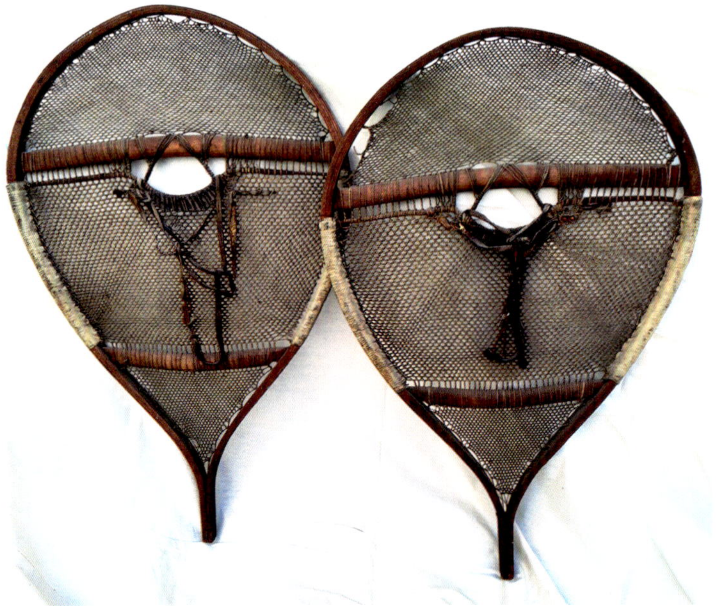
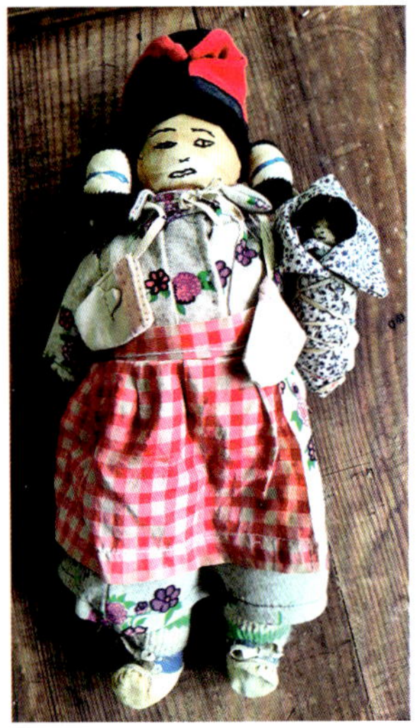
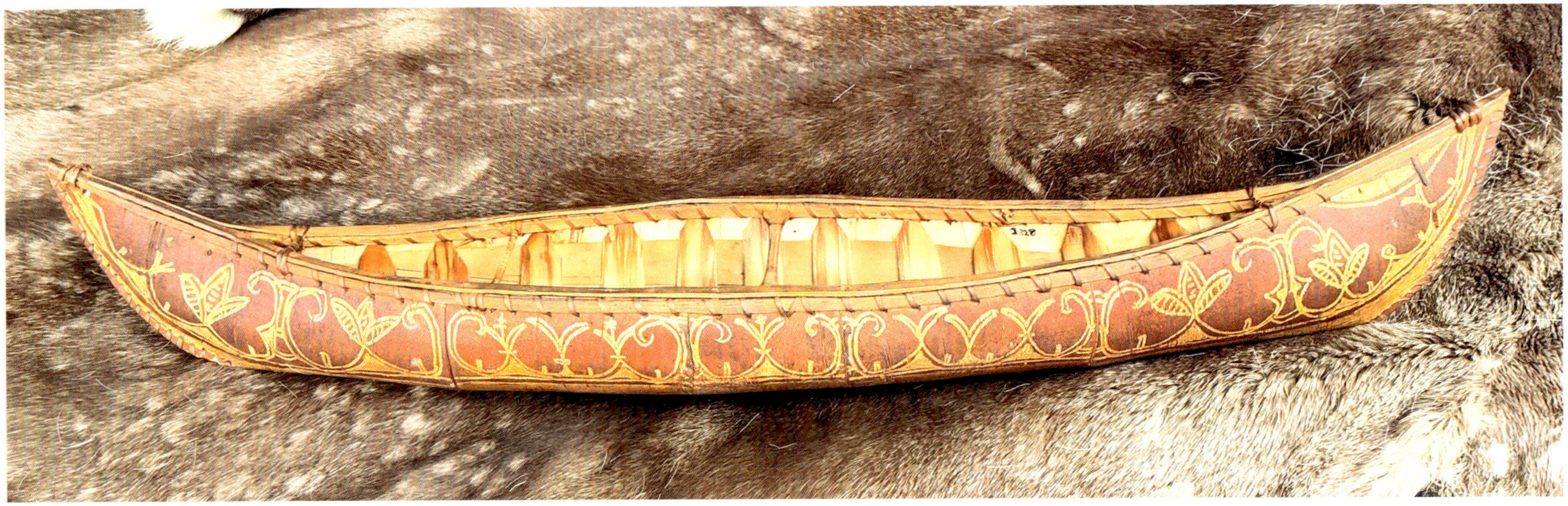

another way, as told to the anthropologist Barbara Bodenhorn by an Iñupiaq whaling captain, "I am not a good hunter, my wife is." Alluding to the premise that animals and their guarding spirits are aware of their pursuers and that upon seeing that their gifts—the skins, furs, and sinew—used to clothe the hunter have been appreciated and respected, they allow themselves to be captured and consumed (Art as fashion, left).

The art of the Indigenous Peoples of the boreal forest is the material expression of a language and a way-of-knowing that has emerged over the centuries, at one with the land and the animals and deeply at odds with the extractive exploitation and resource consumption of invasive government and industrial concerns. With the passing of the last *Tsheniu Mantushiu Kantuat* (the old Innu hunters with special powers) we are losing one of the last links we have to humanity's once universal hunting and gathering heritage.

Some of the earliest surviving artistic depictions of this worldview linking human beings with the animals of the forest is revealed in the Upper Paleolithic cave paintings in France and Spain from more than 30,000 years ago. And while rock art in North America is not nearly as old, it is on bedrock polished smooth by the glacial advances that we can glimpse some of the oldest surviving testaments of the intimacy and profundity of the relationship between human beings and animals in the boreal forest (Art for the ages, left, and *Old and New*, p. 247).

If, as suggested here, we recognize the primacy that respect for animals and the Animal Masters (the species-specific spiritual agency that governed movement and behavior) figured in the lives of boreal forest inhabitants, then might we look for its manifestation elsewhere? As a medium for artistic expression, most of the materials of the forest—skins, bone, bark, and wood—have not survived the passage of time. It is only stone tools that resist decay and disappearance to survive to the present day and testify to the antiquity and pervasiveness of human lives in the forest. These tools, including projectile points, spearheads, knives, and scrapers, are by dint of their function—the killing and processing of animals—a primary component linking the hunter with his prey and the physical world with the intangible.

Throughout the eastern reaches of the boreal forest, people went to extraordinary ends, journeying hundreds of miles to acquire specific stone—Mistassini quartzite in central Quebec, Ramah chert in northern Labrador—from which to fashion their tools (Art as artifact, above). This, despite the availability of suitable local lithic raw materials. Archaeologists have speculated that the choice of particular stones was predicated on the belief that such were linked to appeasing animal spirits that recognized some form of agency or power that was inherent in the stone's distinctive appearance. Some affirmation of this interpretation is based on the observation that the selected stone was a commodity of ritual exchange, being found in ceremonial contexts hundreds, even thousands, of miles from its source, as well as being featured prominently as grave goods that accompanied the deceased in the transition to an afterlife.

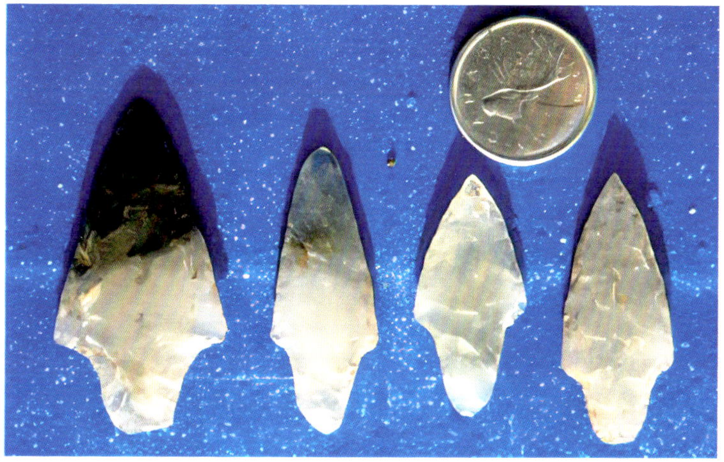

Art as artifact: Ramah chert spear points from Nulliak (left) circa 3800 BP (Before Present) and Kamestastin (right) circa 6000 BP, Labrador. The selection of certain raw materials for crafting stone tools was predicated on showing respect for the animals hunted, in this case caribou. Stephen Loring

FACING, CLOCKWISE FROM TOP LEFT:

Art as fashion: Jodie Ashini (Innu Nation) holds an Innu caribou skin coat made circa 1881, Smithsonian National Museum of Natural History—74457. Stephen Loring

Art for the ages: Middle or Late Woodland petroglyph of a moose, Machias Bay, Maine, 1999. Rick and Martha Jordan, courtesy of Dr. Arthur Spiess and the Maine Historic Preservation Commission

Art as craft I: Cree snowshoes acquired by William Brooks Cabot at Lake Mistassini, Quebec, circa 1899. Great esteem was placed on one's ability to walk far and fast. A hunter's snowshoes, intimately linked to the pursuit of game, formed a reference point that transversed the world of the hunter with a spiritual landscape. Stephen Loring

Art as craft II: Model of an Algonquian (probably Malecite) birchbark canoe, 1867, Smithsonian National Museum of Natural History—2328. Stephen Loring

Innu tea doll: Everyone helped carry supplies while traveling from the trading post. Girls' dolls would sometimes be stuffed with loose tea while at the post and replaced with moss at the winter camps. Stephen Loring

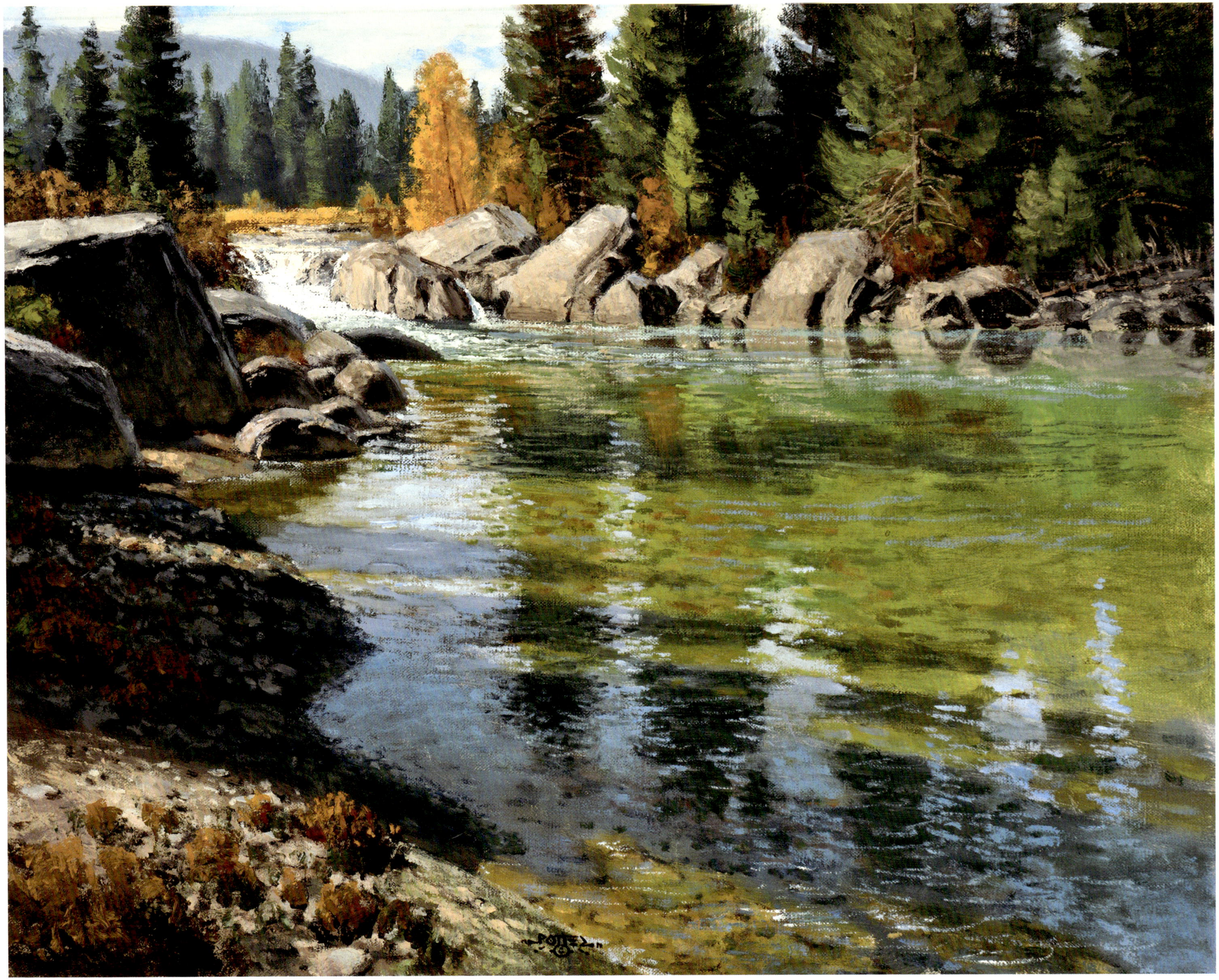

As attested to by the long-distance trade and exchange of lithic raw materials (and doubtless other organic products including bark, babiche, dried meat, and fish), mobility has always figured significantly in the success of boreal forest inhabitants. The ability to move far and fast—with snow-shoes and toboggans in winter (Innu tea doll, p. 244), canoes and pack-straps in summer (Art as craft, p. 244)—enabled access to huge territories in pursuit of game and facilitated a complex web of social interactions that belied the assumption of European trappers, traders, missionaries, and explorers that the land was essentially a wilderness (Art as craft II, p. 244).

There is no denying a precariousness to lives lived in the boreal forest, but the dreaded specter of starvation and cold as manifested by the cannibalistic *windigo* (*witiko*) with his heart of ice belies the comforts and securities that generations of Algonquian and Dene families have crafted from the gifts of the forest. Even as modernization brings roads, dams, and industry, and village life replaces the forest camps of their ancestors, Indigenous communities across the forested North retain a deep abiding attachment to their homelands and a responsibility to the celebration of unique lifeways predicated on pride and place, responsibility and respect.

While canvas and cloth have replaced skins for shelter and clothing, and manufactured guns and motors have replaced spears and paddles, the Algonquian and Athabaskan people of the forest retain a deep, abiding consciousness and commitment to the preservation (and celebration) of their homelands.

FOR A FEW MOMENTS IN 2001, as I imagined blending art, science, and adventure on the Missinaibi River to support conservation, I thought it was an original idea. Of course it wasn't. Thomas Moran's paintings from the Hayden Expedition of 1871 helped inspire Yellowstone National Park, and that's just one famous example. My notion of full immersion

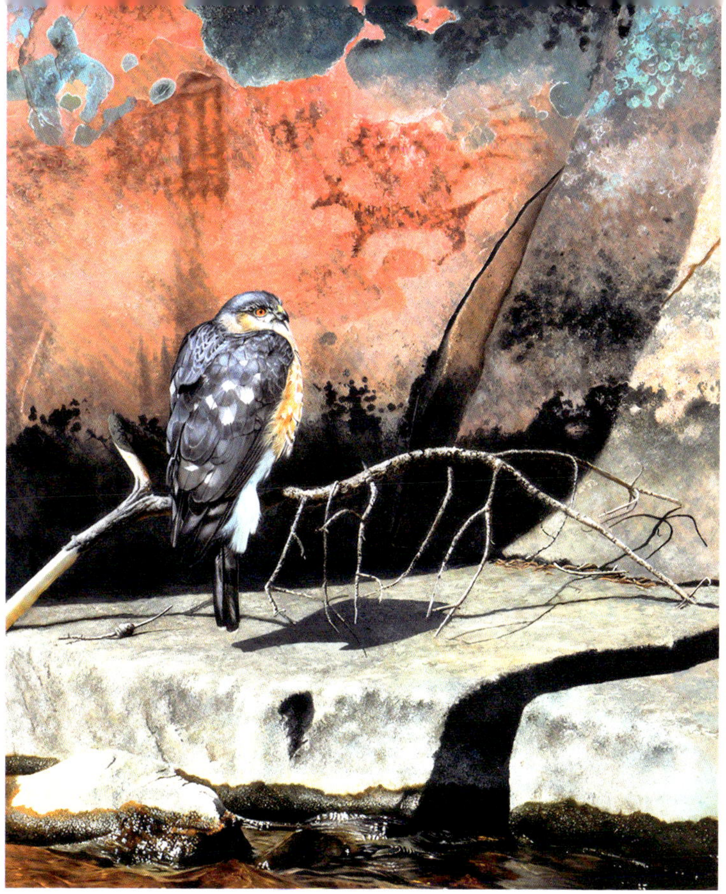

Goshawk and pictograph, Woodland Caribou Provincial Park, Ontario. *Old and New*, oil, Julia Hargreaves

FACING: *Nibi Manidoog* ("water spirit" in Ojibwe), oil, John Potter

in wilderness to shape one's art was simply a voluntary return to the norm before highways, SUVs, and modern cameras. Human creativity almost always builds on the work of others, a continuum stretching back to our hunting-gathering ancestors and forward alongside their Indigenous contemporaries, producing new expressions like Norval Morrisseau's Woodland School or Ojibwe WREAF artist John Potter.

Initially I thought of our predecessors in wilderness art as the Canadian Group of Seven, Thomas Moran, Karl Bodmer, midshipman Robert Hood with Sir John Franklin's first expedition (1819–21), John James Audubon, and George Catlin. But, as Dr. Stephen Loring's essay so eloquently illustrates, art isn't limited to painting and drawing. Alongside it runs music, dance, carved tools, decorated clothing, and birchbark canoes

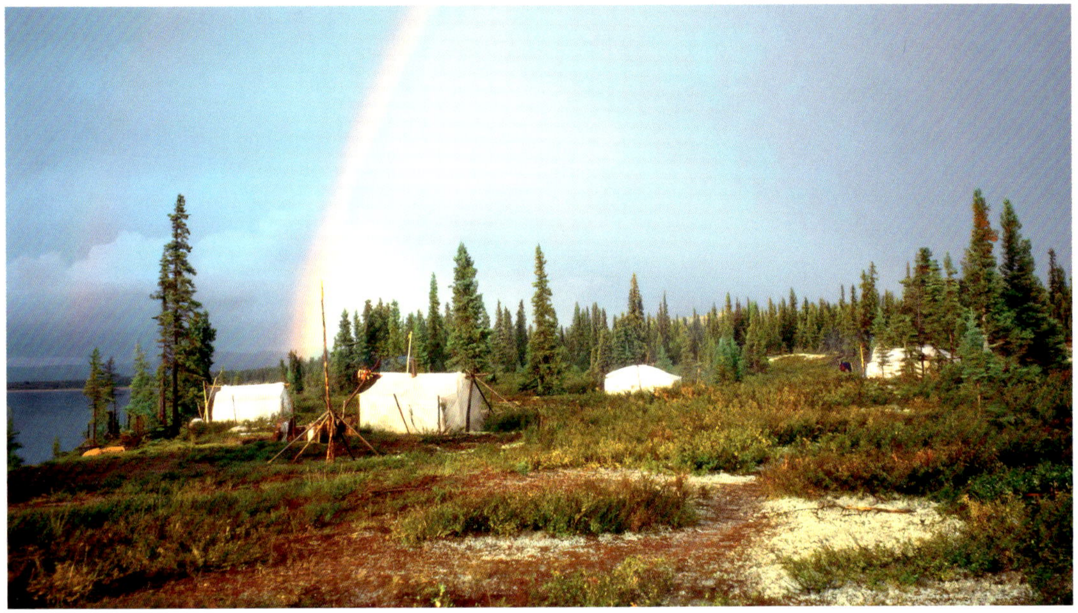

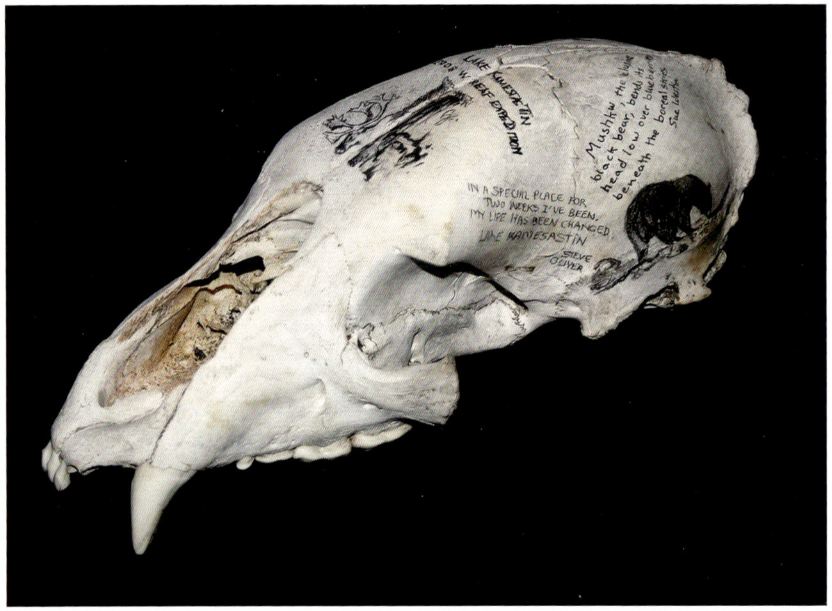

LEFT: Art as a way of life: Waiting for the caribou, Innu camp at Kamestastin, Labrador, 2000. Stephen Loring

RIGHT: Kamestastin black bear skull talisman. Rob Mullen

etched with intricate designs. Personal expressions of identity, reverence, and connection to the land can be found across cultures—just as surely as a First Nations hunter leaving his ocher handprint beside a painted animal on a cliff.

In a bit of a synthesis of older art with our modern styles, the 2008 Kamestastin crew all contributed to this talisman of our shared expedition experience. It is a black bear skull I found while on walkabout in the tundra.

Europeans

When Europeans first ventured to the New World, they mostly hugged the coasts, fishing and trading a few furs. Then began the "Age of Discovery." Of course, First Nations people had explored the interior thousands of years earlier; Europeans almost universally relied on their guidance. "I want to explore that region—show me the way" could have been the start of a Monty Python skit.

In 1534, Jacques Cartier was sent by King Francis I to find gold, spices, and a passage to India. Sailing up the St. Lawrence, he passed the Innu hunting site of Tadoussac, reached the Iroquois village of Stadacona (today Québec City), and then Hochelaga (modern Montreal), where the Lachine Rapids stopped him. Ignoring the locals' views, he claimed the land for France and returned with news of furs and fish from "Canada"—possibly from an Iroquois word for village. Others suggest it came from Portuguese sailors in Chaleur Bay saying *Cá nada* ("Here nothing"), though Canadians, like my wife, prefer the former.

Thirty years later, the English sought the Northwest Passage via Hudson Bay. They didn't find one (it would take global warming to create it), but they established a foothold for England, sparking two centuries of French-English fur trade rivalry. By 1600, European beavers were scarce and demand for Canadian pelts soared. The French, led by Samuel de Champlain, dominated early trade. In 1608, Champlain founded Québec City and made a key conceptual leap past his chauvinistic peers: To move inland, they had to adopt the Algonquian birchbark canoe.

In July–August 2023, we sailed our ketch *Skidbladnir* up Cartier's route from Tadoussac to just below Montreal. Even with charts, weather and positioning apps, and an engine to help the sails, fighting the current, tides, and headwinds earned Cartier a nod of respect. We also sailed Chaleur Bay—far from *Cá nada*, it was a highlight.

In 1659, Médard des Groseilliers and Pierre-Esprit Radisson returned from Lake Superior with a huge fur haul and a Cree tip: ship furs via Hudson Bay to avoid the Iroquois. When Quebec officials, possibly not pleased at the idea of their bypassing the St. Lawrence River, seized their cargo, des Groseilliers and Radisson defected to the English, helping found the Hudson's Bay Company in 1670 and founded Moose Fort (later Moose Factory, where our 2016 expedition ended) and other posts. Then they helped the French capture the posts. Then they helped the English get them back. The Wild West in the North.

By 1763, Britain had won Canada in the French and Indian War, freeing resources to push farther into beaver country. The fur trade became Canada's frontier epic, akin to the US West—only with birchbark canoes, snowshoes, voyageurs, and coureurs de bois instead of horses, Conestoga wagons, pioneers, and mountain men.

As the shortest route between the Great Lakes and James Bay, the Missinaibi was a major transport link for the Hudson's Bay Company for furs and supplies between posts on the ocean and the interior. However, it had also been a major route for Indigenous people for untold millennia.

But the romance masks two tragedies. Beavers, already gone from Europe, neared extinction in North America before silk hats replaced beaver felt (chapter 4). The ecological disruption from losing this keystone species is hard to fathom. The human toll was worse. Metal pots, needles, and firearms were irresistible but impossible to make locally, fostering dependence on European goods and fueling social upheaval. And then came smallpox.

Cultural and social erosion and even extinction (e.g., the Beothuk and the Erie) caused by the commodification of wildlife (trapping), dependence on European metal goods, cultural and religious assimilation, war, and the catastrophic ravages of smallpox devastated Indigenous people throughout the global boreal forest.

In North America, a cruel irony was that one of the most sublimely beautiful technological blends of organic form and function produced by any human culture—the birchbark canoe—helped spread these tragedies among the very people who perfected it and far into the heart of the continent.

Samuel de Champlain was a true Renaissance man: author, artist, cartographer, ethnologist, soldier, sailor, navigator, diplomat, promoter, adventurer, and visionary. He forged New France and died in the nascent city he founded on Christmas Day 1635. His canoe journeys were epic, and he excelled as an observer and chronicler. I fall asleep trying to make short notes in my journal on trips; Champlain wrote books based on his, and they can transport you to a wilderness river over 400 years ago. Highly recommended and available free online.

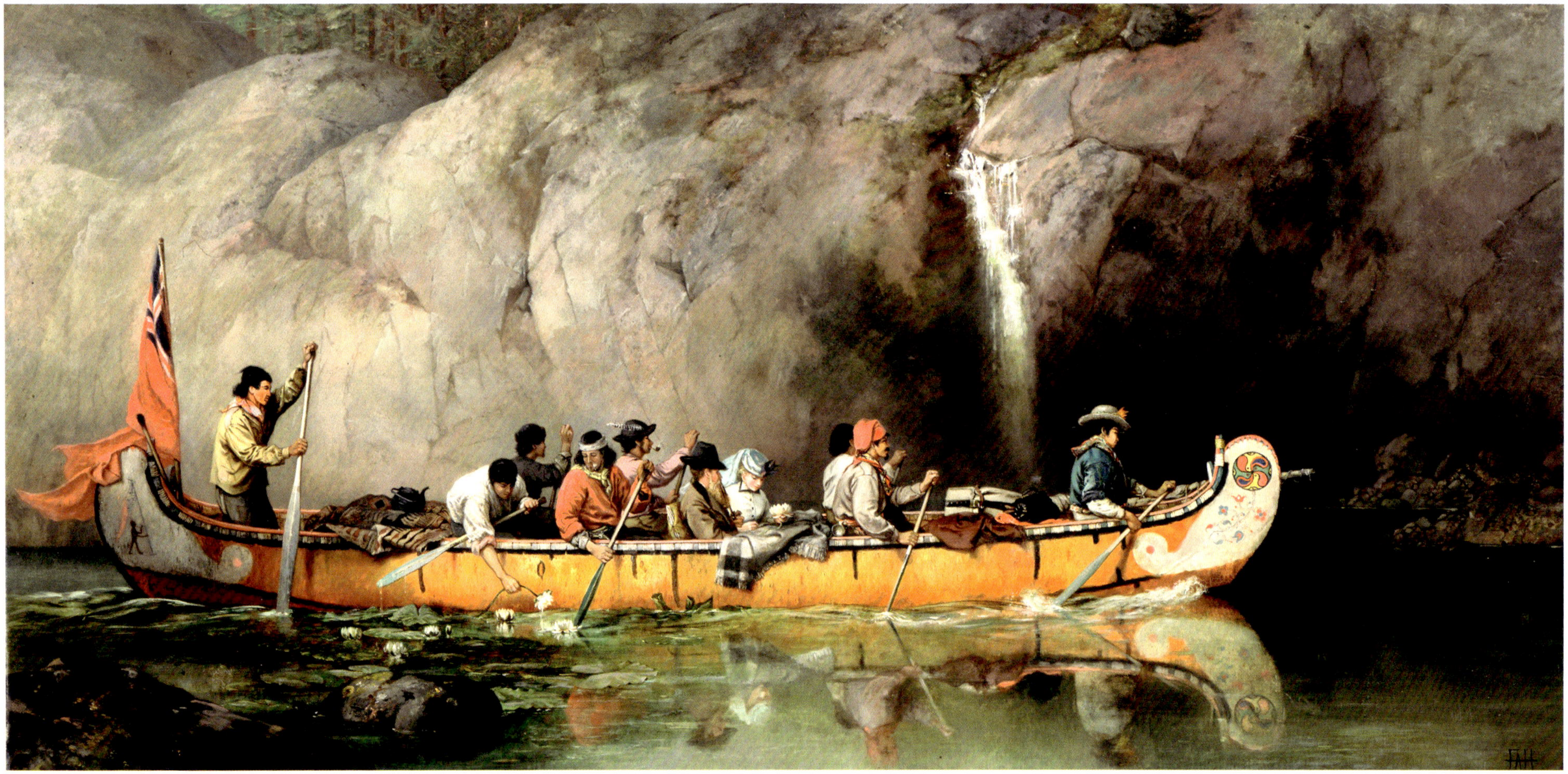

Canoe Manned by Voyageurs Passing a Waterfall, oil, 1869, Frances Anne Hopkins, public domain

The birchbark canoe of the boreal forest

Canoes are the essential backbone of this book, which is appropriate since they are also an essential element in the human story of the boreal forest.

A casual internet search for the origin of birchbark canoes almost invariably yields acres of results claiming they were invented by Algonquian First Nations in North America. The simplicity of dugouts and that the oldest preserved canoes are dugouts in Europe is superficially consistent with a North American origin of bark canoes. However, while birchbark canoes arguably reached their technological and artistic zenith among Algonquian boat builders, birchbark canoe technology almost certainly came east to the Americas from Eurasia.

First, there is direct evidence. Even though dugouts are much more likely to be preserved than birchbark, there are examples and historical accounts of birchbark canoes among many of the Indigenous cultures throughout the current boreal forest of northern Eurasia. They are not common, and not typically more than a couple of hundred years old, but there is supporting, indirect evidence pushing their use back millennia. Moreover, 8,700-year-old paddles and sheets of birchbark were found preserved in a bog in Sweden, strongly suggesting that the use of birchbark canoes among at least the Saami (and likely others) vastly predate the few known examples and historical accounts.

Also, there is compelling circumstantial evidence that the development of bark canoes would not have waited until humans arrived in eastern North America. First, as modern humans moved into the glacial and post-glacial mammoth steppe and the advancing boreal forest of northern Eurasia, glacial meltwater created a need for light, portable boats to travel interior waterways. Dugouts are heavy for portaging and running rapids, and finally, cold-adapted birch would have been increasingly common throughout the biome. Birch, the Tree of Life to the Anishinaabe, was used extensively for a wide variety of purposes, especially for waterproof storage and cooking vessels. I can imagine a young man, possibly with some latent engineering talent, who, just before leaving on a hunting foray, rescued one of his wife's birchbark cooking pots from floating away while she was getting water. Two days later, as he struggled to lift his grandfather's old dugout over a beaver dam, he muttered to himself, "There must be a better way."

As humans made their way across Eurasia and into the Americas, they would have brought their boats and boat-building technologies along. The roots of the birchbark canoe are almost certainly in Eurasia, but it is also quite likely that it was Algonquian canoe builders and the paper birch (*Betula papyrifera*) of North America that brought these wonderful craft to the pinnacle of design and utility.

Nothing could match them in the Northwoods, not the elm bark canoes of the Iroquois and certainly not the dugouts or York boats of the Europeans. Lightweight and portable, strong and capable of carrying heavy loads and maneuvering through rapids, and just gratifyingly beautiful to look at, they were engineering and artistic marvels. Like the rest of Indigenous culture, canoes were of the forest. There was no need to bring spare parts. If a boat was damaged, you stopped and fixed it. Everything needed was at hand.

As Europeans adopted the canoe and morphed into North Americans, canoes were made of wood and canvas, and then cedar strips sheathed in fiberglass. My first canoe was made of aluminum, and WREAF's modern expedition canoes are made of virtually indestructible plastic or tough synthetic fabrics

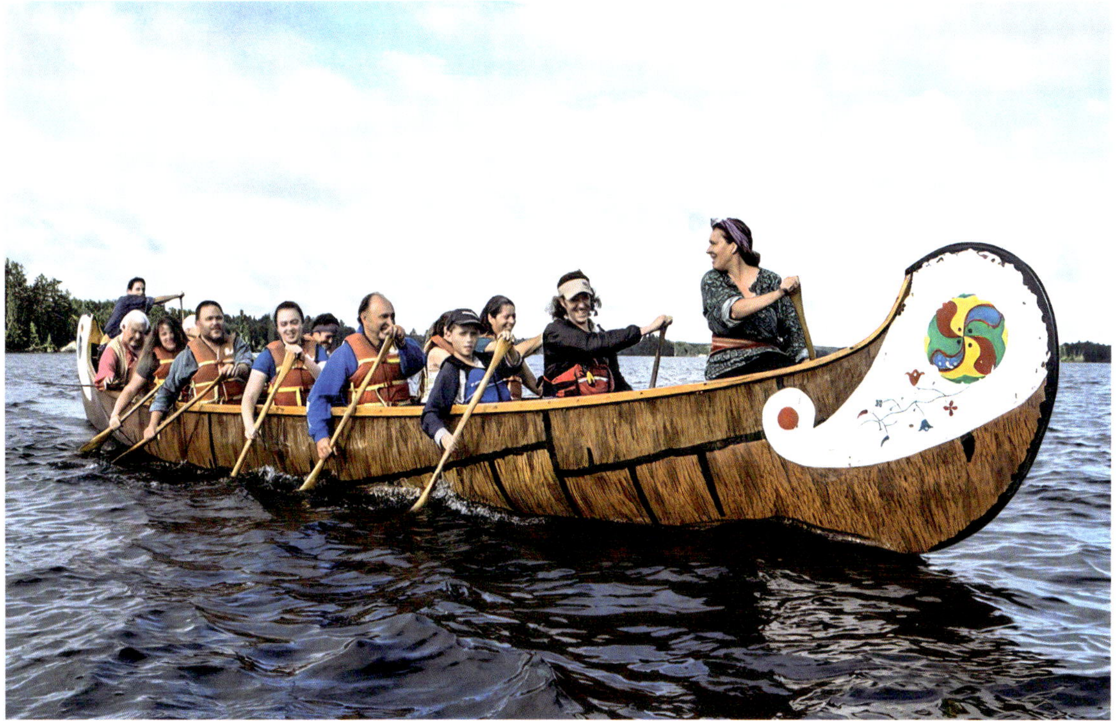

Replica fur trade canoe, 36 feet, Nym Lake, Ontario (Quetico Provincial Park border). The lean, long-locked fellow with the head-band visor near the bow is Scott MacGregor, founder of *Paddling Magazine* and Rapid Media. Steve Bruno

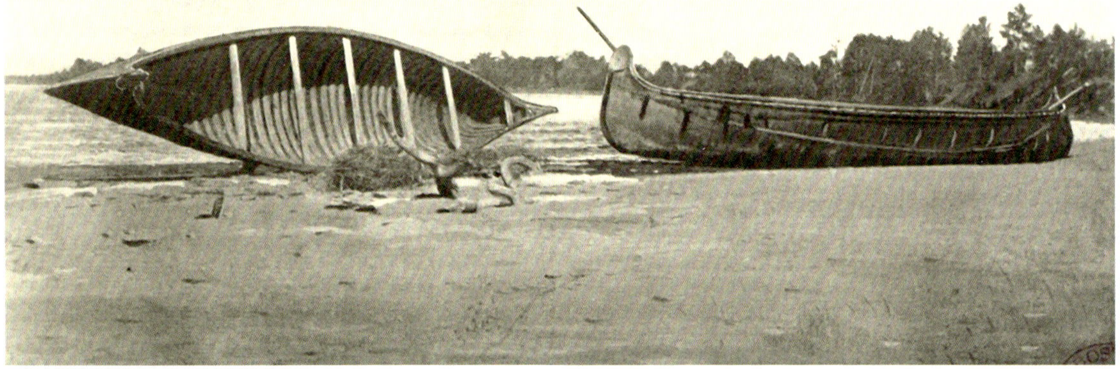

Ojibwe birchbark canoes. Annual report of the Bureau of American Ethnology to the Secretary of the Smithsonian Institution.

(folding PakCanoes). Our utilitarian boats lack the craftsmanship of an antique wood and canvas, and wood and canvas lack the organic grace of a birchbark, but there is one shared attribute that, however attenuated with modern building techniques, still shines through when afloat. Canoes are almost alive—and still the best way to travel in the boreal forest outside of freeze-up.

For a comprehensive overview of North American canoes, read *The Bark Canoes and Skin Boats of North America* by Edwin Tappan Adney and Howard Irving Chappelle. It was an inspiration for Harri Luukkanen and Dr. William Fitzhugh's book on Eurasian watercraft.

The following account of Dr. Fitzhugh's experiences canoeing the Canadian north as a teenager at the famed Keewaydin Temagami camp, how they led to his career as a Smithsonian anthropologist, and, coming full circle, an abridged ethnography of the Anishinaabe of Lake Temagami, offers insights about this land and the people of it that might be the germ of another book.

Note: It was mention of Canadian canoeing being a basis for his career in his bio that led me to call Dr. Fitzhugh at the Arctic Studies Center in 2004. That phone call led to this book. It just took 22 years.

The Teme-Augama Anishnabai canoe: A vessel of knowledge

By Dr. William W. Fitzhugh

My Introduction to the Boreal

A fair wind from the northwest must have been blowing when, at age 15, I met Dartmouth College's Howard Chivers, an Olympic skier, a founder and the first manager of the Dartmouth Skiway, and director of Keewaydin Temagami, then a boys' camp on Lake Temagami, Ontario. At the boundary between the two biomes, Keewaydin has been helping adolescent adventurers from the south discover themselves and their capabilities in an unfamiliar environment: the northern hardwood and boreal forest. In my case, apart from the physicality of canoe tripping, this also meant meeting the Indigenous people who have lived there for thousands of years. J-stroking, axe-bearing, tumpline-strapping canoe guides like Elie Seymond introduced me to ways of living and knowing during my introduction to the Canadian bush.

This part of the boreal, which borders the mixed northern hardwood forest, is one of its most elegant landscapes. Glaciers scoured the lake's 12,000 islands and more than 1,000 kilometers of shoreline to their geological roots. The landscape gleams with sparkling granite outcrops smoothed and contoured by eons of glaciation into rolling curves that emerge like whales' backs at sea. Onshore, red and white pines, paper birch, maples, and fir perch on bedrock, their roots emerging from the topsoil like sunburned knuckles.

At the ends of their lives, the pines become roosts for woodpeckers, eagles, and ravens. Bears and moose prowl the muskeg swamps along with lynx, fox, porcupine, and scores of other species. Small headwater streams and bogs are havens for beavers, muskrats, and fishing birds, while on the lake, ducks and geese crowd watery landing strips, and the eerie cry of red-throated loons echoes from the hills.

The Keewaydin experience: Anishinaabe heritage

Into this landscape stepped I, taking the bow seat for sternman Peyton Conger, the staff man of our 14-person Keewaydin canoe section. Ours was one of several Wabeno tripping sections for 15- and 16-year-olds. Our section had seven canoes, with the more experienced second-year campers in the stern. Ritual governed the portage process: the bowman carried the food wanigan (from the Ojibwe *waanikaan* for storage pit), a wooden box, to a halfway point and returned for the personal gear, packed in duffels lashed together by a leather tumpline with a broad head-strap, and carried it to the portage end. Meanwhile, the sternman carried the canoe, rigged with a tumpline and paddles lashed to the thwarts, and returned to the halfway point for the wanigan. Determining "halfway" was often contentious and usually favored the bowman. On the lake, protocol had the staff at the tail end of the section and the guide, responsible for setting the course and selecting campsites, in the lead. On the water, Elie was like a human outboard, his body silhouetted against the horizon, paddle barely moving. Only his axe, standing behind his seat, wobbled side-to-side with each stroke.

Elie was a master woodsman from Mattawa, whose residents included Cree, Ojibwe, Teme-Augama Anishnabai, and other First Nations, and Métis. He was around 40 years old and understood English but rarely spoke. If you asked a question, the answer was usually a grunt accompanied by a nod or headshake. After we arrived at camp, while we were setting up our tents, Elie gathered the first batch of wood, made the fire, and set up the fire irons (pipes with flattened ends). His canoe carried the "jewelry": the wanigan for kitchen hardware. Campers did the cooking, washing, and drying. While the cook and the potato peelers were preparing dinner, the rest of us, after checking for other canoes, jumped naked into the lake. Wabun, then a neighboring camp for girls, traveled in red canoes, while Keewaydin's were green. We frequently found traces of red paint on the rocks at campsites and pull-out places marking a Wabun passage.

The knife was used with the thumb on the butt of the handle, cutting toward oneself.

It seemed to me that Elie spent much of his time ashore sharpening his axe. His was a light three-quarter length one and was clearly a prize possession, along with his "crooked" knife. With just these two tools he could make a canoe, bowls, spoons, ladles, curly fire-starter splints, and more. The crooked knife replaced prehistoric stone scrapers and had a bent tip that could carve concave surfaces. Indigenous people made them from worn-out files, which they annealed, bent, and sharpened to a razor edge. Elie kept his axe and knife in canvas sheaths. No one was allowed to touch them. Staff carried a similar axe, and the trusted among us used it for wooding.

I got pretty friendly with Elie that summer. One time he took me to an old pine snag at the edge of camp. He swung the axe underhand, blade out, and in a looping throw lodged it into the trunk 20 feet off the ground. Pretty neat, I thought, and then wondered how he'd get it down. He left me to figure that out!

This may have been the last summer Elie guided for Keewaydin, as by then most other First Nation, Métis, and seasoned Canadian guides left and the camp began hiring "homebrews" (former campers). In 1962, I was signed on as guide for a Wabeno section. I worked closely with my staff on provisioning and routes and got to carry an axe behind my stern seat. Things got off to a rocky start when I misidentified a portage out of Obabika Lake into Florence Lake, on a circuit called Trout Streams. Whacking my way through dense water grass, I found a trail that led to a stream. After a couple days working up this boulder-filled creek without seeing green or red paint, I got concerned. Our canoes were suffering boulder fatigue, and one was so damaged I had to carve a new rib from an aspen sapling and patch the canvas with a shirt sleeve soaked in Ambroid glue. Next day we retraced our steps to Obabika and found the correct portage—this one with red and green scrapes. When we got back to base camp I had some explaining to do. I sure missed a seasoned guide on that trip! Afterwards, I often thought how lucky I and hundreds of kids were to have known and learned from Keewaydin's original guides.

Many years later when our sons Ben and Josh were Keewaydin campers, my wife Lynne and I brought our own canoes to Temagami and stayed overnight at Camp Ojibway at the north end of Devil Island. Ojibway is an adult camp frequented by campers' parents that was managed by Sandy Chivers. After the last day of camp, we collected the boys and set out on a trip to Lady Evelyn Lake. On the second day out, Ben cut his leg with an axe, and I had to canoe-ambulance him to Keewaydin, where he was patched up and we returned to the bush. That summer I showed Ben some of the Temagami's old Indian rock paintings. Those Keewaydin summers and others in Labrador and Kodiak, Alaska, helped him, like me, choose archaeology as a career.

Boreal connections

Boreal ecology and geography have influenced the movement of peoples, technologies, ideas, and cultures. From 1897 to 1903, Franz Boas investigated cultural similarities among coastal cultures from Japan to British Columbia, including raven mythology, skin boats, harpoon technology, art forms, and oil lamps. Alfred Irving Hallowell pursued links in bear ceremonialism across the circumboreal forest, and more recently the Russian anthropologist and folklorist Yuri Berezkin catalogued similarities in folklore and mythology among the peoples of Siberia and North America.

The Smithsonian's *Crossroads of Continents: Cultures of Siberia and Alaska* exhibition, organized with the Soviet Academy of Sciences in 1988, explored cultural connections across the Bering Strait. Some trans-Beringian similarities are easier to explain than others. Sea mammals surviving wounds, for instance, can carry embedded harpoons hundreds or thousands of miles, providing distant peoples with ideas about new technology. Fourteenth-century Japanese junks disabled by storms have been carried by circum-Pacific currents and winds, bringing iron tools to the mouth of the Columbia River, where they have been found in Indigenous villages. Amur River birchbark canoes display shape and construction similarities with southern Northwest Coast tribes, suggesting canoe-builders from Siberia crossed into North America.

There are many questions about connections through the global and North American boreal zone for which we have no answers. Similarities in canoes, garments, and personal decoration may be especially promising, along with the more definitive studies emerging from language and genetics. One thing is certain: The fluid nature of small-scale, highly mobile boreal societies fueled by subsistence, and social and economic necessity provided conditions that promoted long-distance contacts and exchange of ideas, materials, languages, and genes.

Boreal deep time

These types of similarities among different peoples raise hopes that a more specific history of connections and developments may emerge from future studies. Archaeologists of arctic North America have been able to track specific cultural histories over four millennia, including migrations of several different Inuit groups. Similar movements must have taken place in the boreal zone of North America. The widespread distribution of Algonquian languages suggests their ancestors descended from Eastern Paleo-Indians, who moved north and east with the receding Laurentide ice sheet. Linguistic studies suggest that Athapascans may have been a second wave of Asian immigrants, arriving in America via the Bering Strait 8,000 to 10,000 years ago.

However, unlike resource-rich coastal regions and areas where ceramics and agriculture developed, providing diverse cultural markers, the precontact cultures of the boreal forest, throughout its entire 10,000-year history, have remained resistant to archaeological definition due to small populations, low cultural profiles, highly nomadic settlement patterns, and more fluid cultural boundaries. To these impediments we must add the physical challenges of extensive forest cover, thin and corrosive soils, and a rather desultory history of archeological research. Nonetheless, the research fairly well demonstrates a prehistoric way of life seen among living boreal forest peoples at the time of European arrival. This boreal culture was honed over thousands of years by peoples who perfected an adaptation, using minimal technology, that enabled survival of their cultures and languages and promoted values of sharing and respect for human neighbors, animals, and the lands in which they all lived.

First Canadians: A Temagami profile*

The Teme-Augama Anishnabai (People of the Deep Water by the Shore) live on Bear Island in the center of star-shaped Lake Temagami, whose branching narrow bays connect with hundreds of other lakes, ponds, and waterways. The community numbers about 800 people, who speak a Central Algonquin language. The larger family of Algonquian speakers occupy forested Canada from Manitoba to Newfoundland, and from the Great Lakes to Hudson Bay and Nunavik.

Among these Indigenous Peoples are the Western and Eastern Cree; Nipissing; Ojibwe; Ottawa; the Innu of Labrador and Quebec; the Mi'kmaq of Nova Scotia, New Brunswick, and Newfoundland; and the Beothuk of Newfoundland (extinct since the 1830s). Other Algonquians live in Northeastern and North Central United States. Other than Inuit, theirs in the largest geographic spread of any language family in the Americas. They are part of the group known as "First Nations" in Canada and are the prime citizens of the northern forests where they have lived for thousands of years as "keepers of the game."

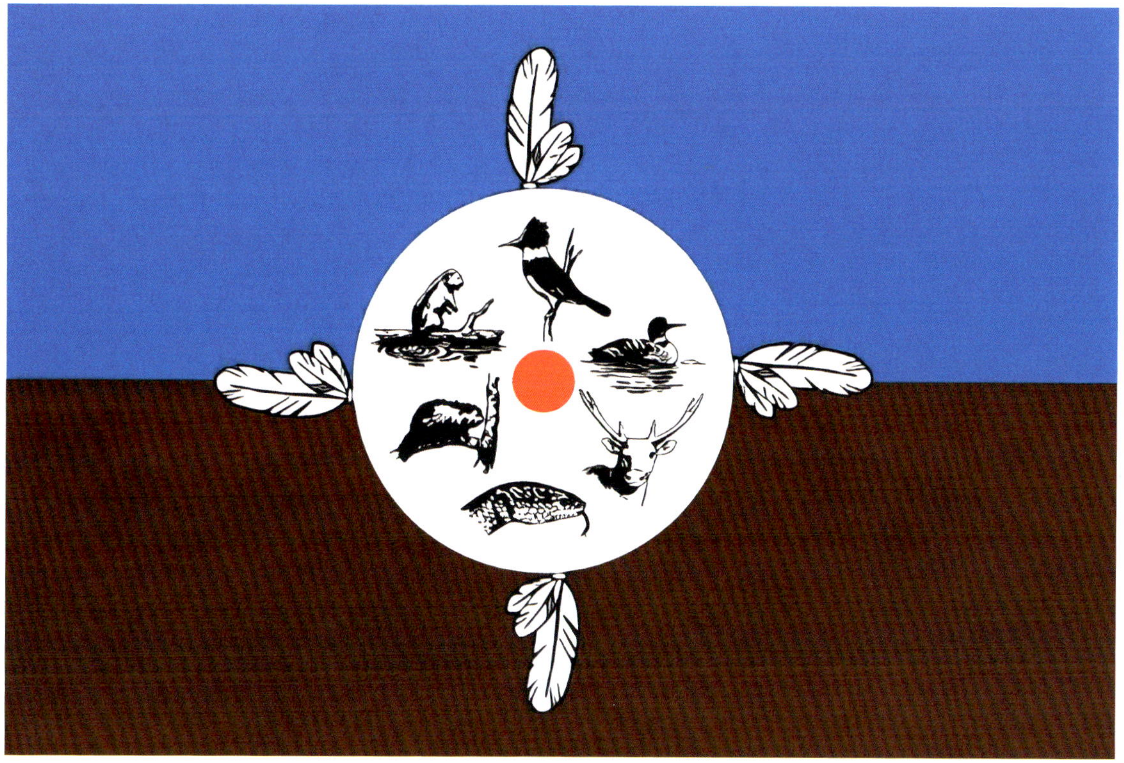

Temagami First Nation emblem. Courtesy of Teme-Augama Anishnabai First Nation

* The full text and photos of this abridged introduction by Dr. William W. Fitzhugh can be found in the online appendix at robmullen.com/boreal-wilderland.

According to the belief of the Teme-Augama Anishnabai, the Creator placed them at Ishpatina Ridge, in N'dakimenan (Our Land), at least 8,000 years ago. They speak a dialect of Central Algonquian and are culturally affiliated with the Ojibwe.

They lived a mobile life in small groups, in tents and insulated winter huts, as a hunting, trapping, and fishing people, making use of the wide variety—but sometimes scarce—game, and gathered plants and berries. The traditional technology was relatively simple for most Algonquian groups: bows, arrows, spears, canoes, toboggans, and traps. Tools and utensils were made of wood, bark, stone, bone, and antler. Clothing was tanned moose or deer hide, often decorated in striking patterns with dyed porcupine quill and later with glass beads. Sinew served as thread, and animal fur embroidery provided highlights.

Like other peoples in the northern forest, the Anishnabai had a social structure that was flexible and based on small family units. During winter when food was scarce, people scattered, living on local game and cached food. Bands of 40 to 80 people came together during the more bountiful summer season. Periodic animal crashes and forest fires required families and bands to maintain long-distance relationships with neighboring groups with whom they exchanged spouses, traded in exotic and needed materials, and shared strategic survival information. Individual hunters and whole families or bands could travel hundreds of kilometers in just a few days, carrying almost nothing but sleeping robes. Unlike in more temperate regions, warfare and inner-societal violence was rare and almost unheard of. The exception was confrontations with the agricultural Iroquois who raided Algonquian groups for their furs to sell to Europeans.

Mobility and unpredictable ecology were major factors accounting for the huge areas occupied by Algonquian and, in western North America, by Athabascan language families. Widespread distribution of mythology, shamanism, shaking tent ceremonies, bear rituals, and other religious concepts are other indicators of the cultural, social, and language connections the Anishnabai shared with other northern forest peoples. For more information about the Algonquin and Athabascan cultures of the boreal forest region, consult the *Handbook of North American Indians Vol. 6: Subarctic* edited by June Helm.

As with many smaller First Nations communities in Canada, the Teme-Augama Anishnabai do not hold title to the land they live on, even though their ancestors have been there for thousands of years. The coming of 18th-century fur traders and trappers, followed by 19th- and 20th-century miners, timber harvesters, and road and dam builders, saw the community gradually lose control of their lands and resources. On New Year's Day 1879, in a speech at the Hudson's Bay Company post on Bear Island, Ignace Tonené warned his people that "white men [are] coming closer and closer every year, and the deer and fur is becoming scarcer and scarcer... so that in a few years more Indians could not live by hunting alone."

Heritage revival

Facing massive changes in contemporary life, including language loss, cultural erosion, and the ongoing struggle for land claims and traditional rights, the Teme-Augama Anishnabai initiated programs to reinforce their traditions. They chose birchbark canoes as the vehicle for rejuvenation. In 2012, they conducted a birchbark canoe-build in partnership with Temagami's Wabun Camp. Robin Potts, who grew up on Bear Island and attended Wabun, helped organize a second build in 2017.

Canoe-making as a nexus for community building, cultural revival, and education has spread to other Anishinaabe First Nations.

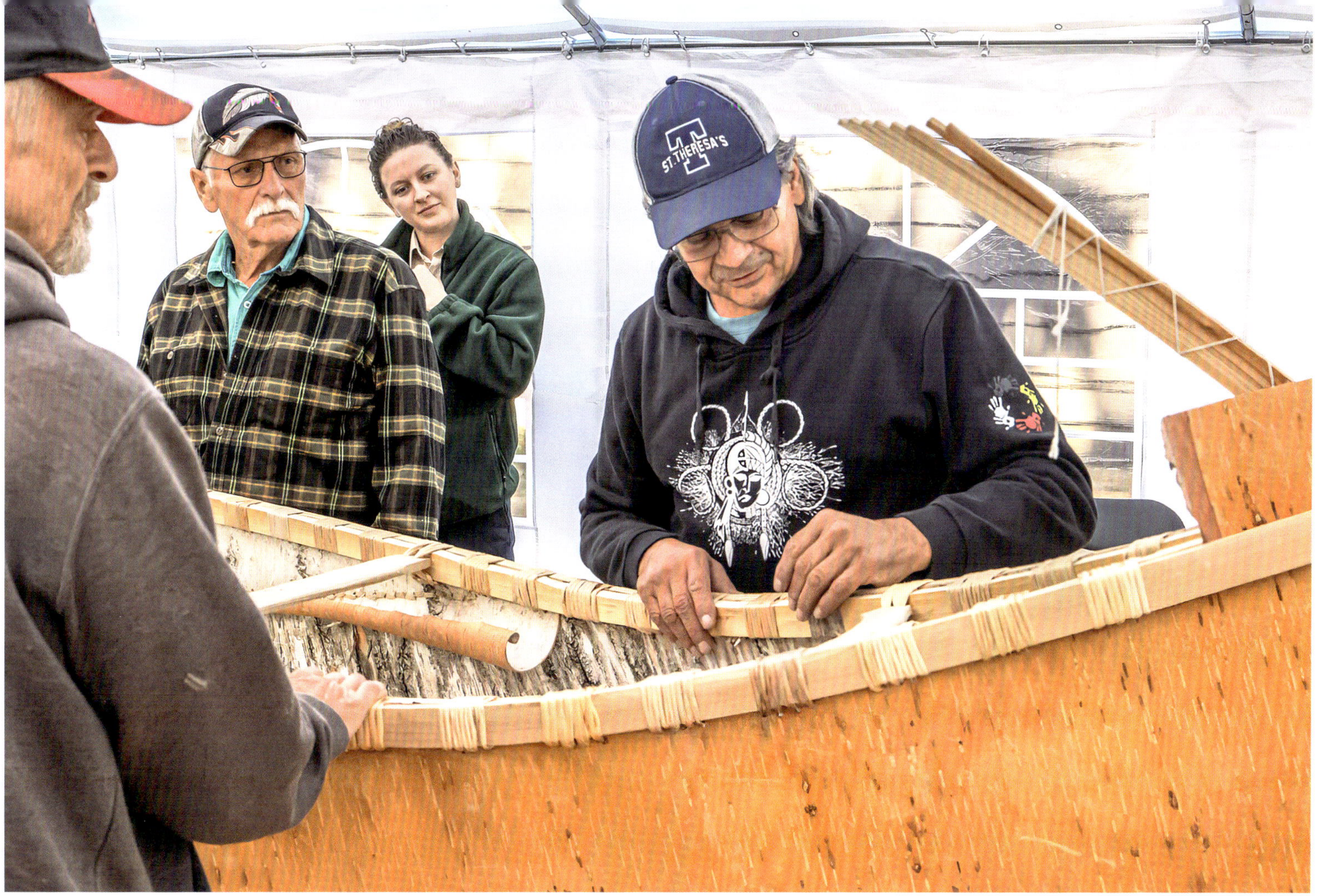

Birchbark canoe-making, Opeongo Lake, Algonquin Park. Steve Bruno

As the world becomes more complex, populated, and interconnected, straining our capacity to maintain respect and balance with each other and our environment, we would be wise to learn from the example of forest people who have been managing to do so for thousands of years. Toward this end, I am reminded of what I and many other young people learned from their youthful wilderness experiences and First Nations guides at Keewaydin and Wabun. These values are well exemplified by the Anishnabai canoe-builders, so well expressed by David Neel in words worth repeating:

Today in its renaissance the canoe carries the knowledge of a millennia-old culture as well as the dreams and aspirations of a younger generation. It is a vessel of knowledge, symbolizing the cultural regeneration of many nations as they struggle to retain and rebuild following a period of systematic oppression and of rapid social and technological change. The great canoe has come back from the abyss a vital symbol for First Nations.

And for the rest of the world.

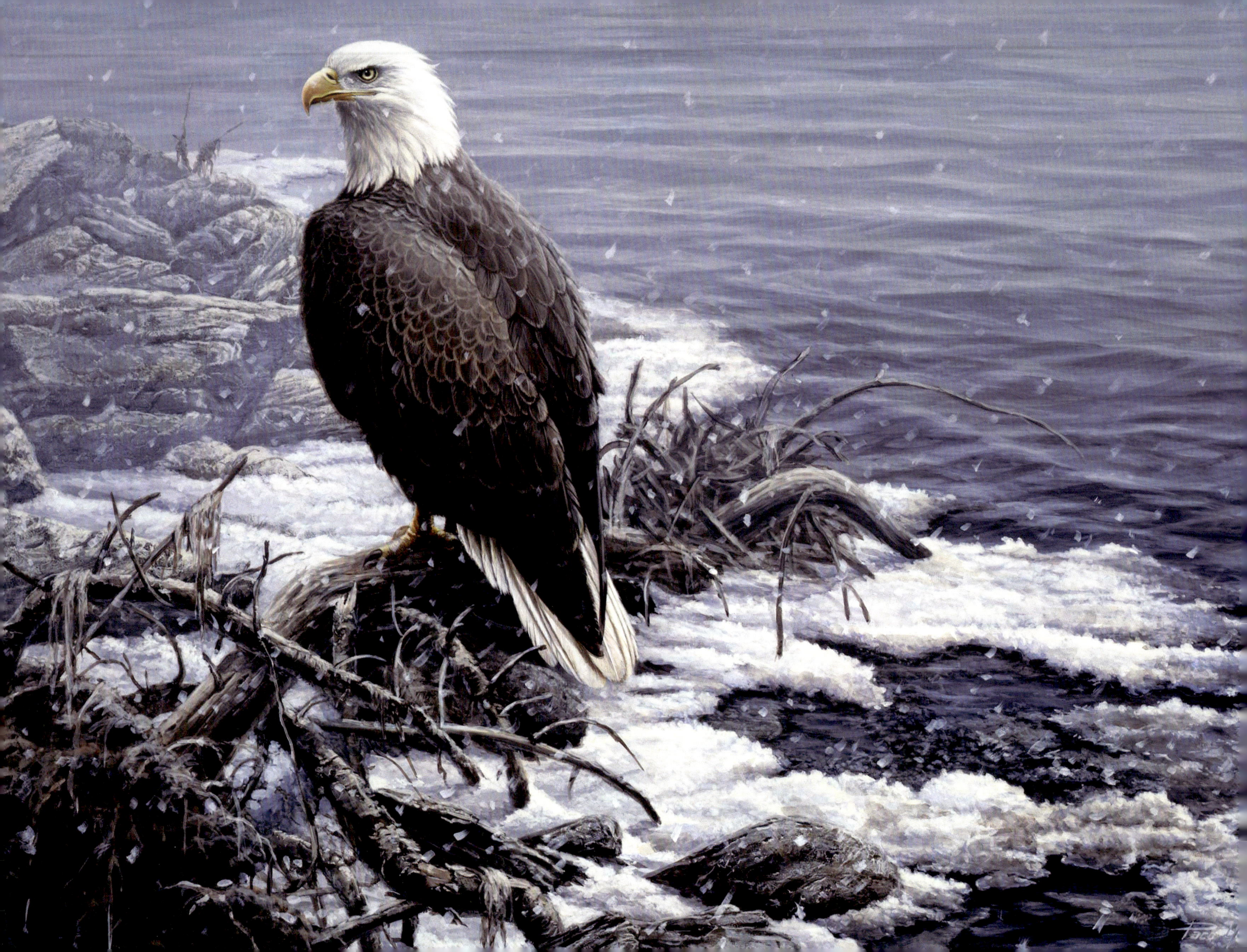

FACING: Bald eagle, Alaska. *Shoreline Flurries*, acrylic, Paco Young

8

Ecosystem Services: The Other Ecological Value

September 13, 2001: Headed home on the train south from Moose River Crossing, I tried to put worries of family and 9/11 aside. The scenery helped, although it was as flat as a lumpy pancake. On both sides of the track, endless bogs, dotted with lonely spikes and clusters of drunken black spruce, fascinated me with the allure of a strange and eerie landscape.

Those nation-sized bogs are one reason the boreal forest matters. Its value lies in its bigness, wetness, and wildness.

Bigness means the boreal operates on an immense scale, turning countless small events into global phenomena. Wetness is self-evident; water is life. Wildness provides both an ecological baseline for study and the possibility of conservation. But the largest wilderness on Earth is shrinking daily, under assault from destruction, fragmentation, and rapid climate change. A warming world is a perilous place for a "snow forest."

In a classic case of not seeing the forest for the trees, many people value the boreal forest only as timber, an industrial wastebasket, or wildlife-watching haven. Yet, the greater worth of any functioning ecosystem is the support it provides for life—its ecosystem services. Note: These are what large intact ecosystems provide on a global scale that major environmental groups in past decades ignored by focusing on small regions with high biodiversity (biodiversity hotspots), as mentioned in chapter 1. Both are critical and a balanced approach is necessary.

These services include oxygen, soil, and food production, pollination, seed dispersal, water filtration, nutrient cycling, waste decomposition, flood control, erosion prevention, pest control, population balance, carbon sequestration, climate buffering, and fisheries enrichment. Unlike home grocery delivery, these services are not optional—they keep the world livable. Even the smallest of them would bankrupt any human economy, yet natural ecosystems do them all for free—and the boreal forest is the largest single land-based provider.

In 2005, the Canadian Boreal Initiative report *Counting Canada's Natural Capital* estimated the annual value of these services at more than Can$700 billion, far exceeding timber, energy, and mining combined. Carbon sequestration was the largest single contributor.

Carbon

August 30, 2018: Rivière aux Mélèzes, Ungava (far Northern Quebec)

Somehow, the 200-pound cast-iron Dutch oven ended up in my canoe (maybe 10 pounds, but still). Fresh bread every night eased my suffering. We had to wait for the wood to burn down to coals—nearly pure carbon—which burn hotter and steadier than wood. One night, staring into the fire, it struck me: We were going through a complex process to transform carbon from trees and wheat into something edible. It would be so much easier if we could just eat the wood or charcoal.

Carbon gets bad press—"CO_2 pollution," "carbon footprints"—but it's the basic building block of all life. Bioavailable carbon comes as organic tissue (living or dead) or CO_2 in the atmosphere or ocean. Photosynthesis consumes CO_2, producing glucose and oxygen. Respiration—photosynthesis in reverse—uses glucose for energy, consuming oxygen and releasing carbon dioxide (or methane and other gases in anaerobic respiration). This exchange is the active loop of the carbon cycle. Slower, geologic loops operate on timescales so long they're effectively "inactive," with no effect on atmospheric CO_2 on a human timescale.

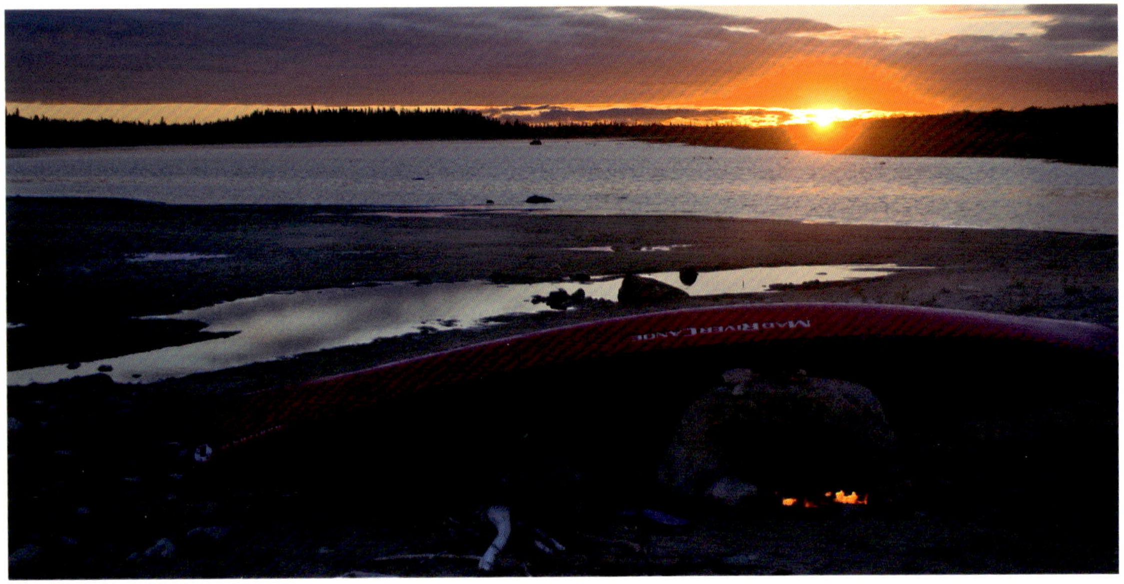

Fin Isle sunset with low campfire, Moose River below James Bay.

The greenhouse effect and greenhouse gases

Our atmosphere isn't just for breathing; it also regulates and distributes heat.

The moon's average temperature is an uncomfortable -10°F (-23°C). However, averages can be tricky, and if you plan to visit, a parka won't help. You'll boil at noon and freeze at midnight. On Earth, the atmosphere changes all that.

Our atmosphere lets sunlight in because no atmospheric gases absorb high-frequency visible light well. However, when that energy is absorbed as heat and radiated back toward space, it doesn't get a free pass like on the moon. Carbon dioxide and a few other gases do absorb low-frequency heat energy. So, as the heat makes a break for it, some is absorbed and reradiated out in all directions. This creates a heat reservoir that is spread around the globe, warming Earth to a balmy average of 59°F (15°C). That is the greenhouse effect, and carbon dioxide and company (e.g., methane, nitrous oxide, and even water) are the "greenhouse gases." They constitute less than 2 percent of the atmosphere yet warm the entire planet (oxygen and nitrogen absorb very little). They are so effective that small changes in their relative abundance can trigger disproportionately large changes in global temperature and climate. Because of its abundance and persistence, carbon dioxide is an important greenhouse gas.

Carbon and climate

Carbon dioxide is essential to life. Without it, Earth would be far colder; with more, it would likely get warmer—and warmer can be good. For most of history, the planet's climate

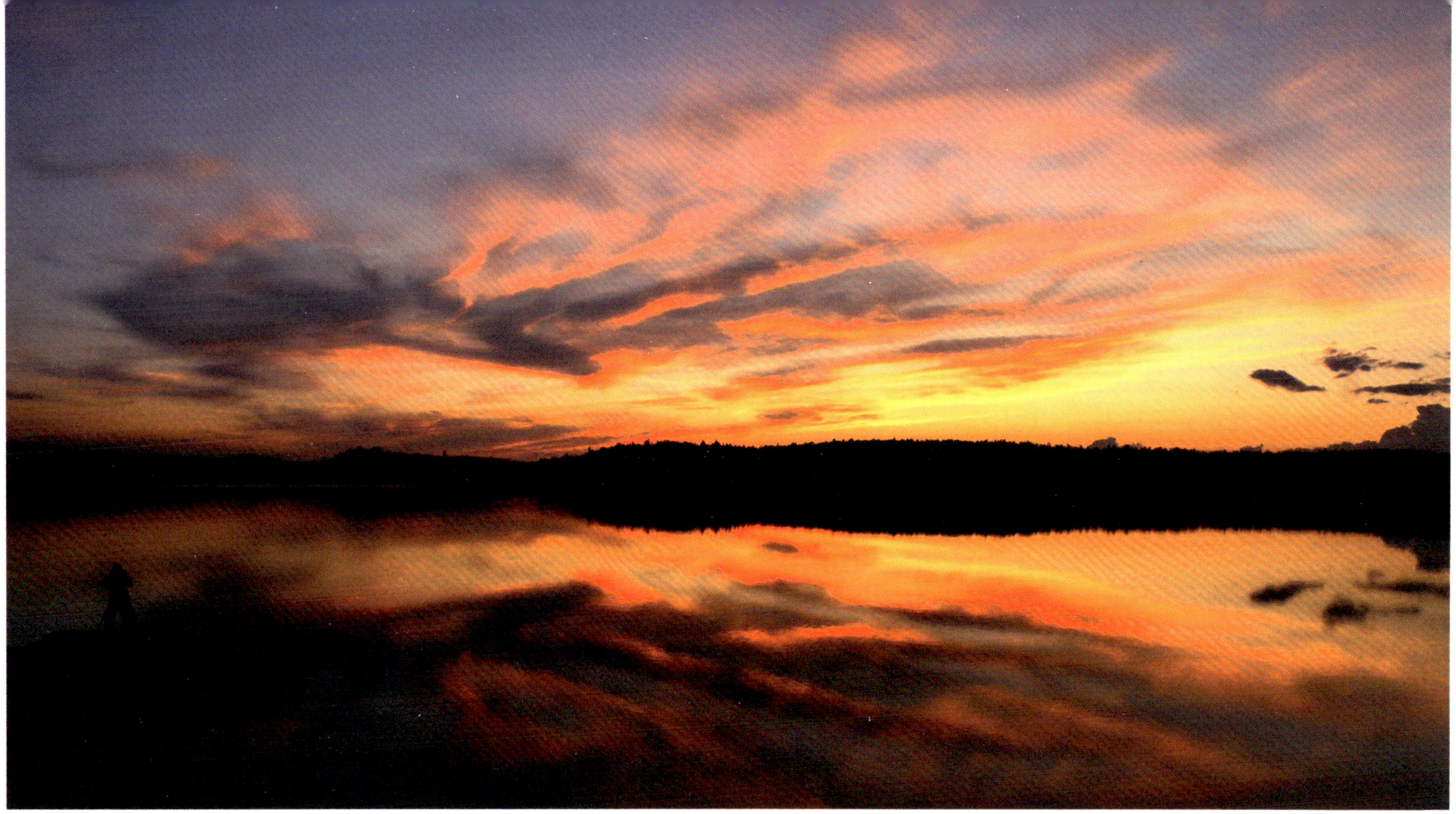

Mexican Hat Lake sunset,
Woodland Caribou Provincial Park.

has been tropical (we are in an ice age now), but such changes usually occurred over hundreds of thousands or millions of years. Rapid shifts, by contrast, have often triggered mass extinctions.

While the distribution of active carbon among the biosphere, atmosphere, and oceans has always fluctuated, the total amount stayed relatively stable for long periods. Since the industrial revolution, however, burning oil, gas, and coal has released carbon long locked underground, warming the atmosphere unusually fast.

Antarctic and Greenland ice cores, fossil pollen, and other "proxy data" allow CO_2 and temperature to be reconstructed across hundreds of thousands—and in some cases hundreds of millions—of years. Recent records show a clear, accelerating rise in both atmospheric CO_2 and global temperature, that when graphed out form the famed "hockey stick" curve: nearly flat until about 1900, then sharply upward.

Some misread this millennium-long snapshot as evidence that current levels are unprecedented. Others point to ancient climates when both CO_2 and temperatures were far higher. This fuels a false debate between "we must save the planet" and "global warming is nothing new." George Carlin got closer to the truth: "The planet will be fine . . . but humans are #@$*%!"

The real concern is not the absolute level, but the speed of change. Our industrial society developed during the graph's flat portion. Now, rapid warming threatens to raise seas and disrupt rainfall faster than we can adapt, jeopardizing food production, infrastructure, and social stability. After the last ice age, rising seas didn't faze our hunter-gatherer

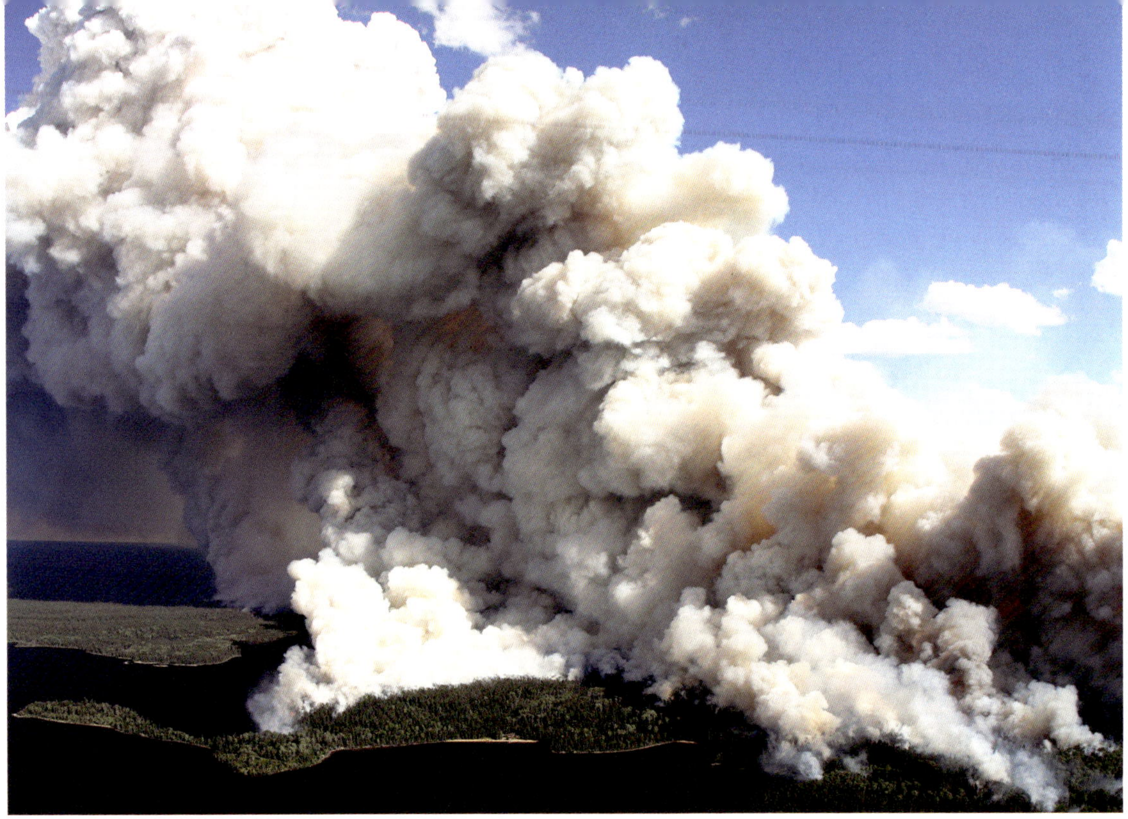

Red 124. Mitch Miller, courtesy of OMNR

ancestors—but they could pack everything they owned and move in an hour, and there weren't eight billion others competing for prime real estate.

Carbon sequestration, semi-sequestration, repositories, and sinks

As noted earlier, for much of Earth's history, its global climate was temperate to tropical, with far higher atmospheric CO_2. So where did all that CO_2 go? The extra carbon has since been "sequestered": locked in sediments or limestone or converted to fossil fuels. Sequestration occurs when organisms fail to fully decompose, often in low-oxygen environments like the deep sea, bogs, or volcanic ash. Over ages, heat and pressure incorporate this carbon into rock or convert it into coal and oil. Once buried, it typically has no effect on climate for millions of years.

Semi-sequestration: Between sequestration and the active carbon cycle, the boreal forest's wetlands and tundra store carbon as partially decomposed tissue in bogs or permafrost. Bogs and permafrost cover a significant share of the boreal forest and tundra around the northern hemisphere.

Repositories: Vast carbon stocks are stored in trees; twice as much in ground cover, soils, fungi, peat, and permafrost (about 30 percent of global terrestrial storage). This "carbon repository" is mostly active, meaning it can be released as CO_2 or methane within the lifespan of a spruce or patch of lichen. If vegetation regrows, the carbon balance remains in a roughly steady state—over time.

Sinks: Historically, the forest has also acted as a "carbon sink," absorbing excess atmospheric CO_2 and buffering climate. That role is threatened as warming drives more fires (see chapter 2).

A 2021 *Scientific Reports* paper found the boreal may already be a carbon source due to wildfires (the US federal agency NOAA reported the same for the Arctic in 2024). A *Science* paper in 2021 projected recovery to sink status in a century as birch and poplar replace burned stands—but both agree that for the next 50 to 100 years, the boreal will emit more carbon than it absorbs.

Bill McKibben, co-founder of 350.org, once pointed me to a paper noting the flaw of heating with wood pellets—even if carbon-neutral over time—to fight warming. We may not have a century to wait for regrowth.

Note: The boreal remains a massive repository, like a large bank account from which withdrawals now exceed deposits. However, continued net losses risk detonating a "carbon time bomb": a feedback loop where warming releases more carbon dioxide and methane, which in turn accelerates warming.

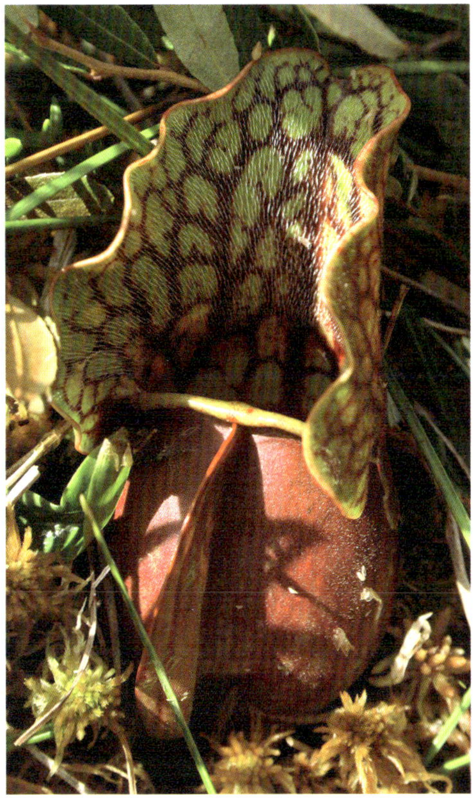

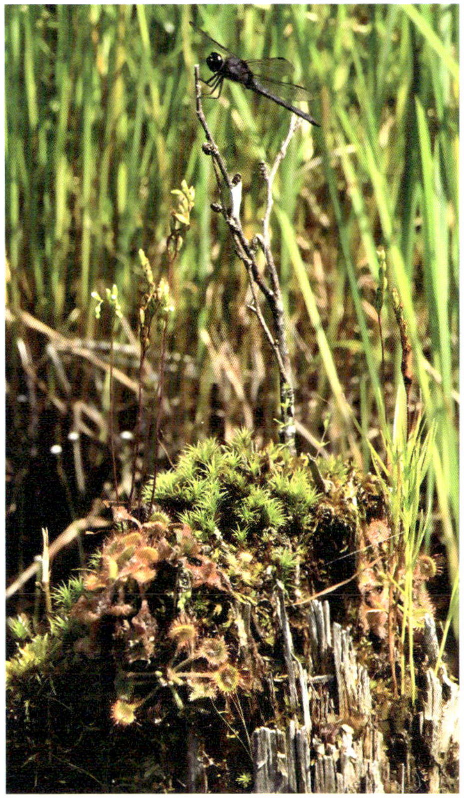

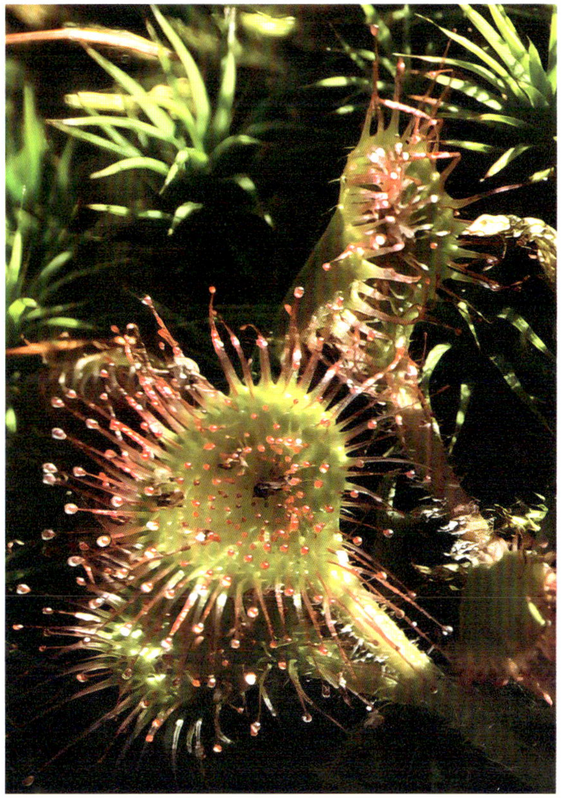

LEFT: Pitcher plants have modified leaves that form a fluted pitcher filled with a sweet-smelling (to insects anyway), sticky fluid that is a concoction of digestive enzymes. Once attracted to the pitcher's colorful, slippery surface, down-curved hairs lead to the pool of doom and make it difficult for those with second thoughts to escape.

CENTER: Sundew plant and dragonfly on a beaver stump in Preston Pond, including its about-to-open flowers. Bonnie Rowell

RIGHT: Sundew close-up showing the traps, one of which is curled around a victim, Preston Pond, Vermont. Bonnie Rowell

Water

August 8, 2011: A bog near Hansen Lake, Woodland Caribou Provincial Park

Threading through reeds between lakes, we reached one of the trip's 57 short portages. Along the carry lay a classic bog: a thick mat of sphagnum moss floating where a lake once was. Jumping up and down gave it a bounce. These unusual habitats harbor unusual life, and this one offered hundreds of carnivorous purple pitcher plants (*Sarracenia purpurea*). On moss mats over deoxygenated water, nutrients are scarce, so pitcher plants and sundews supplement with insects.

The boreal forest holds more than a third of the world's 50 longest rivers and two-thirds of the longest undammed ones, delivering full nutrient loads to lakes and seas. Boreal forest rivers don't just deliver nutrients and minerals to the sea though. Free-flowing rivers also shelter sea-run fish such as salmon.

Every year, young anadromous fish, including salmon (smolts), leave the rivers they were born in and head to sea. The young fish grow big, mature, and return to their freshwater spawning sites, bringing their ocean-derived biomass back with them, and hence to the forest (such as when bears eat the salmon and then do what bears do in the woods),

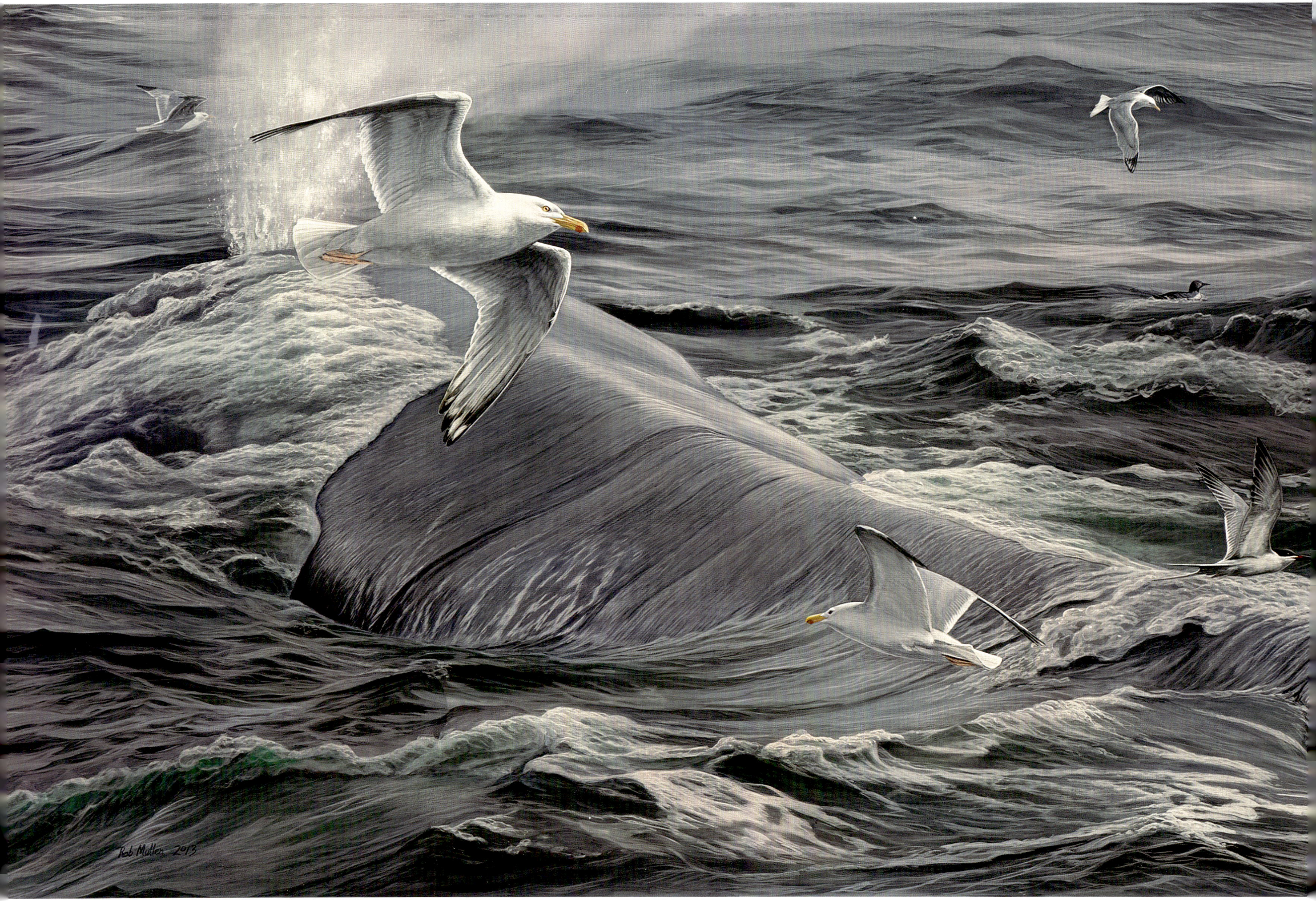
Rob Mullen 2013

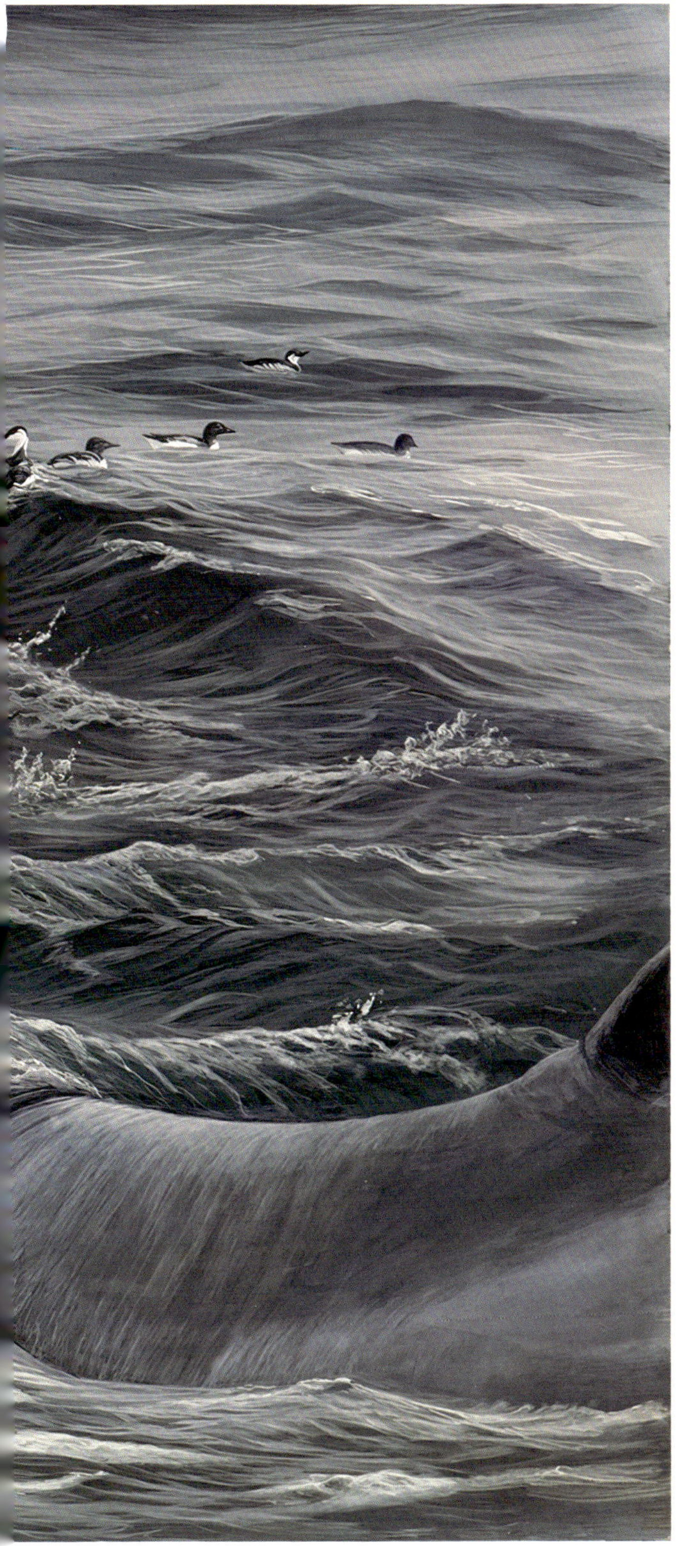

Orcas, Alaska. *The Fog Cutter*, acrylic, Paco Young

FACING: Great blue whale, herring gulls, common murres, common tern, and a dovekie, Tadoussac, Quebec, St. Lawrence River Estuary. *Confluence Cornucopia*, acrylic, Rob Mullen

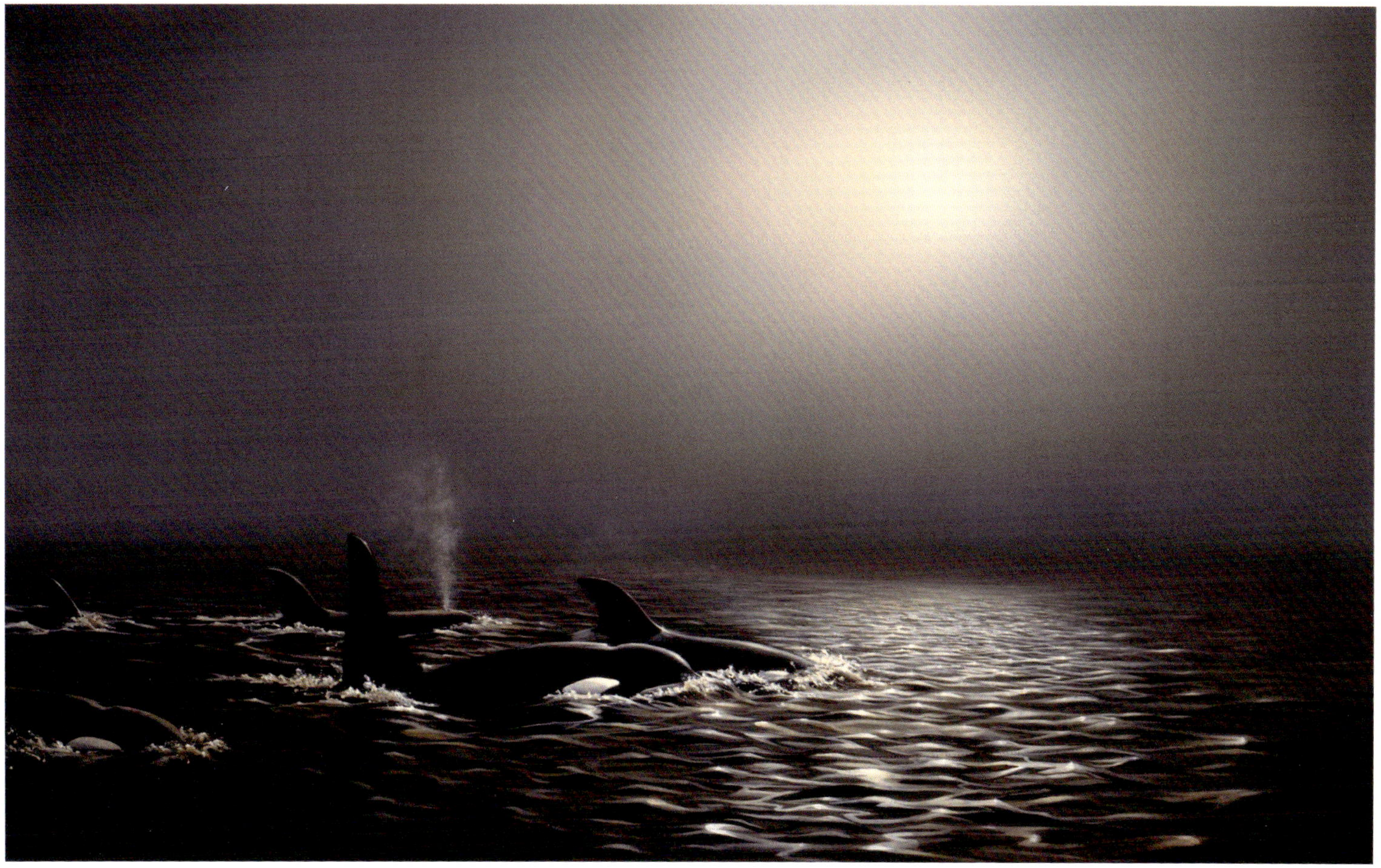

LEFT: Chum, pink, and coho salmon, Alaska. Susan C. Morse

RIGHT: Big Mama, Alaska. Susan C. Morse

FACING: *Ursus maritimus* (polar bear), oil, Linda Besse

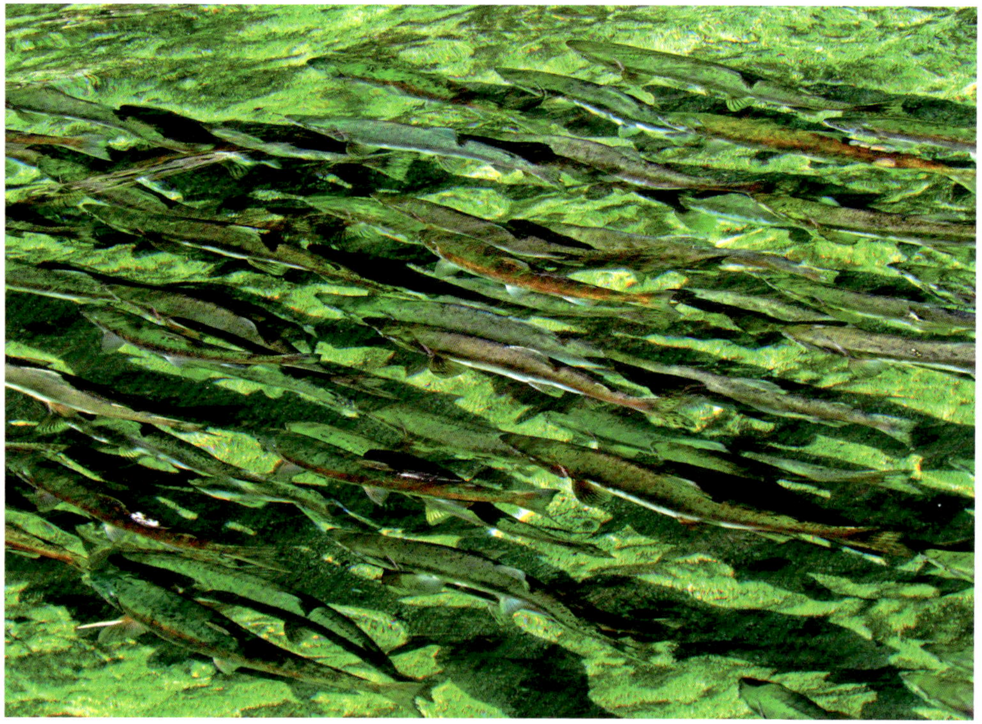

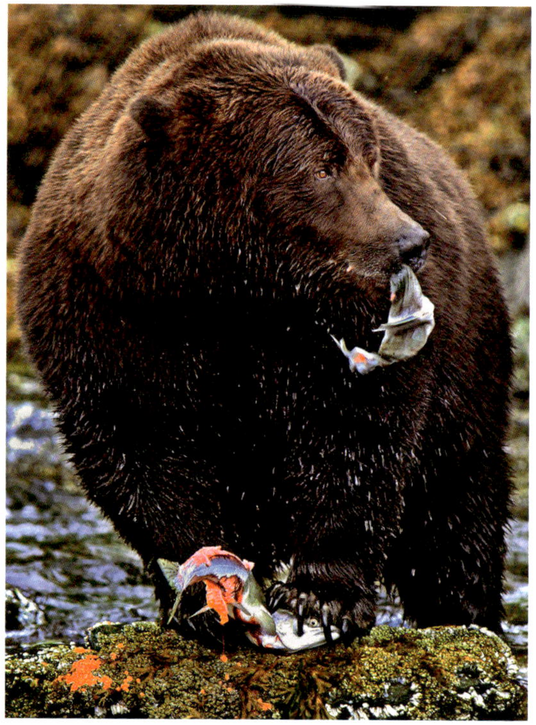

making them keystone species, especially at their historic population levels. Salmon were (and still are to a diminished degree) keystone species not only for their ecosystems—their migratory spawning runs were so enormous (especially Pacific Ocean species of the genus *Oncorhynchus*) that entire human cultures and economies were built on them. Across the coastal ranges from the boreal forest, Pacific Northwest Indigenous cultures elevated salmon to a level akin to caribou among more interior peoples such as the Innu. Athabascans and Inupiat in Alaska can access both in the boreal forest.

These nutrient flows enrich coastal fisheries worldwide. In Alaska, rivers from the boreal—Yukon, Noatak, Kobuk, Selawik, and others—help make the Bering Sea one of Earth's most productive marine ecosystems, supplying half of US seafood. Freshwater inflow drives ocean currents and promotes sea ice formation, which reflects sunlight, cools the planet, and provides habitat for the entire coastal arctic ecosystem, whose apex predator is the polar bear.

Well, along with these cetaceans (orcas; see previous page).

The vast wetlands also support many of North America's waterfowl, plus shorebirds, gulls, and terns. Their migrations, like those of salmon, transport nutrients across continents.

Water shapes the boreal in less visible ways: Just by living, trillions of trees release enough vapor to cool regional and hemispheric weather and climate.

Water is so elemental that its ecological services could fill volumes. Let's sum it up in one watery word: "titanic"—hopefully only in the sense of size, not fate.

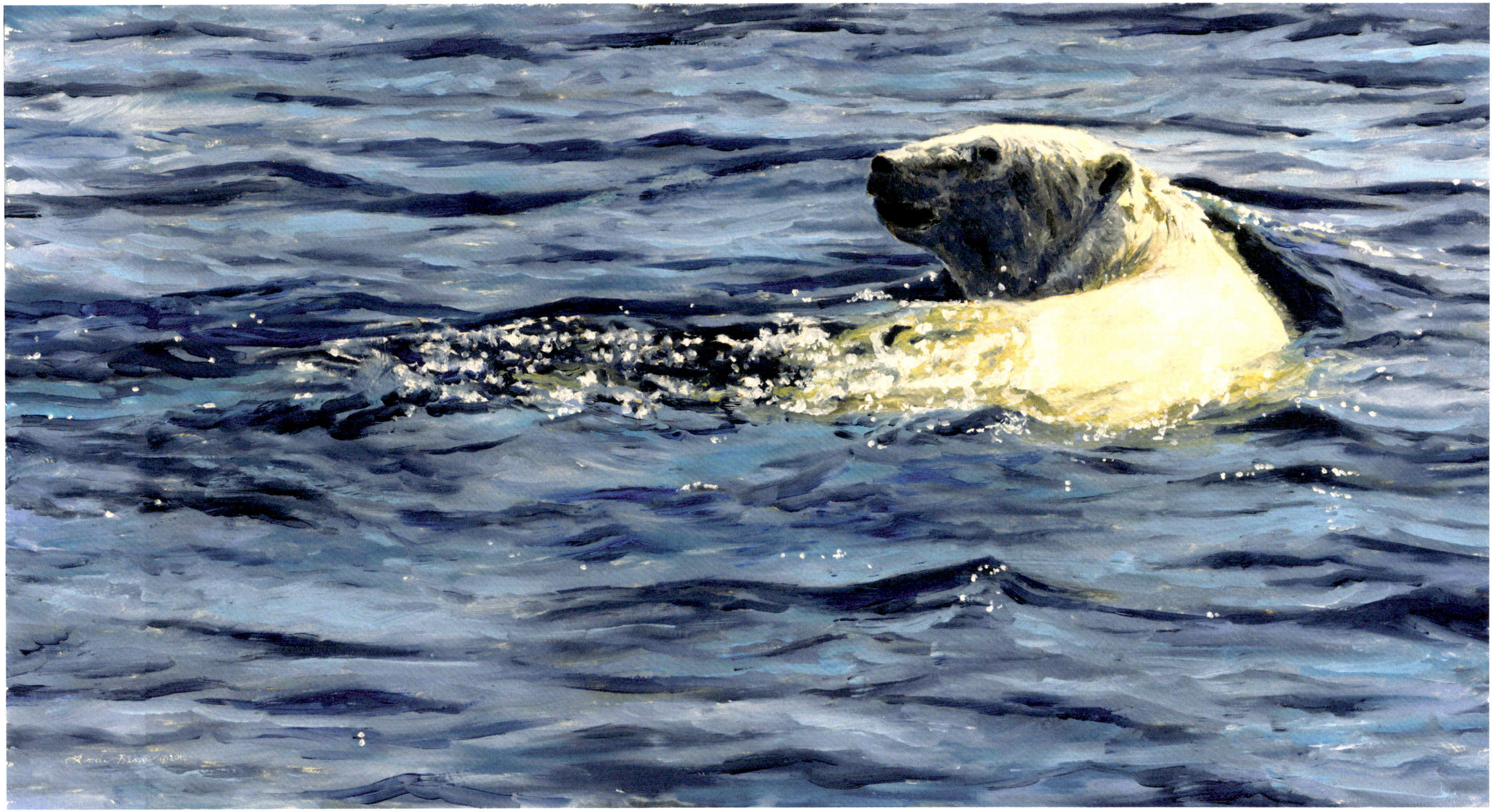

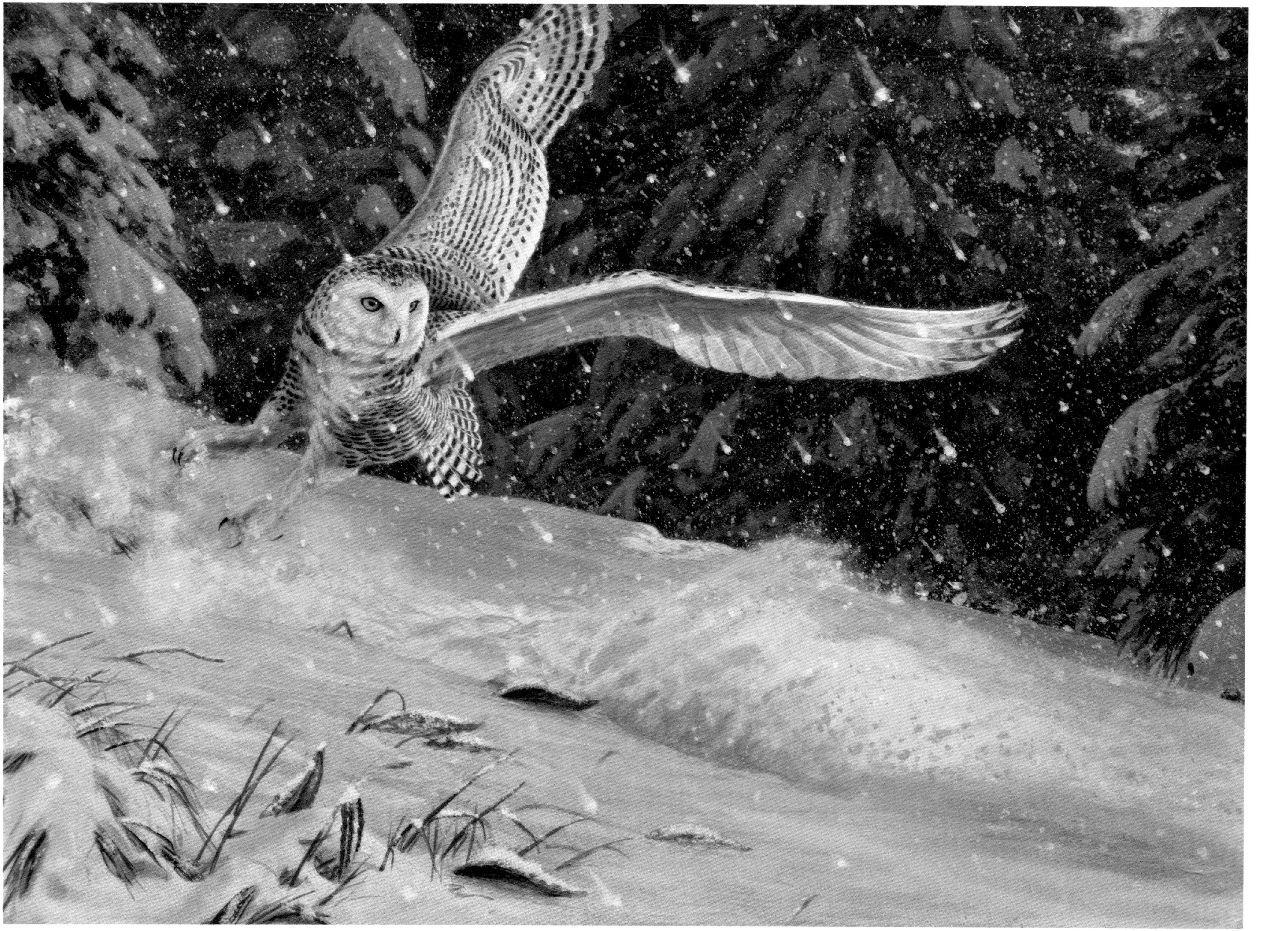

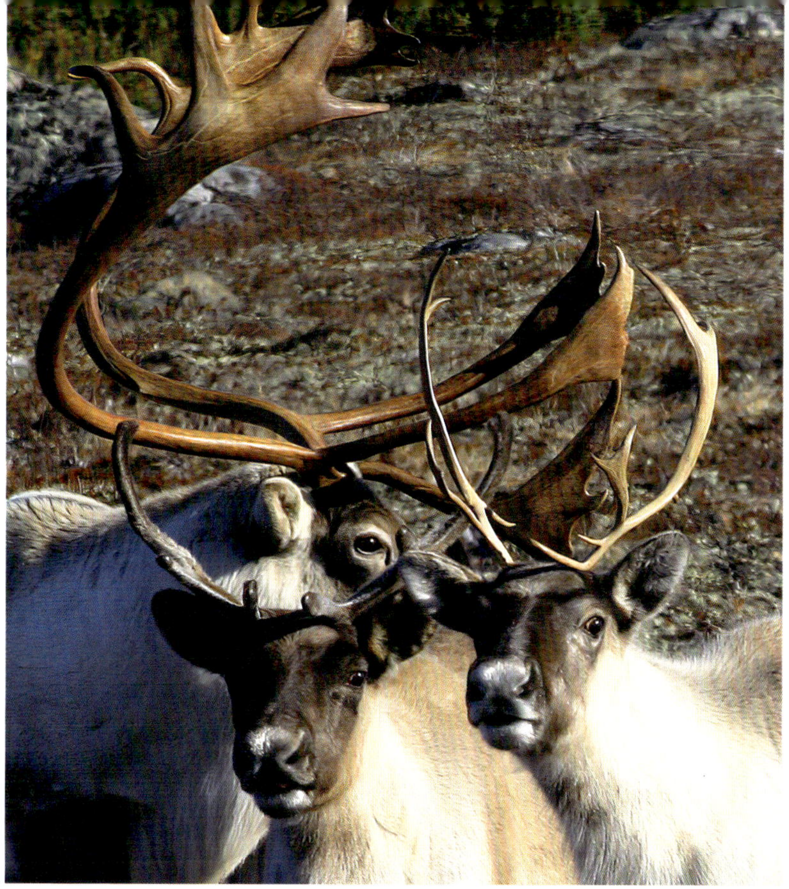

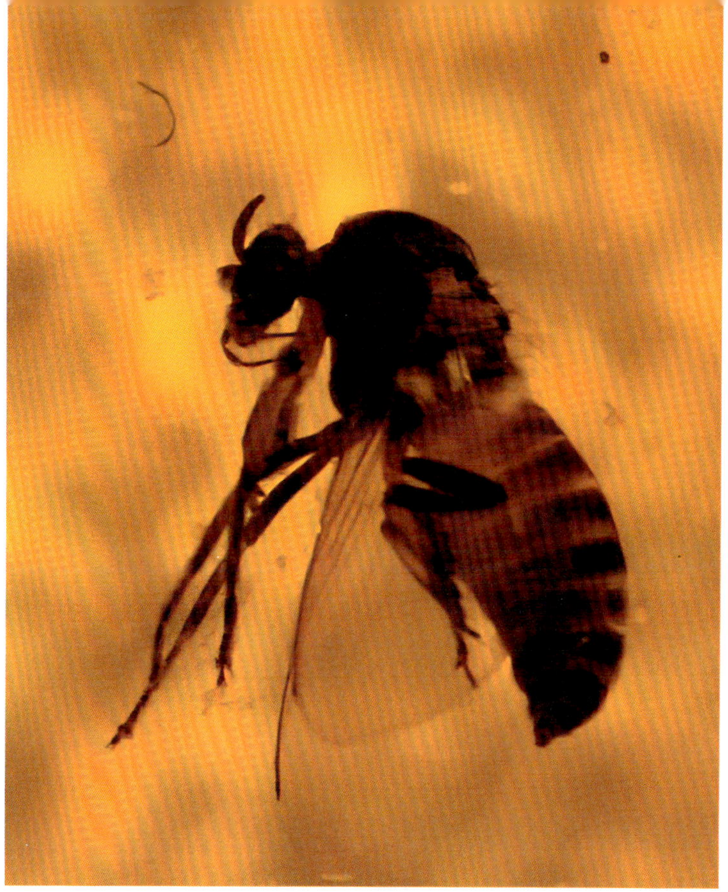

LEFT: Bull's-eye: caribou on the highlands above Kamestastin Lake, Labrador.

RIGHT: Blackfly in amber. (She may have fed on a dinosaur—or, truth be known, got stuck in the bottom of my rum flask.)

FACING: Snowy owl and "essence" of snowshoe hare. *Snow Squall (aka RUN!)*, acrylic, Rob Mullen

Wildlife

Winter 2018: Senate Committee on Natural Resources and Energy meeting room, Vermont State House, Montpelier The Vermont Fish and Wildlife Department testified against a bill banning coyote killing contests, citing license fees as key to funding management. A colleague countered that wildlife watching generated more economic value. The debate felt wrong—not in the facts, but in trivializing wildlife. That got me thinking.

February 22, 2024 (same senate committee; sometimes my thinking takes a while): Testifying on a wildlife governance bill, I closed with the following statement:

On the Fish and Wildlife Department's website, and even those of advocacy groups, wildlife's value is usually summed up as consumptive use (hunting, fishing, and trapping) or "wildlife watching." As if animals matter only as food or entertainment. For a clearer view, I recommended Joe Roman's Eat, Poop, Die: How Animals Make Our World. *Wildlife's environmental services keep our world livable. Valuing them only for flavor or aesthetics misses the big picture.*

(The committee chair called it the best book title he'd heard.)

Plants have a publicity edge. We see them everywhere; they outweigh all other multicellular life combined and are hailed as climate allies. Even fungi are getting the star treatment with Suzanne Simard's *Finding the Mother Tree*. Wild animals,

White-lipped wood snail feeding on a mushroom, Quebec. Danielle Champagne

FACING: Peregrine falcon and willow ptarmigan, Labrador. *Tundra Tailchase*, charcoal, Cole Johnson

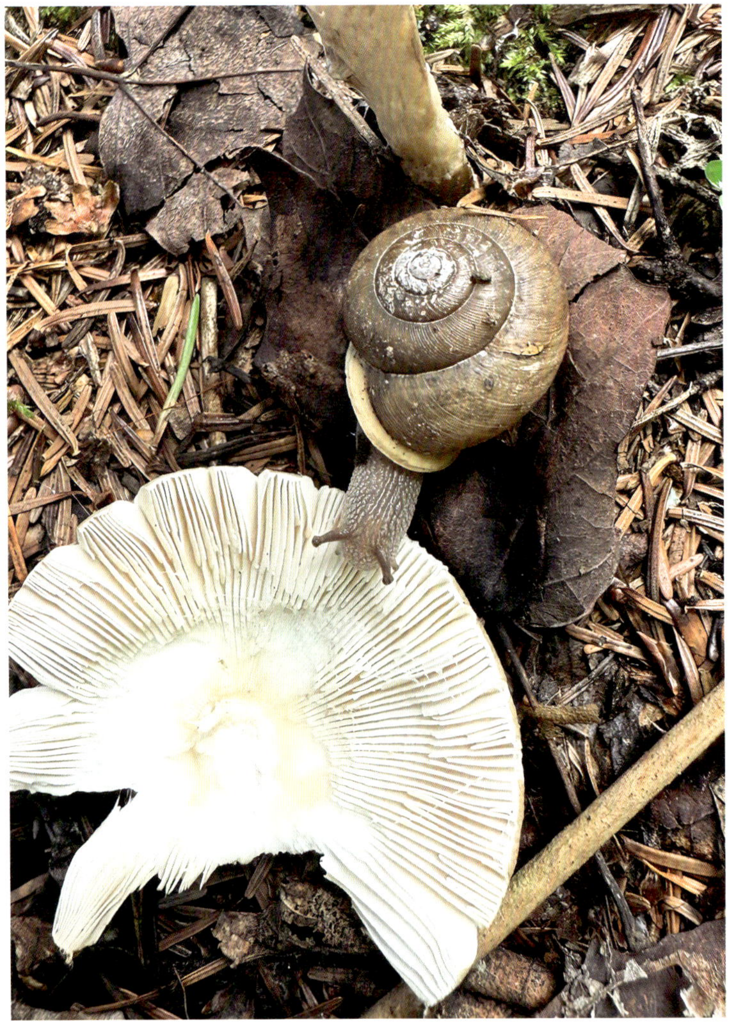

meanwhile—from protozoa to megafauna—are ecologically underappreciated. We think we know them from childhood stories, yet even when urged to save endangered species, the reasons offered are moral, not functional. Few of us could explain to a planning board why preserving biodiversity might justify altering a shopping mall project.

The truth is, we can't live without wildlife. They keep ecosystems productive, and biodiversity assures they can do it reliably. Ten thousand years ago, no one had to be told: Wild animals were food, clothing, shelter, tools, and occasionally predators. Today, most of us don't depend on or fear them, so they fade into the background. They shouldn't. Food and clothing were never more than the flashy tip of the iceberg of what wildlife provide.

If plants are the biosphere's fuel tank, animals are the moving parts, distributing and cycling life's essentials. They prevent plants from exhausting local resources, pollinate flowers, disperse seeds, and, through eating and excreting, feed the chain of smaller organisms that return nutrients to plants. Without animals, Earth could stagnate back into a reeking, primordial mat of microbes.

A word about invertebrates

While we passed over invertebrates in the bestiary of chapter 4 they deserve at least a nod. Insects alone outweigh vertebrates and are just as vital for moving energy and resources. We've mentioned the infamous blackflies and mosquitoes but the dragonflies, damselflies, butterflies, and legions of others are wondrous (and don't bite unless provoked even if they can).

But countless others species move mountains of nutrients, toiling out of sight and largely out of mind.

Biodiversity and ecosystem construction and maintenance

When life first appeared, again as it colonized land, and today when recolonizing new habitats, each new species finds a niche—a way to survive—and, in doing so, alters conditions in the environment, creating opportunities for others. Wind and rain break rock into finer material, enabling bacteria, fungi,

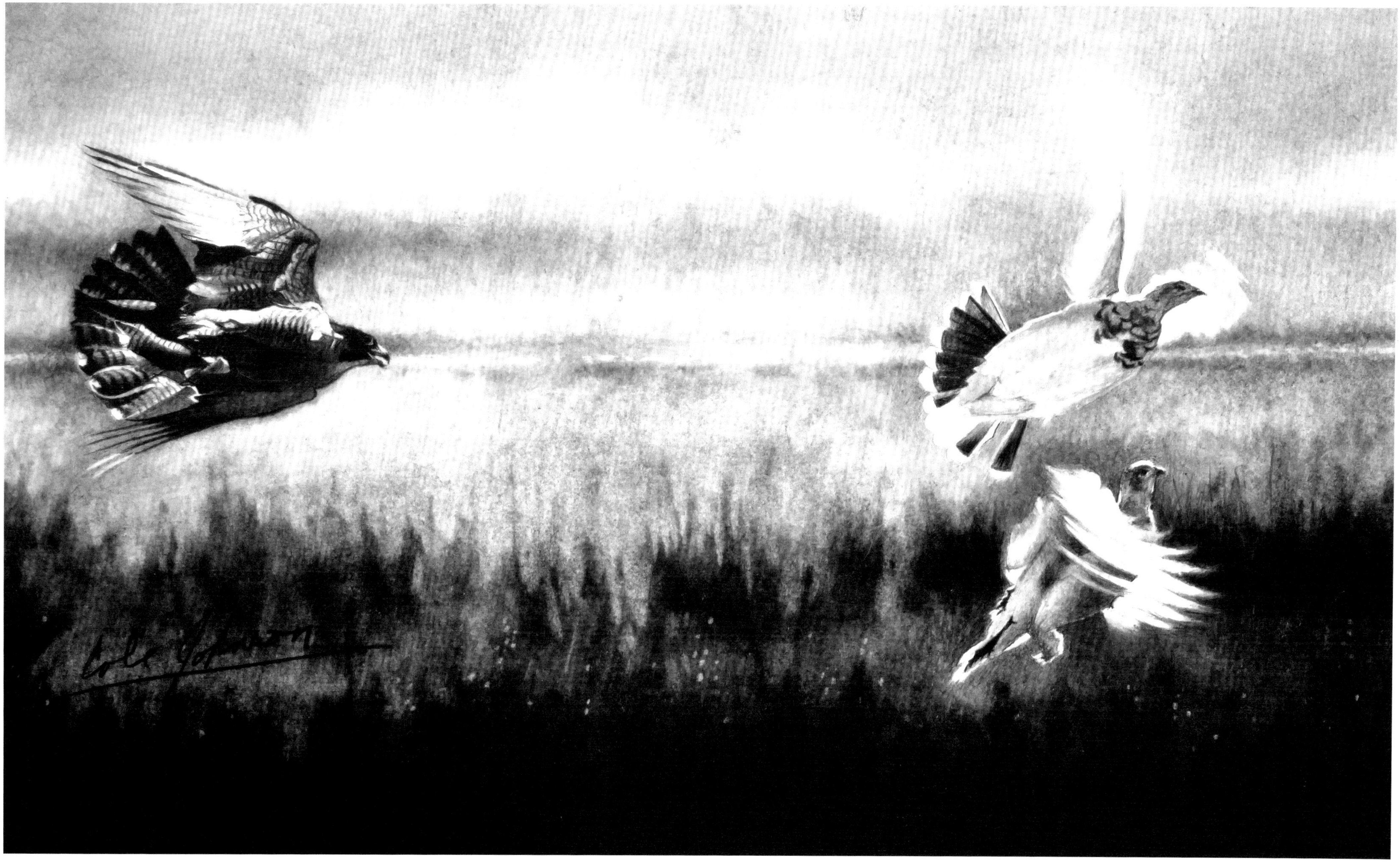

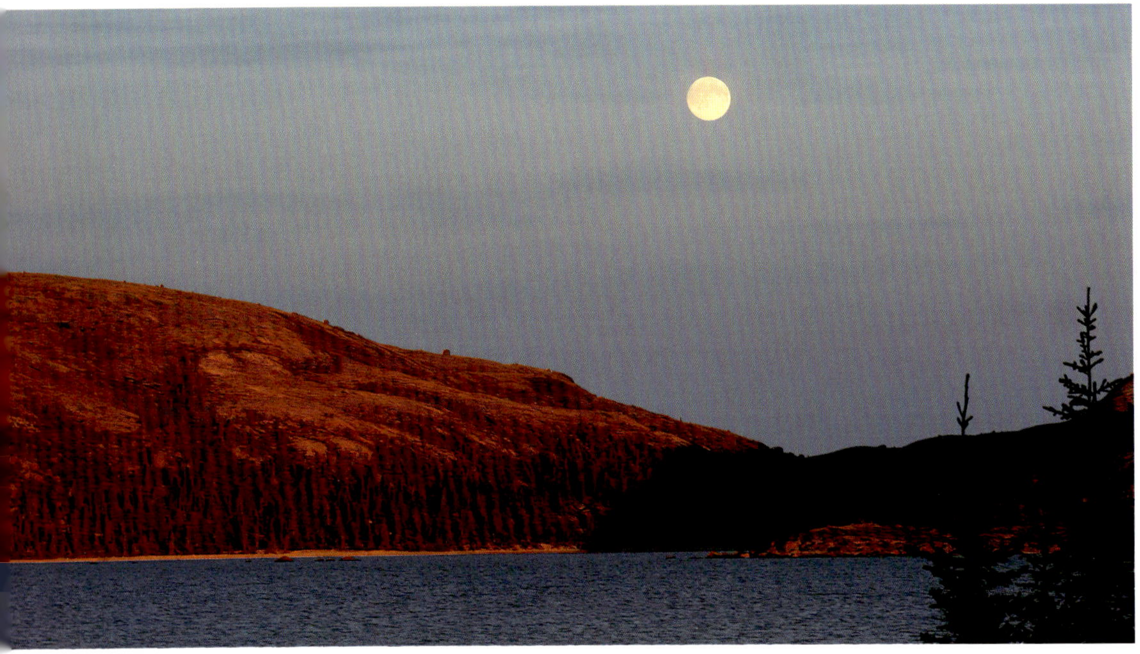

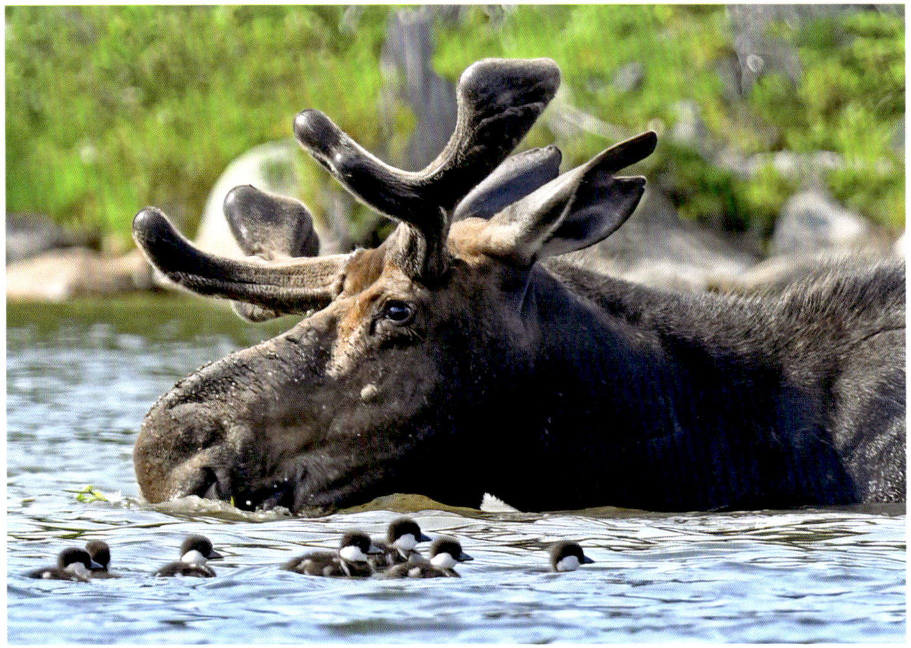

LEFT: Moonrise over Kamestastin.

RIGHT: Biodiversity in wetlands: a moose feeds and finds some relief from the flies as a flotilla of goldeneye ducklings paddle by, Baxter State Park, Maine. Daron Tansley

FACING: Juxtaposition: mining equipment at rest, Labrador. Neil Simmons

and lichens to extract nitrogen and phosphorus. Their presence draws microinvertebrates, and as generations live and die, organic matter accumulates into soil for plants and the animals that follow.

With the ball rolling, mosses, small animals, and vascular plants move in, followed by predators. Each arrival weaves new threads into the growing web. As ecosystems develop toward a climax stage, niches overlap in purpose but differ in method, so essential functions such as oxygen production and nutrient transfer have multiple pathways. This redundancy stabilizes ecosystems, letting them adapt to disturbances—climate shifts, natural disasters, parasites, or disease—that might eliminate a species, but don't destroy the fabric of life.

That's why biodiversity (the number and variety of species) is critical. Reduce it, and you reduce productivity, stability, and the delivery of ecosystem services.

We depend on healthy ecosystems for food, water, and air. Healthy ecosystems depend on biodiversity to endure change, and biodiversity is eroded by species loss. This isn't just a concern for conservationists: Ecosystems have maintained balance for hundreds of millions of years, recovering from massive volcanism, meteorite impacts, tsunamis, and ice ages. If we want our way of life to last, we must regain our balance, lest we become one of those calamities.

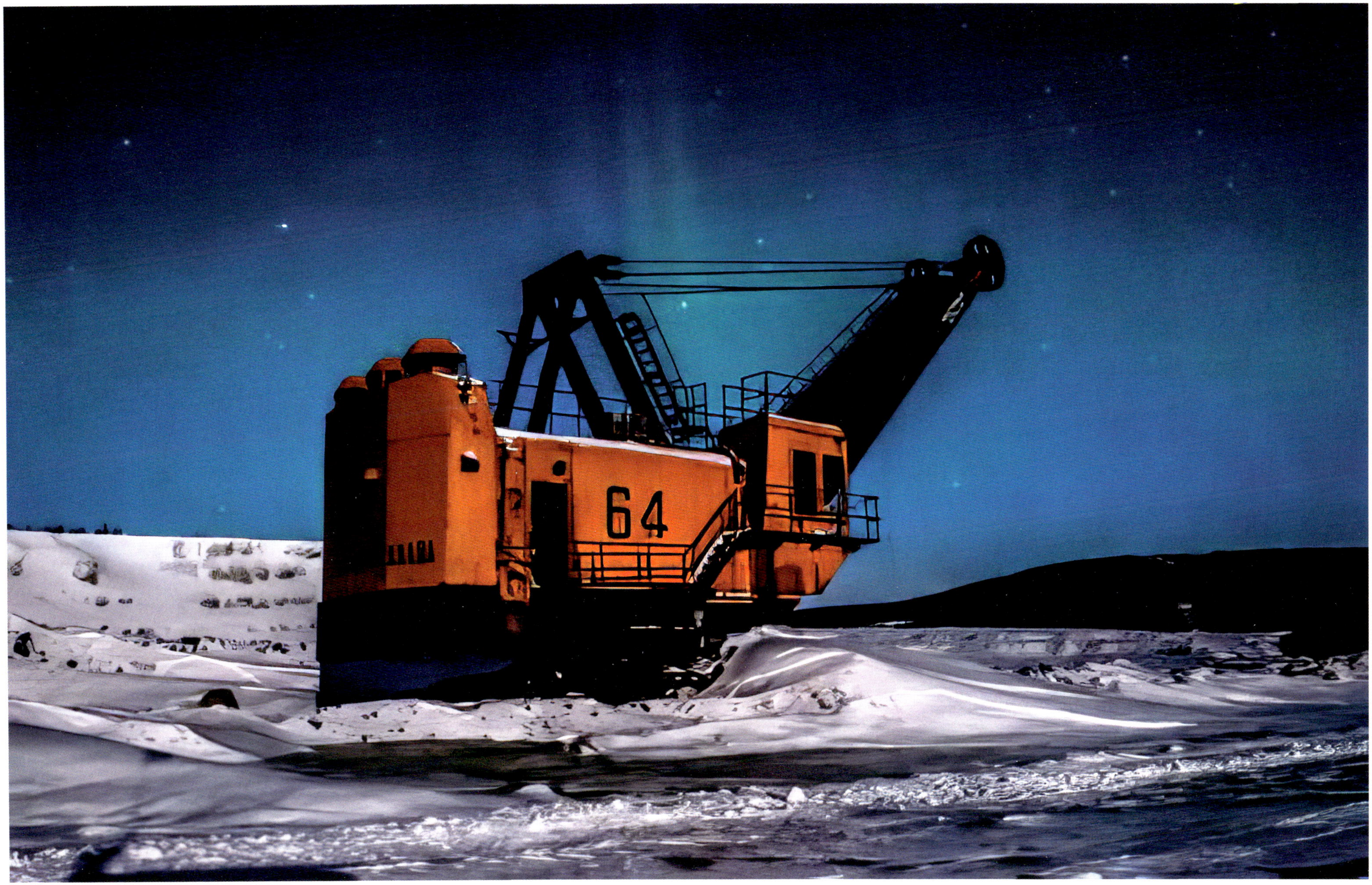
64

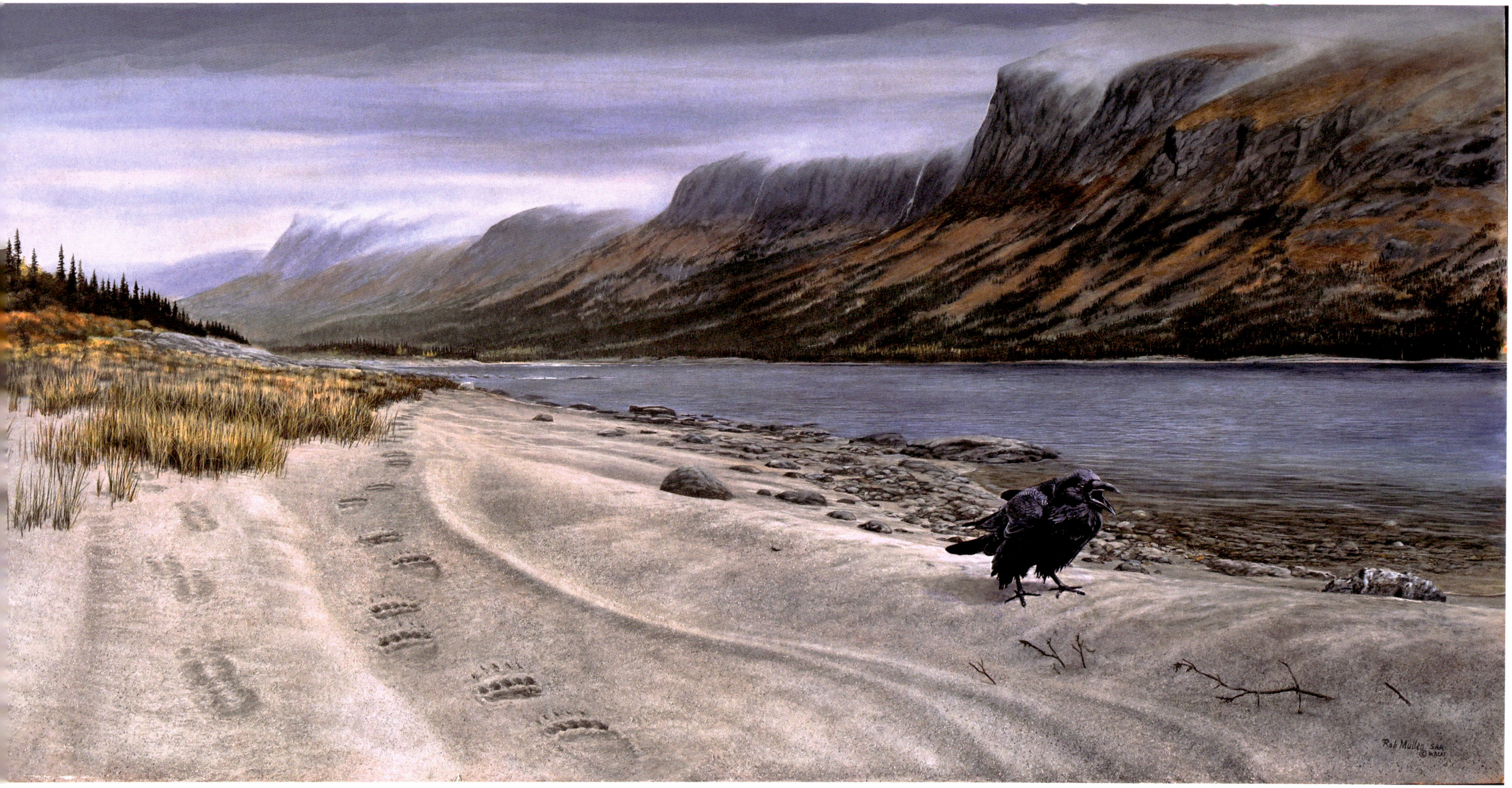
Rob Mullen SAA

FACING: Barren-ground *matashu* (grizzly) tracks and *kakatshu* (raven) on Mushuau shipu (George River). *Matashu* may have persisted into the early 20th century. *Nevermore: Passage of Matashu*, acrylic, Rob Mullen

9

Conservation Threats and Action

EVERYTHING ABOUT the boreal forest is big, including human destruction. The critical ecosystem services the forest provides depend on its enormous size. Massive, horizon-to-horizon clear-cuts, hydro dams flooding forested areas the size of not-so-small countries and leaching tons of mercury into the water, and sprawling mines digging out and processing nickel, iron, gold, uranium, diamonds, rare earth metals, and energy (oil shale and tar sand) are examples of our direct destruction of the biome. On our trips, we have seen slag heaps from iron mines near Fermont (Quebec) and Labrador City (Labrador) that were the size of mountains and glowing red slurry ponds that were visible on the horizon from many miles away. And on my first journey on the Missinaibi River, we encountered something even more shockingly insidious.

June 1986: Missinaibi River near Brunswick Lake Portage

Generally, the clear-cuts in Ontario are masked from canoeists by a mandated 200-meter erosion buffer of trees along riverbanks, but as Dad and I set camp, this one was visible, and it had bright yellow warning signs posted at regular intervals. The signs advised that the area had been sprayed and gave a chemical formula that meant little to me other than it was some organic molecule. However, Dad was a high-school chemistry and biology (and geology, etc.) teacher and simply said, "Agent Orange."

Agent Orange was a broadleaf herbicide that, along with agricultural uses, was employed by the United States Army to destroy cover and food resources for the Viet Cong during the Vietnam War. However, a disastrous component of Agent Orange was dioxin, an infamously devastating and persistent toxin. While it was big news in the US and banned in 1971, in the out of sight, out of mind wilds of the boreal forest, things were different. While I did not note when the spraying had been done as posted on the signs Dad and I saw on our 1986 trip, presumably it had been the year before; Agent Orange was not banned in Canada until 1985. When I returned the next year, I thought to take a photo, but the signs had been updated. The new signs stated that the area had been treated July 14–15, 1986 (shortly after Dad and I had been there), with the safer half of Agent Orange, 2,4-D herbicide.

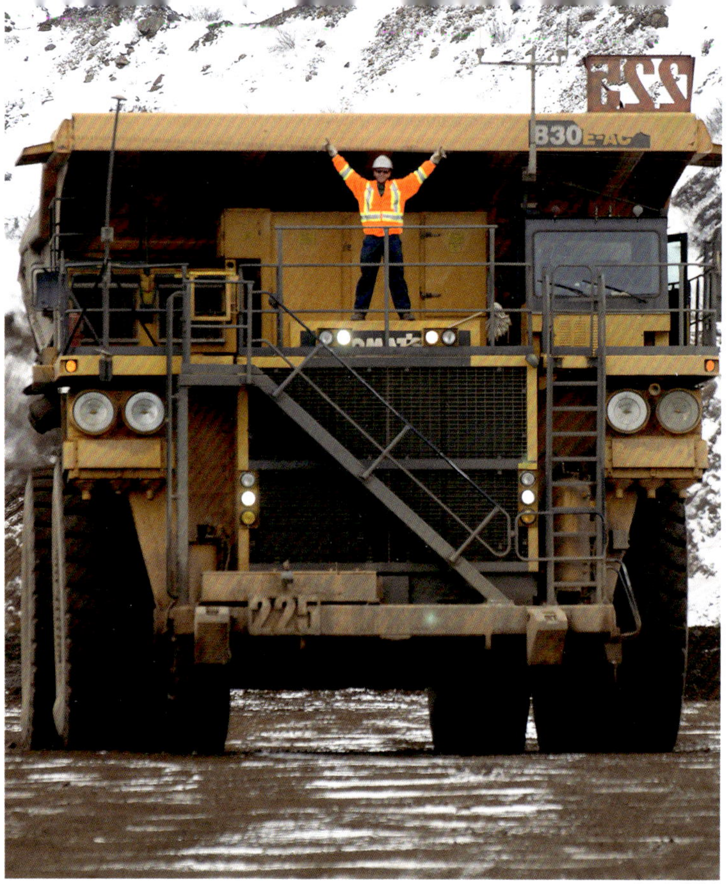

Love your job: iron ore truck in Labrador City. Neil Simmons

(Satellite images of the growing clear-cuts in the area of Thunderhouse Falls from 2015 and 2025 can be found in the online appendix under "Maps and Satellite Images.")

As little information on the boreal forest as was available online when I returned from the Missinaibi in 2001, I was just behind the dawn of an awakening. In 1999, The Pew Charitable Trusts had partnered with the Hewlett Foundation, Ducks Unlimited, World Wildlife Fund, and others to establish the International Boreal Conservation Campaign (IBCC), and subsequently formed the Canadian Boreal Initiative and the Boreal Songbird Initiative in 2002 (WREAF worked closely with both, starting in 2005). Recognizing the depth of knowledge and direct stake of the people who have lived in the boreal forest for millennia, the Indigenous Leadership Initiative, founded in 2013, assumed leadership of the IBCC in 2023.

Now, a quick internet search yields hundreds of pages of boreal forest information. There are many boreal forest programs at US and Canadian NGOs and governmental agencies, a growing body of research, and international organizations that provide data and guidance for conversation efforts and governmental policies. Their work over the last 30 years has focused on conserving large tracts of the North American boreal forest and especially acknowledging Indigenous leadership. As Dr. Jeffrey Wells and his colleagues write in the paper referenced below, "Indigenous governments across the boreal forest biome of Canada are leading in many of the most modern, cutting-edge land and wildlife management plans and models in the world." An example of this is the Pimachiowin Aki UNESCO World Heritage site established by the Anishinaabe mentioned in chapter 3.

Another, more insidious threat to the boreal forest's healthy function is fragmentation. The roads that support industrial operations are perniciously destructive on an even larger scale, reducing nesting habitat (some species depend on contiguous forest blocks far from edges that can draw predators), exposing trees to increased storm and fire damage, and rapidly changing predator and prey hunting and avoidance behavior (and success).

One example is of caribou just north of Québec City. Logging roads have essentially changed the landscape for wildlife so dramatically and so rapidly that caribou have not had time to adapt. Due to roads allowing increased access for predators to smaller, fragmented areas of habitat, the time-tested strategies caribou have used for avoiding wolves drive them into the claws and jaws of bears. Bears are not good at taking adult caribou, but they are highly proficient at killing calves and are stalling any possibility of caribou recovering from their current population crash.

As may be apparent even with this cursory overview, boreal forest conservation must consider many moving and interacting parts. As the authors of the article on caribou in Quebec confirm, "Humans are altering the composition and structure of ecosystems... so rapidly (i.e., mainly within the last century) that prey may have been unable to adapt to the cumulative effects."

And this reality introduces a relatively new scientific management tool: cumulative effects analysis.

In an article on Ungava caribou, Susan C. Morse wrote:

The dramatic increase of human intrusions may have combined with natural stresses to create a kind of tipping point for many herds. Indeed, a new science called Cumulative Effects Assessment should oblige us to acknowledge and measure the aggregate of combined and often interacting and cascading human-caused perturbations that negatively impact wildlife and their habitat.

In November 2024, the U.S. Environmental Protection Agency, U.S. Geological Survey, and Bureau of Land Management all announced new policies to improve the use of cumulative effects assessments (CEAs) in environmental management. A paper on similar Canadian efforts echoed the need for a more holistic approach: "Indigenous-led CEAs have started to foster considerable momentum and authority... Indigenous worldviews, knowledge systems, and ancestral law instinctively understand cumulative effects and are intrinsically better aligned with integrated, ecosystem-based framings than Western science paradigms."

For an eye-opening discussion of the impact of roads on wildlife and ecosystems, read Ben Goldfarb's *Crossings: How Road Ecology Is Shaping the Future of Our Planet.*

In the paper referenced above, Dr. Wells and his colleagues summed up protection of the boreal forest:

The biome will also become increasingly important as a place of refuge for species forced northward by inhospitable climate further south. Further, the best insurance for maintaining resilience of plant and animal communities to climate change will be the maintenance of intact ecosystems and robust populations. Species that must shift ranges northward to survive will have their best opportunity to so do when unimpeded by fragmented habitat full of human-made barriers. Careful land-use planning now that conserves very large parts of the North American Boreal Forest biome will provide the best likelihood of survival for countless species, including humans. The most significant land-use planning and conservation proposals underway across the biome are led by Indigenous governments. Governments, non-governmental organizations, academics and indeed the public at large, should be finding ways to support and encourage Indigenous-led land-use planning, Indigenous guardians and Indigenous protected and conserved areas.

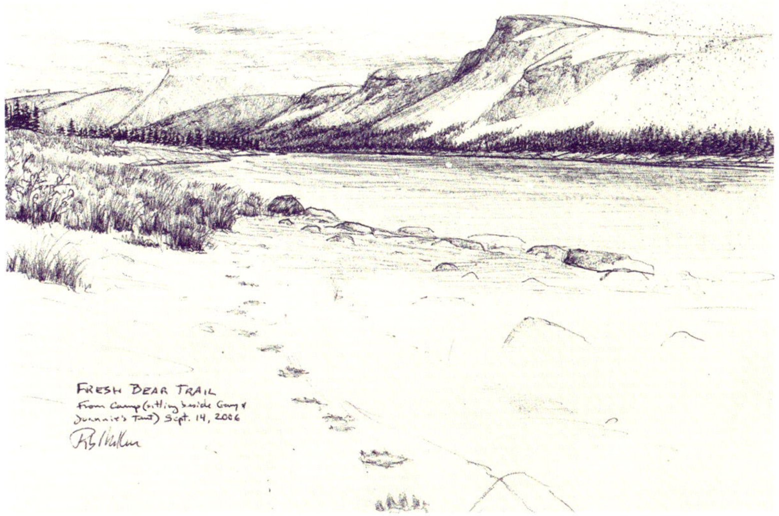

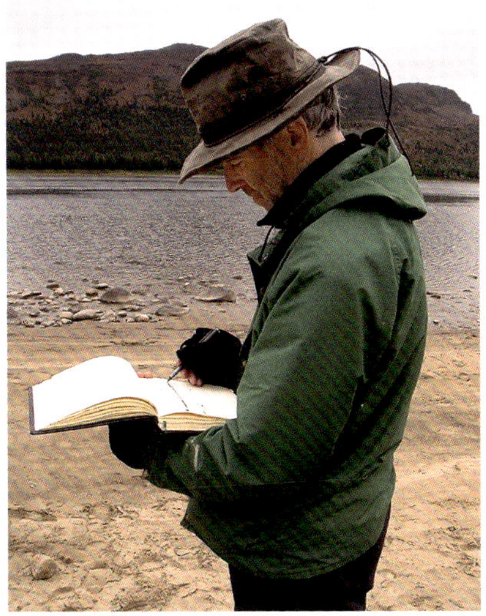

LEFT: Field sketch of black bear tracks on a beach where we camped (the bear departed as we came in), September 14, 2006. As I worked, I thought of the departing bear, the passing storm, the flowing river, and the passage of glaciers, people, and time over this spot for millennia. Then the bear tracks and Stephen Loring's paper on the eastern grizzlies came back to mind, and I imagined the last grizzly departing, and the idea for the painting on page 274 popped out. Ballpoint pen, Rob Mullen

RIGHT: *Rob Mullen au travail sur la Rivière George* ("Rob Mullen at work on the George River"). Jean-Louis Courteau

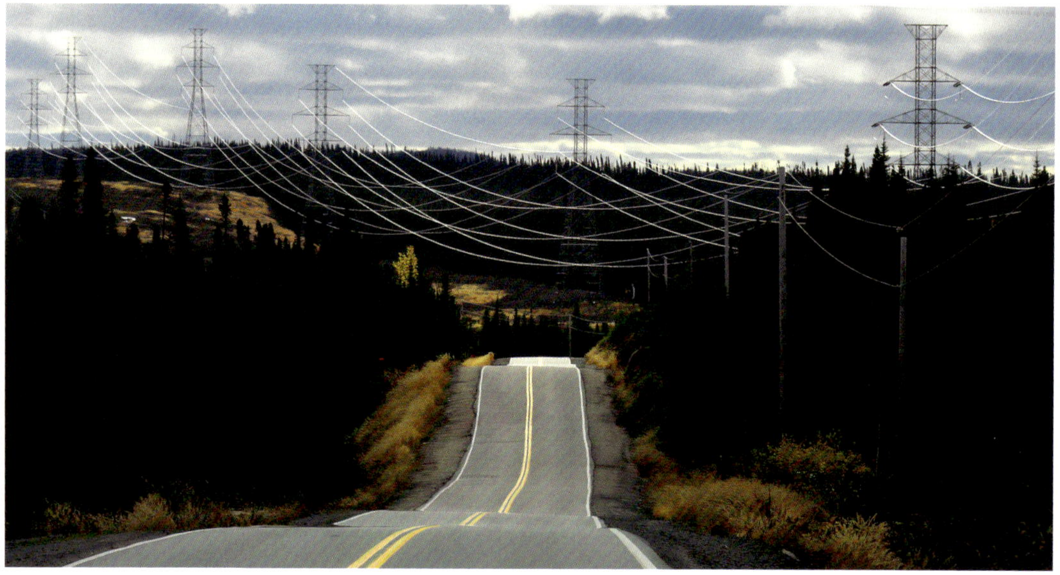

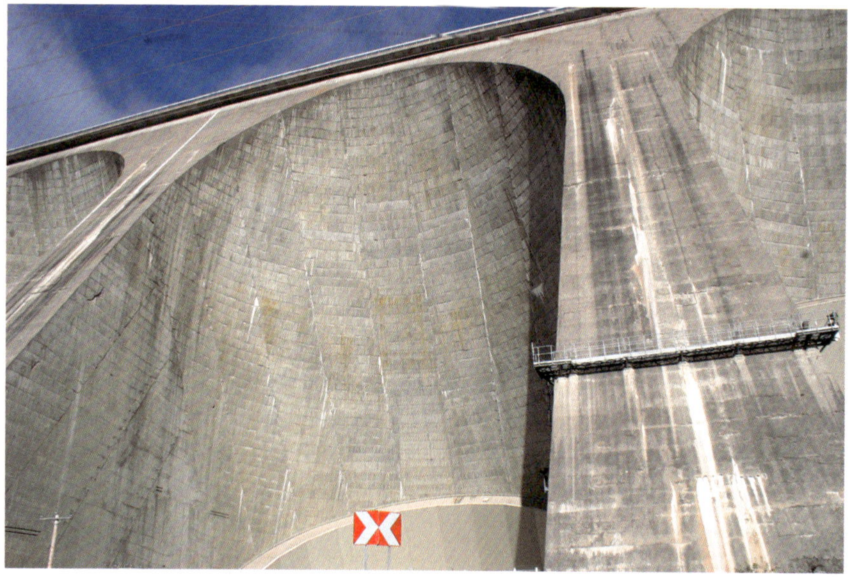

LEFT: Québec 389 and Hydro-Québec power lines from the massive Manicouagan Reservoir.

RIGHT: The Daniel-Johnson dam (known as Manic-5 when the photo was taken), the largest multiple-arch-and-buttress dam in the world creating the Manicouagan Reservoir in a meteorite crater. The resulting 40-mile-wide ring of water is often called "The Eye of Quebec," Kamestastin Lake, Labrador expedition.

FACING: A self-portrait in Swamp Rapids, Missinaibi River. *A Change of Mind*, acrylic, John C. Pitcher

What we can do

While increasing protected areas is a crucial, direct, remedial step in protecting the boreal forest, the root causes of its threatened destruction and the solutions for them lie to the south. To repurpose Walt Kelly's *Pogo* quote for the original Earth Day, "We have met the enemy, and he is us."

An indelicate facet of basic biology is a leading culprit: Americans use more toilet paper than any other nation on Earth, both in total and per capita. Most of the pulp for toilet paper, facial tissue, and paper towels in the US comes from the Canadian boreal forest and accounts for 59 percent of Canadian boreal forest pulp production. Prosaic as it may seem, this is nothing to sneeze at. In 2019, the Natural Resources Defense Council (NRDC) produced the report *The Issue with Tissue: How Americans Are Flushing Forests Down the Toilet*. Stemming from that report, and to give people a way to do something about it, the NRDC produces an annual scorecard that rates paper companies based on how sustainably their tissue paper products are made. Past and current report cards can be found at nrdc.org/resources/issue-tissue.

There are other familiar steps we can take: reduce electricity and gasoline use, weatherproof our homes, ease up on our rabid consumerism, and adopt a philosophy of reuse, repair, and recycle. Be a little leery of getting a new phone every time some billionaire offers a glitzier one to you. If it reliably makes phone calls and locates you for the transporter beam, you are right up there with Captain Kirk. You will help stop the rare earth metals mine planned to be cratered into the Barrens north of Kamestastin.

There is a symmetry in these actions that reflects the cumulative power of trillions of trees in the forest. The part any one of us can play is small, almost to the point of seeming pointless. However, in our hundreds of millions, and in our self-interest, we can have a global impact in saving this greatest wilderness in the world.

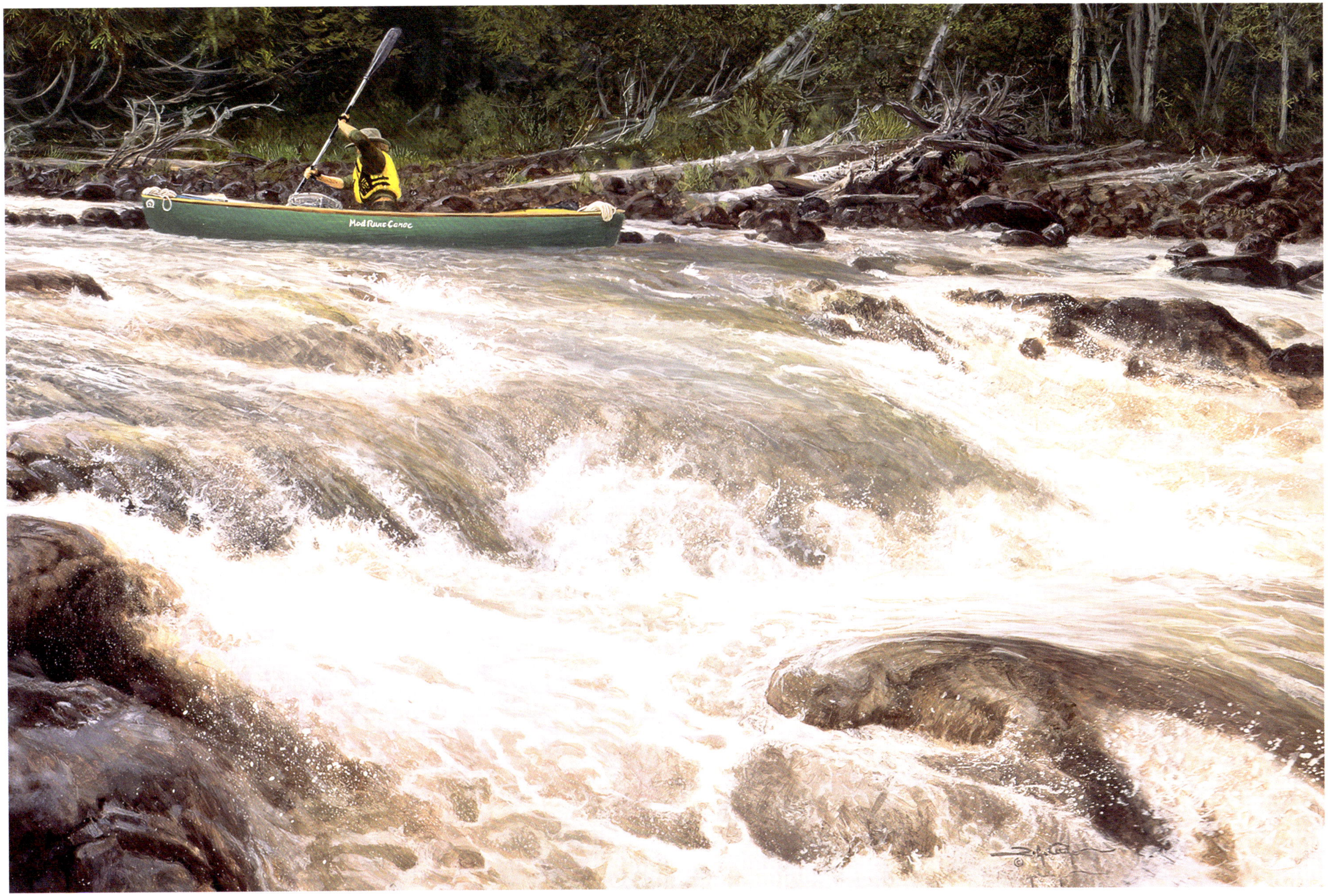

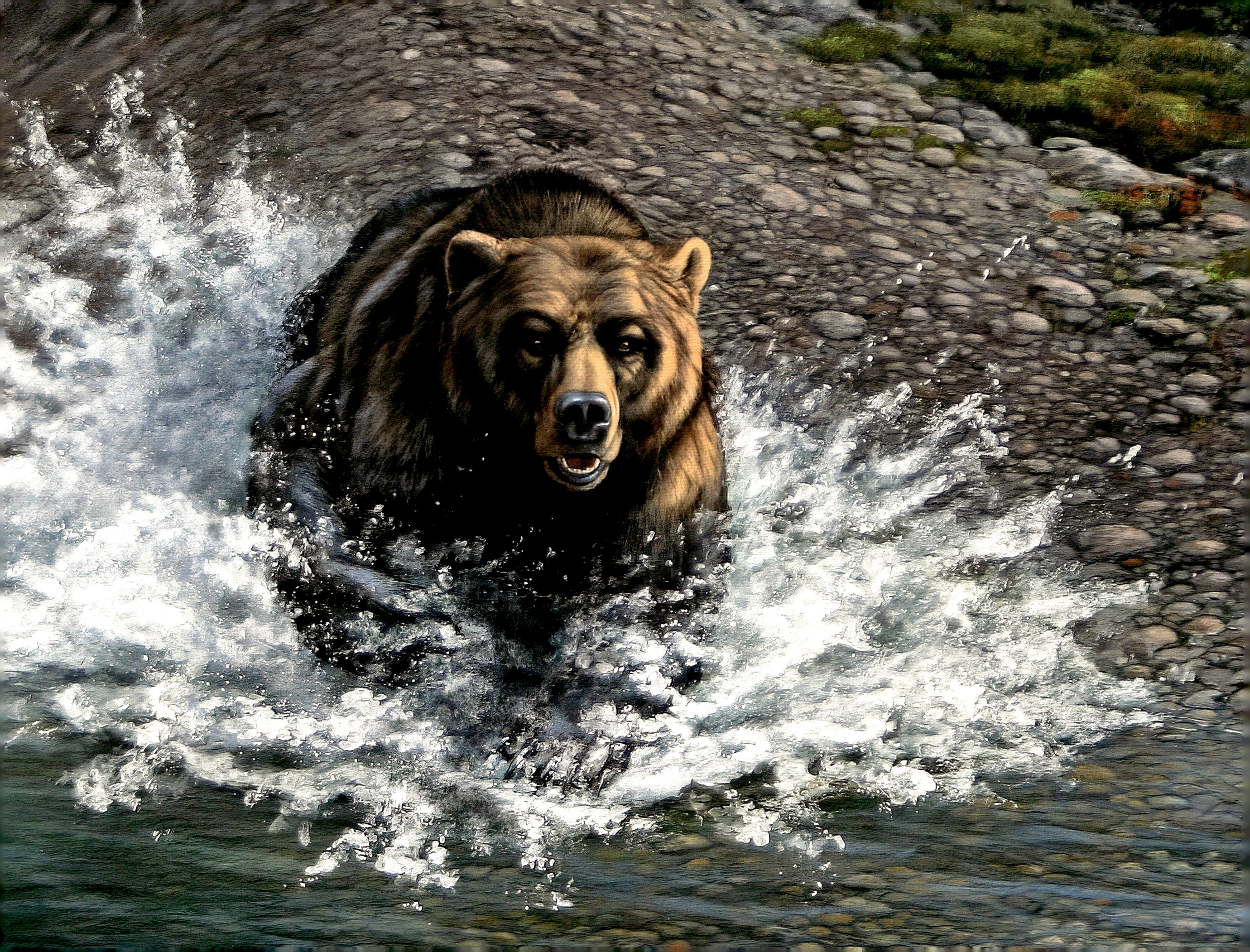

FACING: Grizzly charge, upper Reed River, Alaska. *Alaska Cold Rush*, detail, acrylic, Rob Mullen

A Last Word About the "Wild" in Wilderness

August 17, 2013: Reed River, high in the Brooks Range

Crossing the Brooks Range with a canoe had taken three weeks (I'd naively figured on 10 days) and been a surreal challenge at times, but it was worth it. The forested upper Reed River ran beneath mountain peaks on a crystal current with easy rapids, and hundreds of salmon and graylings flashed under my hull. And the canoe was carrying me rather than the other way around! I wasn't much concerned about grizzlies. Nonetheless, alone on a narrow river teeming with fish between wooded banks, and with the pepper spray long forgotten back at the airfield (it happens), I kept my Marlin .45-70 carbine on top of my pack with the magazine full but the chamber empty (a safety measure, as the weapon cannot accidentally fire without a round in the chamber, but it is instantly ready for an emergency). It was 12:30 p.m. as I passed a cut bank river-left and continued leisurely into the transition to the next bend with the sand and gravel of the shallow bottom clearly visible a few inches beneath the canoe.

I was just thinking it couldn't get better when from behind, there was a rough snort. I knew instinctively without looking and a nauseating cold rush of dread shuddered through me. As I looked back, a grizzly (my first of the trip, as a fracturing part of my mind unhelpfully noted) peered from the brush and instantly plunged down the bank. Stubborn disbelief was blown away as, without discussion, the big bear launched headlong into the shallows toward me at full speed. The canoe spun, time stretched, dumb shock was hammered by spasms of ballooning panic.

Then out of the chaos, as another, unheroic part of me was hoping for a fast end that wouldn't hurt too much, a steely calm took hold and shoved everybody aside. "Yes, idiot, get the gun."

Pulling the rifle free seemed to take long seconds focused on not catching the sight hood on the pack handle. I was certain that if it caught . . . Clear! Chambering a heavy cartridge, I oddly noted how sharp and square the Marlin's lever was compared with my old Winchester .30-30. Bringing the rifle to bear (sorry, couldn't help it), I was surreally (and

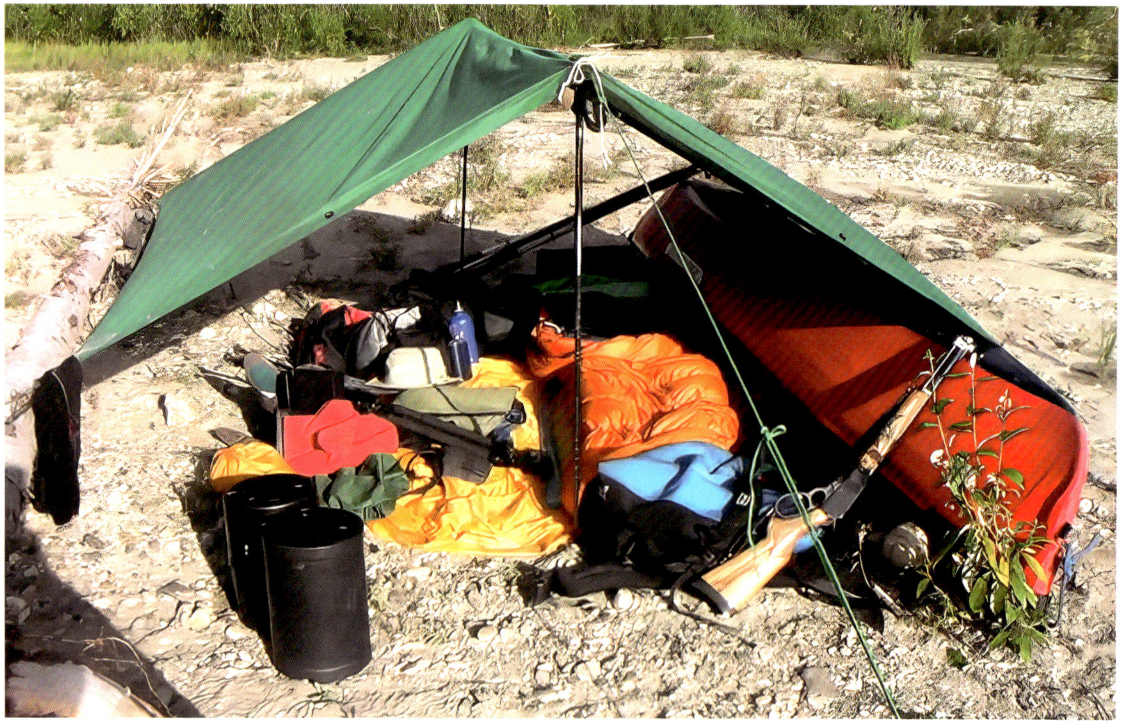

Camp on the lower Reed, near the Kobuk River the morning after the grizzly charge, August 17–18, 2013. The bear barrels with my food (left foreground) were not near the tarp shelter overnight.

FACING: *Off to Rap and Tap*, oil, Dustin Van Wechel

possibly incorrectly) certain I had time for two rounds, and that one would kill him. I could risk a warning shot. But I also "knew" that he had to feel it: just a loud noise would not do, but I mustn't hit him. Simultaneously, I was reassured to see a clear shot to his chest, but I desperately wanted to avoid a fight to the death. Inspiration (and all of that thinking in a split second)! As he rapidly closed, I shifted my aim up and right to just past his left ear—and fired.

The .45-70's recoil can break your nose (or knock you out of a canoe). I hardly felt it, levering the kill shot in as the rifle bucked, and had locked on his chest—when I realized that he'd diverted full power to the brakes. My relief was beyond profound a terror-tinged euphoria. I held fire, even allowing myself to note with grim humor the sudden enthusiasm he had for his new plan. He sloshed to a halt perilously close to the canoe. We stared each other down while part of me recalled some bit of advice about not staring bears in the eyes (Nuts to that!) and the current opened some breathing room. Then he broke for the far shore. Chillingly realizing it might be the dumbest thing I'd ever done, I set the rifle down, dug out my pocket video, and filmed his retreat as he shot me dirty looks over his shoulder. That's a wildlife artist.

He exploded up the bank, spraying sheets of water and continued energetically for parts unknown. As the sound of his flight subsided and the river carried on as though nothing had happened, I was suddenly alarmed that I didn't know where my paddle had gone. I had no recollection of doing anything with it. To my relief, I was surprised to see it with the blade neatly and properly stowed between the bilge and my gear pack with the grip on top of the thwart. That's a canoeist.

(The light spot next to his left ear is where I put the warning round.)

Most grizzly charges on humans are bluffs. This certainly might have been one, but my every instinct in the moment yelled "Predator!" There was no noise, no roar, and no intimidating demonstration: just an all-business charge from behind. Aside from the excitement, this event underscores a more general point made much more subtly on previous expeditions: that the wild non-human earth underlies everything we think we know. The bear seemed a dramatic embodiment of that sense of latent threat I had felt on other journeys: that under the veneer of our clever creations, the world is still wild. And in the crux of the moment, when the prospect of being eaten (possibly while still alive for part of the meal) presented itself, the steely calm, and unequivocal certainty of action that overtook me hints that there is still wild in us.

Absorbed with daily life, living as most of us do in cities or suburbs, it is easy to forget. I lived and worked in New York City for years. While there, I came to see my wilderness canoe trips not as escape from the real world, but as a return to it. They sensitized me to the wild that underlies even midtown Manhattan. Standing at The Battery or on Fifth Avenue, even as the city buzzed around me, the wind blew, the tide and the leaves turned, and birds and bugs kept at their incessant work to survive (and rats hunted for pizza). If the delivery trucks stopped dropping stuff off, and the garbage trucks stopped picking stuff up, the birds, bugs, rats, and plants would take over, and soon be joined by an entire ecosystem of new life. Over time, Manhattan would be hard to find. Eventually, there would be little trace that one of the greatest cities in history had ever existed.

Hand in hand with the stubborn notion that we are separate from nature is the even more foolish idea that we can defeat it. Civilization may help most of us live longer, safer, and more comfortably, but even in the penthouse, we have no more defeated nature than beavers have after they've built a dam.

The wild is where we all come from. As the ice started retreating 18,000 years ago, every human on Earth lived in a hunting-gathering culture, and every human was a wilderness traveler. Indigenous people still link us to that past and the human story of the boreal forest. Their unique and deep knowledge can lead the way in preserving this critical biome. And as previously noted, Indigenous people are leading the way in preserving and conserving this superlative biome. But Indigenous people can't do it alone. We need to use our ballyhooed intelligence to adapt our consumptive ways and our reactive solutions, innovate, pay attention to biodiversity coldspots as well as the sexier hotspots and remember that no matter how clever, we depend on large, intact ecosystems to maintain a liveable planet—and the boreal forest is the largest and most intact.

The rest of us may or may not go adventuring or on nature retreats there. Still, the possibility remains, and, as a matter of spirituality, it is invaluable to know that primal places such as the northern wilderness of the boreal forest and tundra—the Boreal Wilderland—are still here. As a matter of survival, it would be best to keep it that way.

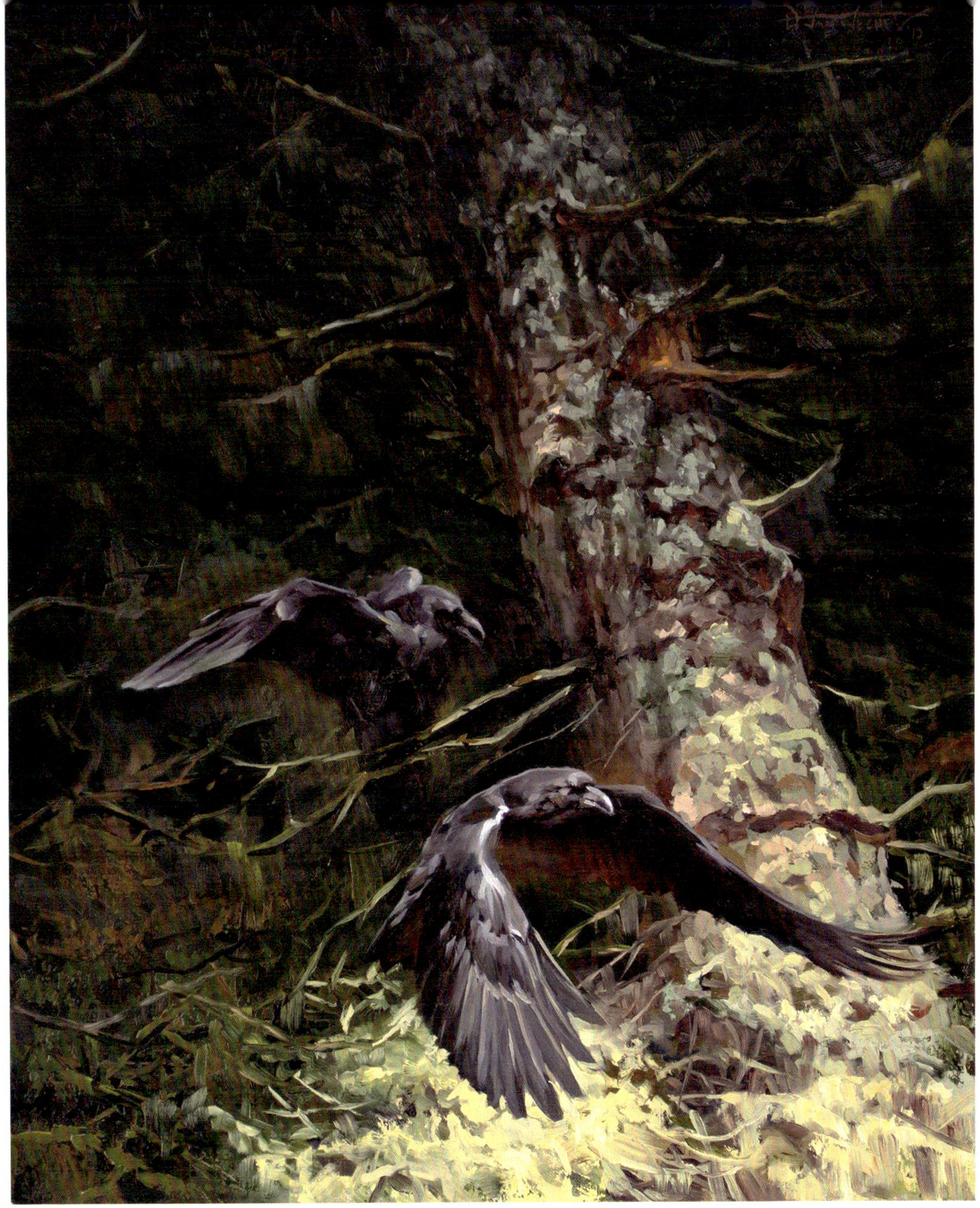

Acknowledgments

FIRST, a special acknowledgment to Luc Walter and the Luc Walter Foundation. After a long, strong run with the book and on the verge of a breakthrough, we hit a few hurdles and had run low on momentum by 2022. With perfect timing, Luc stepped in and provided the funding that has pushed this project to completion. Generous support was also provided by Burlington City Arts.

After nearly a quarter century, there are a LOT of other people to thank. To economize space, I will name only those not mentioned in the main text. Roughly in chronological order, I would like to thank the following:

Gordon Colby, brand manager for Mad River Canoe Company in 2001 and more recently, MTI Life Jackets

Susan Casey-Lefkowitz, Anthony Swift, and a host of people at the Natural Resources Defense Council (NRDC)

Marilyn Heiman, Cathy Wilkinson, and others at the Boreal Songbird Initiative

Larry Innes, Kelly Acton, Tanya Moore, and others at the Canadian Boreal Initiative (CBI)

Buff Grubb, Mad River Canoe Company

Roselyne Hébert and many more at Bonjour Québec

Jean Paquet, Norpaq Aviation

The Innu leaders of Tshiuetin Rail Transportation

Lisa Koperqualuk and others at Makivvik Corporation

Johnson Outdoors and Eureka!

Chota Outdoor Gear

Wendy Barber, EyeWearCliCs

Claude Ménard

Luba Mycio-Mommers and others at the Canadian Wildlife Federation

Trish Manning, Destination Ontario (retired and married to the late Steve Bruno)

Albert Rogalinski, Goldseekers Canoe Outfitting and Wilderness Expeditions

Gary Youngblood, former chief ranger at the Gates of the Arctic National Park and Preserve

Victoria Hust, The Center for Circumpolar Studies

Amy Souers Kober, American Rivers

Staff at the Bureau of Land Management

Gary and Kathy Grant in Hubbards, Nova Scotia. Dear friends. Their unbridled hospitality and patience while I refitted *Skidbladnir* were above and beyond.

Darrin Swinimer and everyone at Gold River Marina

Keith Campbell, a habitué of Gold River Marina, an old salt whose experience and advice pulled me out of the depths

Captain Gregg Colp and Vicki Piersig who own Below The Salt Distillery near Port Hawkesbury, Nova Scotia, who after coming by to admire *Skidbladnir*—they love wooden boats—invited Frank and me to their house to sample their products and see Vicki's photos. All great art, including award-winning rum, whiskey, and gin. Walking distance fortunately.

Steve Megonigle, a boater at Ballantyne's Cove Marina, Nova Scotia, who replaced our broken throttle cable AND bought us a case of Anchor beer

Bobby Burns, another friendly and generous Ballantyne's Cove boater

More boaters than I could possibly fit in here, all along the coast of the Gaspé Peninsula, though I will single out Matane, Quebec, marina manager and master brewer Rachel Lacombe

Finally, Captain John Mott at Barcombs Marina in Rouses Point, New York, on Lake Champlain; Bos Dewey, who helped crew down Lake Champlain late in October; and Chip Taube, owner of Chipman Point Marina in Orwell, Vermont, all helped enormously during a tragic time at the end of the voyage. *Skidbladnir* now lives at Chipman Point and, after a year of neglect while I finished this book, will be treated right this fall and thereafter.

Discerning readers may have deduced that you meet far more people on coastal sailing trips than on wilderness canoe expeditions.

AND... finally finally, because they were naturally the last stage of this endeavor, the personnel at Page Two, those I know and probably many I don't: Co-founder and co-owner Jesse Finkelstein, project manager Natassja Barry, substantive editor Scott Steedman, copy editor Adrineh Der-Boghossian, designer Peter Cocking, and marketing specialist Tessa Eisenberg

Notes

1: The Boreal Forest and "Wilderland"

p. 28 *"biodiversity coldspots":* Peter Kareiva and Michelle Ann Marvier, "Conserving Biodiversity Coldspots: Recent Calls to Direct Conservation Funding to the World's Biodiversity Hotspots May Be Bad Investment Advice," *American Scientist* 91, no. 4 (July–August 2003): 344–51.

p. 35 *"when mushrooms become plentiful":* Olaus Johan Murie, *Alaska-Yukon Caribou*, North American Fauna 54 (United States Department of Agriculture, 1935), 41.

2: Architects of the Boreal Forest

p. 53 *"I did not see one cart-load of earth":* Jacques Cartier, *The Voyages of Jacques Cartier*, trans. and ed. H.P. Biggar (Public Archives of Canada, 1924), 12.

3: A Tour of the Boreal Forest

p. 68 *Some excerpts from "Across Labrador's Spine":* Rob Mullen, "Across Labrador's Spine," *Che-Mun: The Journal of Canadian Wilderness Canoeing*, Outfit 139 (Winter 2010): 6–9, 11.

4: Non-Human Mammals

p. 101 *a paper proposed splitting* Rangifer tarandus*:* Lee Harding, "Available Names for *Rangifer* (Mammalia, Artiodactyla, Cervidae) Species and Subspecies," *ZooKeys* 1119 (August 2022): 117–51.

p. 121 *In a study conducted in Finland:* Katarina Meramo, Ville Vasko, Tia-Marie Pietikäinen, Veronika N. Laine, Otso Ovaskainen, and Thomas M. Lilley, "Tree-Volume and Forest Age Increase Bat Species Diversity in Boreal Urban Landscape," *Landscape Ecology* 40, art. 23 (January 2025): 1–18.

p. 137 *One Arctic fox was radio-tagged:* Eva Fuglei and Arnaud Tarroux, "Arctic Fox Dispersal from Svalbard to Canada: One Female's Long Run Across Sea Ice," *Polar Research* 38 (2019).

p. 137 *the interactions of Arctic and red foxes:* Sandra Lai, Chloé Warret Rodrigues, Daniel Gallant, James D. Roth, and Dominique Berteaux, "Red Foxes at Their Northern Edge: Competition with the Arctic Fox and Winter Movements," *Journal of Mammalogy* 103, no. 3 (June 2022): 586–97.

p. 137 *all native boreal forest red foxes in North America:* Jan Kamler and Warren B. Ballard, "A Review of Native and Nonnative Red Foxes in North America," *Wildlife Society Bulletin* 30 (June 2002): 370–79.

p. 139 *found evidence only for native foxes:* Mark J. Statham, Benjamin N. Sacks, Keith B. Aubry, John D. Perrine, and Samantha M. Wisely, "The Origin of Recently Established Red Fox Populations in the United States: Translocations or Natural Range Expansions?" *Journal of Mammalogy* 93, no. 1 (2012): 52–65.

5: Birds

p. 149 *the whooping crane, the trumpeter swan, and the bald eagle:* The percentages of populations breeding in the North American boreal forest are from the Boreal Songbird Initiative's "Comprehensive Guide to Boreal Forest Birds," based on research and publications done by Dr. Peter Blancher and Dr. Jeffrey Wells. See borealbirds.org/comprehensive-boreal-bird-guide for more information.

p. 149 *his beak can hold more than his bellican:* "Quote Origin: A Gorgeous Bird Is the Pelican, Whose Beak Will Hold More Than His Bellican," Quote Investigator, June 20, 2020, quoteinvestigator.com/2020/06/20/pelican.

p. 179 *Cooper's hawk being prepped for release:* Rob Mullen, "Cooper's Hawk Down" [the editor's title, not mine], *Canadian Wildlife Magazine*, June 2010—I think.

6: Lower Vertebrates

p. 223 *They grow to six feet, and researchers cite reports:* Arcadi N. Matveyev, Nikolai M. Pronin, Vitali P. Samusenok, and Charles R. Bronte, "Ecology of Siberian Taimen *Hucho taimen* in the Lake Baikal Basin," *Journal of Great Lakes Research* 24, no. 4 (1998): 905–16; Juraj Holčik, Karol Hensel, Josef Nieslanik, and Ladislav Skácel, *The Eurasian Huchen, Hucho hucho: Largest Salmon of the World* (Dr. W. Junk Publishers, 1988).

p. 223 *conservation efforts and research on taimen are on the rise:* "International Taimen Initiative," Wild Salmon Center, accessed November 2, 2025, wildsalmoncenter.org/campaigns/international-taimen-initiative.

p. 232 *The humble mink frog is an excellent example:* Amphibian Species of the World v. 6.2, "*Aquarana septentrionalis* (Baird, 1854)," by Darrel R. Frost, American Museum of Natural History, accessed November 3, 2025, amphibiansoftheworld.amnh.org/Amphibia/Anura/Ranidae/Aquarana/Aquarana-septentrionalis.

7: Cultures of the Boreal Forest

p. 242 *Truth and Reconciliation Commission of Canada has deemed a cultural genocide:* "Reports," National Centre for Truth and Reconciliation, accessed September 22, 2025, nctr.ca/publications-and-reports/reports.

p. 242 *the Chukchi kept the Cossack major's head as a trophy:* James Forsyth, *A History of the Peoples of Siberia: Russia's North Asian Colony 1581–1990* (Cambridge University Press, 1994), 146.

p. 249 *Champlain wrote books based on his:* Samuel de Champlain, *The Voyages and Explorations of Samuel de Champlain (1604–1616)*, trans. Annie Nettleton Bourne (The Courier Press Limited, 1911).

p. 251 *examples and historical accounts of birchbark canoes:* Harri Luukkanen and William W. Fitzhugh, *The Bark Canoes and Skin Boats of Northern Eurasia* (Smithsonian Books, 2020).

p. 251 *use of birchbark canoes among at least the Saami:* William Fitzhugh and Harri T. Luukkanen, "The Indigenous Watercraft of Northern Eurasia," *Vestnik of Saint Petersburg University History* 64, no. 2 (January 2019): 474–98.

p. 252 *Annual report of the Bureau of American Ethnology to the Secretary of the Smithsonian Institution:* Bureau of American Ethnology, "Birch-Bark Canoes of Wild Rice Gathering Ojibwa," plate LXXIV, *Nineteenth Annual Report*, 1065, repository.si.edu/bitstream/handle/10088/91702/Wild Rice Gatherers of the Upper Lakes.pdf.

p. 252 *Harri Luukkanen and Dr. William Fitzhugh's book:* Luukkanen and Fitzhugh, *The Bark Canoes and Skin Boats of Northern Eurasia*.

p. 253 *The Keewaydin experience: Anishinaabe heritage:* For Keewaydin history, see Brian A. Back, *The Keewaydin Way: The Story of the World's Oldest Canoe-Trip Camp* (Roy Waters Scholarship Fund, 2004).

p. 254 Crossroads of Continents: Cultures of Siberia and Alaska *exhibition:* William W. Fitzhugh and Aron Crowell, eds., *Crossroads of Continents: Cultures of Siberia and Alaska* (Smithsonian Institution Press, 1988).

p. 254 *Fourteenth-century Japanese junks disabled by storms:* George I. Quimby, "Japanese Wrecks, Iron Tools, and Prehistoric Indians of the Northwest Coast," *Arctic Anthropology* 22, no. 2 (1985): 7–15.

p. 256 *Ignace Tonené warned his people:* Bruce W. Hodgins and James Morrison, "Tonené, Ignacia (Nias)," in *Dictionary of Canadian Biography*, vol. 14, University of Toronto/Université Laval, 2003–, accessed October 1, 2025, biographi.ca/en/bio/tonene_ignace_14E.html.

p. 257 *The great canoe has come back from the abyss:* David Neel, *The Great Canoes: Reviving a Northwest Coast Tradition* (University of Washington Press, 1995), 1.

8: Ecosystem Services

p. 259 *Canadian Boreal Initiative report:* Sara Wilson, *Counting Canada's Natural Capital: Assessing the Real Value of Canada's Boreal Ecosystems* (Canadian Boreal Initiative, 2005), pembina.org/pub/counting-canadas-natural-capital.

p. 261 *Recent records show a clear, accelerating rise:* Michael E. Mann, Raymond S. Bradley, and Malcolm K. Hughes, "Northern Hemisphere Temperatures During the Past Millennium: Inferences, Uncertainties, and Limitations," *Geophysical Research Letters* 26, no. 6 (March 1999): 759–62.

p. 261 *"The planet will be fine . . . but":* George Carlin, "Jammin' in New York," recorded April 24–25, 1992, at the Paramount Theater, Madison Square Garden, New York City, and aired November 10, 1992, on HBO.

p. 262 *the boreal may already be a carbon source:* Bailu Zhao, Qianlai Zhuang, Narasinha Shurpali, Kajar Köster, Frank Berninger, and Jukka Pumpanen, "North American Boreal Forests Are a Large Carbon Source Due to Wildfires from 1986 to 2016," *Scientific Reports* 11, art. 7723 (2021): 1–14.

p. 262 *NOAA reported the same for the Arctic:* "2024 Headlines and Overview," NOAA Arctic Report Card, National Oceanic and Atmospheric Administration, accessed November 3, 2025, arctic.noaa.gov/report-card/report-card-2024.

p. 262 *recovery to sink status in a century:* Michelle C. Mack, Xanthe J. Walker, Jill F. Johnstone, Heather D. Alexander, April M. Melvin, Mélanie Jean, and Samantha N. Miller, "Carbon Loss from Boreal Forest Wildfires Offset by Increased Dominance of Deciduous Trees," *Science* 372, no. 6539 (April 2021): 280–83.

p. 262 *Bill McKibben . . . once pointed me to a paper:* Jamie Fanous and William R. Moomaw, *A Critical Look at Forest Bioenergy: Exposing a High Carbon "Climate Solution,"* Climate Policy Brief no. 8 (Global Development and Environment Institute, Tufts University, March 2018). See also K.B. Matthews, Doug Wardell-Johnson, Dave Miller, Nuala Fitton, Ed Jones, Stephen Bathgate, Tim Randle, Robin Matthews, Pete Smith, and Mike Perks, "Not Seeing the Carbon for the Trees? Why Area-Based Targets for Establishing New Woodlands Can Limit or Underplay Their Climate Change Mitigation Benefits," *Land Use Policy* 97, art. 104690 (September 2020).

9: Conservation Threats and Action

p. 276 *"Indigenous governments across the boreal forest biome of Canada":* Jeffrey Wells, Natalie Dawson, Nada Culver, Frederic A. Reid, and Shaunna Morgan Siegers, "The State of Conservation in North America's Boreal Forest: Issues and Opportunities," *Frontiers for Global Change* 3, art. 90 (July 2020): 1–18. Dr. Wells has been a leader in boreal forest conservation for years. He was the senior scientist at the Boreal Songbird Initiative when WREAF started working with them in 2005 and is now a senior vice president at the Audubon Society, heading up their boreal forest programs. He has been a strong supporter of this project for many years.

p. 276 *Bears are not good at taking adult caribou:* Mathieu Leblond, Christian Dussault, Jean-Pierre Ouellet, and Martin-Hughes St-Laurent, "Caribou Avoiding Wolves Face Increased Predation by Bears—Caught Between Scylla and Charybdis," *Journal of Applied Ecology* 53, 4 (August 2016): 1078–87.

p. 277 *"Humans are altering the composition and structure":* Leblond et al., "Caribou Avoiding Wolves Face Increased Predation."

p. 277 *In an article on Ungava caribou, Susan C. Morse wrote:* This quotation is from a presentation Susan gave at a caribou conference in 2018.

p. 277 *In November 2024, the U.S. Environmental Protection Agency:* United States Environmental Protection Agency, "EPA Announces New Draft Framework to Advance Consideration of Cumulative Impacts on Communities," news release, November 21, 2024.

p. 277 *"Indigenous-led CEAs have started to foster considerable":* Corrie Greaves and Lael Parrott, "Re-Grounding Cumulative Effects Assessments in Ecological Resilience," *Environmental Impact Assessment Review* 105, art. 107403 (March 2024).

p. 277 *The biome will also become increasingly important:* Wells et al., "The State of Conservation."

p. 278 *"We have met the enemy, and he is us":* "Earth Day: A Story of Us," National Air and Space Museum, April 22, 2021, airandspace.si.edu/stories/editorial/earth-day-story-us.

p. 278 *the Natural Resources Defense Council (NRDC) produced the report:* Jennifer Skene, with significant contributions from Shelley Vineyard, *The Issue with Tissue: How Americans Are Flushing Forests Down the Toilet* (National Resources Defense Council, February 2019), nrdc.org/sites/default/files/issue-tissue-how-americans-are-flushing-forests-down-toilet-report.pdf.

p. 278 *the NRDC produces an annual scorecard:* With over a million acres of boreal forest clear-cut for pulp (NRDC), most for tissue paper and most of that for the United States, another way to dramatically reduce deforestation, reduce water waste (over 20 gallons per roll), and provide personal health benefits at the same time is to join the growing movement toward using various forms of bidets. Inexpensive conversions are now available that save water and substantially reduce or even eliminate toilet tissue use.

A Last Word About the "Wild" in Wilderness

p. 282 *dug out my pocket video, and filmed his retreat:* "Grizzly Attack Aftermath Aug 17, 2013," posted September 6, 2013, by Rob Mullen, YouTube, 29 sec., youtu.be/2otTBEtkWpg.

Index

Artwork and photographs are indicated by page numbers in italics

Agent Orange, 275–76
Alaska, 266. *See also* Angiaak Pass, Brooks Range; Kobuk River; Noatak River; Ogriveg River; Reed River
alder flycatcher, 199
alders, 29
algae, 32
Algonquian peoples, 53, 242, 243, 247, 250–51, 255, 256
Algonquin Wildlife Research Station, 84
American bittern, *148*, 173
American dipper, 204, *204*
American ermine, 140, *194*
American goldfinch, 210, *211*
American goshawk, 181, *181*
American larch (tamarack), 29, 56
American pipit, 205, *206*
American robin, 205
American three-toed woodpecker, 192, 195
American toad, 232
American wigeon, 162, *162*
amphibians, 48, 231–32
Amur falcon, 187
Amur leopard, 122
Angiaak Pass, Brooks Range (Alaska), 24, *25*, 91, 92, *92*, 112, 206
animals (wildlife): approach to, 98; ecosystem importance, 269–70; hibernation, 48; migration, 48, 50–51; pelage and plumage, 46–47; scientific names, 232; size, 46. *See also* biodiversity
Anishinaabe peoples, 84, 239, 241, 251
Arctic char, 220
Arctic fox, 47, *48*, 137, *137*
Arctic grayling, *221*, 221–22
Arctic ground squirrel, 48, 113
Arctic hare, 120, *120*
Arctic loon, 154
Arctic tern, 156
art, 243–45, *244*, 247–48
aspens, 29
Athabascan peoples, 242, 243, 247, 256, 266
Atlantic salmon, *218*, 219
Atlantic sturgeon, 218

Baikal seal, 142
bald eagle, *62*, 94, *95*, 150–51, *151*, 185, *258*
balsam fir, 29
balsam poplar, 42
Banning, Dawn: *Blaze of Dawn*, *55*; *Red Canoes*, *14*
barred owl, 188, *189*, 191
Barrens, 38. *See also* Labrador Barrens
Barrow's goldeneye, 163
bass, *146*, *214*, 226, *227*
Bateman, Robert, 10–11, 21, 64, *65*; *Lynx in Snow*, *124*; *Mountain Monarch*, *186*; *Winter Snowshoe Hare*, *40*
bats, 121, *121*
Baughan, Kalon, *52*; *Reflections*, *172*
bay-breasted warbler, 206, *207*
bears, 48, 128–31, 276. *See also* black bear; grizzly bear; polar bear
beavers, 114, *117*, 117–19, *118*, 249
belted kingfisher, 192, *192*
Benuen, Jordanna, 236, *236*
Bering Sea, 266
Besse, Linda, *90*, 91; *Guardian*, *183*; *Onlooker*, *127*; *Outgunned*, *70*; polar bear (oil), *267*; *A Reintroduction*, *113*; *Ursus maritimus*, *267*; willow ptarmigan (oil), *177*; *Winter Solstice*, *109*
biodiversity, 21, 59, 98, 270, 272; coldspots, 28; hotspots, 27, 259. *See also* animals; ecosystems
birchbark canoe, *244*, 249, 250–51, *252*, 254, 256–57, *257*
birches, 29
birds, 28, 48, 50, 147–49
black-backed woodpecker, 195
black bear: about, 128; encounters with, 24, 73, 128, 130, 144; omnivorous habits, 131; paintings and sketches, *126–27*, *128*, *129*, *277*; photographs, *25*, *31*, *131*, *143*; skulls, *243*, 248, *248*
black-billed capercaillie, 176–77
black-billed magpie, 200
blackbirds, 210
blackburnian warbler, *207*, 209
black-capped chickadee, 196
blackflies, 38, *269*, 270
blackpoll warbler, 206, *207*, 209–10
black spruce, 29, 38, 41, 56, 59, *77*
black woodpecker, 195
Blakiston's fish owl, 191
Bloodvein River First Nation, 84, 241
blue-headed vireo, *198*, 199
bluethroat, 199
blue-winged teal, 162
bobcat, 122, 124, *124*
bogs, 36, *37*, 53, 259, 262, 263
Bohemian waxwing, 206, *206*
Bonaparte's gull, 156
Bonaventure Island (Quebec), 76
boreal chickadee, 196, *196*, *197*
boreal chorus frog, 48, 232, *233*
boreal forest: about, 27–28; art and, 243–45, *244*, 247–48; author's introduction to, 17–18, 20–21; carbon storage, 31, 262; conservation, 276, 277, 283; ecosystem services, 259; extent, 28, *29*; fire and, 41, 56–59; Indigenous Peoples and, 242, 243, 244, 247, 254–55; migration and, 48, 50–51; precursor to, 61–63; threats to, 259, 275, 276; trees, 28–29, 41–42, 56, 62–63; understory, 29, 31; water and, 41, 51–53, 263, 266; as wilderness, 36, 38, 236; winter and, 41–42, 46–48
boreal owl, 191
Boreal Songbird Initiative, 276
broad-winged hawk, 185
Brody, Bill, *87*, *90*, 91; *Maidens in Cloud*, *89*; waterfall sketch (ink wash), *87*
Brooks Range (Alaska), 87, *89*, 205, 281. *See also* Angiaak Pass; Kobuk River; Noatak River; Reed River
brook trout, *220*, 221
brown bear, 128
brown dipper, 204
brown trout, *217*, 219
bryophytes, 31–32, *33*
bufflehead, 63, *63*, 163, *166*
buteos, 182–85

Canada, 242, 248–49
Canada goose, 147, *148*, 157, *159*
Canada jay, 47, 84, *84*, 200, *201*
Canada lynx, 122, *122*. *See also* lynx
Canadian Boreal Initiative (CBI), 84, 276; *Counting Canada's Natural Capital*, 259
Canadian sandhill crane, 170, *170*
Canadian Shield, 38, 53, *77*
canoe, 53, *244*, 249, *250*, 250–52, *251*, *252*, 253–54, 256–57, *257*
Cape May warbler, *193*, 206, *207*
carbon and carbon storage, 31, 36, 260–62
caribou: about, 64, 101–4; curiosity, 144; encounters with, 73, *75*; fungi and, 35; paintings, *2*, *57*, *70*, *85*, *100*, *103*, *105*, *106–7*; photographs, *30*, *34*, *49*, *50*, *71*, *74*, *99*, *102*, *104*, *107*, *144*, *269*; threats to, 276, 277; winter adaptations, 47. *See also* reindeer
carnivorous plants, 32, 263
cedar waxwing, 206, *206*
Champlain, Samuel de, 249
Chapleau River (Ontario), 21, 83
chestnut-sided warbler, *208*
chickadees, 47, 196
Chinook salmon, 218, 219–20
chipmunks, 114
chum (dog) salmon, 218, *219*, 220, *266*
climate and climate change, 35, 62, 103, 109, 120, 137, 260–62, 277
clubmosses (Lycopods), 31, *33*
common eider, *169*, 170
common goldeneye, 162, *167*
common grackle, 210
common kestrel, 187
common kingfisher, 191
common loon, *146*, *152*, 154
common merganser, 163, *163*, *165*
common snapping turtle, 229, *231*
common tern, 156
common yellowthroat, 209, *292*
conifers, 29, 41, 42, 56, 58
Connecticut warbler, 206
conservation, 276–77, 278, 283. *See also* Red List
consumers, 35
Cooper's hawk, 179, *179*
cougar (mountain lion), 122, *122*

Courteau, Jean-Louis: *Boreal Watch*, *5*; *Diving Loons*, *153*; *Eagle's Peak*, *26*; *Life's Blood*, *51*; *Tenace*, *45*; *Water Dance*, *161*
coyote, 134–35, *135*, 269
cranes: sandhill, *170*, 170–73, *171*, *172*; whooping, 149, *149*
Cree, 20–21, 81, 83, 112, 218, 239, *244*, 249, 255
cumulative effects assessments (CEAS), 277
cyanobacteria ("blue-green algae"), 32

Dall (thinhorn) sheep, *96–97*, 112, *112*
dark-eyed junco, 210, *211*
decomposers, 35
deer. *See* caribou; elk; moose; red deer; Siberian musk deer; Siberian roe deer
Dene. *See* Athabascan peoples
De Pas River (Quebec), *52*
dippers, 204
dog (chum) salmon, 218, *219*, 220, *266*
doll, tea, *244*
double-crested cormorant, 155
downy woodpecker, 192–93, *194*
ducks, 161–70

eagles, 150–51, 185–86
eastern blacknose dace, *147*, 228, *228*
eastern newt, 231, *233*
eastern phoebe, *198*, 199
ecosystems, 209, 259, 266, 270, 272. *See also* biodiversity
elk (wapiti), 109, *110*
ermines, 139–40
Eurasian bittern, 173
Eurasian eagle-owl, 191
Eurasian ermine, 140
Eurasian goshawk, 181
Eurasian lynx, 122
Eurasian northern wren, 204
Eurasian nuthatch, 200
Eurasian perch, 226
Eurasian pygmy owl, 191
Eurasian three-toed woodpecker, 192, 195
Eurasian wigeon, 162
Eurasian wryneck, 192
European grayling, 222
European pond turtle, 231
European robin, 199
Europeans, 248–49
evening grosbeak, *211*, 212

falcons, 186–87
Farshmin, Jon, *44*, 84
ferns, 31, *33*
finches, 210–12
fire, 41, *56*, 56–59, *58*, *59*, *262*
fish, 214–29, 263, 266
fisher, 140, *141*
Fitzhugh, William W., 21, 252–57
Florida, 216–17
flycatchers: old world, 199; tyrant, 196, 199
Foggett, Lindsey, 32, 80–81; *Autumn Trail*, *126–27*; *Peaceful Solitude*, *108*; *Sun Catcher*, *123*; *Tundra Beauties*, *158*
foxes, 137–39
fragmentation, habitat, 276
frogs and toads, *214*, 231–32
fungi, 28, *34*, 35, 269, *270*

gadwall, 162
garter snake, common, *230*, 231
geese, 147–48, 157, 161
George River (Nunavik): black bears, 131, 144; caribou, 64, *100*, *102*, 102–3; confluence with Nutillilik River, 113; expedition, 38, 64; foxes, 137; Indigenous Peoples, 239; La Colline Misurtuq, *26*, *60*; northern hawk-owl, 191; northern lights, *39*; photographs, *65*, *66*; sculpins, 229; short-tailed weasel, 139; wetlands, *52*
glaucous gull, *155*, 156
golden-crowned kinglet, 204, *204*
golden eagle, *60*, 150, 185
Gorman, Stephen (Steve), 69, *74*, *83*, 93, 142–43, 224
goshawks, 181, *247*
gray-headed chickadee, 196
gray heron, 173
graylings, 221–23
gray wolf, *7*, *13*, *45*, *131*, 131–34, *132*, *133*, *134*, *135*, *145*
great black-backed gull, 156
great blue heron, 173, *173*, *174*
great blue whale, *264–65*
greater white-fronted goose, 157, *160*
greater yellowlegs, 174
great gray owl, 188
great horned owl, 188, *188*, *190*, *191*
grebes, 155
greenhouse effect and gases, 260
green-winged teal, 162
grizzly bear: about, 128; Arctic ground squirrels and, 113; encounters with, 91, 92, 93, 94, 130, 281–82; omnivorous habits, 131; paintings, *127*, *280*; photograph, *266*; tracks, *86*, *274*, *277*
grosbeaks, 212
grouse, 47, 176–78
gulls, 155–56
gyrfalcon, *186*, 186–87

hairy woodpecker, 192–93
hares, 119–20
Hargreaves, Julia, *83*, 93, *223*; Boreal chickadee (acrylic), *196*; *Old and New*, *247*; *Voyeur*, *193*; yellow-rumped warbler (acrylic), *208*
Harlequin duck, *166*, 167, *168–69*
Harricana River (Quebec), *77*, 80, 124
hawks, 179–85
hermit thrush, 205
herons, 173
herring gull, 156, *264*
hibernation, 48
Holocene, 41
hooded merganser, 163, *164*, *165*
Hopkins, Frances Anne: *Canoe Manned by Voyageurs Passing a Waterfall*, *250*
horned grebe, 155, *155*
hornworts, 31
horsetails, 31
Howe, Nancy: *Above Us Only Sky*, *112*; *Oomingmak on the Rise*, *46*; *Of a Secret Nature*, *124*; *True North*, *238*
Hudson Bay, 64
Hudson Bay Lowlands, 36
Hudson's Bay Company, 81, 83, 120, 249, 256
humans, 242. *See also* art; canoe; Europeans; Indigenous Peoples

ice ages, 41, 61, 62
ice sheets, 53, 63, 241
Indian pipe, 32, *33*
Indigenous Peoples: art, 243–45, *244*, 247–48; author's interactions with, 239, 241; birchbark canoe, *244*, 249, 250–51, *252*, 254, 256–57, *257*; boreal forest and, 242, 243, 244, 247, 254–55; colonialism and genocide, 242, 249; conservation and, 276, 283; salmon and, 266; Teme-Augama Anishnabai, 255–57. *See also* Algonquian peoples; Anishinaabe peoples; Athabascan peoples; Cree; Innu; Inuit
Innu, 12, 64, 65, 68, 102, 114, *201*, 241, *244*, 245, *248*
insects (invertebrates), 21, 209, 270
International Boreal Conservation Campaign (IBCC), 276
International Union for the Conservation of Nature (IUCN), 98
Inuit, 68, *74*, 104, 131, 239, 241, 255
iron mining: ore truck, *276*; slag heaps, *28*, *275*
jack pine, 29, 41, 59, *77*
James Bay Aboriginal People, *240*
Japanese quail, 176
Japanese waxwing, 206
jays, 200
Johnson, Cole, 51, 68, 104, 131, 133, 134, 144, 179, 239; black bear (graphite), *129*; brown trout (graphite), *217*; buffleheads (charcoal), *166*; *Dangerous Lunatic*, *145*; *Darkness There and Nothing More*, *200*; *Drummer*, *178*; *Flight of Pintails*, *160*; *Ice Bear*, *130*; *Labrador Bull*, *103*; peregrine falcon (charcoal), *187*; photographs, *69*; red-tailed hawk (charcoal), *182*; *Ruh Ro*, *135*; *Spring Snow*, *164*; *And Then There Was One*, *214*; *Trumpeter Swan*, *150*; *Tundra Tailchase*, *271*; walleyes (charcoal), *225*
Johnson, Jay J.: *Near Kamestastin*, *105*
junco, dark-eyed, 210, *211*

Kamestastin Lake (Labrador), *36*, *57*, 64–65, *70*, 101–2, 241, 248, *248*, *272*
Kapuskasing Lake (Ontario), 46–47, 131, 133
Keewaydin Temagami (camp), 252–54
kingfishers, 191–92
kinglets, 204
Kobuk River (Alaska): about, 84, 87; Arctic char, 220; Arctic loon, 154; bogs, *37*; expeditions along, 91, 92–94; field painting of, *86*; glaucous gulls, 156; merlin, 187; Onion Portage, *93*, *94*; trumpeter swans, 150; whitefish, 222, *223*. *See also* Noatak River; Reed River
Kogaluk, Labrador Barrens, *52*, 68, *68*, 69, 104
Kromschroeder, Lee: great horned owl (acrylic), *191*; *Mama's Boy*, *132*; *Winter's Warmth*, *13*

Labrador, 273. *See also* Innu; Kamestastin Lake
Labrador Barrens, 38, 51, 52, 64, 68, 98, 133, 179, 278
Lac des Loups Marins (Quebec), 73, 142–43
Ladoga ringed seal *ladogensis*, 142
lake sturgeon, 216, 216–17, 218
Lake Superior: Nipigon Bay pictographs, 77; north shore expedition, 41, 83–84
Lake Superior National Marine Conservation Area, 42, 44, 45, 83–84
lake trout, 220, 221
larches, 29, 36, 42, 63
Lawrence, Rod, 52; *In the Shadows*, 138; *Tea Time*, 197
least flycatcher, 198, 199
least weasel, 139
lemmings, 50, 116, 188
lesser sandhill crane *canadensis*, 170, 170, 173
lesser yellowlegs, 21, 174, 175
lichens, 32, 33
little brown bat, 121
liverworts, 31
long-eared owl, 190, 191
long-tailed weasel, 47, 140
loons, 152–54
Loring, Stephen, 21, 27, 64, 69, 73, 74, 102, 142–43, 241, 243, 245, 247
Lycopods (clubmosses), 31, 33
lynx, 46–47, 59, 120, 120, 122, 123, 124, 125

magnolia warbler, 206, 207
magpies, 200, 201
mallard, 162
mammoth steppe, 61–62, 242
Manicouagan Reservoir (Quebec), 278
marbled godwit, 174, 175
marsh wren, 203
marten, American, 140, 141
McGuire, Jan Martin: *On the Fence*, 201; northern harrier (acrylic), 180; purple finch (acrylic), 210; red-winged blackbird (acrylic), 209; white-breasted nuthatch (acrylic), 202
McHuron, Greg: *Brooks Range Haven*, 157; *Golden Sunset Retreat*, 167
mergansers, 163
merlin, 187, 187
mice, 113–14
migration, 48, 50–51
mink, 42, 140, 141
minke whale, 73
mink frogs, 73, 74, 232, 233
minnows, 226, 228
Missinaibi River (Ontario): Agent Orange and, 275; Barclay Bay to Moose River Crossing (2001), 77–80; black bears and, 128, 130; as historic transport link, 249; Mattice to Moose Factory and Moosonee (2016), 81, 83, 83; Mattice to Moose River Crossing (2003), 80–81; moose and, 17, 108; paintings, 18, 19, 20, 22, 78, 79, 81, 237, 279; sandhill cranes and, 18; solo expedition (2001), 18, 20–21; spruce grouse and, 177; sturgeon and, 215–16; Thunder Falls camp, 38; walleye and, 224, 226
moose, 16, 17, 82, 104–9, 105, 106, 107, 108, 109, 111, 144, 272
Moose Factory (ON), 81, 83, 239, 249
Moose River (Ontario), 18, 218, 260
Moose River Crossing (ON), 20, 83, 218, 239
Morse, Susan C., 35, 103, 104, 122, 277
mosses, 31, 33
mottled sculpin, 228, 228–29
mountain lion (cougar), 122, 122
Mount Igikpak (Alaska), 25, 87, 92
mourning warbler, 206
Mullen, Frank (brother), 17, 18, 21, 38, 75, 75, 134, 226
Mullen, Jerry (dad), 17, 46, 75, 75, 76, 134, 275
Mullen, Rob: *Alaska Cold Rush*, 280; *Approaching Rain Squall*, 94; *Back to Business*, 21; *Belly of the Beast*, 57; *Birds of a Feather*, 148; black bear tracks (ballpoint pen), 277; *Brook Bath*, 213; *Chapleau River Dawn*, 82; *Coming Ashore*, 2; *Condo on the Water*, 117; *Confluence Cornucopia*, 264–65; *Conjuring House Rock*, 22–23; *Crossing Paths*, 16; *Curious George I & II*, 115; *Cutting a Wake*, 165; *Damsel in Distress*, 227; Devil Shoepack Falls, Missinaibi River (acrylic), 79; *Echoes of the Ice Age*, 91, 96–97, 112; *Into Every Life*, 95; *Eyes in the Night*, 189; *Fire and Ice*, 194; *First Light*, 18; *Flowing North*, 77, 80; *Forest Flickers*, 193; *Gravel Bar Blues*, 88; *Homeward Bound*, 6–7, 80; Kobuk River (acrylic), 86; *Leaf Gleaner*, 197; *Loon Rise*, 146, 215, 228; *Marsh Missile*, 164; *Marsh Morning*, 203; *Migration Moon*, 100; *Missinaibi Evening*, 81; *Neck Deep*, 118; *Nestled In*, 190; *Northern Night*, 39; *Nunavik Gold*, 60; *Off to Breakfast*, 184; *Old Dam Family*, 159; *Passage of Matashu*, 275; porcupine (ballpoint pen), 114; *Red Sails at Night—Fulmars Take Flight*, 75; red-tailed hawk (gouache), 185; *River's End*, 18–19; *Shoreline Hunter*, 231; *Snow Ending This Evening, Turning Colder, Chance of Rodents*, 189; *Snow Squall (aka RUN!)*, 268; *Solo*, 21; *Spring Peeper*, 292; *Trouble Bruin*, 128; *Turbulent Currents*, 168–69; *Veery Light*, 205; *View North*, 20; wildlife art, 18; *Wing Light*, 72
musk oxen, 46, 96–97, 112
muskrat, 119
mycoheterotrophic plants, 32
Nainette (dog), 68–69, 69
Natural Resources Defense Council (NRDC): *The Issue with Tissue*, 278
neotropical migrants, 48
Noatak River (Alaska): about, 84, 87; Arctic char, 220; expeditions along, 91–92; merlin, 187; musk oxen, 112; paintings, 88, 94; robin, 205; visit with Elder Ricky Ashby, 220, 239. *See also* Kobuk River; Reed River
northern Eurasian wren, 204
northern flicker, 193, 193, 195
northern flying squirrel, 116
northern gannet, 76, 155, 155
northern harrier, 180, 181, 181–82
northern hawk-owl, 191
northern lights, 38, 39, 64
northern nutcracker, 200
northern pike, 224, 224
northern pintail, 160, 162
northern raven, 199, 199, 200, 274
northern saw-whet owl, 189, 191
northern shrike, 199
northern waterthrush, 209
northern wheatear, 199
Norway spruce, 29
nuna (the land), 61
Nunavik. *See* George River; Nutillilik River
nuthatches, 200
Nutillilik River (Nunavik), 38, 113

Oberg, Ralph: *Eagle Rock Along the Missouri*, 64
Ogriveg River (Alaska), 143–44
Oliver, Steve: *If Looks Could Kill*, 141; *On the Move*, 106–7
olive-sided flycatcher, 199
osprey, 151, 184, 185–86
otters, 139, 140, 226, 226
ovenbird, 209
owls, 188–91
Pacific loon, 152
painted turtle, 229, 229
palm warbler, 206, 207
Pépin, Patricia: *Boreal Chorus Frog*, 233; *Boreal Morning*, 111; *Snails*, 54
perch, 224–26
peregrine falcon, 143, 144, 151, 186, 187, 271
permafrost, 262
Philadelphia vireo, 198, 199
pictographs, 77, 77, 242, 247
pied-billed grebe, 155, 155
pileated woodpecker, 193, 195, 195
Pimachiowin Aki (World Heritage site), 84, 276
pine grosbeak, 212, 212
pipits, 205
Pitcher, John C., 18, 20, 32, 104, 177, 192; American goshawk (watercolor), 181; canoe and tundra (watercolor), 67; *A Change of Mind*, 279; *Chilkat Sentinel*, 151; *Land of the Canoe*, 237; *Missinaibi Blues*, 78; northern harrier (acrylic), 181; *The Path Less Taken*, 234–35; *Sharing the River*, 139; spruce grouse (ink and acrylic), 176; wolf track in mud (acrylic), 134
pitcher plants, 263, 263
plant communities, 35–36
plovers, 173–74
polar bear, 64, 128, 130, 131, 239, 266, 267
porcupine, 114, 114, 117
Potter, John, 247; *Nestled In*, 171; *Nibi Manidoog*, 246; *Play Date*, 133; *Soul Searching*, 191
primary producers, 32
ptarmigan, 47. *See also* willow ptarmigan
purple finch, 210, 210

raccoon, 140
rapids, 52, 53, 77, 279
raptors, 179–91

ravens, 199, *199*, *200*, *274*
Red 124 (fire), 56, *56*, *58*, *59*, *262*
red-breasted merganser, 163
red-breasted nuthatch, 200, *204*
red deer, 109
red-eyed vireo, *198*, 199
red-footed falcon, 187
red fox, *136*, 137–39, *138*
Red List, 98
red-necked grebe, 155
redpoll, 212, *212*
red squirrel, 113, *115*, *116*
red-tailed hawk, 182, *182*, *185*
red-throated loon, 152, *152*
red-winged blackbird, *209*, 210
Reed River (Alaska), *16*, *25*, 92, 222, 281–82, *282*. *See also* Kobuk River; Noatak River
reindeer, *238*. *See also* caribou
reptiles, 48, 229–31
Rich, Andrea: *Merganser Brood*, *163*
ring-billed gull, 156
Rio Grande, 63
Rivière aux Mélèzes (Quebec), 69, 73, *143*, 144, 150, 260
Rivière du Chef (Quebec), 17–18
rock ptarmigan, 176
rodents, 113–19
rose-breasted grosbeak, *211*, 212
rough-legged hawk, 182–85, *183*
Rowell, Bonnie, 6, *6*, 68–69, 93, 121
ruby-crowned kinglet, 204
ruffed grouse, *178*, 179
rusty blackbird, 210

Saimaa ringed seal, 142
Sakhalin taimen, 223
salamanders, 231, *232*
salmon, 218–20, 263, 266, *266*
sandhill cranes, *170*, 170–73, *171*, *172*
sandpipers, 20, 173–74
scientific names, 232
Scotch pine, 29, 62, 63
Scott, Sandy: *Height of Land*, *152*
sculpins, 228–29
seals, 142–43
semipalmated sandpiper, 174, *175*
sharp-shinned hawk, 181
sheefish, *90*, 222, *222*, *223*
short-eared owl, 191
shrikes, 199
Siberia, 242
Siberian grouse, 176
Siberian jay, 200
Siberian musk deer, 109
Siberian nuthatch, 200
Siberian roe deer, 109
Siberian rubythroat, 199
Siberian taimen, 222–23, *223*
Siberian tiger, 122
Simard, Suzanne: *Finding the Mother Tree*, 35, 269
Skidbladnir voyage, *75*, 75–76, 249
skunk, striped, 139
slag heaps, *28*, 275
slimy sculpin, 228–29
smooth green snake, 231
snails, *54*, *270*
snakes, 231
snow goose, 161, *161*, 173
snowshoe hare, *40*, 47, *119*, 120, *120*, *268*
snowshoes, *244*
snowy owl, 50, 188, *188*, *268*
solitary sandpiper, *21*, 174, *175*
song sparrow, 210, *211*
sora, 173
sparrows, 210
spearheads, *245*
sphagnum moss, 31, 36
spotted sandpiper, 174, *175*
Sprague's pipit, 205
spring peeper, 48
spruce grouse, *176*, 177, *177*
spruces, 29, 62. *See also* black spruce; white spruce
squirrels: arctic ground, 48, 113; conifers and, 114; northern flying, *116*; red, 113, *115*, *116*
sturgeon, 216–18
sundew plants, 263, *263*
sunfish, 226
Swainson's thrush, 205
swamp sparrow, 210, *211*
swans, 149–50, 156–57

Tadoussac (Quebec), *76*
tamarack (American larch), 29, 56
tea doll, *244*
Teme-Augama Anishnabai (Temagami First Nation), *255*, 255–57
Tennessee warbler, 206
terns, 155, 156
thinhorn (Dall) sheep, *96–97*, 112, *112*
thrushes, 48, 205
thrush nightingale, 199
toilet and tissue paper, 278
towhees, 210
treecreepers, 200
trees, 28–29, 35, 41–42, 56, 62–63
trout, 219, 221
trumpeter swan, 149–50, *150*
tundra swan, *97*, 157, *157*, *158*, *159*
Tursujuq National Park, 69, 73–75, 239
turtles, 229, 231, *231*
two-barred (white-winged) crossbill, *211*, 212

understory, 29, 31
Ungava Peninsula: Tursujuq National Park and trans-Ungava expedition, 69, 73–75, 144, 150
Ungava seal, *74*, *142*, 142–43
Upper Missouri River Breaks National Park, 63, *64*, 121, 200
Ural owl, 191

Van Wechel, Dustin, *90*, 92; *CariBOO!*, *85*; *Off to Rap and Tap*, *283*
varied thrush, 205
veery, 205, *205*
Vermont, 134, 147, 179, 186, 193, 205, 269
vireos, 48, 199

wagtails, 205
walleye, 224–26, *225*
wapiti (elk), 109, *110*
warblers, wood, 48, 206–10, *207*, *208*
water, 41, 51–53, 263, 266
waxwings, 206
weasels, 139–40
Wells, Jeffrey, 96, 276, 277
western capercaillie, 176
Westin, Sue: *The Gift Tree*, *201*
whales, 76, 266
white-breasted nuthatch, 200, *202*
white-crowned sparrow, 210, *211*
whitefish, 222
white spruce, 24, 29, 38, 56, 77, 87, 91, 156
white-tailed sea eagle, *184*, 186
white-throated dipper, 204
white-throated sparrow, 210, *213*
white-winged (two-barred) crossbill, *211*, 212
whooper swan, 157, *160*
whooping crane, 149, *149*
Wicks, Derek, *90*; chestnut-sided warbler (acrylic), *208*; *Winter Sonata*, *110*
wilderness, 12, 236, 282–83
willow ptarmigan, *48*, *72*, 176, *177*, *179*, *271*
willows, 29
willow tit, 196
winter, 41–42, 46–48
winter wren, 204
wolf. *See* gray wolf
wolverine, 59, 140, *141*
wood bison, 112–13, *113*
wood duck, 162, *162*
wood frog, 48, 231, *233*
Woodland Caribou Provincial Park, *37*, *43*, *56*, *58*, 84, *247*, *261*, 263
woodpeckers, 192–95
wood thrush, 205
wood warblers, 48, 206–10, *207*, *208*
wrens, 204

yellow-bellied flycatcher, 199
yellow-bellied sapsucker, 192, *193*
yellow-billed loon, 154
yellow perch, 226, *226*
yellow rail, 173
yellow-rumped warbler, *207*, *208*, 209
yellow warbler, 206, *207*
Young, Paco: *The Fog Cutter*, *265*; *Shoreline Flurries*, *258*
Young, Steven, 61–63, 84, 241, 242

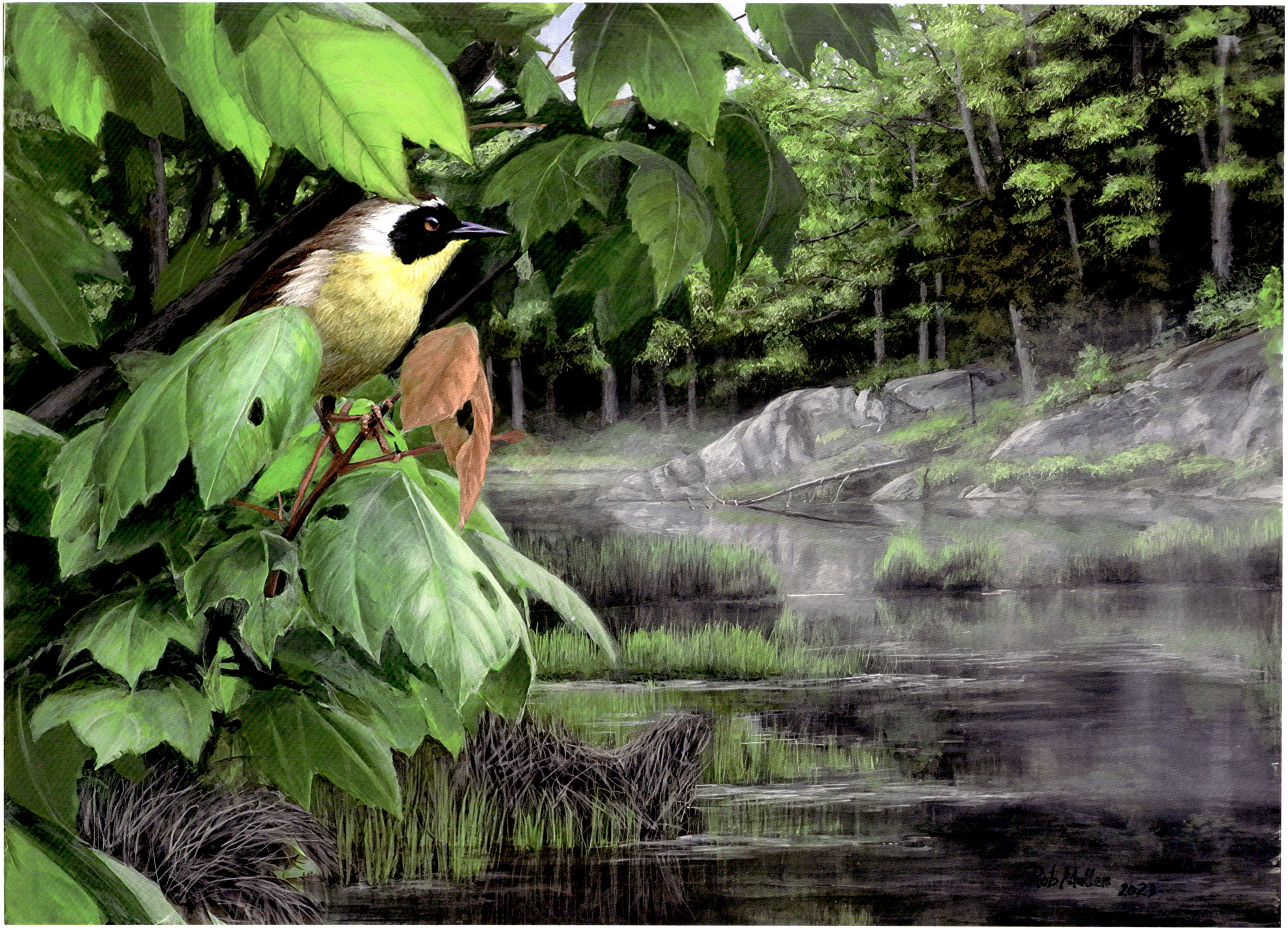
2023

About the Author

1964: One fine summer day, a friend of Dad's brought his new canoe to Preston Pond. The magical craft felt alive on the water. Three years and a Boy Scout canoeing merit badge later (my first of many), I bought a canoe with my paper route money.

Canoe and growing outdoor skills—check.

1966: While I was drawing in fifth-grade art class (I got in trouble when I drew in other classes), Mr. Kennedy stopped at my desk and said, "Rob, you could be a professional." Nine years later, my professors had me illustrating slime mold sporangia, spider genitalia, and other arcane, biological esoterica. My entomology professor suggested a career in biological illustration. I graduated and departed for New York City to become an artist.

Art and science—check.

1970: Entering high school, I had recently shot up to six feet, and my mother and the basketball coach got ideas. Despite the "how to play basketball" book Mom bought, the tryout was humiliating. Crushed, I walked home through the woods. Slowly. I hadn't thought about the kerfuffle my disappearance would cause. The state police were about to become involved when I showed up at the door. Undeterred, the maternal and coaching cabal coerced me into the wrestling team tryout. It was déjà vu all over again until the coach told us all to run laps around the gym. That I could do. And did until I was the only one left running. I was a competitive (and occasional champion) middle- and long-distance runner through high school, college, and on the New York City road-running circuit.

Confidence, strength, and stamina—check.

1971: I was home alone with my brother Frank one fine summer day, admiring the new 10-speed bike I bought with earnings from timber stand improvement work during vacations. Excited to try it, I suggested surprising our grandparents in Proctor, Vermont, 70 miles to the south. It worked beyond our expectations. In our excitement, we had neglected to leave a note.

A penchant for big adventures—check. (Yes, we made it. The parental ire was tempered by grudging respect.)

1978–1994: Built a successful art business in midtown Manhattan and had it destroyed by computers.

Work ethic and the fortitude to roll with big punches—check.

1994–2006: Back in Vermont, dove into wildlife art, exhibiting in nearly 100 art shows from coast to coast and overseas.

Logistics, planning, and a long list of professional friends—check.

2001–2023: Started these canoe trips, formed the Wilderness River Expedition Art Fellowship, and married Bonnie.

2026: We're done with the book (mostly). We can get back to other work and enjoy visits from my daughters, Toni and Lena, our three grandsons, and two step-grandchildren while living on Preston Pond with my 94-year-old father, our dogs, and a stray cat who is still here.

FACING: Common yellowthroat warbler. *Spring Peeper*, acrylic, Rob Mullen

Contributors

Writers

Rob Mullen, Wilderness River Expedition Art Fellowship (WREAF)
Dr. William Fitzhugh, Smithsonian Arctic Studies Center—National Museum of Natural History
Dr. Stephen Loring, Smithsonian Arctic Studies Center—National Museum of Natural History
Susan C. Morse, Keeping Track®
Dr. James Vogelmann, United States Geological Survey (retired)
Dr. Jeffrey Wells, National Audubon Society
Dr. Steven Young, The Center for Circumpolar Studies

Artists

Robert Bateman, British Columbia
Dawn Banning, Ontario
Kalon Baughan, Montana
Linda Besse, Washington
Bill Brody, Alaska/Florida
Jean-Louis Courteau, Quebec
Lindsey Foggett, UK/California
Julia Hargreaves, British Columbia
Nancy Howe, Vermont
Cole Johnson, New York
Jay J. Johnson, Massachusetts
Lee Kromschroeder, California
Rod Lawrence, Michigan
Jan Martin McGuire, Tanzania
Greg McHuron (in memoriam), Montana
Rob Mullen, Vermont
Ralph Oberg, Colorado
Steve Oliver, Pennsylvania
Patricia Pépin, Quebec
John C. Pitcher, Vermont
John Potter, Montana
Andrea Rich, California
Sandy Scott, Wyoming
Dustin Van Wechel, Colorado
Sue Westin, Vermont
Derek Wicks, Ontario
Paco Young (in memoriam), Montana

Principal Photographers

Steve Bruno (in memoriam), Ontario
Stephen Gorman (Steve), Vermont
Susan C. Morse, Vermont
Rob Mullen, Vermont
Peter Riley, Vermont
Bonnie Rowell, Quebec/Vermont
Neil Simmons, Labrador
Daron Tansley, Vermont
Robin Tapley, Ontario

Art Expeditions

Missinaibi River (August 24–September 12, 2001)

Harricana River (October 3–12, 2002)

Missinaibi River (September 26–October 3, 2003)

Chapleau River (September 4–14, 2004)

George River Recon (August 30–September 12, 2005)

George River (September 1–18, 2006)

Rio Grande (December 22, 2006–January 12, 2007)

Lake Superior North Shore (July 15–31, 2007)

Lake Superior National Marine Conservation Area Winter Walk (February 2–13, 2008)

Slate Islands, Lake Superior (September 2–8, 2008)

Kamestastin Lake, Labrador (September 29–October 16, 2008)

Quebec Trans-Labrador (August 29–October 12, 2009)

Smoke on the Water: Woodland Caribou Provincial Park, Ontario (August 2–29, 2011)

Northern Rim of the Boreal: Part 1—Noatak River: Lucky Six Creek to Kotzebue, Alaska (July 29–September 2, 2012)

Northern Rim of the Boreal: Part 2—Lucky Six Creek: Kobuk River via Angiaak Pass and the Reed River (July 25–August 24, 2013)

Northern Rim of the Boreal: Part 3—Kobuk River to Kotzebue, Alaska (August 12–30, 2015)

Missinaibi River to James Bay (July 29–August 14, 2016)

Upper Missouri River Breaks National Monument, Montana (May 2017)

Algonquin Wildlife Research Station (March 5–12, 2018)

Trees and Tundra: Top of the Boreal Forest on an Ancient Route Across Nunavik, Rivière aux Mélèzes (August 1–September 7, 2018)

Hike on the Long Trail (Québec–Massachusetts): Boreal Forest Along the Spine of the Green Mountains (September–October 2020)

Voyage of the Artful Otter: Six Weeks on Lake Champlain (non-continuous: June–October 2021)

Skidbladnir's Journey Home: The Boreal Forest from the Sea—Chester Basin, Nova Scotia, to Lake Champlain, Vermont (July 12–August 6, 2023)

Note: WREAF was formed to create a body of field and studio artwork created by artists who have lived their art, roughing it on wilderness expeditions. The work in this book is not from armchair naturalists.

Cataloguing in publication information is available from Library and Archives Canada.
ISBN 978-1-77458-587-0 (hardcover)

Page Two
pagetwo.com

Jacket and interior design by Peter Cocking
Interior photos by Rob Mullen unless otherwise indicated
Map by Eric Leinberger
Printed and bound in Canada by Friesens
Distributed in Canada by Raincoast Books
Distributed in the US and internationally by Macmillan

This book is printed on FSC®-certified paper. The paper comes from responsible sources, is certified by the Forest Stewardship Council®, and supports sustainable forestry. It is acid-free, ancient forest friendly, and has been processed chlorine free.

FSC www.fsc.org MIX Paper | Supporting responsible forestry FSC® C016245

26 27 28 29 30 5 4 3 2 1

robmullen.com

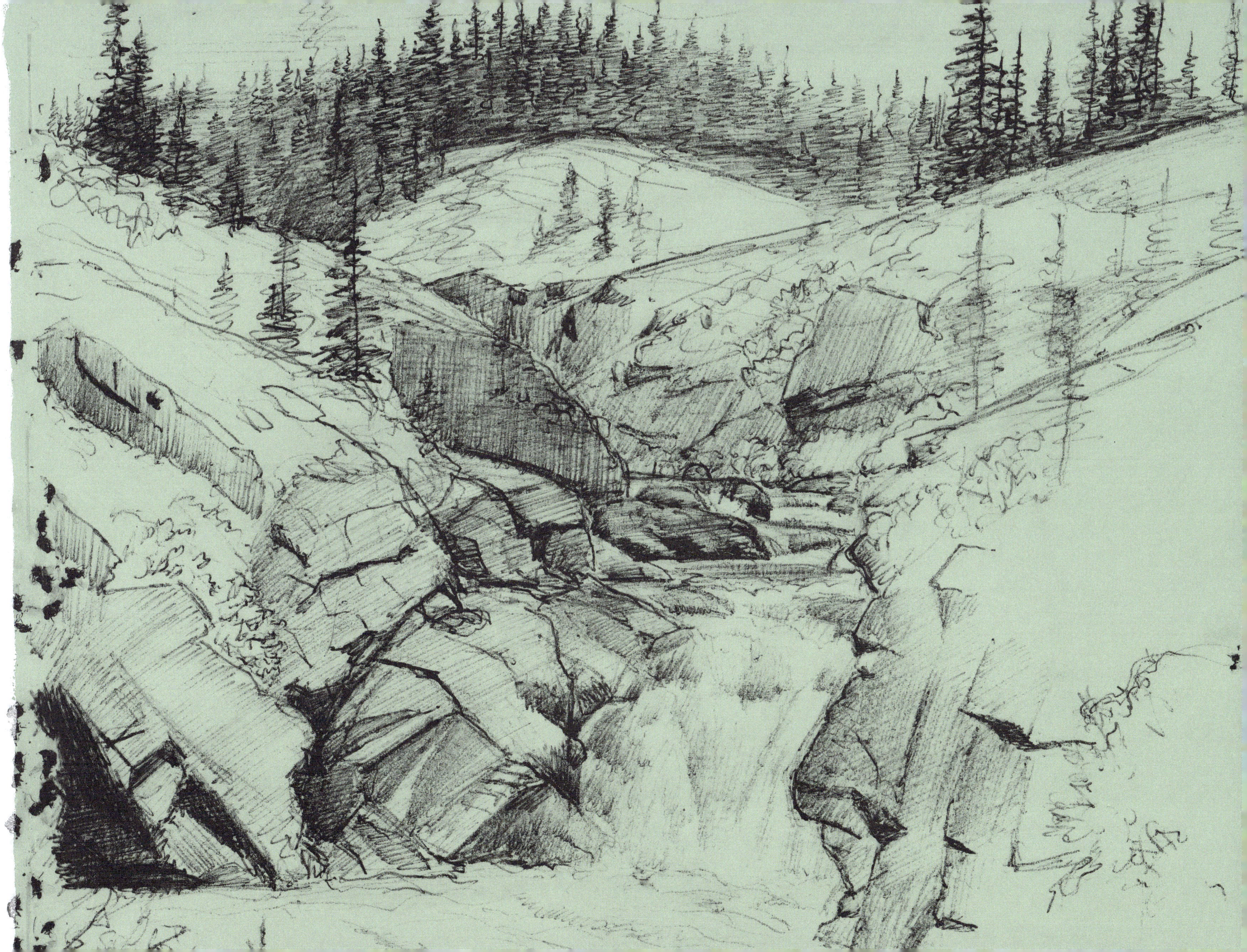

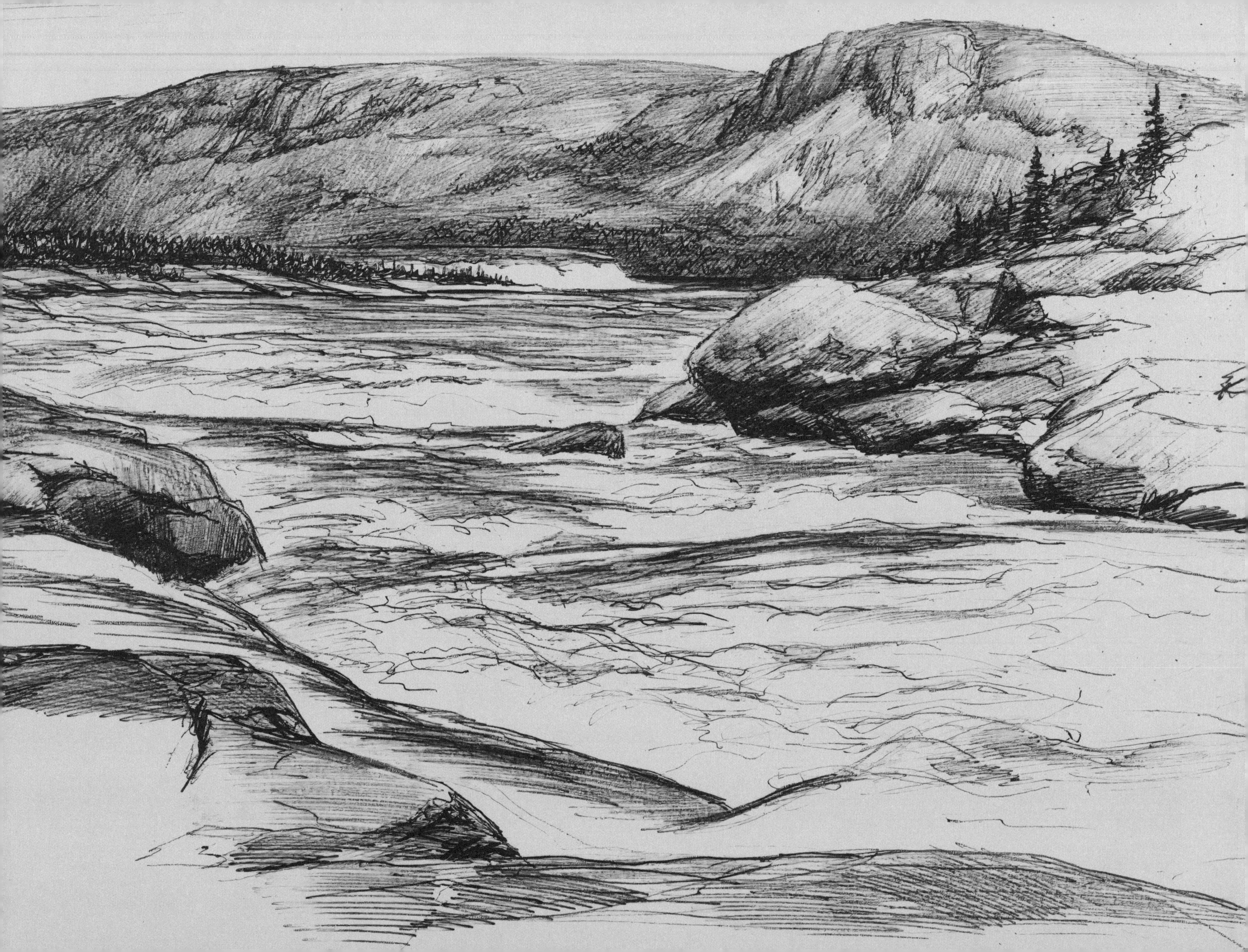